Food Production Operations

Third Edition

Chef Parvinder S. Bali

Corporate Chef—Learning and Development
Oberoi Centre of Learning and Development
New Delhi

OXFORD
UNIVERSITY PRESS

OXFORD
UNIVERSITY PRESS

Oxford University Press is a department of the University of Oxford.
It furthers the University's objective of excellence in research, scholarship,
and education by publishing worldwide. Oxford is a registered trade mark of
Oxford University Press in the UK and in certain other countries.

Published in India by
Oxford University Press
22 Workspace, 2nd Floor, 1/22 Asaf Ali Road, New Delhi 110002

First Edition published in 2009
Third Edition published in 202••
Second impression 2021

ISBN-13 (print edition): 978-0-19-012479-3
ISBN-10 (print edition): 0-19-012479-2

ISBN-13 (eBook): 978-0-19-099218-7
ISBN-10 (eBook): 0-19-099218-2

Typeset in Avant Grade Std and Baskerville BE
by B2K-Bytes 2 Knowledge, Tamil Nadu
Printed in India by Nutech Print Services India

Cover image: © Yuliya Gontar/Shutterstock

For product information and current price, please visit www.india.oup.com

I dedicate this book
to
my two lovely children, *Ojas* and *Amora*,
my nephew and nieces, *Guranjan, Jaijeet, Harshita, Kanan, Vriti,* and *Dhwani*,
my parents, *Late Major Ranjit Singh Bali* and *Gominder Kaur Bali*,
and my wife, *Shalini Bali*.

Let us ride along on this path of knowledge,
eat well and stay fit

— Chef Bali

Preface to the Third Edition

An experienced chef is known by many names—a chef instructor, a culinary educator, a training chef, or a culinary professor. It is so because his/her passion lies in passing on his/her skills to the culinary students aspiring to become a chef one day. These educators teach and train students through lectures and demonstrations and prepare them to face the upcoming challenges in the hospitality industry. *Food Production Operations* will help teachers bring forth the various facets of professional kitchen and familiarize students with the basics of food production operations.

In my experience as a chef instructor, I have come to realize that it is important for culinary educators to keep themselves updated with the prevalent technology. The present students or the millennials are tech savvy and the instructor should make maximum use of technology to keep their interest engaged. Some futuristic ideas that can be implemented in the new age culinary schools are:

- Record the demonstrations carried out in the kitchen to allow students to access any of their classes in the entire year's curriculum at any time
- Orient the chef trainers in various 5 stars hotels continuously so that they are updated in their knowledge and skills
- Set-up training kitchens that replicate a hotel's kitchen
- Divide the kitchen into departments such as cold kitchen, butchery, hot kitchen, and banquet kitchen and deliver the modules accordingly so that the students get to understand inter departmental coordination
- Make the restaurant serve an à la carte menu and a buffet menu, wherein for a few hours the entire F&B class functions like a live hotel
- Capture the entire simulated meal experience on camera so that the students can understand the concept of time and motion study by looking at the recordings and also learn to deal with real guests

As in any other stream of education, the future is to use as much of technology as possible as very soon artificial intelligence will storm the culinary arts education as well. A software that will teach cognitive cooking by analysing the data, will revolutionize the way chefs marry flavours and textures. Many chefs around the world have already started to collaborate with scientists and IT professionals to create an interface between AI and human cooking so that more can be achieved in the culinary field.

About This Book

This third edition continues to provide a comprehensive coverage of the techniques of food production and operations. Based on the National Council for Hotel Management and Catering Technology (NCHMCT) syllabus, it introduces students to the concept of cookery, basic principles of food production, bakery, and production of Indian and Western cuisines. To keep the students and the teachers updated about the new happenings in the culinary world, a new chapter on cheese has been added and a few modern methods and equipment have also been added.

Coverage and Structure

The entire book is divided into four parts—Introduction to Professional Kitchens, Basic Food Production Operations, Basics of Bakery and Confectionery, and Basics of Indian Cooking.

Part I, *Introduction to Professional Kitchens,* has five chapters that build a strong foundation for those who wish to start their career as chefs.

Chapter 1 introduces the students to the culinary world, thus giving them a glimpse of professional kitchens. It also outlines the health, hygiene, and safety procedures followed in the kitchens.

Chapters 2 and 3 discuss kitchen organization structures and various layouts of the kitchen and related areas. They also talk about the various customers the chef has to interact with, for smooth functioning of the hotel.

Chapter 4 talks about the various types of fuels and equipment used in a hotel's kitchen. It also mentions safety guidelines, which must be kept in mind while performing various operations.

Chapter 5 deals with basic menu planning. It focuses on menu, menu engineering grid, and wine and food pairing.

A surge in the number of good hotels in the country and developments in the field of food production operations have made it fundamental for students to develop a keen understanding of food production operations.

Part II, *Basic Food Production Operations,* comprises Chapters 6 to 17.

Chapters 6 and 7 deal with the most common food items–vegetables and fruits used in day-to-day operations in the kitchens. The chapters discuss the various ways to process vegetables and fruits respectively, and the methods of selection and storage. Chapter 6, in particular, describes the scientific approach to cook vegetables to create various textures and variety in the dishes.

Chapters 8, 9, 10, and 11 discuss the basic stocks, soups, sauces, and salads used in the kitchens. Many cooking techniques are featured in the online resources, which will enhance the learning of the students.

Chapters 12, 13, and 14 discuss meats, fish, and eggs, respectively. The selection of meat, its usage and storage is dealt with in detail.

Chapters 15 and 16 discuss nuts, seeds, Western spices, and different kinds of grains such as rice, cereals, and pulses.

Chapter 17 deals with the various methods of cooking such as blanching, poaching, and microwave cooking.

Part III, *Basics of Bakery and Confectionery,* contains Chapters 18 to 22.

Chapter 18 provides a foundation of the topic of bakery and confectionery by discussing various commodities used in bakery and pastry such as flour, raising agents, oils, and milk.

Chapter 19 discusses the steps in bread making and the various kinds of breads made around the world.

Chapters 20 and 21 deal with the production of various sponges, cakes, pastes, creams, fillings, and sauces.

Chapter 22 focuses on laminated pastries and the methods of preparing them.

Part IV, *Basics of Indian Cooking,* comprises chapters 23 to 27 in the book.

Chapter 23 provides an introduction to Indian cooking. It traces the history of Indian cuisine and discusses the influence of travellers and invaders on the same.

Chapter 24 deals with the various condiments, herbs, and spices used in Indian dishes.

Chapter 25 discusses various masalas and pastes that give variety to Indian food.

Chapter 26 deals with commodities such as thickening agents, souring agents, and colouring agents and their usage in Indian cooking.

Chapter 27 discusses Indian gravies, their preparation, usage, and storage techniques.

Chapter 28 is a new chapter in the book. It discusses the process of making cheese, selecting and storing cheese, and its uses in cooking. It also lists some of the most famous cheeses from around the world.

About the Online Resource Centre (ORC)

The book is accompanied by an online resource centre (india.oup.com/orcs/9780190124793) consisting of 55 videos and 365 recipes. The videos may be divided into the following broad categories.

1. *Various cuts of vegetables and fruits*
2. *Methods of using kitchen equipment*
3. *Preparation of various sauces*
4. *Processing of fish, lamb, chicken, and beef*
5. *Preparation of cakes, pastries, and breads*

The recipes are designed on Excel sheets with built-in macros. They also allow users to calculate the right amount of ingredients required for preparing the respective dishes.

New Additions

Besides the categories of videos mentioned above, the following new categories have also been added in the videos.

1. *Mock Interviews*
   a. What are the interviewer's today actually looking at while hiring their future leaders? (MT Preparation Journey)
   b. Culinary Interview, Situation Handling, and Culinary Presentation (Video in Mock interview format)
   c. Culinary Interview Content Building–Zooming Ahead the Right Way
2. *Live Kitchen In-process*

The ORC also contains Assessments–Multiple Choice Questions (chapter-wise) to help students assess their learning.

Acknowledgements

I would like to mention certain people and organizations who have either directly or indirectly contributed towards this book. First and foremost I would like to mention our Chairman *Mr. Prithvi Raj Singh Oberoi* and *Mr. Vikram Oberoi*, Managing Director and CEO under whose able guidance I have been able to collect all the knowledge pertaining to this book. I would like to thank the *Oberoi Centre of Learning and Development* for letting me use their resources for research. A special mention to all my colleagues and friends who have lend their encouragement and support in this venture of mine. A big thank you to the whole Oberoi Group for their support in enabling me to fulfil this undertaking.

I would also like to thank Mr Sameer Kaul, Founder, *Conquer Next* for helping us with the transcripts of the new videos uploaded on the ORC.

My thanks would be incomplete, if I did not mention the academicians and the reviewers, who reviewed the book and gave corrective feedback that helped me to frame the contents of the book. I would also like to thank the editors and the team at Oxford University Press India for their constant follow-ups and all the support that motivated me to accomplish this project.

I would like to thank all the near and dear ones and the professionals in the industry who have in some ways influenced the development of this book. Last but not the least I would like to appreciate the support of my wife Shalini and my children Ojas and Amora, who have shown their immense patience whilst I compiled the third edition of my first book.

Parvinder S. Bali

The publisher and the author would like to thank the following reviewers for their feedback:

- Amit Joshi (School of Hotel Management and Tourism, Lovely Professional University)
- Aparna Deshpande (Maharashtra State Institute of Hotel Management and Catering Technology)
- Sunder Srinivasan (AISSMS College of Hotel Management & Catering Technology, Pune (Affiliated to Savitribai Phule Pune University))
- M. K. Dash (Institute of Hotel Management, Gwalior)

Preface to the First Edition

Globalization has changed the mindset of many youngsters in India who wish to start their career as chefs in five star establishments. Earlier there were only a few established hotels and everyone wanted to be associated with them. Now, with many international chains coming to India, more and more job opportunities exist in the kitchen as there is a shortage of skilled culinary professionals.

A chef's job comprises performance of all kitchen activities including preparation of food and the operations involved in it. Being the heart of a hotel, the kitchen demands high precision and dedication by its staff. For beginners, it would usually be a very stressful place to work in. However, professionals who have spent decades in the kitchen, look upon it as a place to relieve stress.

The rising demand for experts in the culinary department has made food production operations a much sought-after course. Today, it is common to see students studying specialized culinary art all around the world. Thus, the subject of food production has gained significant popularity among young students. Food production is mostly taught at the undergraduate level in the first two semesters in hotel management institutes as also in diploma and foundation courses.

Food Production Operations introduces students to the various facets of the kitchen—ranging from the layout of the kitchen department to menu planning to production of Indian, Western, and pastry food items. This book would, therefore, familiarize students with the basics of food production operations.

About this Book

Recognizing the need for comprehensive books for students pursuing their career in hotel management, I decided to write books for students and everyone interested in cooking. The first volume, *Food Production Operations*, educates on basic commodities, methods of cooking, and basics of pastry and Indian cuisine. The second volume, *Quantity Food Production Operations and Indian Cuisine*, touches base upon volume cooking and Indian cuisine. The third volume, *International Cuisine and Food Production Management*, is an extension of the first two volumes and it is expected that students have read them to be able to grasp the third one better.

This book has been developed keeping in mind the changing trends in modern kitchens. As there is a myriad of differences in the commodities and technology used across the world, it is important that one should be aware of the dynamics of kitchen operations.

Food Production Operations has been specially designed to meet the requirements of students aspiring to become chefs. It would also be helpful for trained professionals in the industry.

This book discusses the basic day-to-day operations that are performed in professional kitchens and more so, it gives readers insights into why a particular thing happens in a particular way in a kitchen, attempting to educate about the various ways to perform a particular task.

This book also brings in my sixteen years of experience with Oberoi Hotels and Resorts. This professional knowledge percolates down through the chapters in the form of 'chef's tips' that are rarely mentioned in books, but are always followed in kitchens. These were handed down from earlier generations and chefs have always carried the same with them. For example, when there is no time to clean an oil spill on the floor, chefs sprinkle salt over it to avert accidents. There are many such tips, which are mentioned all throughout the book.

Pedagogical Features

Each chapter of the book begins with learning objectives, which give an introduction to the various topics discussed in the chapter. Important information is listed in tables, which are numbered for easy reference and accessibility. The concepts and skills are explained through illustrations, figures, and chef's tips. The illustrations provide added information to readers such as English, Hindi, and scientific names of fruits and vegetables, various cuts of vegetables and fruits, equipment used in Indian kitchens, etc. Each chapter is summarized in the form of conclusions and has a list of key terms, concept review questions, and project work.

Other key features of the book can be listed as follows:

- Discusses the basics of all kitchens—Western, Indian, and pastry.
- Includes chapters on menu planning, vegetable cookery, meats, bakery and confectionery, basic Indian gravies, etc.
- Discusses methods of cooking such as sautéing, steaming, braising, microwave cooking, etc.
- Includes sections on food safety, ergonomics, internal and external customers, modern cooking equipment, and game and poultry.
- Includes projects at the end of each chapter, which can be useful for chef instructors and students alike.

Acknowledgements

I would like to mention certain people and organizations who have either directly or indirectly contributed towards this book.

First and foremost, I would like to mention our Chairman Mr Prithvi Raj Singh Oberoi, under whose able guidance I have been able to compile the information needed for this book. I would like to thank Oberoi Centre of Learning and Development (OCLD) for letting me use the resources for research. I gratefully acknowledge the support of my Dean OCLD Dr Paul Zupan, for allowing me to complete this task. I would like to make a special mention of our chef de cuisine Chef Soumya Goswami, who has lent his encouragement and support in this venture of mine. I would also like to make a mention of the kitchen management associates of OCLD 2008–10 batch—Aditya, Ashutosh, Ayushee, Omkar, Neelabh, Shiv, Sagar, Hitesh, Gaurav, Indranil, Parag, Anurag, Hasan, Saket, Mahesh, and Mandar, who have lent their help and support in the compilation of the videos. I would like to thank my colleague Chef Nishant Bhatia, for organizing the workflow for the video shoot. I would like to thank Chef Salaria from The Oberoi, New Delhi for doing butchery videos for the online resources centre of this book. I would also like to thank the entire Oberoi Group for letting me use the layouts and designs from various Oberoi hotels and resorts.

I would also like to thank my friend Mr Jaydeep Patil, who provided the beautiful photograph for the book's cover. I am grateful to professionals in the industry, who have in some ways influenced the development of this book. I would like to thank the academicians and the reviewers, who reviewed the book and gave corrective feedback that helped frame the contents of the book. I would like to thank the editors and the team at Oxford University Press India for their constant follow-ups and support that motivated me to accomplish this project.

Last but not the least, I would like to thank my parents—Late Major Ranjit Singh Bali and Gominder Kaur Bali—who have given me education and made me worthwhile to write a book which would benefit many, and my wife—Shalini Bali—who has given her full support in this venture of mine and has shown immense patience while I sat late at nights to compile this work.

Parvinder S. Bali

Detailed Contents

PART I: INTRODUCTION TO PROFESSIONAL KITCHENS

PART II: BASIC FOOD PRODUCTION OPERATIONS

PART III: BASICS OF BAKERY AND CONFECTIONERY

PART IV: BASICS OF INDIAN COOKING

*Online Chapter: DEALING WITH
PANDEMICS IN FOOD PRODUCTION
OPERATIONS*

▶

CHAPTER 1

INTRODUCTION TO COOKERY

Learning Objectives

After reading this chapter, you should be able to:

- understand the basic operations of a professional kitchen with regard to safety procedures and hygiene
- appreciate the usage of knives and learn how to take care of them
- claim an insight into the basic hierarchy in the kitchen and their placement in the brigade with regard to their skills and experiences
- enumerate the safety procedures while handling basic equipment and how to lift heavy equipment
- comprehend and appreciate the grooming standards to be followed in the kitchen
- familiarize with Food Safety Standard Acts Of India and understand its importance in our food industry

INTRODUCTION

Cooking has neither been a discovery nor has it been an invention; it has been an evolution and food has changed with times and societies. Food is one of the basic requirements of survival for humans and in the past all wars have been fought for mere survival only. In the prehistoric times cavemen killed animals for food and this led them to develop crude tools made of stone so that they could hunt with ease. The accidental discovery of fire changed the way we eat food today. We can only guess how the first cooked food evolved. Maybe one day some piece of meat accidentally landed in fire and it tasted good; it could also be that a whole animal fell into a fire and got chargrilled accidentally to create the world's first barbecue. The old pictures of cavemen also depict whole animals being spit roasted. Even till this day the cuisine of nomads and tribes is essentially whole spit roasts.

After the colonies were built and the civilizations set in, societal structure started developing based on the type of work being done by different people. Man started to demarcate food as well. Food began to be classified as food for warriors, royal cuisine, and poor man's food. With the advent of religions, religious barriers prevented eating of pork for Muslims and beef for Hindus. Kosher laws[1] of Jews, royal cuisines of Thailand, Emperor's cuisine of China, and many others around the world have been segregated according to caste and social status of the people.

Food grows naturally in the forms of grains, nuts, and vegetables; but with the advent of technology in the sphere of agriculture, various types of pesticides and insecticides and hybrid technology gave birth to new kinds of vegetables and fruits which even had different colours and shapes. Japan has started producing tomatoes which are square-shaped like bread loaves, so only a single slice of such a tomato will cover an entire piece of bread in a tomato sandwich.

Now with awareness of the harmful effects of pesticides and insecticides, people are restoring back to naturally produced vegetables and fruits and now we call them 'organic food'. Organic food has become a fashion statement and the same food which was once available naturally, is being purchased at thrice the price and so only the rich and famous can afford it.

In the prehistoric times and until very recently men went hunting, while the women would stay back to cook meals and look after kids and it became a norm that woman of the household will cook, while the man earns and brings in the commodities. However, in the modern era, women have been competing with men in all the spheres of life. Lack of time to cook at home has given a new dimension to food and a whole range of ready-to-eat meals have hit the shelves of markets all over the world. After a hectic day of work, all one has to do is empty the contents of a ready-to-eat meal packet into a boiling pot of water and after a few minutes, one can relish any dish from around the world.

Thus, we see how food has evolved from centuries, and we do not really know where the future will lead us. But there is one thing for sure, for which chefs do not have to bother and that is, whatever form food takes, it will still be chefs who will create the delicacies.

LEVELS OF SKILLS AND EXPERIENCES

As discussed above, women are competing with men in all spheres of life and kitchens are no exceptions as it can be seen that more and more female chefs are becoming world-renowned chefs.

To understand a professional kitchen let us talk about the four 'P's.

Product

To any customer coming to wine and dine in a restaurant, food and service are of paramount importance. The food has to be hygienically prepared and presented in a modern way, so that the restaurant stays ahead of its competitors. All the aspects of food need to be considered while preparing and serving it in a professional kitchen. The taste, serving temperature (hot food served hot and cold served cold), eye appeal, aroma, and hygiene can never be ruled out in order to delight the customer.

Process

Standardization is one of the biggest challenges faced today in professional kitchens. Imagine being in a restaurant where you do not know how the food will taste. To achieve the standard dish each and every time, the processes have to be in place. The way the ingredient is processed to get the maximum usable product, the standard equipment to be used each time to produce that dish, the modulation of heat applied to the food, and many other processes form an integral part of food business.

---

[1] Kosher food laws are very strict food eating laws of Jews.

Profit

A food service provider should be able to offer the most enjoyable dining experience to the customers. At the end of the day, apart from customer's satisfaction, one has to operate a profitable business for further growth and development of the organization. The chef behind the scene contributes greatly to the financial viability of the business. He/she has to be in control of all aspects of food operations such as purchasing, storing and portioning, and wastage control. The old saying 'customer or guest (as we call him/her today in our hotels) is God' means that the guest is always right; but we must remember that financial viability is of the utmost importance to stay in the food business.

Fig. 1.1 Organizational chart

People

The first three 'P's—process, product, and profit—cannot happen unless a human's touch is involved to give finishing touches to the food. The basic organizational chart of a kitchen looks as shown in Figure 1.1.

In the following chapters we shall discuss the job responsibilities of each position in the kitchen. This organizational chart or the hierarchy of the kitchen will be different for each business depending on the volume and type of operations. Most of the hotels today follow a very flat kind of hierarchy known as 'lean hierarchy'. It is very common to see these days that modern hotels do not have commis classified as I, II, and III; the joining level after an apprentice programme is commis.

An apprentice is a person who wishes to make his/her career in the professional kitchen as a cook and finally, with experience and knowledge, wishes to become a chef in the future. An apprentice joins an organization at the lowest level of the kitchen and has no skill and experience. Only a few apprentices become chefs, because apart from skill and experience it requires passion to become a successful executive chef. Working in a kitchen is quite complex as it requires long hours on the feet amidst the heat and noise. Apprentice programme is usually governed by the labour laws and the duration of the course is around two to three years. In India, it is a three year programme. 'Commis' is a French word which means a cook. After completing the apprenticeship programme, one is employed as a commis. Based on the needs of the operations and the style of business, the commis may be restricted to one area, thereby specializing himself/herself into a saucier, butcher, baker, staff cook, soup cook, vegetable cook, roast cook, etc. as the operations demand. This kind of segregation of commis was earlier done in most of the hotels, but now with the labour becoming expensive, it is advisable to follow the modern trends of staffing, whereby, multi-skilling has not only become necessary but it has also become the first choice of hotel operations. Multi-skilled cooks thus have better chances of promotions.

The next level of promotion is demi chef de partie. As we can see, the titles of the chefs are all in French. This is because France was the first country to classify this profession into various levels and the world thereafter followed suit. 'Demi' in French means 'half' and 'chef de partie' means 'a part of chef', and if we try and put this together, then chef de partie would mean a supervisor or a person who, with his/her sheer hard work and skills is preparing to take on the job of a chef. In this role, a chef de partie takes on other managerial responsibilities as well. A demi chef de partie would be an entry level into a chef de partie and can be referred to as an assistant supervisor. In kitchen operations, the main job of any person at any level is to cook on the range for guests; therefore, with promotion the kitchen personnel get another set of responsibilities with addition to the basic job of cooking that they are doing. These days, however, the position of demi chef de partie is being overlooked and people can become chef de partie from commis. There is no particular time span to become a chef de partie from commis, but usually depending on the skills, minimum work experience of at least three years is required before being promoted.

The next level in the hierarchical grid of kitchen is sous chef, which literally translates to 'under chef'. This person would be a kitchen manager of a section, heading a team of chef de partie, demi chef de partie, commis, and apprentices. Depending on the size of the operations, there could be as many sous chefs as the departments in the kitchen. The modernistic view, however, is to have two or three sous chefs overlooking more than two to three kitchens, depending on their levels of skill and experience. The sous chef reports to the executive sous chef. The executive sous chef is in charge of the overall kitchen operations. His/her main job is to assist the executive chef in planning menus, implementing the standard operating procedures, doing portion control, motivating his/her workforce.

Depending on the levels of skill, a person would become an executive sous chef in 10 years time and then may become an executive chef in another two to three years.[2]

In the early times, being a cook was looked down upon and this profession was regarded as a profession for the poor and failures. But today there are many famous chefs around the world and now being a chef is considered to be a very respectable career and hence, more and more people are opting for the same. Now prestigious universities have specialized courses on cooking and catering, thereby preparing people to be successful hotel professionals. There are three-year diploma courses and even four-year degree courses in hotel management, which have an academic component and also industrial exposure, where a student has to work in hotels to get hands-on experience of what has been taught. These students are known as industrial trainees or internship trainees. It is very important for budding professionals to see for themselves, if they can manage to face such challenges in their life, as hotels work 365 days, 24 hours a day. During festivals chefs in the hotels are the busiest, as they have to create a memorable experience for their customers.

Other important chefs seen in the kitchen are master chefs. The master chefs have to perform particular jobs and it usually takes around 12 to 15 years of hard work for a cook to specialize as a master chef. This concept has come from South-East Asia.

ATTITUDE AND BEHAVIOUR IN THE KITCHEN

Kitchen essentially is quite a dangerous place to work. In many areas, accidents can occur, if the basic safety rules of the kitchens are either not known to the staff or are ignored. The three main causes of accidents in the kitchen are:

1. Distraction
2. Haste
3. Failure to observe safety rules and regulations

---

[2] Here it refers to operations in a five star establishment. In small scale operations one can be head of the kitchen brigade in less than five years also.

Distraction

Distraction in the kitchen is usually caused by other personnel working in the kitchen. In the kitchen, a chef works with tools and equipment, such as sharp knives and intricate machinery, which help him/her to do his/her job efficiently and diligently; but if he/she loses concentration while working with any of these, it can prove fatal. For example, while slicing onions with speed, if people talk to someone over the counter it can prove dangerous.

Haste

One of the major enemies of kitchen personnel is 'haste'. A cook or a chef must complete the work which has to be done, that very moment itself. The customer can wait for his/her food for a considerable time only; but that does not mean that chefs should run around in the kitchen and try to do things fast. Therefore, it is very important to set up the workstation in a way that everything is within reach. All the *mise en place* should be in place and the placement of commodities should be thought through by the executive chef after doing time and motion studies. The golden rule in the kitchen is not to run, no matter how urgent the task is.

Failure to Observe Rules and Regulations

The last but not the least is the failure to observe safety rules and regulations. In a busy operational kitchen, sophisticated machineries are provided to perform jobs effectively and all these machines come with safety and handling manuals. Any deviation could result in fatal accidents. For example, while extracting juice from a juice machine, one has to use the pusher to push the fruit in. Using bare hands can cause loss of fingers or hand. Figure 1.2 shows a sample of operating instructions used in hotels, usually found hung near equipment in the kitchens.

HOTEL XYZ
Operational Instructions
3-deck Oven

- Isolate power supply when not in use and while cleaning.
- Always use safety gloves when placing or removing objects from the oven.
- Ensure the oven is at the correct temperature prior to using and preset temperature.
- Ensure to open damper prior to opening the door to exhaust steam within the chamber.
- Use a soft brush to clean the templates in the chamber when the oven is completely cold.
- Clean oven with a wet cloth. Avoid excess water.

Authorized personnel to utilize machine: Sous chef positions
Chef de partie positions

Fig. 1.2 Operating instructions format

PERSONAL HYGIENE AND FOOD SAFETY

Hygiene is the science that deals with sanitation and disinfection. For successful food operations, hygiene and food safety form the backbone of a successful business. News has shown the closure of famous food industries after the outbreak of food poisoning. Personal hygiene and food safety go hand in hand because one is complementary to the other. Most of the contamination of food takes place by improper handling of food by the food handlers and so it is imperative for us to know about these.

Food Safety and Kitchen Hygiene

Most chefs would say that food safety and kitchen hygiene relate to the cleanliness of the kitchen. Clean pots and pans being used each time someone starts cooking, wearing a clean uniform each time someone reports to work in the kitchen at the start of a shift, and food prepared in the right manner using the best quality ingredients ensures food safety and kitchen hygiene. Food and all activities related to food is why a kitchen exists, whether at home or in a hotel.

Food that is produced in the kitchen has to be safe and hygienic so that when a guest eats it, he/she does not fall ill due to food poisoning. Safe food is food that is free of contaminants and will not cause illness or harm. Persons involved in food poisoning investigations often remark about the cleanliness of the premises involved in such cases. Hygiene is more than cleanliness. It involves all measures necessary to ensure the safety and wholesomeness of food during manufacture, distribution, transportation, receiving, storage, issuing, processing, preparation and handling, holding, sales and service. From the time raw ingredients come into the receiving department until the time the food is eaten by the guest, it is the hotel's responsibility to ensure that the food is absolutely safe and free from all contaminants.

CONTAMINATION OF FOOD

Contamination of food is a major hazard that entails the occurrence of any objectionable matter in or on the food. Lamb legs may be contaminated with faecal matter; high-risk food (food high in protein content) may be contaminated with spoilage or food poisoning bacteria; and flour may be contaminated with rodent hair and excreta, weevils, etc.

There are three types of contamination, namely:

1. Physical contamination
2. Chemical contamination
3. Microbiological contamination

Physical Contamination

Foreign bodies found in food (other than what is supposed to be consumed) is termed as physical contamination. The list of foreign bodies is endless—stalks, stones, pebbles, insect bodies, metal particles, nuts and bolts, glass, rodent droppings, metal screws, cigarette ends, plastic pieces, plastic packaging material, stapler pins, pieces of corrugated box material, sawdust, wood chips, human hair, pins, clips, etc. This type of contamination can easily be brought under control if the establishment has sound systems and procedures in place, and a dedicated team of people who ensure that nothing unwanted comes into the kitchen area.

Chemical Contamination

This takes place when unwanted chemicals enter the food during:

- growth, for example, veterinary drugs, excessively-used fertilizers, pesticides, and environmental contaminants such as lead or dioxins;
- food preparation, for example, oil, cleaning chemicals (residues found in pots and pans that have not been cleaned thoroughly), or insecticides;
- processing, for example, excessive addition of preservatives in foods such as sausages, salamis, etc.

Chemical food poisoning can cause long-term illness such as cancer, and many such cases of food poisoning have come to light.

Microbiological Contamination

It is surprising to know that when one talks about contamination, most of the time one refers to physical contamination or chemical contamination. However, the type that can cause real havoc in a food business is seldom spoken about—microbiological contamination. This is by far the most significant type as it results in large amounts of spoilt food and unacceptable instances of food poisoning.

Microbiological contamination takes place due to the presence and multiplication of food poisoning bacteria. Bacteria are single-celled organisms found everywhere—on raw food and people; in the soil, air, and water. Bacteria are microscopic and vary in size—from around .001 mm to .003 mm. While most of the bacteria found in nature are harmless to humans, a few are harmful and can cause damage to mankind. These bacteria are classified as 'pathogens'.

Food poisoning bacteria may be brought into the food premises from sources such as:

- food handlers, people working in related departments such as kitchen stewarding, service personnel, and to a certain degree, guests;
- raw food including poultry, meat, eggs, milk, fish, shellfish, and water, especially when polluted with sewage or animal faeces (vegetables or fruits may become contaminated by manure or polluted irrigation water);
- insects, rodents, animals, and birds;
- the environment, including soil and dust.

Bacterial multiplication takes place in many ways, and more often than not, it is the food handlers themselves who unknowingly contaminate perfectly safe food, thereby making it hazardous to eat.

Vehicles and Routes of Bacterial Contamination

Sometimes bacteria pass directly from the source to high risk food, but as bacteria are largely static and as the sources are not always in direct contact with food, the bacteria have to rely on other vehicles to transfer themselves to food. The main ones are:

- hands of food handlers (many of them carry *pan masala*, tobacco for chewing, *bidis*, etc. in their pockets even when they are working in the kitchen);
- clothes and equipment (when a food handler goes to use the conveniences, he/she is supposed to take the apron off and keep it in the kitchen itself, in pigeon holes made exclusively for holding aprons while the staff go to the lockers; after the work is done he/she is supposed to deposit the soiled uniform in the uniform room);
- hand-contact surfaces (when kitchen staff use the conveniences and do not wash their hands properly; ideally a nailbrush should be provided at the wash hand basin at all times);
- food-contact surfaces.

In the kitchen, one of the most common causes of bacterial contamination is 'cross-contamination', defined as the transfer of bacteria from contaminated foods (usually raw) to other safe foods. This includes the direct contact, drip, and indirect methods.

Direct Contact Method   It takes place when a commis goes to the food store to pick up ingredients, and loads all the butchery items such as vegetables and fruits into one basket or trolley, rather than into separate baskets. Many endeavour to follow good practice, but most of the time it is the junior-most in the kitchen brigade or the industrial trainee who goes for picking up the ingredients and he/she does not know about cross-contamination.

Drip Method This type of contamination takes place when there is inadequate storage (small walk-ins, just one walk-in), where the kitchen staff have to make do with whatever little space they have to store the food. When frozen meat is stored above cooked food and it is thawing, a certain amount of liquid may fall onto the cooked food, thereby contaminating it.

Indirect Method It takes place when a chopping board is used to cut raw meat and then cooked food is cut on the same board without cleaning or disinfecting it.

The above mentioned methods of food contamination happen all the time in most hotels. Even staffs, who have been trained, forget about best practices in hygiene over the years. As the leader of the team, it is the responsibility of the chef to reinforce best practices at all times. It is important to train all the kitchen personnel for maintaining hygiene in the kitchen. Also, there has to be a visible commitment from all the kitchen personnel regarding the hygienic practices to be followed in the kitchen.

Here is a check list for contamination control that may be of value to all.

1. Purchase food and raw materials from known and reliable suppliers.
2. Accept deliveries only if transported in clean, properly-equipped vehicles, whether refrigerated or not.
3. Inspect deliveries immediately on arrival; reject or segregate damaged, unfit, or contaminated items; wherever relevant, check temperature, codes, and date markings, and reject food that have passed the expiry date.
4. After checking, remove deliveries immediately to appropriate storage, refrigerator, or cold store.
5. Ensure adequate thawing of foods and separate the thawing foods from other foods.
6. Make suitable provision for cooling food prior to refrigeration.
7. Use only proper containers for storing food.
8. Keep high-risk foods apart from raw foods, in separate areas with separate utensils and equipment; colour coding is useful.
9. Keep food covered or otherwise protected until it is actually processed or prepared, in which case bring the food out only when needed and do not leave it lying around.
10. Keep premises, equipment, and utensils clean and in good condition, and remember to repair the same; disinfect food contact surfaces, hand contact surfaces and where appropriate, hands.
11. Ensure that all empty containers are clean and disinfected prior to filling with food.
12. Control cleaning materials, particularly wiping clothes; keep cleaning materials away from food.
13. Remove waste food from food areas as soon as possible; store in appropriate containers, away from food.
14. Keep any chemical away from stored food.
15. Maintain scrupulous personal hygiene at all times and handle food as little as possible. A person who has just recovered from an illness should not be given kitchen duties.
16. Maintain an active pest control programme.
17. Control visitors and maintenance workers in high-risk areas.
18. Ensure that hygiene disciplines apply to all personnel, including management.

In spite of modern machines, manual work cannot be avoided when it comes to food preparations and thus personal hygiene becomes the most important thing for food handlers. They should have the highest level of personal cleanliness and clean protective clothing. Before we understand personal hygiene and food safety, let us first know what contamination of food is and what the main causative agents of the same are. Consumption of contaminated food results in acute illness, and symptoms could vary depending upon the kind of food poisoning. The most elementary ones being nausea, sharp gripping pains in the lower stomach, diarrhoea, and sometimes, fever. Food poisoning is mainly caused by the following.

Bacteria

Bacteria are the main causative agent of food poisoning. In some countries, the bacteria are often referred to as 'bugs'. Bacteria are so small that they cannot be seen with the naked human eye. They are everywhere—they live in the air, in soil, in water, on and inside people, in and on the food that one cooks. Not all bacteria are harmful; some bacteria are friends of chefs as they help in production of cheese and yoghurt. Such friendly bacteria are known as 'commensals'. Some bacteria help in decaying and rotting of food and are called 'spoilage bacteria'. Many are pathogenic and the most common types found in the food are as follows:

- *Salmonella*—found in eggs and poultry
- *Staphylococcus aureus*—found on human skin, nose, ears, and hands
- *Clostridium perfringens*—found in the faeces and sewage
- *Clostridium botulinum*—found in fish intestines, soil; this bacteria is a concern in the bottling and canning industries, as it can withstand high temperature
- *Escherichia coli (e-coli)*—found in manured vegetables, raw milk, and intestines of animals

Bacteria enjoy temperatures between 5 and 62°C. This is usually termed as 'danger zone'. So cold food should be stored and served at a temperature lower than 4°C and hot food at above 63°C. Most bacteria will get killed in food that is held at 70°C but to ensure that food is cooked thoroughly we must cook the food till the internal temperature of food reaches 74°C.

Viruses

Viruses are even smaller than bacteria. For survival they must live in a host body. Viruses are responsible for illnesses as common as the common cold or as dangerous as smallpox and polio. Hepatitis is the most common problem in the food industry today. Viruses have the ability to change form and this ability has led to a number of known strains of hepatitis, and in order to track them, the medical profession has given each strain an identifying letter, such as A, B, C, D, and E. Virus usually spreads through the faecal-oral route. One has to be very careful when purchasing seafood; it should be procured from trustworthy sources as most of the seafood is bred in sewage polluted water. However, thorough cooking will kill the virus. The vegetables grown in sewage polluted water are also a prime cause of viral food infections.

Chemicals

Chemical poisoning, metal poisoning, and poisonous plant poisoning are all caused by the carelessness or ignorance of those handling such commodities. Always read the instruction labels for use of cleaning materials and store them away from food in a specific storeroom or an area designated for the same. Many a times, it has been seen that the cleaning chemical is stored in empty mineral water bottles and sometimes during busy operations one can easily mistake it for water, especially if the chemical is colourless and odourless. It is the prime responsibility of the executive chef to ensure that the kitchen is always maintained with regard to repairs and maintenance. Ripping off paint on the ceiling can also be disastrous if it falls on the food.

Metals

Metal poisoning is not very common; however, occasionally finding a piece of metal in food was quite common until a few years ago. These metals were found in commodities such as rice, wheat, pulses, etc. Some foods react to metals such as copper, for instance, while deep-frying food, using a copper tool in the hot oil will make people sick. Similarly, cooking of yoghurt or acidic food in a copper vessel should be avoided.

Poisonous Plants

Toadstools, red kidney beans that have not been cooked thoroughly, rhubarb leaves, deadly nightshade, and many other plants and their related products cause food poisoning.

PERSONAL HYGIENE AND ITS IMPORTANCE IN THE KITCHEN

As a food handler, one must ensure that the food provided to the customer is free of all of the above mentioned contaminants. Food handlers should remember that customers place great trust in them, and that carelessness on their part could make customers ill, or at times even lead to death.

Our normal body temperature, which is around 37°C, is favourable for bacteria to dwell and grow. This is probably the source of most cases of food poisoning. Personal hygiene should be important to everyone but to a food handler, it is of paramount importance.

The food handler has a moral and legal responsibility of having good standards of personal hygiene. The bacteria on the human body are usually found on hands, ears, nose, mouth, throat, hair, and groin. One must wash hands after touching these areas, otherwise the pathogen will enter into the food and then with favourable conditions the bacteria will grow and multiply and will cause the risk of contamination.

Bacteria are transferred to food mainly through hands. So one must ensure washing hands:

- when first entering the kitchen
- when coming back from a break
- after going to the toilet
- after handling raw meat
- before handling cooked meat
- after handling raw vegetables and other dirty foods
- after handling garbage
- after handling cleaning equipment—mop, buckets, clothes
- after touching or blowing one's nose
- after touching one's hair
- after licking one's fingers
- at regular intervals throughout the day

This list can be endless. Hands should be washed in a basin meant for hand wash only and never in a sink. Always use hot water to wash hands and clean with a germicidal soap. Hands should be cleaned all around and between the fingers also. Use nail brushes, while cleaning the hands and apply a disinfectant to keep your hands free from germs.

One must keep the fingernails short as bacteria might grow in the dirt under the nails. Nail varnish should not be used as it may chip and contaminate food.

No jewellery (bracelets, watches, earrings, etc.) should be allowed in food areas, as they also harbour dirt and bacteria. One can wear an important ring such as a wedding ring on a chain around one's neck.

In case of food poisoning it is always advisable to:

- report one's illness to one's employer or supervisor;
- not handle food until given clearance to do so;
- tell the doctor that one is a food handler;
- get medical clearance to start work again.

Food safety is governed by strict food safety laws. Health inspectors can take food samples at random and in case, if a sample fails, prosecution can follow. Every food establishment is being Hazard Analysis and Critical Control Points (HACCP)[3] certified.

The first step is to form a HACCP team, who shall then identify any step in the activities of the food business which is critical to ensuring food safety and ensure that adequate safety procedures are identified, implemented, maintained, and reviewed on the basis of the following principles.

- Analysis of the potential food hazards in a food business operation.
- Identification of the points in those operations where food hazards may occur.
- Deciding which of the points identified are critical to ensuring food safety (critical points).
- Identification and implementation of effective control and monitoring procedures at those critical points.
- Review of the analysis of food hazards, the critical control points and the control and monitoring procedures periodically and whenever the food business operations change.

UNIFORM AND PROTECTIVE CLOTHING

The uniforms for chefs were invented centuries ago; however, these have been developed and modernized as per the requirement and availability of new fabric. In early times the prime job of the uniform was to make a cook look like a cook, but today the uniforms are designed keeping in mind that these keep the workers safe, as they all operate in a potentially dangerous environment with lots of sophisticated machinery and tools around. Most people take the chef's uniform for granted, but there are good reasons for each piece of clothing. These are discussed here.

Chef's Jacket

The typical chef's jacket or chef's coat is also called *veste blanc* in French (Fig. 1.3). It is made of heavy white cotton. This fabric is important as it acts as insulation against the intense heat from stoves and ovens and is also fire resistant. The white colour of the jacket repels heat and thus keeps the worker comfortable. Also, a white uniform will get soiled quickly and

Fig. 1.3 Chef's jacket

---

[3] Hazard Analysis and Critical Control Points, a programme started by scientists in National Aeronautics and Space Administration (NASA).

a cook would have to change it, since personal hygiene is very important in the kitchen. The jacket is always double-breasted as the thickness in the cloth will prevent the chef from being scalded by hot liquids or spattering hot oil and thermal shocks as the chef constantly shuttles between the cold storage areas and the hot kitchen areas. Since there are two rows of buttons, the chef can rebutton the double breasted jacket to change sides whenever a side gets soiled during the course of work during a shift.

Chef's Trousers

Chefs wear either black pants or black and white checked pants. The traditional checkered pants were so designed to camouflage spills and the colour of the pants in some organizations also denotes the seniority of the chef. A black pant is usually worn by sous chefs and other senior chefs, while cooks and apprentices would wear checkered pants. Just like the coats, kitchen pants are designed to provide comfort and protection. The kitchen pants should be straight and without cuffs, which can trap debris and any hot liquid spills. It is advisable to have a snapped fly and elastic waist band and the kitchen trousers should be worn without a belt, so that it can be removed easily in case of hot liquid spills or even fire.

Chef's Hat

The most interesting part of the uniform is the tall white hat, called *toque blanc*. The toque dates back to mid-seventh century BC. Cooks during that time were required to wear hats similar to those worn by royalty of that time so that it resembled the crown and segregated them from the common people. The main purpose of the hat is to prevent hair from falling into the food and also help in absorbing sweat. Along with the other conveniences, disposable paper hats were invented to look like cloth so that they can be thrown away when they are soiled.

Scarf/Neckerchief

Chefs wear white neckerchiefs, which are knotted in the front. These were originally designed to absorb perspiration. Nowadays, chefs wear the neckerchiefs to keep the tradition and finish the look of their uniforms. In some cases scarves are used to represent various levels in a kitchen hierarchical grid.

Apron

It is usually made of thick cotton fabric and is worn around the waist with the help of a long string. The apron should reach below the knees to protect the chefs from spilling hot liquids. The string of the apron helps hold the chef's kitchen towel in place. The loose ends of the same should be tucked under or else they can be trapped in machinery and can cause accidents.

Kitchen Towel/Duster

They are used to pick up hot pots and pans and also to wipe hands in order to keep them dry. Usually two dusters should be kept with the chef—one to wipe the wet hands, and the other (dry one), to pick up hot pans, as a wet duster can scorch the hands. Considering the modern hygiene trends, it is advisable to use disposable paper towels for wiping and cleaning. The kitchen dusters should be used only for handling hot equipment.

Shoes

The shoes should be black and well polished. To prevent slipping, the sole should be made of rubber. Black cotton socks, preferably the sweat absorbing cotton variety should be worn. The shoes should be closed, to prevent the feet from scorching in case of spills. The shoes have to be comfortable, as we know that cook will have to stand for long hours.

GROOMING IN KITCHEN

Those days are gone when people felt that chefs do not need to focus a lot on their grooming as they will be only working in the back area of the kitchen and will rarely ever come in front of the guests. This thought has now changed worldwide. All food establishments now lay a lot of emphasis on the grooming of chefs. A clean uniform and personnel with good personal hygiene is of utmost importance in the kitchen.

We have already explained the different parts of a kitchen uniform and the importance of personal hygiene, let us now discuss what should be the basic norms of grooming in the kitchen for both men and women.

1. The chef uniform must be clean and well ironed. It is a symbol of pride for all chefs and a clean and well-kept uniform represents a professional chef with passion towards this industry.
2. Chefs should wear inner clothing of cotton under their chef's jacket, which can absorb sweat and keep them clean at all times.
3. They should wear the right sized uniform as too tight or too loose a uniform can cause hindrance in their work.
4. The sleeves must be folded just below the elbow and there should be no loose threads hanging from the uniform. Any loose clothing can get caught in a machinery and can cause fatal accidents.
5. Gentlemen should be well shaved with no moustache or maybe trimmed moustache. People who cannot shave due to their religious beliefs should wear a beard net for hygiene purposes.
6. The hair should be trimmed and short and side burns should not be below the middle of the ear. For ladies working in the kitchen, the hair must be tied up and pinned and both ladies and gentlemen should wear head nets before wearing the chef hats.
7. No jewellery should be worn, though the wedding ring is allowed, its design should be simple and conservative with no embedded or protruding gems. Body piercing, etc. must not be done.
8. Tattoos on skin must be avoided as having them can restrict your appointments in large five-star hotels and their equivalents due to strict grooming standards.
9. The shoes should be well polished and clean at all times. Safety shoes with closed heels and top should be worn in the kitchen.
10. Chefs should wear a neat and clean apron and its strings must be concealed as loose hanging strings can get stuck in machinery.
11. One should look crisp and attentive at all times.
12. For women working in the kitchen conservative ear studs are allowed, but no bangles should be worn as whilst working in hot areas such as tandoor and ovens, the bangles become hot causing burns on skin.
13. The ladies may be allowed to wear light make up depending upon the company policies, however the makeup should be resistant to hot and humid conditions in the kitchen.
14. Black cotton socks must be worn under the shoes. Nylon and other material can be dangerous in the kitchen environment.

IDENTIFICATION OF KNIVES AND HOW TO SHARPEN THEM

The importance of knives to a chef cannot be overstated. It is the most important piece of equipment in the kitchen. Knives come in various shapes and sizes and each is meant for a specific use though some knives can be used as multi-purpose knives. Let us now familiarize ourselves with different parts of a knife (Fig. 1.4).

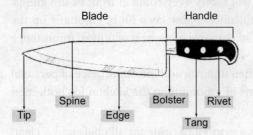

Fig. 1.4 Parts of a knife

Blade   The blade is usually made up of a metal compound called high carbon stainless steel. It combines the property of carbon of being sharpened easily and the non-corrosive properties of steel.

Tip   The tip of the knife is the pointed edge where the knife blade ends. The tip is generally used for scoring patterns and working with meats or carving.

Spine   The spine of the knife is the topmost, thick edge of the knife, which gives strength to the knife.

Bolsters   In some knives there is a collar known as a bolster, at the point where the blade meets the handle. It reinforces the structure of a knife.

Cutting Edge   The cutting edge is the most important part of the knife. It should always be kept honed and sharpened.

Handle   The handle of a knife should be easy to grip and should be non reactive to most cleaning agents. The different materials used to make handles are wood, plastic, plastic fibre, or even metal.

Tang   The tang is the continuation of the blade and extends into a knife's handle.

Rivets   These are metal fasteners that hold the handle and the tang together.

Many types of knives are used in the kitchen that make up an essential kit for a successful chef. Since these are the most valued possessions of a cook, utmost care needs to be taken to keep them in good working condition. Table 1.1 shows the basic knives used in a professional kitchen.

Table 1.1   Types of knife

Knife	Description	Photograph
Chef's knife	Also known as French knife, it is usually an extension of a chef's hand. This is so because it is the most common knife used for various operational jobs in the kitchen such as chopping, slicing, etc. The length of the blade is usually 8 to 12 inches.	
Paring knife	Also known as fruit knife, it is usually used for small jobs such as paring of apples, taking wedges of lemon, hulling strawberries, etc. The blade is usually 3 to 4 inches long.	
Tourne knife	Also known as bird's beak knife, is of the same size as a paring knife, only difference is that the cutting blade is slightly curved to facilitate the cutting of a vegetable into a barrel shape.	
Boning knife	Boning knife has a thinner and shorter blade than a chef's knife and is used to cut meat away from the bone. The heel of the knife is slightly curved so that the knife can rest on the bone whilst deboning.	

Contd

Table 1.1   (Contd)

Knife	Description	Photograph
Filleting knife	Similar to boning knife but has a flexible blade for the ease of filleting a fish.	
Bread knife	A long serrated knife used to slice bread or sponge cakes. The blade is usually 12 to 15 inches long.	
Carving knife	Thin sharp blade usually used for carving cooked meats or big joints of roasts on the buffet.	
Cleaver	It is generally used in the Chinese kitchen for cutting and chopping. It has a large wide blade and is heavier than a chef's knife. A multipurpose knife which can also be used as a mallet, to flatten a piece of ginger before chopping.	
Palette knife	It is not a knife for cutting purposes; but is a flexible spatula with a rounded tip and is widely used in confectionery to decorate cakes.	
Tomato knife	It is usually serrated and has a round or a forked tip to remove the eyes from the tomato or to pick up the slices of the tomato. Though a sharp blade can also be used for slicing a tomato, a serrated knife allows the slicing of the tomato even when the knife is blunt, without crushing the flesh of the tomato.	
Asparagus peeler	There are many designs of an asparagus peeler, however the most common one is the one in the shape of Y. The handle allows a strong grip and the sharp blade helps to peel the asparagus whilst keeping it flat on the chopping board.	
Vegetable knife	Though one can use the chef's knife for cutting vegetables, the knives in the kitchen are meant for specific ingredients to avoid contamination and also for the ease of cutting that particular ingredient. The vegetable knives have a straight edge and a less wide blade for the ease of cutting vegetables.	
Wavy knife	This knife has a wavy blade and is used for cutting fruits and vegetables to enhance their appearance.	
Cheese knife	This is a knife with a unique design. The blade has a ribbed line on the cutting blade and has holes with a fork tip. This unique design allows to cut through crumbly cheeses like feta, without breaking the cheese. The fork tip helps to pick up the sliced cheese.	
Santoku knife	The Japanese chef knife, which literally means 'three virtues' is a knife used for fish, meats, and vegetables. This knife has a unique design with a flat edge and its tip curving like the foot of a sheep. Many designs have horizontal indentations so as to cut through soft fish such as salmon and tuna, without hampering their flesh.	

Contd

Table 1.1 (Contd)

Knife	Description	Photograph
Sashimi knife	This is a slender and long knife that is sharpened on one side only. This unique style of the knife allows it to sharply cut a slice of meat, by dragging the knife on the meat in one direction only. This allows the meat to retain its natural design of marbling, thereby enhancing its appearance.	
Oyster shucker	As the name suggests, this knife is used for shucking oysters. The term shucking is used for opening a fresh oyster shell. The unique design allows the chef to open up the oyster with ease.	

Sharpening a Knife

Sharpening a knife involves two processes—occasional honing and regular sharpening. A dull blade should be honed or ground against a carborundum stone. Long sharpening steel is used to smoothen the blade after honing and to sharpen the knife each time you use it. The various kinds of sharpening tools are shown in Table 1.2.

Table 1.2   Types of sharpening tools for knives

Sharpening Tool	Description	Photograph
Sharpening stone	It is essential for the proper maintenance of a knife that it is sharpened by passing its edge over the stone at the correct angle. The grit and its degree of coarseness abrade the surface, creating a sharp cutting edge. While sharpening, start with the coarse surface and then move to the finer one. Most stones are moistened with water or oil but if they are moistened with oil once, then the process has to be continued every time. There are three basic types of sharpening stones namely: 1. Carborundum stones (most commonly used) 2. Arkansas stone 3. Diamond impregnated stone	
Steel or sharpening rod	The steel should be used immediately after sharpening a knife, it helps in alignment of the sharp edge of the knife, length of the working surface can vary from 3 to 14 inches and usually the material is made up of hard steel. This steel can have diamond impregnation.	
Commercial sharpeners	These are available in various shapes and sizes, some are manual and some are machine operated. The machine has small rollers made out of the same material as the sharpening steel. There are grooves through which the knife can be slid in and out for sharpening.	

There are five steps in sharpening a knife. They are as follows:

1. Use a good medium to fine steel.
2. Ensure to maintain the same honing angle during all strokes. Twenty degrees make a good edge.
3. Blades should be drawn with the edge first, across the face of stone or steel.
4. Count your honing strokes. Then turn blade over and hone the other side with an equal number of strokes.
5. Start strokes with heavy pressure. Then ease off to lighter pressure and finish off with a lighter stroke.

Safety Instructions Regarding Knives

There are many safety instructions which one must follow, to ensure longevity of the knives and to prevent any accident.

1. Always handle knives carefully, especially when cleaning and sharpening.
2. Never leave knives in a sink of water. This is bad for the blade and presents a safety hazard to anyone who puts his/her hands into the water. Always clean and wipe the knives dry before storing in a knife box. A knife box has a magnetic strip inside, on which the knives can be stuck and the same can be locked to ensure that there is control on the knives, as most of the knives are very expensive.
3. As a general rule, hold the knife away from the body.
4. When carrying a knife in the kitchen, make sure the blade faces downward to avoid injury to self or to others.
5. When cleaning a knife ensure that the cutting blade is facing away from your wiping hand. Always focus when handling your utensils.
6. Do not attempt to catch a falling knife. Let it fall and get your feet out of the way. It is important to wear leather footwear for protection.
7. Do not hide a knife under anything.
8. Do not hand a knife to anyone else. Put it down on the table and let him/her pick it up.
9. When a knife is placed down onto a bench, ensure that the blade is flat.
10. Do not let knives hang over the table edge.
11. When using a knife keep your mind and eye on the job.
12. Use the right knife for the right job.
13. Always keep knives sharp. There is an old saying that 'sharp knives cut vegetables and blunt knives cut hand'.
14. Always keep the handle free from any grease.

SETTING UP OF WORKSTATION

In a busy kitchen there is a lot of hustle and bustle and noise. The guest's orders on the pick-up counter keep the service staff on their toes. Each dish has to be consistent and the hot food has to be served hot to the guest. One always has to be prepared even if the entire restaurant is full and the orders are waiting on the counter, commonly known as 'pass' and is usually handled by the sous chef or the chef in charge

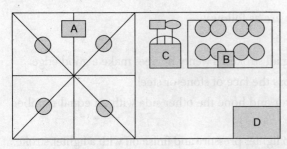

Fig. 1.5 Workstation set-up for basic cooking

of the meal service. In order to meet deadlines, it is mandatory for a cook to be well organized.

Let us understand a basic workstation and its components (Fig. 1.5). The following can be seen in the Fig. 1.5.

A: Gas burner range

B: Mise en place (ingredients kept for cooking)

C: Pot to hold cooking spoons and ladle

D: Chopping board

Gas Range

The cook has to ensure that all the ranges are in proper working condition. He/she must light all the burners and physically check them. In case of any malfunctioning, engineering must be informed for necessary repairs. A cook must light the small protruding pipes known as 'pilots'. Pilots should be kept lit as they emit a very small amount of flame and one does not have to light the flame every time an order comes in.

Mise en Place

This is a French word, which literally translates to 'putting in place'. A pre-preparation of a dish is also referred to as *mise en place*. Before being able to put finishing touches to a guest's food, the entire *mise en place* has to be in place and within reach so that the cook does not have to move from one place to another as this would tire him easily.

Pot Full of Water

Pot full of water to hold ladles and spoons kept near the cooking range saves a lot of time while cooking. The water in the pot helps rinse the ladles and spoons; but the water should be changed at regular intervals. Some people add a small amount of chlorine into the water so that the knives are constantly sanitized; however, care has to be taken as we do not want the smell of chlorine getting into the food.

Chopping Board

Wet the duster and squeeze out excess water, fold to the size of chopping board and place the chopping board firmly on it. This will prevent the chopping board from slipping. Usually it is advisable to fix the chopping board on the corner of the work table so that the debris can be collected easily in the bin and processed vegetables can be scraped into a bowl. As per hygiene laws, colour-coded chopping boards are used for various food commodities.

Red chopping boards are used for raw meats, and yellow for pork, green for vegetables, blue for fish, brown for cooked meat, and white for dairy products.

Setting up of a workstation can be different for each workstation. A cooking range for pasta cooking (Fig. 1.6) will be very different from grill section or a tandoor section (Fig. 1.7) to the Chinese cooking set-up. The following can be seen in Fig. 1.6.

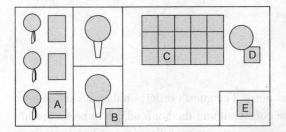

Fig. 1.6 Workstation set-up for pastas

A: Pasta cooker—This equipment is like a bain-marie, with dipping baskets to blanch pastas

B: Cooking pans for tossing pastas

C: Condiment tray for *mise en place* such as seasoning, olive oil, herbs, and vegetables
D: Pot for keeping cooking equipment such as slicers and ladles
E: Chopping board

The following can be seen in Fig. 1.7.

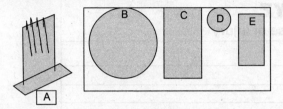

A: Stand for holding hot iron *seekhs*
B: *Tandoor*
C: Wooden plank for flattening Indian breads
D: Condiment tray for oil, dry herbs, etc.
E: Tray for basting the kebabs

Similarly, there can be many such set-ups but one has to ensure that each set-up provides a smooth workflow.

Fig. 1.7 Workstation for tandoor

SAFETY PROCEDURES IN HANDLING EQUIPMENT—ERGONOMICS

This section is very important for students aspiring to be chefs in hotels. Working in kitchens demands long hours on the feet and lifting of heavy pots and pans are normal way of life in the kitchens. Most of the students and also aspiring cooks suffer from back pain and spinal injuries in their industrial training because of their wrong postures. This section will suggest students how to work efficiently and safely as safety is the biggest concern in the kitchens.

Most people will agree that food quality, safety, and hygiene is of utmost importance as it satisfies the guests and makes them happy, which leads to more revenue generation as guests love to come back to such places. So strict control measures and systems should be in place for serving and cooking of food to the guests. It is seen that many a times some employers do not give too much importance to the health and safety of those people who serve or cook the food. The health and safety of workers should be a part of the induction and orientation process of every organization.

Kitchens can be potentially dangerous places to work and cooks and staff face a range of health and safety risks. According to surveys, slips, trips, and falls are some of the many causes of injuries; however, there are uncountable number of risks in lifting and carrying heavy baskets of food commodities, pots, pans, and heavy food pans containing food items. High temperatures in the kitchens and smoke, which is not good for patients of asthma, the use of chemicals, equipment such as knives, food slicers, and many other sophisticated machinery (if not handled carefully), can cause an injury.

Working conditions in many kitchens are not very conducive to work due to high temperatures and poor ventilation, which can be hazardous for the people working there.

As we read above that the main causes of accidents in the kitchens are slips, trips, and falls; but sometimes stray kept trolleys, protruding pots and pans, or being struck by falling objects or even at times being exposed to hazardous and toxic chemicals are also major causes of accidents in kitchens.

The employers are legally bound to report all the injuries and accidents to the management. It is also advisable to talk to the supervisor and take leave from work if a person is suffering from any food poisoning or infection, as the person may be the carrier of germs which will affect the others if he works in the food area. If this does not happen, then it should be assumed that the health and safety procedures in that workplace are not in place and it demands a serious review by the management. In case of an accident while on duty, report should be duly filed in. An incident report will look like the one shown in Fig. 1.8. The employer is legally liable to remunerate the worker if the injury results in death or partial or permanent disability.

HOTEL XYZ
Incident Report in Case of Injury
To be filled in by department head
1.   Name of the person injured:
2.   Address:
3.   Telephone number:
4.   What type of injury:
5.   Which part of the body:
6.   Where did the injury take place:
7.   What was the person doing at the time:
8.   Who witnessed the incident:
9.   Who attended to the person:
10.   Who administered first aid:
11.   What first aid was given:
12.   Was the person taken to hospital:
13.   Name of the hospital:
To be filled in by human resources/house doctor
14.   Room number:
15.   Name of doctor:
16.   Is the person still in hospital:
17.   Number of days admitted:
18.   Is the person insured:
Reported By/Date   Department Head/Date
To be sent to human resources department within three days of the incident

Fig. 1.8 Incident report format

All the employers should assess all risks to the health and safety of employees as it is not only their legal but also their moral responsibility. This assessing is known as 'risk assessment'. If the risk assessment report reveals that the work cannot be carried out safely, then the management must ensure that other arrangements are made for the same. The employers must appoint a team of safety representatives who would assist them to assess the risks. Employers should also provide the following to the employees as a safety measure.

- Facilities for health check-ups and first aid on duty. This is liaised with the human resource department.

- Suitable equipment to be provided and also its training should be imparted so that the employees can work safely.
- Training on fire safety and how to handle various types of fire. Fire extinguishers need to be placed in the right places and engineering department is responsible for checking the equipment on a periodic basis.
- Systems in place so that it is ensured that all the electrical equipment are checked regularly and the records updated and kept in a place for audit purposes.

First of all it is advisable to set up a team in each department that will act as safety inspectors. The team should do a risk assessment on the shop floor level and sometimes a format such as Fig. 1.9 can be duly filled and given to the concerned people to fix the problems.

HOTEL XYZ
Hazard Spotting Format
Date: _____

Hazard Spotted	Area and Time	Accident if Any	Recommendation	Action Taken	Engineering Comments

Manager's Name : _____

Executive Chef : _____

Engineering Leader's Acknowledgement : _____

Fig. 1.9   Hazard spotting format

When doing the risk assessment, wide range of jobs should be taken into account such as moving, lifting, and carrying materials and equipment (such as pots and pans, or cooking ingredients) and the chemicals used for cleaning and washing. The assessment should also take into account the layout of the work area, assessing whether it is crowded and should also take into account the storage areas, which are regularly used by the kitchen staff.

It should also be noted that even the ancillary staff in the kitchen such as kitchen cleaners, often referred to as 'kitchen stewards', are also prone to similar kinds of accidents on duty and sometimes it can also be said that probably they are the worst affected, as they deal with cleaning chemicals and toxic materials.

Slips, trips, and falls are the most common causes of injury in kitchens. More than a quarter of these result in major injuries, such as a broken arm or other injuries requiring hospitalization. Carrying loads or pushing/pulling trolleys increases the risk of slips and should be avoided or reduced. In order to prevent accidents in kitchen areas, the following should be kept in mind.

- Kitchens should have non-slip flooring which should be easy to clean and maintain.
- Floors should always be mopped and kept dry and there should be no obstruction in the walkway to the kitchen or in the kitchen itself.
- Use only the recommended cleaning materials from a reputed company and train the staff to use them in right dilutions, as if a chemical is used incorrectly or in wrong quantity it can cause the floor to lose its slip-resistant properties.

CHEF'S TIP
If oil falls on the floor and there is no time to clean, immediately sprinkle salt on it. The friction will prevent slips.

- If there are any spillages such as oil, water or food substance, it should be cleaned up immediately.
- Signage should be put on wet floor while cleaning kitchens.
- Proper storage should be available to keep floors clear.
- Ensure that non slippery and covered footwear is worn by the kitchen and ancillary staff working in the kitchen.

Hazardous Chemicals and Other Substances

One can find all sorts of chemicals in the kitchens and here we are not talking about condiments and spices; but cleaning materials such as:

- Dish washing liquid
- Detergents
- High-acid oven cleaners
- Disinfectants
- Drain and other cleaning products

Exposure to these is likely to be through contact with the skin or eyes, breathing in or swallowing. Many of the chemicals used in the kitchens are hazardous in nature because they are corrosive and can cause skin and eye burns if they are accidentally touched by hand or come in contact with the body. Some can simply cause irritations on the skin, whereas some can be fatal. Some substances can also cause breathing problems if they are sprayed in large quantities on hot surfaces, for example, high acid oven cleaners and this can happen especially where there is no proper ventilation.

We must ensure that the employees are properly informed about such hazards and should train them to ensure safe practices. Even after training is imparted the supervisors must ensure that the staff is supervised at all times, especially the new ones on induction.

Sprains and Strains in the Kitchen

Cooking involves a lot of lifting of commodities and equipment and also manual work such as decorating pastries or garnishing large amount of salads, which can be very repetitive in nature. It is seen that lifting, handling, and carrying account for more accidents than cuts, slips, and falls and such accidents are more fatal because they usually happen over a period of time by doing the same job over and over again. Back injuries can be very painful and most of the time they are difficult to treat and can lead to partial disability, but they can be prevented by being aware of the right posture to work, etc. Lot of manual handling tasks happen in the kitchens and these include pulling trolleys of food commodities and carrying food stores as well as lifting heavy crates of vegetables and fruits.

Unplanned work methods and inadequate training can lead to manual handling injuries to the kitchen staff. Most common strenuous tasks in kitchens include the following.

- Lifting, pushing, pulling, folding, or moving tables around, especially when the kitchens have to be deep cleaned.
- Setting up equipment and workstations.
- Moving food stores from storage and sometimes from the kitchens to the larger storage areas such as walk-ins and deep fridges.
- Filling and carrying large food containers with liquids or prepared foods.

Employers are required by law to prevent manual handling injuries. In brief, the regulations state that employers must do the following.

- As far as possible avoid the need for handling the work manually. For example, use well-oiled trolleys to move around with ease, when they are loaded with the permissible weight limits.
- Where manual handling cannot be avoided, identify and assess the risks and provide the necessary equipment and support systems.
- Ensure permissible load limits are marked and followed by the staff.
- Staff should be trained and well informed about the safety rules and regulations.
- Overalls should be supplied, especially when entering into deep freezers where a considerable time would be spent, comfortable safety shoes and uniforms should be provided. Refer back to the section on kitchen uniform and its uses.
- Training should be provided for ergonomics and other safety procedures such as evacuation in case of fire. But in spite of all this, one must remember the old phrase 'prevention is better than cure'.
- It should be ensured that the staff is trained when new equipment is being introduced and also if there are changes being made to the work methods which are directly related to the safety of the workers in the kitchens.

Certain jobs in the kitchens, such as chopping large amount of onions for a banquet operation, where along with repeated movements of limbs one also needs to apply pressure, can cause aches, pain or stiffness of muscles in the neck or shoulders. These disorders are known as WRULD in medical language, which stands for 'work-related upper limb disorders'. Kitchen staff should thus be trained to take care of the following.

- They should avoid too many repetitive movements and tasks, especially the ones which are of fast pace.
- They should take sufficient rest by taking breaks as kitchen involves long hours on the feet.
- They should not use unsuitable tools or equipment, for example, blunt knives.
- They should use the work areas of sufficient height for jobs such as chopping.
- They should do variety of tasks at the same position; doing the same job in the same position can cause a muscle to strain and stress out.

Fatigue or stiffness in the neck can result in neck pain and can cause headache at the same time. This situation will occur when the neck is in one position for a long period of time. This mostly happens when cooks do intricate jobs such as decorating cakes or salads. Always maintain a straight posture and avoid keeping neck in one position for long.

Cooks always forget the basic rules of lifting heavy things and end up having back problems. Figure 1.10 shows some right and wrong positions of working in the kitchen.

The following five general rules should be applied while lifting a weight.

- Plan the lift. Both squat and stoop lifting is now considered acceptable for jobs requiring repetitive lifting. The term used to describe this is 'free form lifting'. No matter what type of lift is used, it is never permissible to exceed the maximum acceptable load of the worker.
- Squat on the floor and access the weight.
- Keep the load as close to the body as possible.
- Lift the load with a smooth body motion (avoid jerking).
- When turning, do not twist. Turn with the feet.

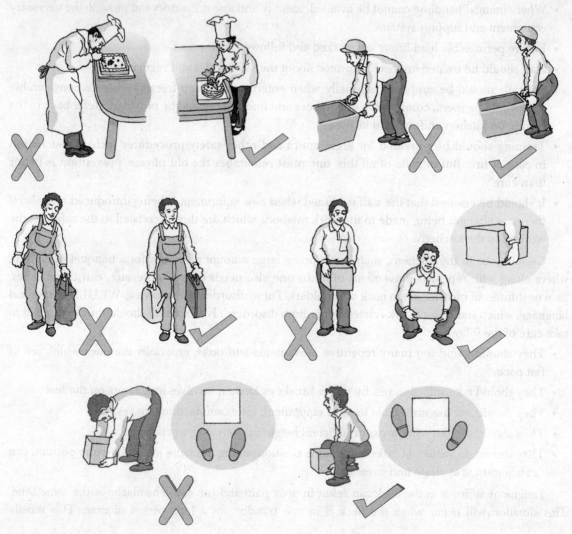

Fig. 1.10 Right and wrong positions of working

Temperature

High temperatures and humidity are very common in kitchens because of the fuel and heat used in cooking and the need for food to be served hot. But high temperatures can sometimes have an adverse effect on the cooks.

Working in high temperatures can cause many side effects such as lack of concentration, irritability, cramps in muscles, and sometimes fainting. Cooks should ensure that they keep drinking lots of fluids while on duty, especially in summers, to avoid dehydration. Some women are more at risk when working in areas that have high temperatures, especially pregnant women or lactating mothers, as pregnant women are more prone to heat stress and fatigue in high temperatures and breastfeeding for lactating mothers may be affected by dehydration.

On the flip side, cold temperatures can also cause discomfort, loss of concentration, irritability, and tiredness. Cold conditions can also cause fatigue since the body would use more energy to keep itself warm. More accidents can also occur as the cold may make the fingers numb and hence, an increased risk of accidents while working with sharp equipment.

The employers must ensure that:

- temperature inside a working kitchen is reasonable and this can be done by installing air handling units known as AHU and should have proper exhaust and ventilation systems;
- a warm overall coat is placed outside areas of deep fridges, so that one can wear it and enter deep freezer if one is going to take time inside the same.

Burns and Scalds

Most scalds and burns in the kitchens are caused by spillage of hot oil from the griller or the deep fat fryer or food spills from the pots and pans. In Indian kitchens, these occur mainly from hot *tandoor* while making bread and kebabs. One can see the scars on hands of every cook as, most burns or scalds occur on the hands, arms, and feet. These gave rise to the use of knee-length aprons and closed protective shoes to protect the sensitive parts of the body. These accidents can be avoided by ensuring that:

- the staff do not lift or carry heavy pans of food or hot water;
- oil and fat is filtered or moved only when it is cool;
- appliances are allowed to cool before being cleaned;
- special oven gloves are used when opening ovens and dry dusters to be used while handling hot pots and pans and utensils while cooking (wet dusters would emit steam, when they come in contact with the hot pans and this can scorch the hands).

All staff including those who work in kitchens should be given health and safety training. This is in addition to food hygiene training. Employers have a legal duty to give health and safety information and training to all employees. But for kitchen staff this is very important, as the cooks are exposed to potentially dangerous equipment in kitchens. Apart from all this basic training, the staff should also be trained to handle small fire in the kitchens or any risk that employees are exposed to.

FSSAI

The Government of India in the last one decade has realized the importance of establishing strict guidelines for food establishments. Under the Food Safety Standard Acts of India, 2006, the government has established the *Food Safety and Standards Authority of India*, abbreviated as FSSAI. This body is responsible for framing various regulations that would direct a food establishment towards better food safety standards. Strict audits would be done and defaulters would be penalized or forced to close down, if they default more than once on the regulations.

FSSAI prepares and circulates guidelines for the food establishments so that they can comply and obtain accreditations and certifications from the nominated bodies. It not only looks into the present, but is also responsible for calculating any potential threat that can occur in future with regards to food contamination. In regards to this, FSSAI advices the Central Government and state governments to put right practices into place. It also conducts training programmes for people involved in the food business. The aim of the FSSAI is to contribute to the development of international technical standards for food.

FSSAI is largely responsible for the following.

1. **Food Licencing and Registration Systems:** FSSAI lists down strict guidelines for the design and facilities of any food establishment in order to procure licence. The directive comprises of areas such as:

   (a) **Location and surroundings-** This involves placing or locating food away from any potential sources of environmental contaminants such as smoke and chemical or biological odours that can cause a health hazard to an individual.

   (b) **Premises and rooms-** The establishment should be designed in such a way that it facilitates the processes of cleaning and disinfection, and prevents any biological, physical, or chemical contamination. It lists down various parameters that should be adhered to so that no contamination of food can take place. Some of these guidelines suggest usage of specific fixtures and fittings and non-toxic materials in the food area. Important points such as proper sealing of the floor, ceilings, and walls to prevent entry and harbouring of pests are also mentioned.

   (c) **Equipment and containers-** FSSAI also lays down strict guidelines for the materials used in construction of equipment for preparing and storing food. The equipment should be designed in such a way that it facilitates cleaning and should be made of non-corrosive materials to prevent chemical food poisoning.

   (d) **Facilities and utilities-** The food unit must have an adequate supply of potable water. Even the ice machines should use potable water for forming ice for food usage. Waste disposal should be done in a hygienic way in accordance with the local rules and legislations as amended from time to time.

   (e) **Personal hygiene and employee facilities-** The premise should be designed in a way that it provides facilities for employees to maintain personal hygiene. Space should be provided for bathing and washing to maintain proper health and sanitation. The air quality, ventilation, and lighting of the kitchen should also be maintained as per the standards to aid in the good health of the employees and the designed food premise.

2. **Food Safety and Hygiene:** Food should be prepared in the most hygienic manner free from all types of contaminants. There should be a system of testing food in laboratories and the reports of such tests should be made available to the food inspector in case of any audit. The procurement of raw material should be free from any hazardous contaminant and the raw material should be stored in the correct container under correct temperatures. Even the packaged products should

be checked for expiry dates. Any allergen that can cause a medical threat should be clearly mentioned in the menus and on the packed food label.

3. **Food Import Clearance Systems (FICS):** The FSSAI also regulates and advices on the ingredients and food imported into our country. The manufacturers of these products abroad, must adhere to the guidelines laid down by the FSSAI and should clearly mention the ingredients on their labels. One must only use ingredients that are specified by FSSAI and considered fit and safe for human consumption.

CONCLUSION

This is the first chapter of this book that introduces the concept of professional kitchen. The things mentioned here are very basic and prepare the learner to understand the jobs done in a professional set up. This chapter discussed about the history of cooking and evolution of food with respect to the societies, castes, and status. It was necessary to write about the basic elements of the kitchen with regards to the 4 P's namely people, product, process, and profit. We saw how the role of chef is crucial in making sure that the food service organization meets the budgets and controls the bottom line. We also read about the basic organization in the hotel; however we shall discuss more about the same in the next chapter.

Few key elements were stressed upon such as professionalism in the kitchen, where every person works in a methodical and disciplined manner. Kitchen can be a dangerous place to work in as the modern kitchens deal in machines and equipment that are very sharp and sometimes to a layman this place can look very confusing. Hence it is important to be vigilant and display high levels of alertness and passion to survive in these odd working conditions.

We also discussed the most sensitive areas of kitchen and that is health, hygiene, and safety. We discussed about the importance of personal hygiene, food safety and the many ways to protect food from spoilage and contamination. We read about different contaminants such as microbiological, chemical, physical, and poisonous plants. Since the job in the kitchens requires odd and long working hours and physical labour, the aspect of safety such as the correct way of lifting weights cannot be ignored.

We discussed the different parts of the chef's uniform and understood the usage of each element in the same. We discussed about the knives, their parts and also the ways of sharpening them. We discussed about various equipment used for sharpening of the knives and learnt about the basic set up of the workstation to carry out the job effectively. Right equipment such as knives and chopping boards play an important role in the safe and hygienic preparation of nutritious food that will keep bringing happy customers back.

FSSAI is playing an important part in today's kitchen and food industry, it was important to introduce students to the basics of Food Safety Standard Acts of India also abbreviated as FSSAI.

The next chapter will deal with the organizational set up of various establishments and we shall talk about the hierarchy of kitchen department.

KEY TERMS

Apprentice   Person under training to become a cook.

Arkansas   A type of sharpening stone for knives.

Asparagus   A kind of vegetable used in European cooking.

Bacteria   Unicellular microorganisms.

Bain-marie   An equipment with hot water used to hold hot food.

Barbecue   Literally means barbe (beard) cue (tail). So a whole animal roasted from head to tail.

Bugs   Another name of bacteria and germs.

Carborundum   Type of stone for sharpening knives.

Chargrill   Method of cooking where a commodity is cooked on high heat.

Chef de partie   Supervisor in the kitchen. Also called CDP sometimes.

Chef   French word for chief, usually referred to experienced cooks and kitchen managers.

Commensals Friendly bacteria that help humans in many ways.

Commis French word for cooks on the bottom of the organizational structure.

Contamination Presence of harmful microorganisms in food.

Cook One who cooks for professional living.

Cuisine French word for things related to food.

Demi chef de partie Assistant to the kitchen supervisor. Also called DCDP sometimes.

Ergonomics Scientific approach to lifting weights and using equipment for human use.

Executive chef One who heads the kitchen and is overall in charge.

Executive sous chef One who is next in position to the executive chef.

Feta A sharp tasting cheese usually made from sheep's milk.

FICS Food import clearance system is a set of standards for importing food into India.

Food cost Cost of the raw material incurred to prepare a dish.

Food poisoning Sickness caused by food infected with microorganisms.

Food safety Methods of preparing food in a safe and hygienic manner.

FSSAI Food Safety Standards Authority of India.

Generation time Time taken for the bacteria to divide into two for multiplication.

HACCP Hazard Analysis and Critical Control Points is a programme that certifies safe and hygienic practices.

Hierarchy Organizational structure of kitchen.

High risk food Food rich in protein and which does not need further cooking.

Honing Sharpening of edge with a sharpening steel.

Hygiene Science that deals with cleanliness and sanitation.

Incident report Format filled in case of accident on the job.

Industrial trainee Students who are studying hotel management and go to hotels for their training.

Internship trainees Same as industrial trainees.

Kitchen stewarding Department that takes care of cleaning of kitchens.

Kosher laws Food eating laws of Jews.

Lean hierarchy When in an organizational structure there are not many levels of reporting.

Marbling The distribution network of fat and muscle in the meat.

Master chefs Skilled chefs who have gained expertise in one area of kitchens.

Mise en place Arranging things prior to cooking, putting up things in place.

Multi-skilling When variety of skills are learnt by cooks.

Operating instructions Instructions on how to use a machine.

Organic food Food grown without the addition of pesticides and chemicals.

Organizational chart The number of staff in a kitchen and their seniority structure.

Oyster A kind of seafood often eaten raw.

Pass Area where food is picked up from, for service.

Pathogens Bacteria that can cause infection and disease.

Portioning Serving of agreed quantity of food to the guest.

Restaurant A place that offers food and beverage service on charge basis.

Saucier Cook who is responsible for preparing sauces.

Serrated An edge of the knife that has teeth like or saw like sharpening edge.

Shuck A term referred to opening of live oysters with a special knife called oyster knife.

Sous chef Literally means under the chef. He/she is a person reporting to executive sous chef.

Spit roast Method of cooking where the piece of meat is grilled over direct fire.

Staff cook Person who cooks meals in cafeteria for staff.

Tandoor A cylindrical clay oven for baking Indian breads.

Thawing Defrosting the frozen products.

Toque blanc Chef's/cook's hat worn to prevent hair falling into food.

Toxins Poisonous wastes secreted by bacteria.

Veste blanc Chef's/cook's coat.

Walk-in Large fridges where one can walk into.

CONCEPT REVIEW QUESTIONS

1. What are the food laws of Jews known as?
2. List the basic organization of the kitchen department.
3. What is the head of the kitchen department known as?
4. Who is at the bottom of the hierarchy?
5. List the attitudes and behaviour of the kitchen staff in a professional hotel.
6. 'Personal hygiene' in food rooms is of utmost importance. In a chart write down five dos and five don'ts that must be followed by food and beverage professionals.
7. What do you understand by the term 'food safety'?
8. List the three main types of food contamination.
9. Give examples of at least five physical contaminants in food.
10. What are the vehicles and routes of contamination? Give examples.
11. What is the difference between bacteria and virus?
12. How would you ensure that the food is safe from microorganisms?
13. Write at least four points that need to be considered while designing the uniform for the kitchen staff.
14. What is so special about the chef's jacket?
15. What is the role of the scarf in the kitchen uniform?
16. List the various parts of a knife and its uses.
17. What is the difference between a tang and a bolster?
18. List at least three things that can be used for sharpening a knife.
19. List at least five safety considerations while handling knives safely.
20. What is the basic workstation set-up for any kitchen?
21. What do you understand by the term *mise en place*?
22. List the various kinds of chopping boards and their uses.
23. List at least 10 hazards that can take place in the professional kitchen.
24. List at least five safety procedures of handling any kitchen equipment.
25. What is an incident report and how is it useful in the hotel?
26. What is the importance of grooming in the kitchen?
27. What kind of material is used for kitchen uniforms and why?
28. Why should the chef's coat sleeve be folded just below the elbow?
29. What kind of shoes are normally worn in the kitchen?
30. What kind of peeler is used for peeling asparagus?
31. Why is santoku knife used for slicing meats?
32. What is unique about the sashimi knife?
33. Why does the cheese knife has a hole in its blade?
34. What is the FSSAI?
35. What is the importance of FSSAI in the food industry?
36. What are the FSSAI guidelines for design of premises?
37. What are the FSSAI guidelines for importing food into India?

PROJECT WORK

1. In groups of five, do a market survey of hotels of various categories and a few stand-alone restaurants and draw out an organizational chart. Compare the charts and record your findings.
2. Conduct a survey of professional kitchens and institutional canteens and study the kitchens from health, hygiene, and safety point of view and list down the observations made therein. Give suggestive feedback to the person concerned with a detailed report on the same.
3. While visiting fast food restaurants or hotels, study the arrangement of the workstation from the health, hygiene, and safety point of view and operational feasibility. Discuss with faculty and friends and draw out a better plan, if any, and give justifications for the same.

CHAPTER 2

HIERARCHY OF KITCHEN DEPARTMENT

Learning Objectives

After reading this chapter, you should be able to:
- understand the importance of the kitchen department and the role it plays
- know the classical brigade of the kitchen
- figure out the organizational structure of the kitchen department in hotels and food establishments of various types and sizes
- know about the various personnel in the kitchen and their duties and responsibilities
- list the personal attributes required in a kitchen personnel

INTRODUCTION

In the previous chapter we discussed the fundamentals of the kitchen in relation to health, hygiene, and safety. We also discussed about the organization in the kitchen. This chapter deals with the various levels of hierarchy in the professional kitchen for various kinds of establishments. The French term 'chef' is the most popular word used for professional cooks around the world; but there is a small difference—a chef is more like a kitchen manager, whose prime job is to cook good food and also lead the team with a common goal and vision. There was a time when cooks were specialized only in a particular field and so there were large organizational structures where each cook headed a section and had helpers and apprentices under him/her. With modernization and the labour being replaced by machines, a more linear hierarchical design of an organization was needed. The need for multi-skilling replaced the specialized cooks with people who could handle more than one kitchen at the same time and this gave a very structured and lean hierarchy that is followed in most of the hotels till today. This chapter discusses the classical brigade and the latest lean brigades followed in many five star establishments today. It also focuses on the job responsibilities of various chefs in the hierarchy, which helps understand the modern approach to optimize the staffing levels to run the establishment like a well-oiled machine.

CLASSICAL KITCHEN BRIGADE

The word chef is literally translated to 'chief'. Hundreds of years ago, this profession was put in the same category as domestic help; but media, awareness, and professional chefs' organizations have promoted the professional stature of the chef. A great deal of cross exposure is happening around the world and chefs are travelling to different parts of the world to learn and hone their skills. One can never stop learning in the kitchen. Many people choose to cook, but only a few, who reach a certain level of proficiency, are called chefs. We discussed in the first chapter the various jobs a chef has to do throughout his/her career. In this chapter we will discuss in detail the duties and responsibilities of each person in the kitchen department.

The job opportunities available to a chef today are varied and are still expanding with the diversification of societies.

Classical Brigade

Hotels and restaurants come first to mind when a cook wants to start his/her career as a chef. In the European style of wining and dining, the restaurants in the hotels were known as 'fine dining', which were also at times referred to as white tablecloth restaurants or highend restaurants. With changing times, the need for casual dining or 'bistro' came into being. With the expansion of business, people became busy and had less time to spare and this gave rise to family restaurants and fast food restaurants. Today, five-star hotels usually have a number of restaurants such as a fine dining and a coffee shop, where families can spend time together. Chef Auguste Escoffier introduced a system known as 'brigade' in a professional kitchen. This brigade or the kitchen hierarchy has been modified and made lean because of rising costs and availability of modern equipment and machinery to accomplish jobs faster and more accurately. Earlier, chefs were restricted to royal palaces and clubs. With the advent of hotels, it was impossible to do without a structured layout, as the professional kitchen is quite complex. There are many jobs and responsibilities that have to be distributed to the kitchen staff so that there is sense of accountability and ownership.

Following are the key points related to the classical brigade in a professional kitchen.

- The traditional system was founded by Chef Escoffier.
- A professional kitchen operates with a very distinct rank and file and hierarchy just like an army of soldiers in a battlefield; probably this is why Escoffier called it a brigade. Till today the team is lead by a leader—the executive chef.
- Each member of a kitchen brigade has his/her own role to play for the success of the team.
- It was based on various sections in the kitchen (for example, pastry, butcher, bakery, etc.).
- Every section was named in French.
- Every section featured a head of department, with cooks, helpers, and porters working under him/her.

Nowadays the requirements of the kitchen are different and hence, the hierarchy has changed. Manpower is expensive and menus are not elaborate as in classical cuisine of the good old days. Every hotel has a brigade depending on its cuisine, number of outlets, volume of business, and functioning style.

The classical brigade in the kitchen had its own flaws; the cooks had a monotonous life as they did the same job all their lives. There was no creativity involved as the cooks had to do what was told to them.

ORGANIZATIONAL STRUCTURE OF THE KITCHEN

Modern kitchen organizations aim at orienting staff in all the areas of the kitchen, so that a multi-skilled workforce is created. A business organization is defined as an arrangement of people in jobs to accomplish the goals of the operation. The organizational structure of the kitchen reflects the needs of the operation, the job functions, and the various goals.

The jobs and duties of staff members also vary from kitchen to kitchen, and so do the titles attached to the jobs. But certain positions and titles do occur throughout the industry. Here are some of the most common positions with a general definition for each and a place in the typical kitchen hierarchy.

Chef De Cuisines (Executive Chefs or Head Chefs)

This position carries overall responsibility for all aspects of production, for the quality of the products served, for hiring and managing the kitchen staff, for controlling costs and meeting budgets, and for coordinating with departments not directly involved in food production.

Duties also include making new menus, purchasing, costing, and scheduling of employees. They are also responsible for kitchen plant and machinery.

Sous Chefs (Under the Chefs)

They are the principal assistants to the head chefs and aid the chefs in general administration and in particular, supervising food production and overseeing its service. They are the acting head chefs in the absence of the head chefs.

Chef Garde Mangers (Pantry Chefs)

They are responsible for all cold food presentations, which might include hors d'oeuvres, salads, sandwiches, pates, etc.

Butcher Chefs

They are in charge of the butcher shop which prepares meats, fish, and poultry as desired by the user departments of the kitchen.

Pastry Chefs

They enjoy a different status and the work of their department is generally separated from the main kitchen and is self-contained in the matter of cold storage, machinery, and equipment. They are responsible for all hot and cold desserts. These may include cakes, pastry, ice creams, creams, etc.

Boulangers

They are the bakers who work under the pastry chefs and are responsible for all baked products such as bread, breakfast rolls, etc.

Potagers (Soup Cooks)

They are responsible for preparing soups and stocks, which may include cream soups, consommés, bisques, broths, national soups, essences, etc.

Entremetier (Vegetable Cooks)

The *entremets* course is, on the menu, the sweet which is prepared by the pastry chefs. *Entremets de legumes* were the vegetable courses traditionally featured on a menu. The entremetiers are therefore concerned with the preparation of the following.

- All vegetable dishes
- All potato dishes
- All egg dishes
- All *farinaceous* dishes

Chef Rotisseurs (Roasting Cooks)

They are responsible for braised meats, roasted meats, and meat dishes. Their section is also responsible for deep-frying of foods.

Sauciers

They are responsible for all sauces and sauce-related dishes.

Banquet Chefs

They are responsible for all food to be prepared for banquet functions and also for the buffet in coffee shops.

Chef Tournants

They are the reliever chefs who take charge in the absence of the section chefs. They were usually multi-skilled cooks, who would fit into any job in case of emergencies.

Chef De Parties (Section Chefs)

All chef de parties (CDP) are supervisors in charge of a clearly defined set of activities within the kitchen. They are the station heads and must be skilled to cook every dish made by their stations. They should also have a certain degree of administrative skills. They should be able to plan and carry out production schedules for the section.

Demi Chef De Parties (DCDP)

They are also in a supervisory capacity. They take charge in the absence of the chef de parties. They assist the chef de parties.

Commis

There are two types of commis—commis I and commis II; the I being seniors. They are the assistants to the chef de partie. However, in most hotels now, the commis I and II have been classified as commis only.

Apprentices

These are the trainees who help out in day-to-day operations.

The positions defined here are in a classical sense. In the real world, they are combined, altered, and adapted to fit the specific goals of the individual operation.

Figures 2.1–2.5 represent the organizational charts of the kitchen departments of different types of hotels.

MODERN STAFFING IN VARIOUS CATEGORY HOTELS

One can see that the classical brigade of the kitchen (Fig. 2.2) has an exhaustive list of cooks and chefs for a particular job. The classical menus were very elaborate, because in those times, it was not mere dining out, it was a feast. Each section of the kitchen had specialized cooks to do the jobs and were supervised by supervisors, who had been in their positions for years. It was very difficult for everyone to become an executive chef as the promotions depended not always upon one's performance, but on the number of years one had spent in an organization.

With the modernization of the society, the kitchen also had to undergo a complete overhaul and this was necessary because of various factors such as changing patterns of eating, expensive labour, etc.

In early times, people had a lot of time for themselves and for their families, and outings in the evening were lavish and ended late in the night. This style gave rise to two words commonly used in hotels now—'high tea' and 'supper'. A high tea, also written as 'hi tea' on menus, is often confused with afternoon tea. A high tea was meant for children and this would take place between five and six in the

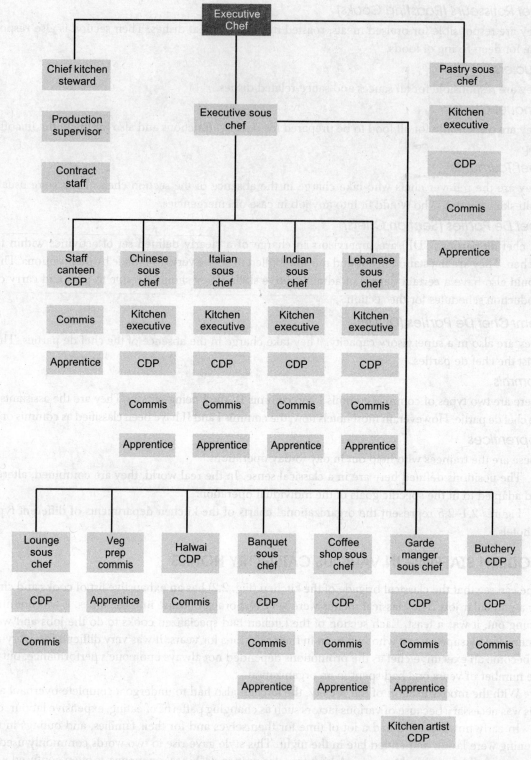

Fig. 2.1 Organizational chart of kitchen department of a large-size hotel

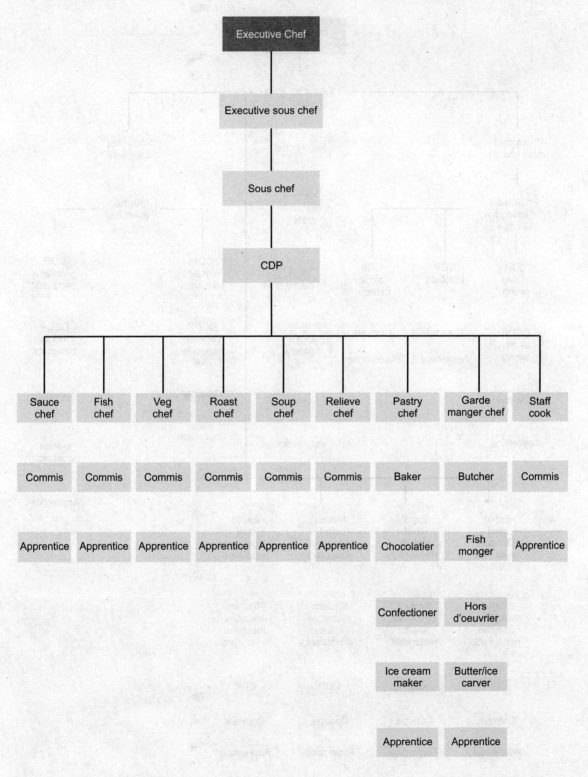

Fig. 2.2 Organizational chart of kitchen department of a large-size hotel (classical version)

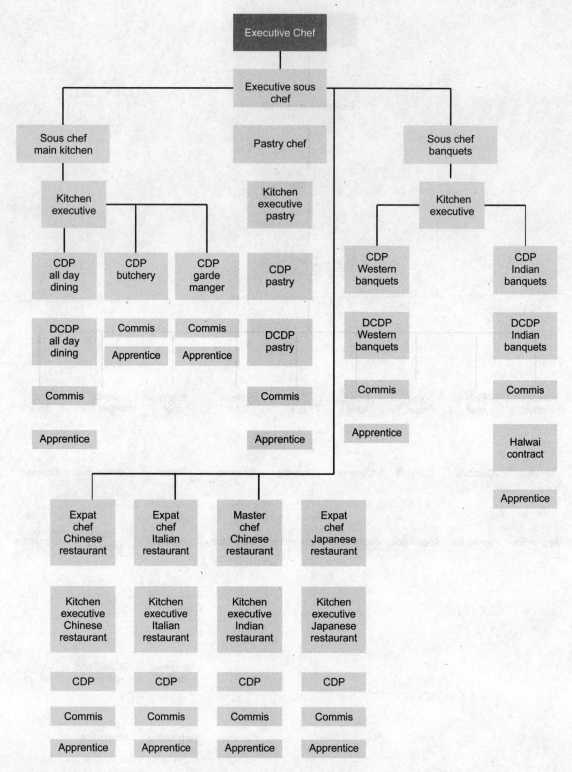

Fig. 2.3 Organizational chart of a modern kitchen department of a large-size hotel

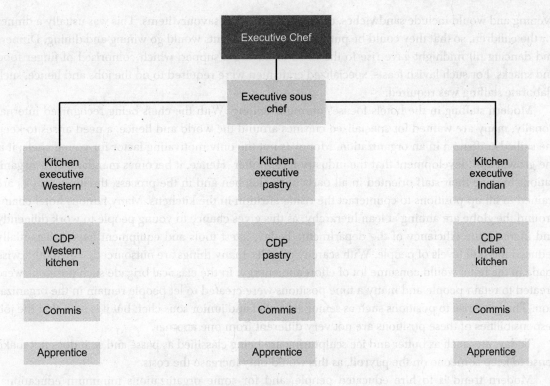

Fig. 2.4 Organizational chart of resort hotel

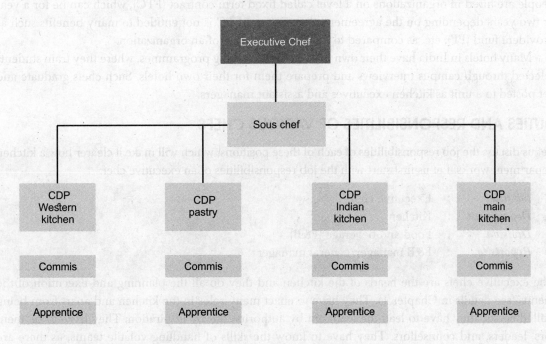

Fig. 2.5 Organizational chart of a medium-size hotel

evening and would include sandwiches, dry cakes, and other savoury items. This was usually a dinner for the children, so that they could be put to sleep, while parents would go wining and dining. Dinners and dancing till midnight gave rise to the need of having a supper, which comprised of finger food and snacks. For such lavish feasts, specialized craftsmen were required to do the jobs and hence, such elaborate staffing was required.

Modern staffing in the hotels focuses on multi-skilling. With the chefs being recognized internationally, many are wanted for specialized cuisines around the world and hence, a need arises to keep the skilled craftsmen in an organization. Money is not the only motivating factor for young chefs; it is the growth and development that the industry has to offer. Hence, it becomes mandatory for organizations to have their staff oriented in all parts of the kitchen and in the process, the junior cooks are trained to fill up positions to counteract the rising attrition in the kitchens. Many famous hotel chains around the globe are aiming at lean hierarchy, as this gives chance to young people to work diligently and increase the efficiency of the department. Today, latest tools and equipment have substantially reduced the skill levels of people. With scarcity of skills, many things are outsourced, which otherwise made in the hotel would consume lot of effort and energy. In the classical brigade such positions were created to retain people and many a time positions were created to let people remain in the organization. This gave rise to positions such as senior sous chef and junior sous chef; but it is seen that the job responsibilities of these positions are not very different from one another.

Today, arts such as butter and ice sculpturing are being classified as passé and so it does not make sense to keep someone on the payroll, as this would only increase the costs.

Modern trend is to hire educated people, and for some organizations minimum educational requirement is a graduation. Many skilled craftsmen, such as Indian sweet makers or dim sum specialists, have learnt their skills when they were teenagers and may not have had any education. Such people are hired in organizations on a level called fixed term contract (FTC), which can be for a year or two years depending on the agreement. A person on FTC is not entitled to many benefits such as provident fund (PF), etc. as compared to a regular employee of an organization.

Many hotels in India have their own management training programmes, where they train students selected through campus interviews and prepare them for their own hotels. Such chefs graduate and get posted to a unit as kitchen executives and assistant managers.

DUTIES AND RESPONSIBILITIES OF VARIOUS CHEFS

Let us discuss the job responsibilities of each of these positions, which will make it clearer how a kitchen department works. Let us first start with the job responsibilities of an executive chef.

Job title	:	Executive chef
Department	:	Kitchen
Division	:	Food and beverage (F&B)
Reports to	:	F&B manager/general manager

The executive chefs are the heads of the kitchen and they do all the planning and execution of the menus (see details in Chapter 1). They have to enact many roles in the kitchen and apart from being skilled cooks, they have to lead the team not by authority, but by inspiration. They have to be mentors, leaders, and counsellors. They have to know the skills of handling volatile teams, as there are always petty issues between the kitchen staff themselves and also with the service teams. The animosity between the food and beverage (F&B) service and the cooks is not hidden from any professional working in the hotels and in spite of trainings and team building exercises, neither of the teams appreciates or is empathetic towards the other. The executive chefs are the role models for the budding chefs and so their job in the kitchen is very stressful.

An executive chef has to direct and guide the food production team in providing a consistent quality of food and beverage in all outlets, as per international standards, in order to achieve the maximum level of guest satisfaction and organizational profitability in an atmosphere of high employee morale. The specific responsibilities of an executive chef are to:

1. Drive the vision and the goal of the company.
2. Reflect the company's philosophy by providing the highest standard of personalized and attentive, but discrete service in a professional and friendly manner, which exemplifies the best of hospitality.
3. Always lead by example, adopting a positive attitude to keep the team spirit at its highest level.
4. Greet with a smile, colleagues and guests at any time or place within the hotel, whether front or back of the house.
5. Anticipate guests' needs and wishes, and surpass their expectations.
6. Look continuously for ways to achieve the hotel's strategic vision and goals by working as a team and being a team player.
7. Be proactive in developing oneself by taking advantage of all learning opportunities, and by striving to achieve the goals of one's personal career development plan and personal mission statements.
8. Be committed to quality and profitability of product to ensure that guests return and to aim to be the best hotel or outlet.
9. Identify and develop new products and equipment, to enhance the product quality.
10. Develop and define quality standards of food preparation and presentation.
11. Define the organization of work within the department including assignments, time schedules, and vacations of staff (Fig. 2.6).
12. Ensure the quality of food preparation and presentation, as per organizational standards.
13. Ensure availability of stock and raw ingredients by proper planning and coordination with purchase and stores.
14. Coordinate with the engineering department to carry out preventive maintenance programme in the kitchen.
15. Establish recipes and methods of preparation, inform the F&B director of significant change in prices affecting the preparation of menu items.
16. Recommend menu pricing in coordination with F&B director/F&B manager/F&B controller/ banquet manager.
17. Be responsible for the hygiene and cleanliness of the kitchen areas, equipment, and staff.
18. Ensure compliance with company and hotel policies by department employees.
19. Constantly monitor key performance indicators for the department and take appropriate action.
20. Analyse and monitor costs, (material, energy, and staff) to ensure high profitability on a regular basis and initiate corrective action whenever necessary.

HOTEL XYZ
Leave Planner
2013–14
Kitchen Department

Name	Kitchen	Apr	May	June	Jul	Aug	Sep	Oct	Nov	Dec	Outstanding—Previous years
Bali	Exec chef										
Vishal Arora	Exec sous chef										
S. Singh	Banqueting										
S. Castelino	But/veg										
A. Jorhda	Rasoi										
G. Kakkar	Pastry										
L. Penpet	Baan thai										
Chef Wei	Taipan										

Fig. 2.6 Leave planner chart of kitchen personnel maintained by an executive chef

21. Ensure that menus are changed on a regular basis, as per corporate guidelines and market needs, in coordination with F&B manager/F&B director.
22. Ensure that the best quality of raw material is procured and used in food preparation.
23. Prepare capital and operational budget in order to achieve desired profitability.
24. Ensure storage of raw and cooked food/raw material as per international standards.
25. Keep oneself updated with market knowledge and trends by conducting regular market surveys in coordination with the purchase department.
26. Ensure department employees are fully trained through continuous on-the-job training.
27. Attend behavioural and vocational training in own and related work areas to enhance skills and develop multifunctionality.
28. Ensure practice of hygiene and safety precautions as well as compliance with hotel and company policies by the kitchen staff through training.
29. Provide career development and succession planning for subordinates through training.

Job title	:	Executive sous chef
Department	:	Kitchen
Division	:	F&B
Reports to	:	Executive chef

An executive sous chef assists the executive chef in directing and guiding the food production team in providing a consistent quality of food and beverage in all outlets, as per international standards, in order to achieve the maximum level of guest satisfaction and organizational profitability in an atmosphere of high employee morale. The specific responsibilities of an executive sous chef are to:

1. Coordinate in detail all food preparation and productions for all the outlets, to ensure total smooth running of the operation.

2. Ensure that all set-ups and prepared food presentations are up to agreed standards with the executive chef.

3. Fully monitor portion control in assigned kitchen area and that par stocks are kept at a minimum with daily stock and product turnover, utilizing leftovers where and if possible.

4. Fully supervise all food tasting, presentations, and food samplings in all production phases.

5. Conduct daily cleanliness checks with the assistant chief steward, with feedback to the executive chef with corrective actions to be taken.

6. Follow up on any issues that are delegated by the executive chef.

7. Recommend changes in menus, methods of preparation, kitchen/service equipment or staff, to improve production standards and quality.

8. Assist the executive chef in developing new cuisine concepts.

9. Ensure implementation and maintenance of standards of food quality, preparation, and presentation.

10. Assist in menu planning and costing.

11. Ensure good physical upkeep, condition of equipment and utensils in all the kitchens, and coordinate with the engineering department for repairs and maintenance.

12. Ensure organization of work within all kitchen sections, including assignments, time schedules, and vacations as per organizational standards.

13. Make sure all the raw material received, in the hotel, for preparation is of high quality, as per organizational standards and product specifications.

14. Recommend changes in systems and procedures to increase efficiency and improve service quality.

15. Ensure prompt and accurate service by kitchen staff to the customer, to achieve a high level of customer satisfaction.

16. Be responsible for maintaining international standards of safety, security, hygiene, and cleanliness in all food preparation and storage areas.

17. See that all the employees in the department adhere to the organization's rules and regulations.

18. Ensure availability of raw ingredients at all times, by proper planning, requisitioning, and storage.

Job title	:	Sous chef
Department	:	Kitchen
Division	:	F&B
Reports to	:	Executive sous chef

A sous chef has to organize, develop, and supervise the food production in the main kitchen as per standards and recipes developed by the executive chef and to handle independently one of the satellite kitchens assigned to him/her. The specific responsibilities of a sous chef are to:

1. Look after the function of the executive sous chef during his/her absence.

2. Train staff on improved work procedures, quality food production, economical usage of food materials, and the attractive presentation of food items.

3. Approve requisitions from stores for stations assigned and in the executive sous chef's absence for the entire main kitchen.

4. Be responsible for all food production in area assigned to him.

5. Be responsible for overall food cost control without affecting standards and specifications as laid out by top management.

6. Account for the usage, consumption, spoilage, and control of food stuff produced or stored under his/her supervision.

7. Be responsible for the preparation of *mise en place* at all stations.

8. Attend the food and beverage meetings and departmental meetings.

9. Maintain all attendance records.

Job title	:	Pastry chef
Department	:	Kitchen
Division	:	F&B
Reports to	:	Executive sous chef

The pastry chef has to organize, develop, and supervise the pastry shop including stations such as bakery, pastry, and confectionery. The specific responsibilities of a pastry chef are to:

1. Be responsible for the *mise en place* and food preparation for all bakery and pastry stations.

2. Account for the usage, consumption, and control of all foods and equipment in the stations supervised by him/her.

3. Train staff in his/her stations on improved work procedures, quality food production, the economical usage of food materials, and the attractive presentation of food items.

4. Supervise and train the chef de parties, commis, and apprentices, and to review staff working in his/her department.

5. Initiate performance reviews of staff working in his/her department.

6. Attend the daily and weekly kitchen chef's meetings and the F&B meetings.

7. Approve requisitions from stores for materials required in his/her stations.

8. Recommend schedule changes and changes in personnel for adequate manning of all stations.

Job title	:	Kitchen executive
Department	:	Kitchen
Division	:	F&B
Reports to	:	Sous chef

A kitchen executive ensures that the food production team provides a consistent quality of F&B in the area/shift under his/her control, as per the corporate preset international standards, in order to maximize guest satisfaction and organizational profitability in an atmosphere of high employee morale. The specific responsibilities of a kitchen executive are to:

1. Ensure adherence to organizational standards of food quality, hygiene, preparation, and presentation in his/her kitchen.

2. Make sure all the kitchen equipment and machinery is in good working order at all times, in his/her area of work.

3. Recommend changes in systems and procedures to increase efficiency and improve service levels.

4. Ensure prompt, courteous, and accurate service to all the guests to achieve high level of guest satisfaction.

5. Be responsible for maintaining of records/documentation in his/her area as per operational/control requirements.

6. Check the quality and availability of raw ingredients at all times for smooth operation.

7. Provide timely follow-up on any sick team member and convey the report immediately to HR and executive chef.

8. Ensure buffets/food displays are set up and maintained professionally, as per organizational standards.

9. Be responsibe for the hygiene standards of his/her kitchen, storage areas, equipment, and machinery.

10. Control food wastage, without compromising on food quality.

11. Check that cleaning schedules by kitchen stewarding department are being followed in timely manner.

12. Ensure level of dry stores and perishables are maintained on daily basis, and also ensure correct store requisitioning.

13. Check attendance and punctuality of every team member.

14. Provide functional assistance to all subordinates and peers of various areas.

15. Ensure excellent relations and professionalism amongst all staff in his/her kitchen and with related departments and staff.

16. Work in close coordination with F&B service team.

17. Maintain appropriate and professional communication with F&B team at all given times, and for any special occasions.

18. Provide constant on-the-job and classroom training for his/her kitchen employees.

19. Personally conduct critical training sessions.

20. Encourage team building through regular informal meetings and keep an open door policy.

21. Coordinate functions and activities with other F&B section, engineering/house-keeping, etc. whenever required.

22. Assist sous chef with on-the-job training and class room training for his/her kitchen and related F&B employees.

23. Attend behavioural, vocational, and skill-related training, to enhance his/her skills and develop multifunctionality.

24. Provide cross training to employees of other departments.

25. Personally conduct critical training sessions.

26. Provide constant on-the-job training and coaching to all the staff in the department.

27. Share his/her skill and knowledge with all employees; follow the company standard operating procedures (SOP) in his/her kitchen.

28. Counsel subordinates in work-related and personal matters.

29. Attend behavioural and vocational training in own and related work areas to enhance skills and develop multifunctionality.
30. Maintain records as required of training in the department.

Job title	:	Chef de partie
Department	:	Kitchen
Division	:	F&B
Reports to	:	Kitchen executive

A chef de partie assists his/her superior in maintaining the highest standard of quality in food preparation by following standard recipes and high level of hygiene standards maintained as per the hazard analysis and critical control points (HACCP) standards in his/her area, in order to maximize guest satisfaction and profitability in an atmosphere of high employee morale. The specific responsibilities of a chef de partie are to:

1. Ensure prompt and accurate service by all kitchen staff under his/her control, to all the customers to achieve a high level of customer satisfaction.
2. Be responsible for implementing hotel standards on food quality, preparation, and presentation in his/her section/shift.
3. Recommend changes in systems and procedures to increase efficiency and improve service levels.
4. Recommend changes at the time of preparation of new menu by introducing new dishes/presentation.
5. Ensure that the hygiene and cleanliness of the kitchen area is maintained as per predetermined standards.
6. Be responsible for controlling food wastage, without compromising on food quality.
7. Ensure proper security and safety of raw and cooked food, and equipment by proper storage.
8. Make sure that all the kitchen equipment is operated, maintained and stored properly and is safe to use.
9. Check that all the kitchen records are maintained properly at all times.
10. Ensure that organizational policies and standards are adhered to by all in the department.
11. Ensure availability of ingredients in the kitchen, at all times, in order to provide a prompt service.
12. Assist the chef de partie/sous chef in implementing standards set by executive chef on food quality, preparation, and presentation in his/her section.
13. Assist the sous chef and higher authorities to define the organization of work within his/her kitchen department including assignments, time schedules, and vacations.
14. Control food wastage without compromising on food quality.
15. Check that inter-kitchen food transfers are accurate and conform to hotel policy.
16. Ensure proper *mise en place* in his/her production sections for speedy preparation and service.
17. Make sure that hygiene and cleanliness of the kitchen area/equipment is maintained as per predetermined HACCP standards.
18. Discuss production planning with his/her commis, demi chef de partie, and concerned higher kitchen authorities.

19. Ensure all the company SOPs are followed by all the team members.

20. Make sure that cleaning schedules by kitchen stewarding department are being followed in timely manner.

21. Ensure par level of dry stores and perishables are maintained on daily basis, and also ensure of correct store requisitioning.

22. Receive daily requirements, from storeroom and get it checked and duly signed by his/her senior kitchen executive.

23. Recommend quality status on all the products in his/her kitchen to senior authority and rectify it as soon as possible.

24. Register complaints regarding improper machinery functioning, or employee's ill behaviour towards his/her kitchen executive.

25. Brief his/her team members on menu changes or introduction of new ingredients/new dishes on the menu.

26. Provide functional assistance to all subordinates and peers of various kitchens.

27. Ensure excellent relations and professionalism amongst all staff in his/her kitchen and with related departments.

28. Maintain appropriate and professional communication with all the team members at all given times.

Job title	:	Demi chef de partie
Department	:	Kitchen
Division	:	F&B
Reports to	:	Chef de partie

A demi chef de partie assists his/her superior in maintaining the highest standard of quality in food preparation by following standard recipes and high level of hygiene standards maintained as per the HACCP standards in his/her area, in order to maximize guest satisfaction and profitability in an atmosphere of high employee morale. The specific responsibilities of a demi chef de partie are to:

1. Ensure prompt and accurate service by all kitchen staff under his/her control, to all the guests to achieve high level of customer satisfaction.

2. Assist the chef de partie in implementing standards set by executive chef on food quality, preparation, and presentation in his/her section.

3. Assist the chef de partie to define the organization of work within his/her kitchen department including assignments, time schedules, and vacations.

4. Control food wastage without compromising on food quality.

5. Make sure that all the kitchen equipment is operated, maintained, and stored properly and is safe to use.

6. Ensure all organizational policies and standards are adhered to by all in the department.

7. Check that inter-kitchen food transfers are accurate and conform to hotel policy.

8. Ensure proper *mise en place* in his/her production sections for speedy preparation and service.

9. Make sure that hygiene and cleanliness of the kitchen area/equipment is maintained as per predetermined standards.

10. Ensure all the kitchen records are maintained properly at all times as per organizational standards in his/her department.
11. Discuss production planning with his/her commis and concerned higher kitchen authorities.
12. Receive daily requirement from storeroom and get it checked and duly signed by his/her senior kitchen executive.
13. Maintain daily log book and register equipment issues, and any critical information to be passed on to higher authority or next shift.
14. Recommend quality status on all the products in his/her kitchen to senior authority.
15. Register complaints regarding improper machinery functioning, or employee ill behaviour towards his chef de partie or senior kitchen executive.
16. Provide assistance to all subordinates and peers of various kitchens.
17. Promote excellent relations and professionalism amongst all staff in his/her kitchen and with related departments.
18. Coordinate with other food and beverage section, engineering/housekeeping, etc. whenever required.
19. Ensure appropriate and professional communication with all the team members at all given times.
20. Attend behavioural, vocational, and skill-related training, to enhance his/her skills and develop multifunctionality.
21. Provide constant on-the-job training and coaching to subordinates.
22. Share his/her skill and knowledge with all employees.

Job title	:	Commis
Department	:	Kitchen
Division	:	F&B
Reports to	:	Chef de partie/demi chef de partie

A commis has to prepare and provide the highest quality food in his/her area by following standard recipes, and high level of hygiene standards maintained as per the HACCP standards, in order to maximize guest satisfaction and optimum profitability in an atmosphere of high individual morale. The specific responsibilities of a commis are to:

1. Prepare food and provide prompt, courteous, and accurate service to all the customers as per organizational standard of quality, as directed.
2. Control food wastage without compromising on food quality.
3. Prepare all *mise en place* in production sections for smooth kitchen operation, as directed.
4. Ensure hygiene and cleanliness of his/her area at all the times.
5. Assist chef de partie in implementing and following organizational standards on food quality, preparation, and presentation.
6. Be responsible for maintaining all kitchen equipment in his/her area in good working condition.
7. Take responsibility for adherence to all organizational policies and procedures.
8. Maintain complete hygiene in his/her work area and adhere to the HACCP standards.

9. Ensure exact collection of perishables, grocery, and meat/fish items as per the storeroom requisition.

10. Ensure timely cleaning and sanitization of all the equipment and tools in appropriate hygienic manner.

11. Recommend daily requirement from storeroom to the demi chef de partie.

12. Maintain daily log book and register equipment issues, and any critical information to be passed on to higher authority or next shift.

13. Recommend quality status on all the products in his/her kitchen to demi chef de partie.

14. Provide assistance to all subordinates and peers of various kitchens.

15. Promote excellent relations and professionalism amongst all staff in his/her kitchen and with related departments.

16. Coordinate with other food and beverage section, engineering/housekeeping, etc. whenever required.

17. Ensure appropriate and professional communication with all the team members at all given times.

18. Attend behavioural, vocational, and skill-related training, to enhance his/her skills and develop multifunctionality.

CONCLUSION

It is important to have a well laid out organizational chart with listed job responsibilities to carry out the various jobs in a kitchen effectively and professionally. Chef Auguste Escoffier was the first person to formulate the hierarchy of a professional kitchen, where he placed the chefs on various levels depending upon their skills and the requirement of the menu. The classical French menu comprised of many courses and hence, many chefs were required to head each section of the kitchen. The modern hotels, however, are drifting away from the classical model due to high labour cost and availability of efficient machinery.

In this chapter we discussed various key points that formed the classical brigade of the kitchens.

Apart from the classical model, we also discussed the hierarchical structure of various hotels such as modern large-size hotel, small-size hotel and also lean hierarchy of resort hotels. Apart from these, we discussed the job responsibilities of all the people in the organizational structure of a modern hotel. We saw how the executive chef is responsible for various other jobs apart from cooking such as hiring, cost controlling, designing new ways of doing jobs effectively, etc. An executive chef has to organize and motivate his/her team by leading with example and also perform many other versatile jobs.

In the next chapter we shall discuss the various kinds of layout of the kitchens and also the concept of show kitchens which forms a part of the restaurant.

KEY TERMS

Afternoon tea It is served between 4 to 5 p.m. and includes mini pastries, finger sandwiches, etc.

Appraisals A tool used by the personnel department to evaluate the performance of an employee at the end of the year.

Auguste Escoffier The famous French chef, believed to be the father of classical cuisine.

Banquet A place let out to guests for private functions and conferences.

Bistro A place which serves fast food or casual food with soft beverages.

Boulanger French word for baker, who bakes breads.

Brigade Organizational structure of a kitchen department.

Capital budget Money allocated for buying heavy duty equipment.

Classical cuisine Usually referred to old French food.

Chef de cuisine French for executive chef.

Coffee shops Small restaurants which serve fast food, coffee, tea, and other non-alcoholic drinks.

Daily arrival list List prepared by front office, of guests arriving in the hotel to stay.

Duty manager A person in charge of the hotel in the night shift.

Entremetier A person who is responsible for preparation of vegetables.

Fine dining A restaurant which serves food and wine as per order in an elegant style and flair.

Food tasting Tasting of food to approve for a particular function or menu.

Garde manger A section in the kitchen which is responsible for cold food preparations such as salads and sandwiches.

Gate pass A format used to take out things from the hotel's vicinity to another location.

Guest history A record maintained by the front office, about the feedback of guests from previous visits.

Hors D'oeuvres French for starters or appetizers.

Inter-kitchen transfer Format used for taking goods from one kitchen to another for internal costing purposes.

Kitchen executive Chef manager who heads the team of supervisors and commis and reports to sous chef.

Market survey Visiting of market so as to compare quality and prices.

Menu A list of food items available to the guest in printed format.

Operational budget Money allocated for buying small equipment for daily operations.

Outlet A place serving food and beverages.

Par stock Minimum level of stock in the stores before reordering is done.

Pate French word for a 'paste'.

Potager Person responsible for cooking soups.

Preventive maintenance Regular servicing of machines to avoid breakdowns.

Recipe A written document for a dish which lists ingredients, the quantities, costs, and the method of preparation.

Requisition A format required for withdrawing anything from the stores.

Satellite kitchen Kitchen attached to a particular restaurant, which cooks food for the same.

SOP Standard operating procedure.

Specifications Standardized points listed for receiving commodities.

Supper Snacks served late in the night, usually mid-night and it takes place after dinner.

Tournant A person who relieves cooks on their off days and in exigencies. He/she is a multi-skilled cook.

USP Unique selling product, or a speciality of a particular hotel/restaurant.

VIP A very important guest who is to be treated with more attention.

CONCEPT REVIEW QUESTIONS

1. Who invented the classical kitchen brigade and why was it necessary?
2. What do you understand by the word 'fine dining'?
3. What is bistro style of dining?
4. Over the years, why has the kitchen brigade been modified?
5. List at least five key points of classical kitchen brigade.
6. What does sous chef mean?
7. What section does a garde manger chef take care of and what is his/her job?
8. Who is a boulanger?
9. What do you understand by the word entremetier?
10. Who is the chef tournant and why is he/she so important?
11. Draw out a layout of classical kitchen brigade in a large metro hotel.
12. Draw out a layout of modern kitchen brigade in a resort hotel.
13. Draw a layout of modern kitchen brigade in a large metro hotel.
14. What is the need of keeping staff on contract?
15. List at least five job responsibilities of an executive chef.

16. List at least five job responsibilities of an executive sous chef.
17. List at least five job responsibilities of a sous chef.
18. List at least five job responsibilities of a chef de partie.
19. List at least five job responsibilities of a commis.
20. What is the procedure of opening stores during its non-operational times?
21. What do you understand by trade test?

PROJECT WORK

1. In a group of three to four, visit hotels of various categories and draw out an organizational chart of the kitchen department. Compare the chart with classical brigade and draw out your conclusions. Justify the staffing levels and give your suggestions to optimize the efficiency.
2. Conduct a survey of various other institutional kitchens (such as hospitals, industrial canteens, etc.) and prepare an organizational structure and list the roles and responsibilities of various team members there in.

CHAPTER 3

LAYOUT OF KITCHEN DEPARTMENT

Learning Objectives

After reading this chapter, you should be able to:
- understand the basic workflow in the kitchen
- know the basic layout of the kitchen and understand the importance of the same
- figure out the layout of the receiving areas of a hotel and understand the importance of the same
- know the layout of the show kitchen and understand the modern trends about the same

INTRODUCTION

It will be right to say that kitchen is the heart of a hotel; just as the heart pumps out blood to all the parts of the body, the kitchen supplies food to all the sections of the hotel. On a busy day when the restaurants are full and room service is also busy, it can get very chaotic if the operations are not well planned. Before we can talk about a well-planned kitchen it is very important to understand good workflow in the kitchen. In the previous chapter we discussed in detail about the various organizational structures and understood the general responsibilities of each chef handling an area. This chapter shall discuss various areas of kitchen and shall stress upon the importance of organized layouts for smooth workflow.

A well-designed kitchen from the architectural point of view might not be the best option for a functional kitchen that would meet the requirements of the operations. Therefore, it is important to understand the workflow of the jobs being done in a kitchen in order to plan them effectively. Time and motion study is carried out to see the feasibility of laying out the machines and equipment so that the staff does not have to move around too much while cooking, otherwise it will cause fatigue and accidents. This chapter deals with many kitchens from around India.

GENERAL KITCHEN LAYOUT

Kitchen is a busy place and cross-traffic can really hamper the operations. There are certain factors that one needs to keep in mind while planning a kitchen. This is usually done by 'facility planning department', which carefully plans the layout of the kitchens. Some hotels commission these services on contract but certain hotel chains, such as Oberoi Hotels and Resorts, have their own facility planning department which is responsible for planning and layout of all the kitchens.

Planning a kitchen entails much more than just placement of equipment in its place. A well-planned operation will always follow a systematic procedure (see Fig. 3.1).

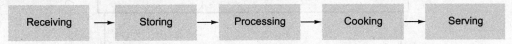

Fig. 3.1 General workflow pattern in a kitchen

When we talk of the design of a kitchen, it generally means the overall planning of the space with regards to the size and shape of the operations. Layout means the detailed arrangement of the floor of the kitchen and allocation of places for the kitchen equipment to be placed where the specific tasks would be carried out. A cluster of such places is referred to as workstation.

A well-planned layout is not only important for the smooth workflow in the kitchen but it also adds to the profitability of the entire operation. Smooth workflow will ensure timely pick up of food for a busy meal period thus creating happy guest and good reputation. Such an operation is also directly linked to the motivation and overall morale of the staff.

Well-planned kitchen operations would always follow a basic three-flow pattern, which would be *back–front–back*. All the raw materials are received at the kitchen (back of the operations) then processed and sent to the restaurant (front) for consumption. The leftover unusable food is brought back and disposed as garbage. A typical workflow is shown in Fig. 3.2.

It would be ideal to have straight lines of production to speed up the service to the guests, but it is rarely achieved. The design and layout of the kitchen should be such that the cooking and the service staff do not intersect or cross each other. The service team usually picks up food from a designated space in the kitchen often referred to as 'pass window'. This space in the kitchen restricts the entry of the service staff beyond this point. This is done for various reasons and one of them is the smooth workflow of the kitchen. During busy times when cooks are under stress and are working with hot pots and pans, they do not want to have any accidents with the intersecting traffic.

Receiving Area

Receiving area is a place where the goods are received into the hotel. This place is not only used for receiving raw food commodities, but also to offload all the supplies of the hotel. This area also restricts the entry of unauthorized personnel into the hotel and is located near the rear entry of the hotel also known as staff entrance. A typical layout of the receiving area is shown in Fig. 3.3.

Various jobs are carried out in the receiving area and this is one place in the hotel where all the supplies are received. From food commodities to engineering supplies, everything lands up at the receiving dock, where the items are checked as per specifications given to the supplier and recorded in various formats. Let us discuss some of the receiving procedures carried out at this place.

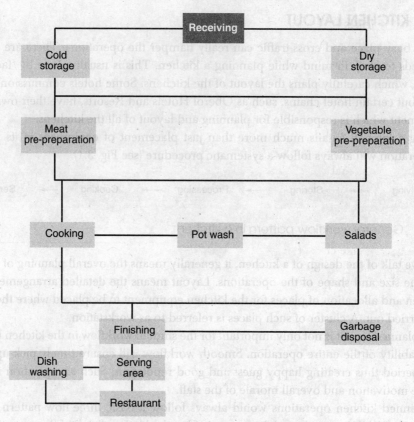

Fig. 3.2 General principle behind layout of the kitchen areas

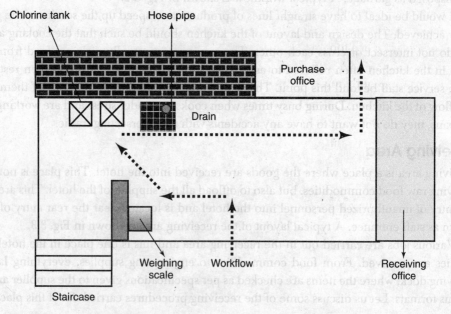

Fig. 3.3 Layout of receiving area

Ordering Vegetables

All ordering is done through the purchase department for proper control. All order sheets should be sent to the purchase department in the evening, at a time agreed by the kitchen and the receiving department. If any order is placed directly to the suppliers due to emergency, a mail should be sent to purchase stating the reason as well as the ordered quantity. The procedure of ordering through purchase ensures that the orders are placed to the suppliers in time and the receiving can be done as per the standards specified by the hotel. When the goods are received, they are entered into a format known as daily receiving report (DRR). The DRR is made daily and a copy of it is forwarded to the chef on a daily basis so that he/she has an idea of cost, short supply, and unavailable items. This report is a check for the quantity received against ordered and also highlights any short supplies. In case of emergencies, the short supplies can be purchased on cash and debited to the contracted vendor.

Quality Check

Receiving personnel take charge of the receiving of vegetable, stacking, and cleaning of the receiving area. The receiving clerk receives the commodities as per the standard specifications agreed by the hotel and the vendor. The receiving clerk then checks all the commodities as per the guidelines and has the authority to reject any commodity which does not conform to the standards as agreed upon.

Sanitization

All vegetables should be sanitized with 50 ppm chlorine. Usually a double sink is used for washing and sanitizing vegetables.

Awareness

A standard purchase specification with photographs of the product should be made available to the receiving clerk so that he/she can make the receiving more accurate.

Food Storage

Once the food is received, it becomes important to store it or hold it before it is issued to the production areas. The food is usually stored in a place known as storeroom. Storeroom is the most valuable and important part of a food service organization. It is designed with care so that it is able to store the food hygienically to process it later. Certain considerations have to be kept in mind while designing storerooms. Some of them are as follows:

- It should be under close supervision of security as it stores expensive commodities.
- The area should be enough to place the items according to the standards and one is able to walk easily in it to access different items conveniently.
- It should be near to the receiving area and the user department.
- It should have walk-in refrigerator and fire extinguishers in the same area.
- FIFO[1] system should be followed in the department.
- The shelves should be 3 inches away from the wall and 6 inches above the ground.
- It should have different shelves for the ingredients which have code numbers and labels.

---

[1] First-in first-out refers to commodities that were purchased and stored first.

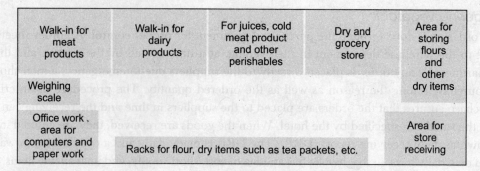

Fig. 3.4 Layout of the storeroom

Layout of the stores (Fig. 3.4) is again very crucial as they should be well ventilated and temperature controlled so that the commodities are stored safely for consumption. They should have enough space for the trolleys to move around.

Workflow in Stores

Meats should be stored in deep freezer at a temperature of 16–18°C. It should also be stored on the basis of FIFO. Separate racks for meats, poultry, and fish are a must. Meats should be kept in colour-coded baskets, with meat tags on them. Vegetables should be stored in walk-ins at a storage temperature of 3–5°C.

For dry stores also all the items should be kept on the shelves with application of FIFO. Before receiving these items manufacturing and expiry dates should be checked. All the items in the dry stores are kept at room temperature until and unless something is specified by the dealer.

All the items come first to the receiving bay before going to the stores. At the receiving bay, the quality and the quantity is checked and items are tallied with the ordering sheet and approved. These items are then sent to the stores, cross checked, and distributed to the various departments as per the requisition.

Fast, Slow, and Non-moving Items

Fast moving items are very critical as their stock has to be maintained correctly at all times as it could be required at any time. These are the items which move fast and hence, are also ordered at the same rate.

There are chances that certain items become non-moving or slow moving and this could happen due to various circumstances such as change in menus and cancellations of a function for which the items were procured. Generally the store informs the executive chef of the items which are non-moving, then the chef checks with various kitchens if any one can use it; if not then he/she suggests an appropriate kitchen which can make use of it and use it to generate profits. The items in the store can get spoilt due to various reasons, and in such cases spoilage report is made for items which get spoilt as they are not used.

LAYOUT OF THE COMMISSARY

A commissary kitchen is the backbone of the kitchen as most of the pre-preparation of food is done here. This kitchen does the basic *mise en place* for large-scale operations, such as banqueting, and so it is very important to have a well-spaced work area as there would be heavy movement of staff. This kitchen also uses specialized machines such as pulverizers, which are used for making pastes and powders for curries, etc. Certain fruits and vegetables that cannot be stored in the walk-in are stored here and provided to the kitchens on basis of requisitions. In many hotels this kitchen forms a part of the store and hence, control becomes easy (refer to Fig. 3.5.).

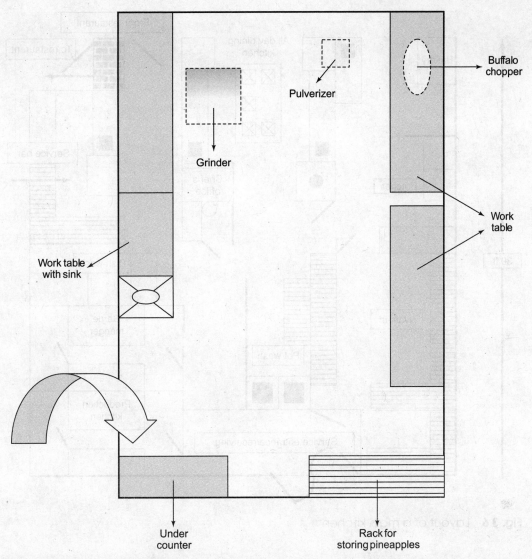

Buffalo chopper

Pulverizer

Grinder

Work table

Work table with sink

Under counter

Rack for storing pineapples

Fig. 3.5 Layout of commissary kitchen

BASIC LAYOUT OF MAIN KITCHEN

Main kitchen can be rightly called the nerve centre of all the kitchens. The kitchen associated with an individual restaurant is known as 'satellite kitchen' and the main kitchen comprises butchery, cold kitchen also known as 'garde manger', coffee shop kitchen, bakery, chef's office, production kitchen, etc. (refer to Fig. 3.6).

LAYOUT OF BUTCHERY

Butchery is the biggest money holding centre of the hotel operations; though it does not make money directly, it is indirectly responsible for maintaining the food cost. This section of the kitchen stores and processes most expensive meats which are both local and imported. Hence, a well-planned butchery

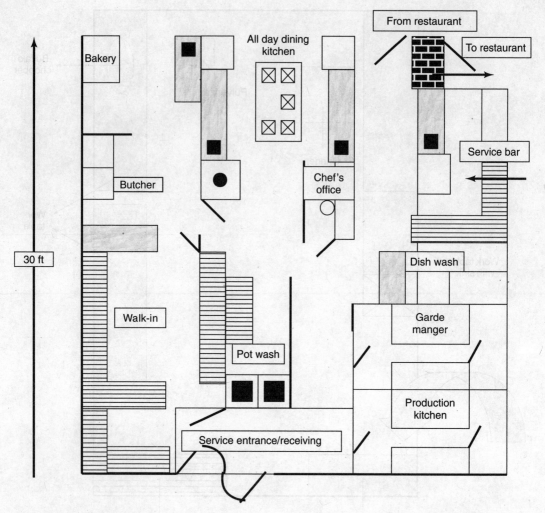

Fig. 3.6 Layout of a main kitchen

operation is a must for a profitable food business. Butchery, in various hotels is part of food stores; as they process all the meats and store them for usage in the kitchen against food requisitions.

Separate walk-ins and separate workstations are provided for separate meats as there could be chances of food contamination. Figure 3.7 shows the layout of a butchery.

LAYOUT OF GARDE MANGER

Garde in French means 'to keep' and manger means 'to eat', so garde manger literally means 'kept cold to be eaten'. This kitchen is called cold kitchen as foods such as sandwiches, salads, juices are prepared and served out from this kitchen.

This kitchen is always a part of the main kitchen as it not only serves à la carte and banquet, but is also responsible for room amenities, welcome drink to the guests and many other support roles for other kitchens (see Fig. 3.8).

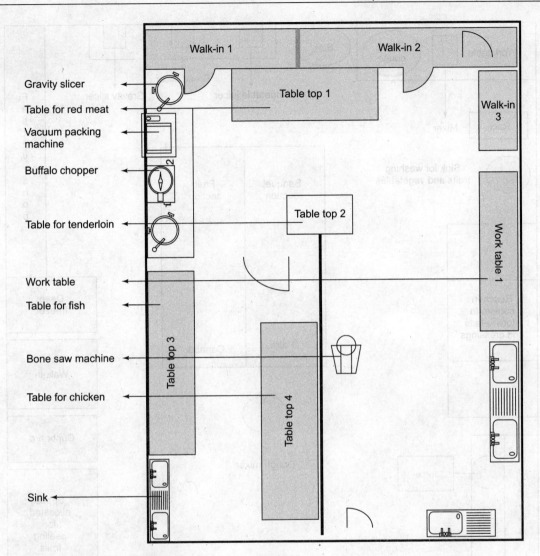

Fig. 3.7 Layout of butchery

LAYOUT OF BAKERY AND CONFECTIONERY

Bakery and confectionery is a very important section of the kitchen, as this department operates round the clock and is the busiest operation. It produces breakfast rolls for the morning breakfast, cakes and pastries for the pastry shop, and also the à la carte desserts of various restaurants and banqueting operations. There is a separate chocolate room which produces chocolates and garnishes (see Fig. 3.9).

LAYOUT OF WESTERN BANQUET KITCHEN

The Western banquet kitchen prepares food for banquets and is also responsible for bulk *mise en place* required by other satellite kitchens such as soup bases, stocks, and sauces.

Banquets also prepare buffet food for all day dining restaurants. The segregation is not done on the basis of amount of work but rather for the flow of work. A Western kitchen (Fig. 3.10) is divided into the following sections.

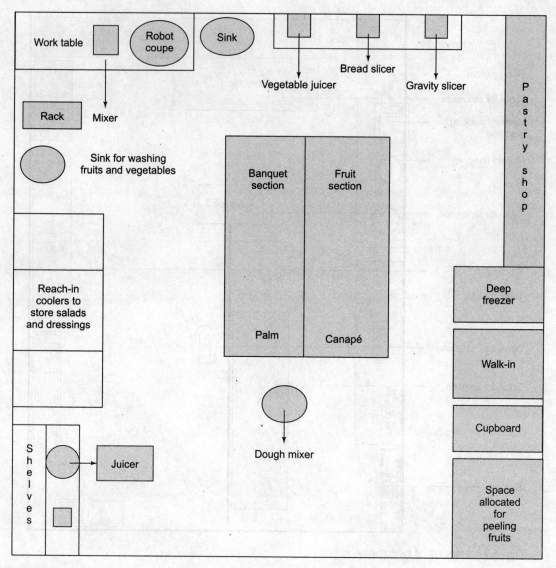

Fig. 3.8 Layout of garde manger

Range This section basically takes care of food which is prepared on the range. The actual cooking such as preparation of dishes is done here.

Grills This section takes care of the grills, hot plate, baking, deep fat fryer and finishing different types of meat in the oven. All the grilled work is done here, such as grilling and searing of meats and vegetables.

Mise en Place Section The bulk *mise en place*, such as cutting and chopping, is done here.

The banquet kitchen also has sophisticated equipment such as brat pans, steam kettles, and combination ovens that are used for cooking, steaming and also reheating of food. As this kitchen can cater up to a size of 3000 guests, it is imperative to have a smooth workflow and storage space available for bulk *mise en place*.

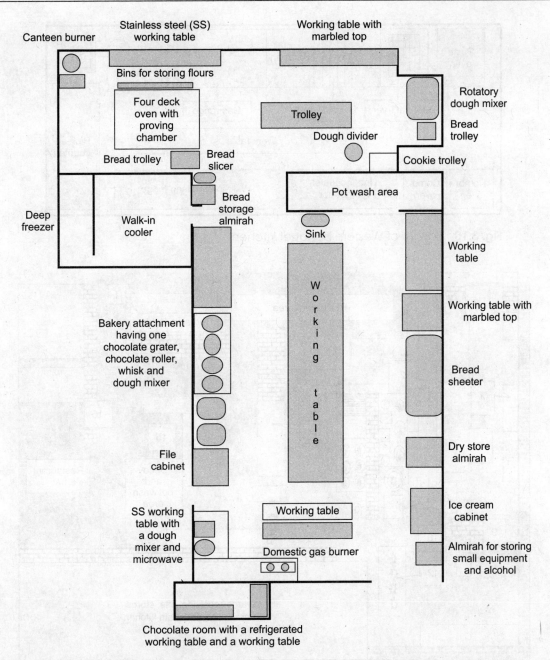

Fig. 3.9 Layout of bakery and confectionery

LAYOUT OF SHOW KITCHEN

Show kitchens also known as 'display kitchens', are the most modern trends in today's restaurants. These kitchens not only add a style statement to the business but also act as a marketing tool, telling the guests that the food prepared here is fresh and hygienic. The guests do not complain about the delay in food, as they can see their meals being cooked in front of their eyes. The idea of having

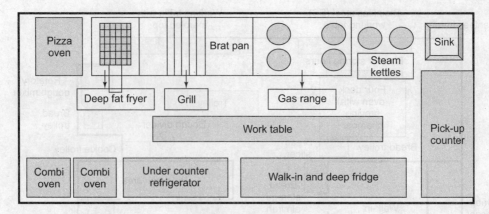

Fig. 3.10 Layout of Western banquet kitchen

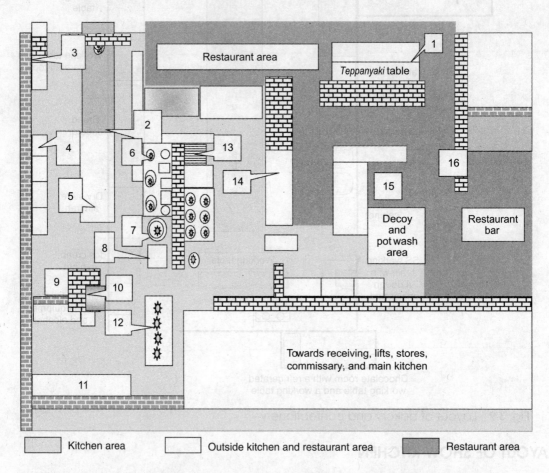

Fig. 3.11 Layout of show kitchen of India Jones restaurant at the Trident, Mumbai

a show kitchen as a feature of the restaurant is very commonly seen in modern hotels, where the live cooking by chefs acts as a USP.

An Oriental show kitchen serves food from entire South–east Asia. Its menu is very elaborate and if not well planned it would be chaotic to cook here. This section serves food from China, Japan, Thailand, India, Malaysia, and Indonesia.

While designing show kitchens it is important to decide what should be visible to the guest and what should be skilfully masked away. For example, a station where a skilled chef is cooking *teppanyaki* or carving a peking duck is a far better sight than someone handling raw poultry or fish.

Show kitchens use best of equipment as the guests can see them and the maintenance cost of the same is very high as compared to the equipment of the normal kitchens.

The staff working here needs to be very skilled and the design of the uniforms must also blend with the entire ambience. Food courts in various malls are examples of display kitchens. Figure 3.11 presents the show kitchen of India Jones restaurant at The Trident, Mumbai. The key to Fig. 3.11 is given in Table 3.1.

Table 3.1 Key to the show kitchen as shown in Fig 3.11

Key	Description
1	*Teppanyaki* table
2	Two mounted tandoors, stainless steel handwash sink
3	Four door reach-in refrigerators for storage of *mise en place*
4	Four door reach-in refrigerators for storage of *mise en place*
5	Stainless steel (SS) table, with an under counter heater/plate warmer for the heating up of dispensing cutlery, a cold bain marie for *mise en place* stacking, a plastic garnish tray for nonperishable garnishes, a chopping board is aligned on the same, an under counter for rice and other condiments
6	Chinese wok range with three high pressure Chinese burners, pilots, emergency safety valve, and the automatic fresh air for the burner, with built-in arrangement for water spout
7	Tilting steam kettle
8	Convection steamer for temperatures above 75°C
9	Dim sum section with a top shelf for storage of dimsum baskets, a tabletop wooden board and under shelves for stacking cold *mise en place*. Next to it is a refrigerated showcase
10	Aquarium; 3 sections one fresh water for scampis and other two salt water for grouper, snapper, crabs, and spiny lobsters
11	4 SS sinks with overhead shelves for storage
12	Thai and pantry section, pantry SS table with under counter Foster's drawers and a plate warmer on the side, hot section has 6 low-pressure burner range, deep fat fryer, grill, a salamander and a peking duck gas oven
13	Pot wash and equipment/restaurant ware storage area
14	Restaurant bar
15	Decoy and pot wash area
16	Bar area

Planning a Show Kitchen

Some important points to be taken care of while planning a show kitchen are as follows:

1. Give emphasis to visually appealing cooking. A crepe station, an open dessert bar, or a well-ventilated dim sum station adds to the feel.

2. Kitchen should be well illuminated as against the restaurant to make a visual impact.

3. There should be areas in the show kitchen where the single guests can sit and enjoy their meals while interacting with the chef and there should also be sitting areas where the guests would have privacy and hear no noise of the cooking ranges.

4. A skilled staff is a must as it will directly influence the guests, so a sense of working hygienically and methodically must be some of the prerequisites while hiring staff.

5. While planning display kitchens keep in mind the changing trends so that renovations are easy if the demand changes over a period of time.

6. Plan a decor that will blend with the restaurant. Spice jars, oil bottles, and traditional cooking equipment add a great deal of enthusiasm.

7. Connect the display kitchen to the rest of the kitchen to facilitate the bringing of stores and disposing of the debris.

8. Focus on the ventilation of the kitchen as it should not let the smoke and temperature hamper the restaurant ambience.

9. Layout of the restaurant should be planned well to accommodate the traffic flow of kitchen as well as service staff.

10. The walls and counters must be made of attractive material but these should be easy to maintain and clean as they will have to be scrupulously clean at all times.

11. The lighting areas of the kitchen should be a mix of both fluorescent and incandescent lighting.

CONCLUSION

This chapter dealt with various kinds of layouts of the professional kitchen that make it functional and cost effective at the same time. We saw the principle behind an ideal layout of the kitchen, where the food comes in and gets stored for processing there and then or later, depending upon the operations. The food is cooked and served and the garbage is disposed off from the receiving area. We discussed the set-up of a receiving area and the various activities that take place there. We also focused on storerooms where the food is stored and the formats that are used as a tool to control the flow of goods from market into the store and then from store into the operations to generate revenue for the organization. We discussed formats such as daily receiving reports, storeroom and purchase requisition format. We also laid emphasis upon the workflow in the storeroom and how it contributes to the profit of the department. The concepts of FIFO and slow moving items to control and maintain the inventory levels were also briefly touched upon.

We then discussed layouts of various kitchens such as commissary, main kitchen, butchery, pastry, and many other kitchens such as show kitchens. There are many factors that need to be considered while planning a show kitchen and the various salient features of the same were also discussed.

In the next chapter we will discuss various kinds of fuels and equipment used in the kitchens, which will help us to begin our culinary journey.

KEY TERMS

All day dining   A 24-hour multi-cuisine restaurant in a hotel.

Bone saw machine   An equipment used in the butchery to saw the meat on the bone.

Brat pan   An equipment used for bulk cooking. It is usually electrically operated and is available in various sizes.

Buffalo chopper   A machine that uses a sharp blade to mince or chop commodities. It is also known as bowl cutter.

Butchery   A kitchen in a hotel, which processes all the meat and fish for usage in the kitchen.

Canapes   Small titbit savouries made in the garde manger.

Chlorine tank   A sink in the receiving area containing chlorinated water to wash vegetables before receiving them into the hotel.

Combi oven   A convection oven usually uses both dry and moist methods of cooking.

Commissary   A kitchen that processes vegetables and makes pastes and powders for bulk kitchen preparations.

Deep fat fryer   An equipment with a container to heat oil, used for frying.

Dim sum   Steamed savoury flour dumplings prepared in China.

Dish washing   Area in the back of the restaurant where all the crockery, cutlery, and glassware are washed and polished.

Dough sheeter   An equipment used in bakery to sheet out dough to desired thickness.

Facility planning   An architectural department in the hotel responsible for designing hotels and kitchens.

Garde manger   French name for cold kitchen. This is the kitchen that produces salads, sandwiches, and juices.

Gravity slicer   An equipment used for slicing commodities. One can adjust the thickness of the slices.

Layout   Detailed arrangement of the floor of the kitchen and allocation of places for the kitchen equipment to be placed where the specific tasks would be carried out.

Main kitchen   Main kitchen is an area that comprises butchery, garde manger, commissary, all day dining kitchen, production kitchen, and bakery and confectionery.

Pass window   A place demarcating the kitchen from service area. This is the place where the service team picks up the ready food.

Peking duck   A Chinese style of preparing duck.

Pot washing   An area in the kitchen where the soiled pots and pans of the kitchens are washed.

Pulverizer   A machine used to make puree or paste.

Range   An area consisting of burners and hot plates used for cooking.

Reach-in   A free-standing refrigerator with shelves.

Receiving   An entry into the hotel where all the goods are received for the usage of the hotel.

Robot coupe   A high-speed blender used for making pastes and purees.

Room service   A food and beverage outlet which is responsible for taking orders from guest rooms and providing service to the guests in their rooms.

Satellite kitchen   The kitchen attached with a restaurant which is responsible only for the food prepared for that restaurant only.

Short supply report   It is a report generated by the receiving clerk stating the items which have not come into the hotel or have not been received as per the quantity specified.

Show kitchen   Also known as display kitchen, it is a kitchen that opens up into the restaurant.

Teppanyaki   A Japanese style of cooking food on hotplates. Lot of skill is involved in cooking on *teppan*, as it involves juggling of food and cooking with showmanship.

Under counter   A refrigerating unit with a work space on top and refrigerated unit below.

Vacuum packing machine   An equipment used for vacuum packing the commodities for storage purposes.

Workflow   A systematic and organized flow of work from the start to finish.

Workstations   A place where the job is carried out to prepare and dispense a dish.

Work table   A table used for *mise en place*

CONCEPT REVIEW QUESTIONS

1. What do you understand by kitchen layout?
2. What is the difference between workflow and workstation?
3. Which department does the planning of kitchens?
4. Draw a flow chart of a well-planned kitchen.
5. What do you understand by design of a kitchen?
6. Describe a workstation.
7. What is a pass counter or pass window?
8. What do you understand by a receiving area?
9. What is the purpose of a chlorine tank?
10. What is the procedure followed for ordering vegetables and meats?
11. What should be done in case of emergency ordering?
12. What is the procedure of sanitizing the vegetables?
13. What is a standard purchase specification?
14. List four considerations to be kept in mind while designing stores.
15. What are fast moving items and slow moving items?
16. How can one use the non-moving items?
17. What is a spoilage report?
18. What is a commissary kitchen and how is it useful?
19. Describe a main kitchen.
20. List the areas of the bakery and confectionary kitchen.
21. What is a chocolate room?
22. List the equipment used in the butchery.
23. List the sections of a Western kitchen.
24. Explain the concept of a show kitchen.
25. List at least five salient features of a show kitchen.
26. List five advantages and disadvantages of a show kitchen.
27. List five parameters to be kept in mind while planning a display kitchen.

PROJECT WORK

1. In a group of four to five, visit a hotel and study the layout of any kitchen with regard to its food style and service. Critique the layout and offer necessary solutions.
2. Visit the stores of a hotel and study the design from the operational point of view. Record your findings and compare amongst each other.
3. Draw out the improvised layout of your cafeteria and give your feedback on the same. If you have any better layout in mind, give justifications and offer solutions.
4. For a simulated event in your institute, plan out the design of a kitchen and prepare the workflow keeping in mind the style of service and smooth operation.

CHAPTER 4

EQUIPMENT AND FUELS USED IN THE KITCHEN

Learning Objectives

After reading this chapter, you should be able to:

- understand the various fuels used in the kitchen, their optimum usage to maximize profits and the storage of the same
- know different types of equipment and their safety operating procedures
- understand the difference between capital equipment and operating equipment
- know the various kinds of modern cooking equipment such as induction ranges and the utensils used for the same
- comprehend and justify the use of technologically advance modern equipment used in modern western cooking

INTRODUCTION

Different types of complicated equipment are used in the kitchen and many of them need fuel to operate. Most of the equipment use electricity, but many others use various other kinds of fuel to run smoothly and efficiently. This chapter deals with various kinds of equipment and fuels used in the kitchen. We must understand that each kitchen can be versatile, with its own set of specialized equipment. The use of equipment in a kitchen could depend upon various factors such as regional cooking, eating habits of people, and cultural diversity. Indian kitchen equipment would be very different from that of a Chinese kitchen or an African kitchen. Though a cooking medium or fuel could be same, there is some specialized equipment that is specific to a particular cuisine. Later in this book, we shall discuss specific equipment used in various cuisines and also their usage. We shall also discuss the classification of equipment based upon their status, for example, capital equipment and operational equipment. Capital equipment are fixed machines such as ovens and weighing scales, whereas operational equipment are the ones that are required to carry out daily jobs in the kitchen such as knives and utensils. This chapter is very important as it forms the base of cooking. In every cuisine, the selection of the utensil and the medium of cooking form the foundation of the same.

FUEL USED IN THE KITCHENS

Various types of fuels are used in hotel kitchens to cook or to operate equipment to aid in cooking. In our homes, we use LPG or electricity; but in professional kitchens it is quite different. Various types of fuels are used in hotels. Some of them are discussed in Table 4.1.

Table 4.1   Fuels used in the kitchen

Fuel	Description	Usage
LPG	LPG or liquefied petroleum gas is the generic name for commercial propane and butane composed fuel used in the kitchen to fuel gas burners. This gas is supplied in industrial cylinders or in bulk storage tanks at the hotel premises. It is supplied to the kitchen through pipelines. The LPG cylinders are stored in a separate place, usually known as 'gas bank' and are usually operated by a department in the hotel called 'kitchen stewarding'. Amount of gas cylinders used in the gas bank depends on the types of operations they are used for. A large hotel may have more than 100 cylinders. Half of these would be installed to supply gas whereas half would always be a back up. A certain gas pressure is maintained by the stewarding department as certain ranges require high pressure for cooking. LPG liquefies under pressure and converts into gas when the pressure is released. It is almost smokeless and easy to handle. LPG should be handled with utmost care as being a transparent gas its leakage is not easily detectable and it is also highly combustible. There are certain government regulations regarding the usage of LPG. For example, LPG cannot be used in basement kitchens or below the sea level and hence, in those kinds of conditions one has to rely on electricity or steam-operated equipment.	It is one of the most essential fuels used in the kitchen and is known for its efficiency. It is used as a fuel for cooking ranges, ovens, and salamanders. Some *tandoors* used in Indian cooking are also fired by LPG.

Contd

Table 4.1 (Contd)

Fuel	Description	Usage
CNG	Compressed natural gas is slowly gaining popularity for its fuel efficiency and environment-friendly properties.	It is used in eco-friendly hotels as fuel in many types of equipment such as ovens and gas ranges.
Coal	Though it is a very crude form of fuel to be used in a modern kitchen, it is still very popular. The smoky flavour which the charcoal imparts, is much desired. Coal should always be stored away from food area, ideally in a cool, dark room away from any moisture. Usually separate areas are built near the receiving area for coal storage as coal is combustible and messy.	Coal is used in a kitchen to light tandoor and grills for barbecue. The coal used in the hotels is wood charcoal only.
Wood	One would have come across wood-fired pizza ovens in the modern restaurants today. Gaining popularity of pizzas has led to the origin of this oven which lends an aesthetic appearance to the restaurant, where chefs prepare pizzas in front of the guests. Though it is also operated by LPG, few logs of wood are placed inside to impart a smoky flavour to the pizzas.	It is used as a fuel for wood-fired pizza ovens. The pizza ovens have LPG fire but wood is also kept inside to impart a smoky flavour to the food.
Electricity	Electricity is also used as fuel to operate many types of equipment. Care should be taken while ordering such equipment as many countries operate on certain voltage. In India equipment work on 220 volts, whereas in the USA equipment work on 110 volts. So care should be taken while importing equipment. Some of the heavy duty equipment use three-phase electric current and some use only single phase. So it is important that the instructions are read before installing the new machinery.	Electricity is generally used to operate most of the equipment in the kitchen. It is popular in the kitchens because it is easy to control electrical equipment.
Steam	Most of the hotels produce steam, which is used to cook or operate equipment. Steam is supplied to the kitchen through insulated pipes.	Steam is used in equipment such as dishwashers and steam jacket kettles.
Solid fuel/ handy fuel	This fuel is made from petroleum jelly and comes in small tins. These are generally used in F&B service areas.	This type of fuel is hardly used in kitchens but is very commonly used in F&B service, where it is used in heating up food in the chafing dishes used commonly in banquets.
Solar energy	The heat from the sun is used as a fuel. This is not a very commonly used fuel in the kitchens, but many eco-friendly hotels have solar cookers that are used in cooking.	Solar cookers utilize solar energy to cook food.

Contd

Table 4.1 (Contd)

Fuel	Description	Usage
PNG	PNG is an abbreviation for Petroleum Natural Gas and it has been recently introduced into the modern Indian kitchens.	It is used in eco-friendly hotels to fire up many equipment such as ovens and gas ranges.
Butane cylinders	These small cylinders that contain compressed butane gas are used as a handy fuel to gratinate certain dishes such as crème brulee, where the sugar is spread on the dessert and torched with this fuel for a caramelized effect that is traditional to this dessert.	It is used for caramelizing sugar on crème brulee or gratinating few dishes.

EQUIPMENT USED IN THE KITCHENS

A professional kitchen set-up can be very confusing for beginners. At times it looks like a science laboratory where people are performing experiments. It looks so because the sparkling kitchens are adorned with sophisticated machines and equipment. Chefs use equipment and utensils in the kitchen to cook meals. Equipment can be further classified as the following.

Capital Equipment

These are also known as large equipment or fixed equipment and they include ovens, gas ranges, and grillers. These equipment are usually fixed and inventoried. Inventory is a procedure followed to audit physical presence of the equipment. The capital expenditure would depreciate over the period of time and hence, it would include items that are expensive to buy. The inventories can be done quarterly or annually depending on a hotel's policy. When the yearly budgeting takes place, the chef in conjunction with the F&B manager prepares a budget for capital expenditure, which is then moderated by accounts and presented to the approving authority through the general manager. These budgets are usually presented during financial review meetings and once they are approved, the purchase department sources the same at best possible prices.

Operational Equipment

These are used directly to operate the basic jobs in cooking and all the utensils, moulds, knives, etc. are categorized under this heading. These equipment are also inventoried to keep a check and discarded when they become unsafe to use. An annual budget is made for the operational equipment based on those to be discarded and those purchased in lieu.

Commonly Used Equipment and Their Operating Procedures

All equipment used in the kitchen have certain procedures of operating and cleaning them. The equipment are very expensive, and the breakdown of the same due to improper handling can cause huge loss to business. So it is very important for chefs to ensure that the operating procedures of equipment are adhered to. During industrial training or internships, students should always ask how to operate any equipment. Table 4.2 shows some of the common equipment used in modern kitchens and includes instructions for operating them.

Table 4.2   Some capital equipment used in the modern kitchen

Gravity slicer and buffalo chopper

The slicer and buffalo chopper are the most dangerous of all kitchen equipment and should always be handled with extreme care.

Gravity slicer

1. Make sure the machine is unplugged before starting your set up.
2. Use the machine according to proper specifications.
3. Turn the machine off and unplug when not in use.
4. Avoid hard materials that could damage the blade.
5. Sharpen the blade regularly.
6. Clean the machine properly after each use.

Buffalo chopper

Food processor and robot coupe

4.1

1. Check whether all the parts of the processor are in place.
2. Insert the items to be processed and secure the lid.
3. Never use any food processor with the lid off.
4. Do not overload and make sure the blade is fitted well.
5. Wash and rinse, and sanitize all the parts and reassemble when finished.
6. Do not send to pot washer.

Robot coupe

Bench saw/Bone saw

Only the butcher or designated butcher chef should use the bench saw.

1. Make sure all areas are clear before using.
2. Make sure the saw and belt are secure before using.
3. Concentrate while working on the saw.
4. After use, unplug the saw and pull apart all pieces.
5. Wash all parts in soapy water with a disinfectant. Rinse all parts and sanitize all areas.
6. Ensure that the saw is clean and sharp at all times.
7. Ensure that the guard is always secured.

Deep fat fryer

1. Fry all frozen foods from the frozen state (do not let thaw).
2. Do not refry any cold fried items.
3. Lightly shake all excess oil from the fried food before plating.
4. Do not mix old and new fat together.
5. Never allow fat to smoke.
6. It should be washed daily.
7. Do not sprinkle salt onto fat.
8. Stop the fryer before draining the fat.
9. Be careful of hot fat when draining.
10. Fat is used at the correct temperature (general 350 °F–375 °F).
11. Ensure that the fat is clean with no debris.
12. Ensure that the fat is filtered after each shift.
13. Dispose off the fat properly.

Contd

Table 4.2 (Contd)

Flat top (grill)	

Flat top (grill)
1. Before using the flat top, make sure the surface is clean and the temperature is correct.
2. Skim all extra grease and fat.
3. Wipe, clean, and scrape after each use to avoid build-ups.
4. Use a chemical to get rid of all the sediments after each meal period.
5. Oil the flat top after cleaning.
6. Turn off if not in use.
7. Use proper tools (no plastic).

Oven
1. Check the inside of the oven before turning it on/off.
2. Preheat the oven at the appropriate temperature.
3. Make sure to open and close doors by the handles and not by kicking them.
4. Use the timer for baking.
5. Shut off when not in use.
6. Check the pilot light and check for gas leaks.
7. Use the proper shelf in the oven.
8. Do not spray water in the oven.

Salamander
1. Do not leave food unattended in the salamander.
2. Turn it off if not in use.
3. Use a kitchen cloth to pick up dishes from the salamander.
4. Preheat the salamander.
5. It is not always necessary to have all the heating elements turned on full.

4.2

Universal dough mixer
1. Isolate power supply when not in use and while cleaning.
2. Clean with a wet cloth only. Avoid excess water while cleaning.
3. Do not overload machine (capacity 1 kg).
4. Do not change speed rapidly.
5. Use relevant accessories with care:
   - whisk for cream, eggs
   - beater for soft butter, eggs, flour, sugar
   - dough hook for dough only

Dough sheeter
1. Isolate power supply when not in use or cleaning.
2. Ensure that all parts are connected properly prior to operating the machine.
3. Ensure that one is briefed prior to operating the machine.
4. Do not use hands or utensils while machine is operational.

Contd

Table 4.2 (Contd)

5. Fully disconnect the machine prior to cleaning.
6. Ensure that the mats are scrapped daily to avoid foreign particles forming.
7. Clean the machine with a moist cloth. Avoid excess water.

Microwave ovens

1. Isolate power supply when not in use or while cleaning.
2. Ensure power supply is connected properly.
3. Do not press any buttons on the machine while in use.
4. Fully disconnect the machine prior to cleaning.
5. Clean with a moist cloth. Avoid excess water.
6. Use only microwave safe utensils. Do not use metal.

Pasta boiler

1. Isolate gas/electrical supply when not in use.
2. Ensure that the equipment is fully operational prior to using and that grid is placed at base of kettle and properly fitted.
3. Be careful of excess steam exhaust while using.
4. Do not tamper with any of the gas lines or valves.
5. Do not use hands in machine while operational.
6. Ensure correct utensils are used at all times.
7. Once food products are removed, ensure grid is stored properly in correct allocated position.
8. When cleaning the machine, use hot soapy water on the inner part and a wet cloth on the outer part.

Steam kettles

1. Isolate steam supply when not in use.
2. Ensure machinery is fully operational prior to using and that grid is placed at base of kettle and properly fitted.
3. Be careful of excess steam exhaust while using.
4. Do not tamper with any of the gas lines or valves.
5. Do not use hands in machine while operational.
6. Ensure correct utensils are used at all times.
7. Once food products are removed, ensure grid is stored properly in correct allocated position.
8. When cleaning the machine, use hot soapy water on the inner part and a wet cloth on the outer part.

Tilting brat pan

1. Isolate power supply when not in use or cleaning.
2. Always use correct utensils when operating.
3. Ensure the pan is at the correct temperature prior to using. Pre-set temperature.
4. Do not let fingers/hands come in direct contact with the pan while operational.
5. Fully disconnect machine prior to cleaning.

Contd

Table 4.2 (Contd)

6. Use hot soapy water to wash the inner pan and a wet cloth to clean the outer.

Vacuum packing machine
1. Isolate power supply when not in use.
2. When switching on the machine, ensure that the correct pressure is set prior to using.
3. When vacuuming food products, ensure that the packets are not over filled.
4. Only use assigned plastic packaging in the machine.
5. When cleaning the machine, make sure it is turned off at the wall socket. Only use a wet or moistened cloth to clean. Do not use water.

These are the most basic equipment used in every kitchen and apart from these, with the modernization and change in food habits, new machines are used in professional kitchens. Walk-in refrigerators and deep freezers are large rooms where one can easily walk inside. Blast chillers are used in flight catering and institutional catering where the food has to be quickly chilled and packed for further usage. Live cooking stations have again given rise to artistically fabricated kitchen concepts that are very appealing to the eye.

Increasing health awareness has given rise to sophisticated steam cookers which preserve nutrition and use minimal oil for preparing dishes.

Some Latest Equipment Used in Today's Kitchens

Induction Cookers

Induction cooking works on the principles of transformers. A coil of wire is placed under a cooking surface usually made of heat-resistant ceramic glass. The electricity when passed through this coil generates a strong magnetic field. When the cooking equipment, which is magnetic in nature, comes in contact with the magnetic field, an electric current is passed to the cooking pot. The pot which is not a good conductor offers some electrical resistance; this results in high-intensity heat which is then used for cooking.

Only special pots of iron or steel, which are magnetic in nature, are used for such a cooking apparatus. These are very expensive, but nevertheless more efficient than normal electrical stoves. The induction cooker can be mounted onto a stainless steel unit or embedded into buffet counters (see Fig. 4.1). It provides flameless cooking and conserves energy, because as soon as the pot leaves the contact surface, the heat ceases due to break in flow of magnetic field.

Fig. 4.1 Induction cookers—flat top and for wok frying

Infrared Cookers

This is by far the most advanced method of cooking food, in which food retains all the natural juices and nutrition.

Heat is always transmitted to food mainly in three ways—conduction, convection, and radiation. Conduction is when the food comes in contact with the heat and the molecules in the food get heated up. In convection, the heat passes through the medium of gas or liquid. The heated molecules heat up these mediums by colliding against them and hence, the heat transfers to the food. Radiation on the other hand is emission of the heat from an external source and does not need any medium. It heats the molecules that come in its way and transfers heat to the food that needs to be cooked.

Electromagnetic waves are of varied wavelengths. The shorter ones are ultraviolet rays, X-rays, and gamma rays and the longer ones are radio waves, microwaves, and infrared waves. This electromagnetic spectrum is separated by the visible light waves starting from red, orange, yellow, green, blue, and violet. That is the reason why the waves above this light spectrum ending with violet are called ultraviolet rays. Similarly, the electromagnetic waves just below the red spectrum are called infrared waves or infrared energy. When these waves strike any organic molecule such as food, it causes the molecules to vibrate and this then generates a very high-intensity heat which is used in modern infrared cookers.

Oriental Cooking Ranges

Chinese and Thai cuisines use high-pressure burners which are powered by air thrust to bring the flame out of the cooking pit, so that the food can be tossed into the flame for the char-grilled effect (see Fig. 4.2).

These are very sophisticated ranges, which have complete arrangement for cooking, washing, etc. The regulation of flame in some of these ranges are operated by knees. As these cooking ranges are of low height they require constant bending to operate with hands.

An oriental kitchen can be a very noisy and busy kitchen. The high-pressure ranges cook the food fast and efficiently. The regulation of the heat requires lots of skill as food can burn very fast on thin woks on these flames.

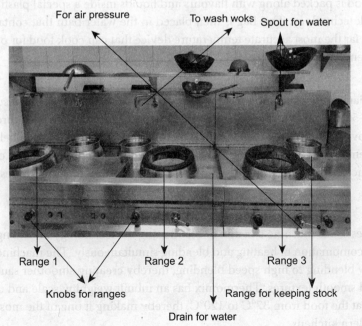

Fig. 4.2 Oriental cooking range

Josper

Joseph Armangue and Pere Juli, both chefs, created the Josper oven around 1970. They gave their name to the company that now produces the Josper ovens with the latest technology. These ovens are a unique combination of grill and an oven encased in a single machine. They are being used in barbeques and grills all across modern and trending hotels worldwide.

The oven operates on charcoal completely. Its design which enables it to function like an oven with a closed door, ensures that there is no moisture loss and that the heat is trapped in the oven to raise the temperatures of the grill up to 400°C, which is the optimal temperature for the juicy BBQ meats.

Combi ovens

These ovens, also known as convection ovens, come in various sizes and work on the principle of circulation of hot air. Some models are also available with roll in trolleys that can be loaded and rolled inside the cabinet. This equipment comes very handy in cooking as well as reheating of food. They are called combi ovens as they have the facilities of moist as well as dry heat. Some modern combi ovens come with inbuilt sensors that help you to programme a certain recipe and then with a click of the button, the oven does its job from start to finish.

Immersion circulators

As the name suggests, this is an electric device that is submerged in water. The heating coils or elements heat up the water whilst propelling it thereby maintaining the temperature of water at a constant degree as selected by the user. This method of cooking allows the food to cook at a precise temperature over a long period of time. In modern cooking methods, sous vide cooking has again emerged as the trending cooking technique. Sous vide cooking, literally translating to cooking under cover, is an ancient art of cooking followed in France and other European countries where a piece of fish was cooked in sealed paper bags.

Today, the food is packed along with flavours and liquids inside a special plastic bag that is sealed with vacuum pack technology. This bag is then placed in the water bath that contains the immersion circulator. It is by far the most accurate temperature device that can cook food for over 24 hours at the desired cooking temperature.

Anti-griddles

This kitchen appliance resembles a box which has a flat metal surface. This metal surface can go up to a temperature of –50°C due to the use of a refrigerant that is continuously passed through a compressor and helps to remove heat from the metal surface thereby taking its temperature below –40°C.

This equipment is used for preparing dishes with frozen exterior and soft and creamy interior, by freezing the product in less than a minute. One could make savoury or sweet dishes using this modern equipment.

Thermomix

As the name suggests, this machine can mix or blend using the technology of heating. Therefore, it can be said that it is a combination of heating and blending simultaneously. The machine blends food from a selection of slow blending to high speed blending, thereby creating smoother sauces, with better air incorporation and smooth textures. Thermomix has an inbuilt weighing scale and unique heating elements that can heat the food from 37°C to 120°C, thereby making it one of the most unique machines to be used in modern kitchens.

Thermomix can be used for a variety of tasks such as grinding, blending, whipping, steaming, mixing, and chopping ingredients for various requirements.

High Speed Blenders

High speed blenders are high performance table top blending machines that are capable of blending ingredients at a very high speed, thereby creating smooth sauces, smoothies, and shakes. Due to its unique blender design, it is possible to blend small amount of ingredients into a very smooth texture. It is used in kitchens, bars and in many fast food outlets across the world.

Air fryers

This equipment has off late been very popular in both domestic as well as commercial food establishments. Its unique design and mechanism enables the food to be cooked by circulating hot air around the food using the convection principle of heat transfer. In this equipment, a circulatory fan moves the hot air around the food creating a crisp layer that resembles the food cooked by submerging in oil.

The deep frying happens when the food is submerged in hot oil at a temperature of 150–200°C. In case of air frying, a thin layer of fat is sprayed on the food product and hot air at a temperature of 200°C or more is circulated over the food, creating the same results.

Dehydrators

As the name suggests, this device removes the moisture from the food, thereby drying it for either preservation or for creating a new texture that can add interesting elements whilst preparing or presenting food. This machine not only dehydrates fruits and vegetables, but it can also be used for dehydrating meats, fruit purees, and yoghurt to create pliable leather like texture.

When the food is allowed to dry up in open air, it changes colour due to oxidation. In case of drying food in dehydrators, a small fan keeps the air in motion, thereby not allowing the oxygen to get in contact with the food, thus maintaining the natural colour of the food and making it look more attractive.

CONCLUSION

The first step to understand any cuisine is to dwell in its history and evolution. It will reveal many unknown secrets that form an integral part of that cuisine and culture. The fuel used in the olden times was wood, as it was the only thing available. Today with modernization and degradation of natural resources, it is becoming mandatory to look at other sources of fuel. Solar energy is also used in many organizations and solar cookers are installed to carry out the cooking jobs in many eco friendly hotels. Necessity and demand have always led to discoveries and inventions and kitchen is no exception. The increasing cost of labour has led to the invention of cost efficient machinery, which is used to carry out the jobs more efficiently. In this chapter we discussed about various kinds of fuels used in the kitchen and the usage of the same.

We also discussed about capital and operational equipment and how they form part of the annual budgets. We also discussed various types of capital equipment used in modern hotels and their standard operating procedures with regards to health, hygiene, and safety. We shall discuss more about these equipment in the coming chapters. In chapter 17 we shall discuss various methods of cooking and there we shall discuss about equipment used in each principle of cooking. Also in chapter 23 we shall talk about some special equipment used in Indian cooking.

In this chapter we have introduced the modern equipment that are being used in fine dining restaurants and hotels in India and abroad. These equipment with improved technology are very efficient in producing high quality products. The next chapter shall discuss about various kinds of menus and basic art of menu planning.

Air fryers An equipment which is used for frying by using minimum amount of oil.

Anti-griddle A machine that has a surface temperature of −50–60°C and is used for making instant frozen desserts and savoury items.

Bar-be –Que grilling meats on a charcoal medium or spit fire.

Blast chillers Equipment used in institutional caterings where hot food is chilled in a very less time.

Capital Equipment Fixed equipment usually a machine used in the hotel.

CNG Compressed Natural gas used for fuel efficiency.

Dehydrator An equipment that is used for removing moisture from fruits, vegetables, and meats, so as to increase the shelf life of the product.

Gas Bank Area designated in the hotel for storage of gas cylinders, from where the LPG is supplied to all the parts of the kitchen.

High speed blender Usually used for making smoothies and shakes.

Immersion circulators An equipment which is used for *sous vide* cooking.

Induction A type of transference of heat through electromagnetic waves.

Infrared cooking modern style of cooking using infrared rays to cook the food.

Inventory Physical count of equipments done on quarterly basis for auditing purposes.

Josper A modern oven and griller that works on electricity and wood fuel at the same time.

Kitchen Stewarding A department in an hotel which offers support to Food and Beverage operations in terms of cleanliness and supplying of crockery and cutlery.

LPG Liquefied Petroleum Gas, commonly available in cylinders.

Operational equipment Basic tools and utensils used for operating a job.

Salamander equipment used in the kitchen which gives radiated heat to the food.

Solid fuel also known as handy fuel comes in pots and is used to heat food in chaffing dishes.

Sous vide French word that literally means 'under cover'. The food is vacuum packed and then cooked in water by using immersion circulators.

Steam Jacket Kettles A cooking equipment which runs on steam.

Thermomix A modern high speed blender that can even be used for cooking.

Woks Cast iron utensil, like a shallow round bowl, used in oriental cooking

CONCEPT REVIEW QUESTIONS

1. List the various types of fuels used in the kitchens.
2. How does LPG work?
3. What is the difference between industrial cylinders and domestic cylinders?
4. What is a gas bank and how does it function?
5. Where is coal used in the kitchen?
6. List few uses of steam in kitchen.
7. What do you understand by handy fuel?
8. What is a capital expenditure?
9. What is the difference between fixed equipment and operational equipment?
10. What is an inventory and how often it is done?
11. List three safety operating procedures of a gravity slicer.
12. List three safety operating procedures of a bone saw machine.
13. List three safety operating procedures of a brat pan.
14. List three safety operating procedures of a deep fat fryer.
15. List three safety operating procedures of a salamander.
16. List three safety operating procedures of a microwave oven.
17. Why should we not use metal in microwave ovens?
18. What is the vacuum machine used for?
19. Explain the process of induction cooking.
20. What is the difference between induction and infrared cooking?
21. What kinds of utensils are used for induction cooking?

22. List three advantages of infrared cooking.
23. Describe an oriental cooking range.
24. Explain the budgeting of food and beverage equipment.
25. Describe the principle behind Josper ovens and why they are unique?
26. What is the relation between thermo circulators and sous vide cooking?
27. What is an anti-griddle and what is the temperature that it provides?
28. What is the difference between a thermomix and a high speed blender?
29. What is the principle behind air fryers?

PROJECT WORK

1. Get a copy of menu of any hotel or restaurant and list the equipment that would be required to run that operation. List your observations and justify your thought process.
2. Make a group of eight to ten students belonging to various states and cultures. List the special cooking equipment or special fuel used in production of your typical home foods. List the observations and share in a group.
3. Make a list of equipment in your organization and list the safety operating procedure of the same. Laminate the findings and paste near the equipment.
4. Large machines and heavy duty equipment use lots of energy for their operation. Liaison with engineering department and calculate the per hour running cost of each machine and get it painted on the machine. This will help you to think on the lines of energy conservation.
5. In your library, pull out the international kitchen journals and periodicals of previous two years and conduct a study of emerging changes in the cooking equipment and fuel efficient procedures. This would help you to understand, how fast the kitchen trends are changing.
6. In groups of three to four, visit the hotels that use modern equipment and understand how they impact the quality of the product.

CHAPTER 5

BASIC MENU PLANNING

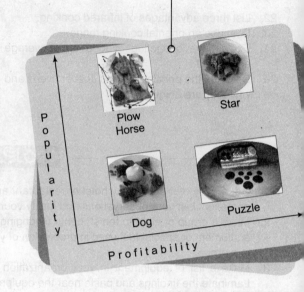

Learning Objectives

After reading this chapter, you should be able to:
- know what a menu is and understand its importance
- understand the various types of menus and the requirement of the same
- understand the concept of balance in a menu
- know the term 'menu engineering' and learn how to analyse it
- know the modern trends in wine and food pairing

INTRODUCTION

This chapter introduces menu, the various ranges of menu available, and the roles of menu and its importance in hotels and food service business. A food service manager identifies the role of menu in food service operations as extremely essential. All the factors of prime concern revolve around the very consideration of menu as a 'bill of fare'. Menu analysis is defined as a range of techniques and procedures that enable decision making with respect to pricing the dish, costing, portion sizing, menu planning, and menu marketing. Customer feedback has become an important aspect of menu analysis as it helps organizations in analysing and refurbishing their existing menus. In order to be able to plan a menu, it is important that we first understand the meaning of the word 'menu' and how it is used as a tool in a food service organization.

MENU

Menu is a list, in specific order, of the dishes to be served at a given meal. Menu is central to the food service concept—it defines the product offering, establishes key elements of financial viability namely price and contribution margin, and provides a powerful marketing tool. The word 'menu' dates back to 1718, but the custom of making such a list is much older. It is said that in the year 1541 Duke Henry of Brunswick was seen referring to a long slip of paper. On being asked what he was looking at he said it was a form of programme of dishes and by referring to it he could see what dish was coming and reserve his appetite accordingly. It is believed that perhaps it was this idea that led to development of menu cards (Philip 2003).[1]

During olden times 'bill of fare' of ceremonial meals was displayed on the walls to enable the kitchen staff to follow the order in which the meal had to be served. Modern menus did not appear until the nineteenth century when the Parisian restaurant Palais Royale provided customers with small, handy reproduction of menu displayed on the door. Mid-nineteenth century saw the placement of menus at the end of the table from where the guests could choose the menu item that they wished to have. However, as time progressed, individualized menus came into being.

The menu is the most significant factor in a food service operation. A menu epitomizes a caterer's F&B intention. People eat away from home for various reasons. However, to many, the food that they eat has the greatest and the most significant impact upon their experience. Therefore, the menu, which proclaims to the guests the choice of food items available, is a major factor in popularizing a restaurant and promoting F&B sales.

Badly composed menu is likely to spoil the best of dinners. Menu plays a competitive role in the commercial industry. Its effect is not only observed in satisfying a client but also in generating sufficient revenue for the businesses. Firms should understand the role of menu and take steps to better it if required.

FUNCTIONS OF THE MENU

A menu has the following functions.

Information   It satisfies a guest's need for information about what food is available, how it is cooked and presented, and at what price.

Order   It presents the dishes in a logical order, usually listing the menu items under course headings, thereby making comprehension of the menu easy.

Choice   It determines the freedom of choice that a guest may have.

Image   Menu helps present the overall image and style of the restaurant.

Sales   It is a means of promoting sales by appropriately describing the dishes which appeal to the guest.

In order for the menu to perform all these functions successfully, it must be informative, accurate, understandable, and well designed. A restaurant manager must ensure that the items mentioned on the menu are available at all times and as per description since it is frustrating for a guest to make a decision only to be told that the dish is not available or to receive a dish that is not as stated.

TYPES OF MENU

Menus are broadly classified into several types. They are as follows:

À la carte   It is a list of all the dishes on offer, which is within the resources of a particular kitchen. It means 'from the card'. From it a guest may select items to compose his/her own menu. The charge of

---

[1] Philip, T.E. 2003, *Modern Cookery for Teaching and the Trade*, Fourteenth Edition, Orient Blackswan.

meal will be the total of the prices of individual dishes served to the guest. This is where the skill of the steward will come into picture, where he/she would do the suggestive selling and let the guests mix their choices in such a way that they enjoy the meal.

Table d'hôte   It literally means 'from the host's table'. It is a meal usually divided into various courses with little or no choice, and is available at a fixed price.

Plat du jour   Food on a *plat du jour* menu is normally the specialty of the day or the chef. Both à la carte and table d'hôte menus are compiled to meet the requirement of the items to be served in the following meals by F&B outlets.

- **Breakfast/*petit déjeuner***   Breakfast is the starting meal of the day and helps boost up the metabolism of the body.
- **Brunch**   A meal had between breakfast and lunch.
- **Elevenses/*goûter***   A light meal usually had at mid-morning hours.
- **Luncheon/*déjeuner***   It is a meal had during the daytime, ideally between 1 p.m. and 2 p.m.
- **Afternoon tea/high tea/*le five o'clock***   It is a light meal, usually had between 4 p.m. and 5 p.m, where tea is served with light snacks.
- **Cocktail**   In this meal small bites are served normally with beverages.
- **Dinner/*diner***   This is the main meal of the day eaten between 7 p.m. and 9 p.m. Many people prefer a light dinner, but for some people it is a lavish fare of wining and dining.
- **Supper**   It is a less formal meal eaten before dinner.

MENU USED AS CONTROL TOOL

Menu is not only a selection of items presented to the guest in a written form, but it is also used as a control tool. There are various things associated with menu that can be used to analyse and control the same. Let us see and analyse a few of them.

Menu Specifications

Menu specification is described as information or description available on the menu, especially if there is use of a foreign language or use of a special type of cooking process. The same is desirable in order for the guests to get clarity on their selection. Some menus also use graphics and illustrations in order to give an idea of the item that the guest can expect to be served.

Menu Composition

A well-balanced meal satisfies guests' appetite, pleases them, and yet leaves them without any feeling of overeating. Hence, the various dishes that constitute a meal should be balanced, otherwise a series of dishes which are excellent individually, may collectively make an indigestible meal. Gastronomic experience has led to an expected order of courses in a meal. The order of dishes is from those that aim to stimulate appetite, such as appetizers, followed by a light meat course onto a main dish, followed invariably by desserts, fruits, or cheese. The last meal is finally had with coffee or tea.

Menu Engineering

Menu engineering is a tool used by modern day managers and chefs to analyse different aspects such as profitability and popularity of dishes on the menu. It is also a very essential tool in making pricing decisions and modifying recipe design, keeping in mind what guests want. In the present scenario it has become a powerful tool in the hands of restaurant managers and chefs alike.

Menu engineering is a unique tool for controlling the food cost in an outlet/organization. This process involves the methodical selecting, costing, pricing, and evaluating of the items in the respective menu. Menu engineering also provides the manager/chef with information about a menu item's profitability, as well as popularity, so that proactive planning, recipe design, and customer pricing decisions can be made.

The limitation of menu engineering is that it is not a substitute for proper purchasing, food rotation, standard recipes, or any of the other basic kitchen controls that can negatively impact costs. Rather it is a method of evaluating every item on a menu relative to its present contribution to the profits, thereby allowing managers to recognize the items they want to sell the most so as to increase the profitability or maximize the profit.

MENU ENGINEERING GRID

A graph can be formulated so as to put in all the menu items and can be plotted with an axis of popularity margin and profitability percentage, as shown in Fig. 5.1.

Stars

Items on a menu that are high on contribution margin and popularity (refer to Fig. 5.1).

Managing Stars

The following has to be done for managing 'stars'.

- Feature the items prominently on the menu.
- Regularly promote these items.
- Increase guest prices in peak season.
- Ensure that rigid specification and quality standards are maintained.

Plow Horse

Items on a menu that are low on contribution margin and high on popularity (refer to Fig. 5.1).

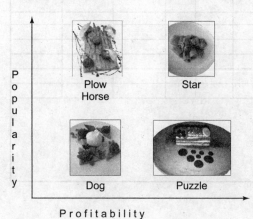

Managing Plow Horse

The following has to be done for managing 'plow horse'.

- Test small increase in selling price.
- Test for the demand of food items.
- Relocate the items to the lower profile of the menu.
- To increase the contribution margin, package it with high contribution items.
- Ensure that strict specification for purchasing, preparing, and presenting the food is maintained.
- Make sure that there is no loss through wastage, spoilage, or pilferage of such food items as these are high in food cost.

Fig. 5.1 Menu engineering grid

Puzzle

Items on the menu which are high on contribution margin and low on popularity (refer to Fig. 5.1).

Managing Puzzles

The following has to be done for managing 'puzzles'.

- Reposition the item on the menu and feature it more prominently on the menu.
- Rename the items and change the presentation.
- Increase merchandising or up-sell the item.
- Give the item a special status.

Dogs

Items on the menu that are low on both popularity and contribution margin (refer to Fig 5.1).

Managing Dogs

The following has to be done for managing 'dogs'.

- Increase the price of the item and raise the status to puzzle.
- Try and up-sell the item.
- If the item is new on the menu, check the amount of time it has been there and accordingly improve that particular dish.
- Remove the item from the menu.

Menu Engineering Worksheet

A menu engineering worksheet is formulated every month and this is used as one of the most important tools in menu engineering. It is represented in Fig. 5.2.

Name of Outlet:													Month:
Sr. No.	Item	No. Sold	MM per-centage	Cost Price	Selling Price	Selling Price (all inclusive)	Total Sale	Total CM	Menu CM	Item	MM Category	CM Category	Category
Fair share													
Fair share percentage													
Popularity index													
Margin													
Food cost													

Fig. 5.2 Menu engineering worksheet

Steps in Formulating a Menu Engineering Worksheet

These are as follows:

Number of Items Sold This column shows the number of dishes sold for that particular month or a given period of time.

Menu Mix (MM) Percentage In this column we get the menu mix percentage by dividing the total number of that particular item sold, by the total number of dishes sold in that particular category.

Cost Price It is the cost incurred to produce a particular dish which is obtained from the standard recipe card of that particular dish.

Selling Price In this column we place the price at which the dish is being sold on the menu.

Menu Mix Rating First the average of the menu mix percentage has to be calculated. If the menu mix percentage of the particular dish is greater than the average then it is termed as high and if it is lesser than the average it is termed as low.

Total Cost The cost price of the particular dish multiplied by the numbers sold.

Total Sales The selling price of the particular dish multiplied by the number sold.

Menu Contribution Margin (CM) It is the difference between the total sales and the total cost.

Item Contribution Margin It is the difference between the selling price and the cost price of a particular dish.

Contribution Margin Category The average of the contribution margin has to be calculated. If the contribution margin of the particular dish is greater than the average then it is termed as high and if it is lesser than the average it is termed as low.

Category The particular dishes on the menu are categorized keeping in mind the following points.

- A puzzle has a high contribution rating and a low menu mix rating.
- A star has a high contribution rating and a high menu mix rating.
- A plow horse has a low contribution rating and a high menu mix rating.
- A dog has a low contribution rating and a low menu mix rating.

Fair Share It is the average of total number of dishes sold by the total number of dishes being sold in the category.

Fair Share Percentage This is fair share divided by number of dishes sold multiplied by 100.

Popularity Index The index is based on the notion of expected popularity. Each item is analysed in relation to those items which are assumed to be equally popular.

MENU BALANCING

The most delectable and well-cooked food might not be appreciated by the guests, if it is not being served in the right portion size and with the right accompaniment. For example, well-prepared fluffy steamed rice teamed with a dry preparation such as 'aloo jeera'[2], will not be appreciated as both the items are dry and will be unpleasant to eat. Before we understand the balance in the menu, let us understand the basic French classical courses often referred to as 'courses of meal'.

---

[2] Potatoes tossed with cumin and Indian spices.

Hors d'oeuvre

This course is usually aimed to simulate appetite and therefore, is composed of tangy and salty dishes. For example, potato salad, caesar salad, Russian salad, caviar, smoked salmon, smoked ham, oyster, etc.

Potage

It refers to soups of two types—clear (consommé) and thick (cream, velouté, or puree). A clear soup on the menu card is generally listed first.

Poisson (fish)

In this course normally poached/steamed/baked fish is served with an appropriate sauce and boiled vegetables.

Farinaceous

Dishes such as risotto, spaghetti, gnocchi, and penne may be served in place of the fish course. Egg dishes, such as en cocotte, sur le plat, may be served in this course, especially during a luncheon menu. They are seldom included during dinner.

Entrée

This is the first of the meat course at dinner; and it is usually complete in itself. For example, sweet breads, vol-au-vent, tournedos, etc.

Remove/Relevée

It is a large joint of meat. For example, saddle of lamb, braised ham, etc.

Sorbet

This course is intended to be a pause during a long meal. A sorbet is supposed to settle dishes already served and to stimulate the appetite for the ones to follow. It is water ice flavoured with champagne or any liquor or delicate wine. It is usually served in a champagne saucer with a teaspoon. Russian cigarettes may be passed around the table and ten minutes are allowed before the next course.

Rôti

This course consists of roast poultry or game, such as chicken, duck, turkey, pheasant, partridge, etc., served with their sauces and gravy. A dressed salad is served along.

Legume (vegetable)

The French customarily served vegetables as a separate course, for example, asparagus served with hollandaise sauce.

Entremets

This may consist of a hot sweet dish such as soufflé, rum omelette, etc. Petit fours are served with this course.

Savoury

A savoury course consists of a titbit on a hot canapé of a toast or fried bread. Cheese platter may also be presented with crackers, watercress, walnuts, etc. as accompaniments.

Dessert

This finale consists of a basket of fresh fruits and nuts. They are sometimes placed on the table as a part of the decoration.

Café

Different types of coffees are served in this course.

The menu for an event is decided by combining the courses listed above. A fine dining menu would be a combination of at least four courses of meals. In formal dinners a fine dining menu will definitely have a sorbet course. State banquets and similar formal events have more than six to seven courses depending upon the type of function.

Selecting Dishes and Courses

A balance in menu is very important because it can leave a guest either satisfied or unhappy. In selecting dishes and in deciding the number of courses, a suitable menu is the one that conforms to the following.

Demographic Choices   While planning a menu, it is important to know the guests who would be eating the meal. It is not a good idea to plan for frog legs for an Indian market segment as it is not relished in India like it is in China. The market segment is to be kept in mind while choosing the same.

Availability of Seasonal Items   A well-planned menu should use seasonal commodities as they will be more nutritious and healthy.

Nature of the Occasion   A menu is planned keeping the occasion in mind. A children's party will have a different menu from an engagement party and a wedding party will be more lavish. A formal dinner is always a course-wise table d'hote and never a buffet.

Resources of the Kitchen Staff and Equipment   Only those items should be put on the menu which the kitchen staff can cook and there should be proper resources and equipment to cook the same. It is not a good idea to put tandoori breads on the menu if one does not have *tandoor* in the kitchen.

Cost and Pricing Policy   A well-balanced menu should be a mix of expensive and inexpensive items. A menu having only expensive commodities will yield less profit.

For Gastronomic Reasons   Balance of menu is not only done for nutrition and digestion. A menu should be balanced in terms of variety and flavour, colour, texture, and consistency. A meal should not contain two dishes which are composed of the same ingredients, for example, if cream of cauliflower is being served then cauliflower should not form a part of the vegetable course to be served later. Two white meats or two dark meats should not follow each other. For example, pork should not follow veal or beef should not be served after lamb. A heavier *relevée* should follow a light entrée. If the menu is long then the dishes chosen in each course should be light and small in portion size and if the menu is short, more substantial dishes may be included to ensure that diners will have a sufficient meal.

WINE AND FOOD PAIRING

There is possibly no better beverage to accompany a meal than wine. As wines are available in infinite variety of tastes, textures, and aromas, they can be paired with several dishes from all over the world.

When the perfect wine has been chosen, the total experience of the meal doubles up; the wine tastes better and so does the food and there is perfect sensory harmony.

In order to match wine and food it requires from the diner, an elevated and educated palate and from the *sommelier*, an in-depth familiarity with both the food and the wine. Wines exist for the pleasure of pairing them with food, and if they are paired wisely, the resulting harmony of smells and flavours will enhance their characteristics.

It is impossible to establish strict rules about wine pairing because both food and wine are complex substances which are easily altered. As the taste of a dish depends on its ingredients and its cooking process, the same holds true for wines. The taste of wines will differ depending on the types of grapes used, the different soils, climatic conditions, and techniques of cultivation and manufacture. The general rule is that the flavour of the wine should not prevail over food and vice versa. In other words, a delicate dish requires a light wine, while a rich dish calls for a well-structured wine.

Here are some of the commonly accepted rules that are by no means hard and fast.

Rule 1 Always drink white wines with white meats and seafood. With a spicy seafood dish, a herbaceous New World *sauvignon blanc* would do well, but so would a chianti (Italian white wine) or new season *beaujolais* (French white wine). Red fish, such as salmon, particularly goes well with wines of its own colour namely a light *pinot noir*, a dry rose, or *blush zinfandel.*

Rule 2 Red wines with red meats this rule probably works well. But choose the right red wine with the right meat. Duck pairs well with *pinot noir* while lamb with *cabernet sauvignon* and most game dishes match perfectly well with peppery, full-bodied *shiraz.*

Rule 3 Sweet wines go well only with dessert. One of the greatest matches is foie gras and sauternes. The sticky sweetness helps to assimilate the fatty richness of the liver on the palate. Same goes with pork.

Rule 4 There are no rules regarding champagne. If one enjoys it one can have it as an aperitive, to finish the meal or drink it throughout the meal.

CONCLUSION

Menu is the soul of a F&B commercial business. It is thus essential for organizations to asses their menus. A menu is an important tool for control and it will be right to say that it is central to F&B operations. In this chapter we discussed the various basic elements of menu engineering. Though this subject is vast in itself, we have got a fair idea of how menu is used as an effective tool to plan and organize a food business. We learned that a menu is much more than just a list of food items with prices mentioned on it. We understood various types and categories of menus used in hotels and other establishments. We also read about the menu engineering grid and the concept of categorizing the dishes with regard to their popularity and profitability on the menu. We also discussed the classical French menu, which helped us to understand the courses of meals. We also learned about the pairing of food with wine.

KEY TERMS

À la carte A list of dishes on the menu card where a guest can choose what he/she wants to eat.

Banquets An area of a hotel designated to celebrate parties and functions for a group of people.

Brunch Meal which combines breakfast and lunch.

Café The last course in the French classical menu which serves coffee with chocolates.

Category   Stars, plow horses, puzzle, and dogs are the categories of items in a menu.

Cocktail   Small savoury titbits usually served with drinks.

Consommé   French classical soup which is made with clarified and reduced beef stock.

Courses of meal   A systematic layout of the meal starting with starters and ending with tea/coffee.

Déjeuner   French for lunch.

Dessert   In English it refers to the sweet course served at the end of the meal; but in French classical menu, nuts, and fruits are served in this course.

Dog   Dishes that are low on profitability and low in popularity.

Elevenses   A light meal usually had at mid-morning hours.

En cocotte   French term to denote dishes served in a casserole.

Entrée   French for 'entry'. It is the first meat course at a dinner or the first entry of a main course.

Entremets   French for sweet course.

Fair share   Average of total number of dishes sold by the total number of dishes sold in the category.

Farinaceous   Fourth course in the French menu which includes starchy dishes such as pasta, risotto, etc.

Game   Wild birds such as pheasant, turkey, duck, partridges, etc.

Gnocchi   A popular Italian dish made with potato, flour, and cheese and poached before being tossed in a sauce.

Goûter   French for elevenses.

Hollandaise   Emulsified sauce of egg yolks and butter.

Hors d'oeuvre   The first course in the French classical menu, usually consisting of salads and tangy dishes to stimulate the appetite.

Item contribution margin   Difference between the selling price and the cost price of a particular dish.

Legume   French for vegetables and pulses.

Menu balancing   Creating a balance in the dishes with regards to colour, nutrition, cost, etc.

Menu composition   A well-balanced or composed menu served as a table d'hôte.

Menu contribution margin   The difference between the total sales and the total cost.

Menu engineering grid   A graphical plot of dishes with respect to profitability and popularity.

Menu engineering   Tool used to analyse the dishes with respect to profitability versus popularity.

Menu mix percentage   Total number of a particular dish sold divided by the total number of the dishes sold in that category.

Menu specification   Information or description of dishes on the menu card.

Menu   A list of dishes usually printed to be served for a meal from where a dish is chosen by the guest and demanded through an order.

Petit déjeuner   French for breakfast.

Plat du jour   French for special of the day.

Plow horses   Dishes that are low on profitability but high on popularity.

Poisson   French for fish. It is the third course in the French menu.

Popularity index   An index based on the notion of expected popularity

Potage   French for soups. It is the second course.

Puzzle   Dishes that are high in profitability but are not very popular.

Relevée   This is usually a large joint of meat and relieves the entry course. It is also called remove.

Risotto   A starchy rice preparation from Italy.

Rôti   French for roasted and in this course, roasted poultry or game are usually served.

Savoury   Third last course in the French classical menu, in which small tit bits are served. Salty foods are also referred to as savoury.

Sorbet   A frozen ice flavoured and sweetened drink, used to wash down the tastes of previous meal and to give a break for other courses to follow.

Sphagetti   String-shaped pasta.

Stars   Dishes high on both profitability and popularity.

Steward   A person who is responsible to take order and serve food and beverage to the guest.

Supper   Less formal meal eaten before dinner.

Sur la plat   French term to denote dishes served on a platter.

Table d'hôte   It is a list of dishes, pre-chosen and written on the card. The entire meal is served when the guest orders table d'hôte.

Velouté   A blond coloured creamy soup made with white meat.

Wine   An alcoholic drink made from juice of grapes which has been fermented.

CONCEPT REVIEW QUESTIONS

1. What is menu? Why is it important?
2. List few roles or functions of the menu.
3. How does menu help to create the image of an organization?
4. List the types of menu and for which meal period you use them.
5. How is table d'hôte menu different from an à la carte menu?
6. What do you understand by the term *plat du jour*?
7. Explain menu specification and its importance.
8. Why is it important to have a well-composed menu?
9. Explain the term 'menu engineering'.
10. How does menu engineering act as a control tool?
11. Draw out the menu engineering grid and plot the categories on the same.
12. How would you manage 'star' category items on your menu.
13. How would you manage 'puzzle' category items on your menu?
14. Explain a menu engineering worksheet and its importance.
15. What do you understand by the term 'menu balancing'?
16. List at least five factors to be kept in mind while balancing a menu.
17. List the French classical courses of menu.
18. List the general rules of pairing wine and food.
19. Which course will be served with champagne?

PROJECT WORK

1. In a group of three to four, select a menu of any establishment and critique it from the point of menu balance and list your findings and share with other groups.
2. Get a menu engineering report of any hotel for a particular month and analyse it from the point of view of a chef and manager and list a plan of action to tackle any unpleasant findings.
3. Divide yourselves into groups of five each. Each group does a menu for a particular function. Do at least six different functions and draw out the menus for the same. Use your discretion to choose the type of menu and give reasons for the same. Draw out the list of wines which can be served with these menus.
4. Do a study on pairing of wine with Indian food and give valid reasons for choosing the same.

CHAPTER **6**

BASIC PRINCIPLES OF COOKING VEGETABLES

Learning Objectives

After reading this chapter, you should be able to:
- identify various types of vegetables used in the kitchen
- know the way a vegetable is processed
- figure out the selection and storage criteria of vegetables
- understand the pigments in the vegetables and the effect of heat on them
- list the cuts of vegetables and their uses in cookery

INTRODUCTION

In the previous chapters we understood the basic work operation of the kitchen in terms of layouts, equipment and fuels, and various kinds of menus. Now let us begin our journey to being professional chefs by understanding about various commodities used in the kitchen. In this chapter an attempt has been made to discuss most of the vegetables used in modern kitchens. It is important to know how heat and temperature affect these commodities as this helps us in creating an array of dishes with different kinds of textures and consistencies. We shall also discuss how each vegetable is processed using the right kinds of tools. Vegetables are rich source of nutrients and are responsible for the metabolism of our body. It is important for chefs to understand the vegetables and how to process them in order to prepare dishes for the vegetarians.

VEGETABLES

Any part of a herbaceous plant that can be eaten, either raw or cooked, is termed as a vegetable. Vegetables contain more starch than sugar unlike fruits and hence, they are used extensively in savoury dishes. Vegetables can be used in a variety of forms such as frozen, canned, cooked, mashed, dried, dehydrated, and fresh. However, only selected parts of some plants are eaten such as flowers, flower buds (globe artichoke), leaves (lettuce), leaf buds (brussels sprouts), shoots (asparagus), shoot buds (cabbage), stems (rhubarb), flower stems (cauliflower), seed pods (green beans), and immatureseeds (broad bean).

Vegetables are eaten in a variety of ways—raw or cooked, as main courses, or as snacks. Different vegetables have different types of nutrients in them. Vegetables have water soluble vitamins such as vitamin B and C, and fat soluble vitamins such as A, D, E, and K, and also contain minerals and carbohydrates. Since each category of vegetable responds to a particular method of cooking, it is important for us to know how vegetables are categorized. Broadly speaking vegetables are put into the categories discussed in Table 6.1.

Table 6.1   Classification of vegetables

Category	Description	Examples
Brassica	Brassica or the cabbage family consists of vegetables used for their head, leaves, or flowers. These are mostly available in winter and hence, are mostly used in broths and braised dishes to accompany meats.	Cabbage, cauliflower, brussels sprouts, and bok choy
Fruit vegetables	These are the fruits of flowering plants. They also contain seeds.	Tomato, avocado, brinjal, and pepper
Gourds and squashes	These are available both during summers and winters. The gourds are classified into summer squashes and winter gourds. There are over 750 varieties of gourds available and they are grown around the world. The long trailing vines of a complex root system bears this vegetable, adorned by large leaves and attractive flowers. The flowers are also used extensively in modern cookery.	Bottle gourd, butternut squash, and ridged gourd
Greens	The term greens is usually referred to vegetables which are leafy and eaten cooked, with the exception of lettuce. Most of the greens are mildly spiced and slightly strong in flavour.	Spinach, watercress, and radicchio
Fungus	Although not a real vegetable, fungus is a plant that has no seed, stem, or flower and usually reproduces from the spores. It is commonly known as mushroom and usually has a stalk covered with an umbrella. One must be careful in selecting a mushroom, as not all varieties are edible and some can cause poisoning and can be fatal.	Button mushroom, shitake, portabella, and porcini
Roots and tubers	Roots and tubers are interchangeable words. However, scientifically speaking, tubers are fat underground stems while roots are the single bulbs which extend into the ground and supply the plant with nutrients. The tubers would be more starchy.	Root vegetables—carrots, radish, and onions Tubers—potatoes, Jerusalem artichoke, and colocasia

Contd

Table 6.1 (Contd)

Category	Description	Examples
Pods and seeds	Some vegetables contain seeds enclosed in a pod. In some cases only seeds are eaten like in case of peas, corn, and pulses. In some cases the plant is eaten as whole such as okra, snap peas, and French beans. The pods and seeds contain the highest source of proteins and carbohydrates.	Green peas, okra, snap peas, and pulses
Stems	Also known as stalk vegetables, they have the highest percentage of cellulose fibre and are usually eaten when they are young and tender.	Celery and rhubarb
Baby vegetables	This is a very modern classification of vegetables that include vegetables which are either created with hybrid varieties or are picked up before maturity. These vegetables are rarely peeled as they are quite attractive and one would not like to lose the appeal.	Tiny turnips, baby cauliflower, baby carrots, and baby squashes

PIGMENT AND COLOUR CHANGES

It is important for chefs to know the various kinds of pigments present in food and how they react to heat and various acidic and alkaline medium, as this would largely impact the style of cooking them. The colour of a vegetable is determined by the pigment it contains. Pigment is the colouring matter within the cells and tissues of the plant. The various types of pigments are affected differently by heat, acid, alkali, and other elements involved in cooking. To maintain as much colour as possible in cooked vegetables, one needs to know about these pigments. Guests may reject vegetable dishes if they do not like the colour of the dishes when presented. Preserving as much of the natural colour as possible is important. Each vegetable has trace amount of acid present in it and we shall see how this acid could be used favourably.

CHEF'S TIP
To retain the original white colour of the flavones pigment, one must cover the white vegetable while cooking. This allows the acids released from the vegetable during cooking, to blend with the vegetable. Use short cooking time and add a small amount of lemon juice, cream of tartar, or vinegar to the cooking liquid to create a slightly acid medium.

Flavones

It is the pigment present in white vegetables. It is present in potato, onion, cauliflower, white cabbage, cucumber, zucchini, and the white of celery. This pigment stays white in an acid medium and turns yellow in an alkaline medium. Overcooking will also turn white vegetables yellow or grey. This reaction is not reversible. Adding acid to yellowed white vegetables will not help in retaining their original white colour.

Anthocyanin

This pigment is present in red vegetables and it gives them their colour. It is present in very few vegetables, primarily in red cabbage, purple peppers, purple potatoes, etc. It is strongly affected by acid and alkali mediums. An alkali medium will turn their colour into blue or bluish green. An acid medium will give them a brighter red colour. This red pigment is easily dissolved in water. The use of excessive amount of water while cooking will leach the colour out of the vegetable.

CHEF'S TIP

To retain the red colour given by the anthocyanin pigment, cook the red cabbage with a small amount of acid, lemon juice, cream of tartar, or vinegar or sprinkle with a food acid at the end of the cooking process. An example of this type of preparation is braised red cabbage with the addition of a tart cooking acid (use only as much as necessary).

CHEF'S TIP

To retain as much of the natural green colour as possible do as follows:
Cook vegetables uncovered to allow the volatile acids to escape.
Cook them quickly until just *al dente*.
Extended exposure to heat will destroy the colour and leach out nutrients.
After boiling, plunge the vegetables into cold water to arrest cooking. This helps to brighten the colours and is known as shocking or refreshing.
Cook the vegetables in small batches; this reduces the cooking and holding time.
Do not hold for long periods of time.
Steam the green vegetables whenever possible. This shortens the cooking time, allows far less acid build up, and retains more colours.

CHEF'S TIP

If beets are not peeled and one or two inches of the stems are left intact, they may be cooked in boiling water, without loss of pigment and colour. They can be peeled after cooking.

Chlorophyll

Green colour in green vegetables is due to the presence of a pigment known as chlorophyll, which is affected by pH. In an acid medium the colour of a green vegetable changes to olive green and in alkaline medium it changes to bright green. This is the reason why some cooks add cooking soda or any other alkaline medium while cooking greens. When a green vegetable is cooked covered, the plant's natural acid is leached into the cooking liquid and is trapped there creating an acidic cooking medium. This, combined with the heat present, destroys the pigment. Green vegetables should be cooked uncovered so that the natural acids present, are expelled into the air.

Carotene

This pigment gives colour to yellow and orange vegetables such as carrot, corn, winter squash, sweet potato, tomato, and red pepper. This is the most stable of the colour pigments. It is only slightly affected by acids or alkalis. However, long cooking periods can make the colour of the vegetables dull. Short cooking times help retain the flavour and vitamins of most vegetables. Carotene pigment is fat soluble and leaches out into the fat used to cook the item. This is why red grease floats to the surface of stews and meat soups which contain tomatoes or carrots.

Betalains

The pigments in the root tissue of red beets are not chemically similar to anthocyanins; they contain nitrogen and are called betalains. Some of these pigments are purplish red, whereas others are yellow. Beets lose much pigment and become pale when they are pared and sliced before cooking, because the pigments are soluble in water and leach from the tissues.

Anthoxanthin

The anthoxanthin pigment changes from white or colourless to yellow, as the pH increases from acidic to alkaline ranges. This pigment is widely distributed in plants and often occurs with anthocyanins. It may combine with some metals, such as iron, to form a dark, complex compound. When it combines with aluminium, a bright yellow colour is produced. The anthoxanthin pigment generally remains quite stable when heated. If the cooking water is alkaline, the pigment may appear yellow. In case of excessive or prolonged heating, the pigment darkens.

Table 6.2 summarizes the effect of various factors on the colour of plant pigments.

Table 6.2 Effects on the pigments of vegetables

Name of the Pigment	Colour	Solubility in Water	Effect of Acid	Effect of Alkali Heating	Effect of Prolonged Ions	Effect of Metal
Chlorophyll	Green	Slightly soluble	Changes to olive green	Intensifies green	Olive green	Changes to olive green in iron
Carotenoids	Yellow and orange; some are red or pink	Slightly soluble	Less intense colour	Little effect	Colour may be less intense	None
Anthocyanins	Red, purple, and blue	Very soluble	Red	Purple or blue	Little effect	Violet or blue with tin or iron
Betalains	Purplish red; some are yellow	Very soluble	Little effect	Little effect	Pale	None
Anthoxanthin	White or colourless	Very soluble	White	Yellow	Darkness	Dark with iron, bright yellow with aluminium

EFFECT OF HEAT ON VEGETABLES

Cooking is the application of heat to food in order to make it safer to eat, digestible, and more palatable. Cooking also changes the appearance of food. Heat has to be applied to food in order to cook it. Heat breaks down the cellulose and the starches present, changes and blends flavours within the food, and also destroys bacteria in order to make the food more digestible for humans.

Vegetables and other foods are composed of proteins, fats, carbohydrates, water and also small amounts of minerals, vitamins, pigments (colouring agents), and flavour elements.

Carbohydrates

Both sugar and starch are carbohydrates and are present in many forms in vegetables, fruits, grains, beans, and nuts.

Heat plays a very important role when applied to vegetables in terms of caramelization and gelatinization. Caramelization is browning of sugars. The browning of sautéed vegetables is a form of caramelization.

When starch absorbs water and swells, the process is known as gelatinization.

Vegetable Fibres

Fibres are a group of complex substances that give structure and firmness to plants. They cannot be digested. The softening of vegetables on application of heat is a result of breaking down of fibres. Alkalis make fibre softer. Vegetables should never be cooked with alkalis, as this would make them mushy and lose the essential vitamins.

Minerals, Vitamins, Pigments, and Flavour Components

Minerals and vitamins are most important for the nutritional quality of the food, whereas pigments and flavour components are important to food as far as the appearance and taste are concerned. Pigments and flavours may also determine whether the food is appetizing enough to eat or not. So it becomes very important to preserve all these elements. On application of heat, all these components may be leached out and dissolved away from food during cooking. Vitamins and pigments may also be destroyed by prolonged cooking.

Proteins

Proteins are present in smaller extent in vegetables as compared to meat, fish, and poultry. When heat is applied to proteins they become firm or they start to coagulate. With an increase in temperature, proteins become even firmer and start shrinking. On being exposed to very high heat, proteins become tough and dry.

Acids present in lemon juice, vinegar, and tomato products help in speeding up coagulation and dissolving few connective tissues.

Heat affects the following characteristics of the vegetables.

1. Texture
2. Flavour
3. Colour
4. Nutrients

Let us see how each of the above four things can be used to the advantage of the chef preparing the vegetables.

CONTROLLING THE CHANGES IN TEXTURE

The changes in texture while cooking vegetables need to be controlled. This can be done by various methods as discussed in this section.

Fibre

Fibre structure in vegetables, including cellulose and pectin, gives shape and firmness to the vegetables. Cooking helps in softening some of these components. The amount of fibre varies in different vegetables, for example, tomatoes and spinach have fewer fibres than turnips or carrots. Even the age of vegetables would determine the amount of fibre in it, for example, young baby carrots would have less fibre as compared to older, tougher carrots. In some vegetables, such as asparagus, the tip would have less fibre content than its stalk.

Acids present in lemon juice, vinegar, and tomato products make fibre firmer and also increase the cooking time.

Sugar strengthens the cell structure and makes the fibre firm. This is more applicable for fruits such as apples. Apple sauce should be cooked and then sweetened, but when the shape of the fruit has to be retained then it should be cooked in heavy syrup.

Fibre is softened by heat, which means the longer one applies heat to vegetables, the softer it becomes. Although vegetables become softer on addition of alkalis such as baking soda, this should be avoided as it makes the vegetables mushy.

Starch

Starch is another component which affects the texture of a vegetable. Dry starchy foods, such as dried legumes, beans, lentils, peas, rice, and macaroni products must be cooked in sufficient amount of water so that starch granules can absorb moisture and soften. Dried beans are often soaked before cooking to replace the lost moisture.

Moist starchy vegetables, such as potatoes and sweet potatoes, have enough moisture of their own, but still they must be cooked until the starch granules soften.

Doneness

A vegetable is said to be 'done' when it reaches a desired degree of tenderness; this varies from vegetable to vegetable and most of the vegetables taste best when they are still firm, which is known as *al dente* in Italian cuisine. At this stage of tenderness not only do the vegetables get the most pleasing texture but also retain maximum flavour, colour, and nutrients.

For proper doneness some rules must be followed.

- Do not overcook.
- Cook as close to service as possible.
- In case vegetables have to be precooked, they should be undercooked, refreshed in cold water, and refrigerated. Then they should be reheated and served.
- Cuts of vegetables should be uniform in order to cook them evenly.
- Different vegetables should be cooked differently.

CONTROLLING CHANGES IN FLAVOUR

Many flavours are lost during cooking, more so if cooked longer. Flavour loss can be controlled by:

> **CHEF'S TIP**
> In some strong flavoured vegetables it is important to dilute the flavours by cooking them uncovered in large amount of water. Examples of such vegetables are onion family, cabbage, sprouts, cauliflower, turnips, parsnips, etc.

- cooking for short time;
- adding salt in boiling water, as it helps in reduction of loss of flavour;
- using as less liquid as possible in order to minimize leaching; and
- steaming, which helps reduce leaching out of flavours and reduces cooking time also.

Overcooking produces flavour changes which are undesirable and results in strong and unpleasant flavour. Younger vegetables have more amount of sugar which changes to starch as and when they ripen during storage. This happens usually in carrots, beets, lettuce, peas, etc.

CONTROLLING NUTRIENT LOSS

Factors responsible for nutrient loss are:

- high heat or temperature;
- cooking for longer periods;
- too much of liquid that causes leaching; and
- use of alkalis (baking soda, hard water).

Some nutrient loss in vegetables is inevitable. Some tips for reducing nutrient loss are given below.

- Use of pressure steam reduces cooking time but at the same time high heat causes some nutrient loss.
- Braising uses low heat but extends the cooking time.
- Boiling is faster than simmering; but high heat can destroy the vegetable.

CUTS OF VEGETABLES

6.1 Vegetables are cut in various sizes and shapes for various cooking purposes, creating different textures, tastes, and mouthfeel. Different cuts would cook differently from one another; all of them will create a different flavour. Flavour is a combination of different things including texture and mouthfeel. For example, shredded cabbage will taste different from diced cabbage in a salad and boiled mashed potato will taste different from boiled chateau potato.

Before understanding the cuts of vegetables, a cook must know the following things.

Know one's Vegetables   The shape, size, and the texture, are different for different vegetables, so the ways of cutting and processing will also be different. Some cuts are specific to certain vegetables, for example, you cannot turn a tomato or artichoke or asparagus or you cannot cut the julienne of asparagus as it will lead to wastage.

Know your Equipment   Equipment are very essential in the processing of vegetables. Knives, peelers, shavers are required to be in prime condition, to obtain neater cuts and prevent wastages and injuries.

Know the Purpose   Vegetable cut for one purpose may not be suitable for another. For example, in mirepoix one cannot use shredded vegetables as long cooking periods in case of roasting will burn them, leading to a burnt flavour.

Know the Effect of Heat   The effect of heat applied during the cooking process will also get affected by the cuts of the vegetables. For example, whole potato will take longer time to cook than cut potato because the surface area in the potato will get increased many fold, due to which absorption of heat will also increase.

A group of internationally accepted cuts of vegetables are termed as 'classical cuts'. The most common among them are julienne, chiffonade, baton, brunoise, dice (small, medium, and large), slices, chop and mince, emincer, and shred. Refer to Table 6.3 to understand them better.

Table 6.3   Cuts of vegetables

Name of the Cut	Description	Purpose	Photograph
Julienne	Thin strips of 1 mm × 1 mm × 25 mm	Garnishes/Chinese stir-fries/salads	
Jardinière	Batons of 3 mm × 3 mm × 18 mm	Sautéed preparations	

Contd

Table 6.3   (Contd)

Name of the Cut	Description	Purpose	Photograph
Macedoine	Large dices 5 mm × 5 mm × 5 mm	Salads/sautéed preparations	
Brunoise	Small dices 2 mm × 2 mm × 2 mm	Garnishes/stuffing	
Lozenge	Cutting obliquely at a slant in equal lengths	Stir-fries/sautéed preparations	
Slicing	Cutting roundels from round vegetables	Salads/roasts/grills/bake	
Paring	Peeling the skin of vegetables using paring knife	Paring fruits such as apples for compotes	
Chiffonade	Shredded leafy vegetables	Sautéed preparations/stuffing/ garnishes	
Paysanne	Geometrical shapes of 1 mm thickness	Sautéed preparations/ garnishes	
Wedges	Round vegetable cut equally lengthwise	Stews/fried/grills/roasts/boil/ poached/braised	
Mirepoix	Roughly cut vegetables, sometimes with skin	Stocks/soups/flavouring	
Tourne/Chateau	Turning of vegetables into barrel shapes	Sautéed/boiled/poached/ roasted/baked	
Fluting	Mushrooms turned for aesthetic appeal	Cut specifically for mushrooms	
Matignon	Evenly cut root vegetables	Used as mirepoix; but served along with the dish, for example, casseroles	
Mincing	Chopping the vegetables very fine	Herbs/stuffing/garnishes	

Contd

Table 6.3 (Contd)

Name of the Cut	Description	Purpose	Photograph
Chopping	Evenly cut vegetables smaller than brunoise	Garnishes/sauces/gravies/stuffing	
Parisienne	Scooped out with a parisienne scooper	Poached/boiled/sautéed/stews/salads	
Segment	Usually done for citrus fruits, where each segment is removed from the fruit	Salads/fruit platters/garnishing	

SOME INDIAN CUTS OF VEGETABLES

Unlike the Western cuts of vegetables, the Indian kitchen does not have a particular way of classification of cuts; but in Indian kitchen too, there are various kinds of cuts which are typically used for certain vegetables. Some of the prominent cuts in Indian cooking are discussed in Table 6.4.

Table 6.4 Indian cuts of vegetables

Vegetable	Type of Cut	Use	Photograph
Okra	Trim from head and tail and slit open lengthwise without cutting through	Used for preparing stuffed okra as in *aamchoor masala bhindi*	
Baby brinjal	Slit into four, keeping the stem intact	Used for stuffed brinjals/*bharwan masala baingan*	
Bitter gourd	Scrape the bitter gourd and keep scrapings for stuffing, slit open without cutting through and remove seeds and inner flesh	Used for stuffed bitter gourd/*bharwan karela*	
Jackfruit	Cut into quarters, remove the skin not wasting too much, remove the centre pith and cut into quarters	Used for kofta, curries, biryani, etc.	
Drumsticks	String the drumsticks and cut into batons	Used for *sambhar*	

Contd

Table 6.4 (Contd)

Vegetable	Type of Cut	Use	Photograph
Banana flower	Peel the banana flower to obtain small flowerets; remove the hard woody style inside the flowers	Used for stir-fires and kofta	
Bamboo shoot	Peel till the white pith is visible and cut into desired shapes	Used for curries	
Chilli	Slit the chilli lengthwise, taking care not to cut through	Used for stuffed peppers, pakoras	
Lotus root	Peel and wash very well to ensure that it is free from sand and mud; cut into one inch chunks or lozenge	Used for curries and kofta	

Let us discuss various kinds of common vegetables used in professional kitchens. Each vegetable listed below is described with its English name, scientific name, and Hindi name. Some of these vegetables are of Western origin and are not cultivated in India and hence, they do not have Hindi names.

Broccoli

Scientific name: *Brassica italica* **Hindi:** *Hariyali gobi*
Broccoli is a member of the brassica family and in a way it is related to cabbage and cauliflower.

Broccoli has been growing for more than 2000 years and was first grown in Italy. *Brocco* means 'arm branch' in Italian. This green vegetable consists of clusters of tiny buds on a stem. The stem is edible but usually not eaten as a vegetable dish; but is used in soups and sauces.

It is available all year round. See Fig. 6.1 to know how to cut broccoli florets. The cauliflower is done in the same way.

How to Select
Broccoli should be bright green and the cluster bud should be firm. The stem should be proportionately smaller than the flower bud. It should be seen that it is not infested with insects. Any hole in the stem would denote that the vegetable is infested with insects.

How to Store
Broccoli should be stored refrigerated at 4 to 5°C, in a perforated basket so that there is lot of room for it to breathe. It can be stored fresh up to three or four days, after which the green buds will discolour to yellowish green.

CHEF'S TIP
- Salt the bitter gourd and keep it aside. Osmotic pressure would drain out the bitterness from the bitter gourds.
- Cut the jackfruit using oil on the knife and also smearing oil on the hands as otherwise its thick milky liquid will stick to the hands.
- After processing the banana flower, boil it in turmeric water to avoid blackening of the same.
- Boil lotus roots in milk and water to avoid the blackening of the lotus roots.

Hold the broccoli firmly; with tip of the knife, cut away the florets.

Cut the florets to the desired size.

Wash the florets under running water or soak in salted water to get rid of insects.

Fig. 6.1 Cut broccoli florets

Cabbage

CHEF'S TIP
Whole cabbage is used in various salads such as coleslaw.

• Shred the cabbage and add salt and leave it to stand for 20 minutes. This would draw out some unpleasant juices out of the cabbage.

• Use whole leaves of cabbage; blanch in hot water to make them soft. This can be used as stuffed vegetable preparation as done in Lebanese dish called dolmas.

Scientific name: *Brassica oleracea*
Hindi: *Bandh gobi*

There are around 70 varieties of cabbage and they all belong to the brassica family.

Cabbages can be of different shapes, sizes, and colour and these characteristics make them useful for different purposes. The common thing in all cabbages is that there would be leaves closing on one another on a short stem. Cabbage is rich in fibre and vitamin C and also has compounds that can fight cancer. Cabbage can be used in various forms. It can be used shredded for Chinese dishes or diced for pickles and curries. Cabbage can be boiled, roasted, or braised. Sauerkraut from Germany is a very popular braised cabbage dish. Red cabbage is very similar to the green cabbage and the only difference is the colour. See Fig. 6.2 to know how cabbage is processed.

How to Select

While selecting cabbage, ensure that its colour is bright. Cabbage should be heavy for its size and firm when pressed. A loose cabbage would indicate the presence of air pockets and hence, chances of infestations.

How to Store

Store cabbage in 4 to 5°C in the refrigerator.

Carrot

CHEF'S TIP
To store root vegetables, do not cut the root ends away. The attached roots help keep the vegetables for a longer time in the fridge.

Scientific name: *Daucus carota*
Hindi: *Gajar*

These root vegetables can be eaten raw as salad or cooked in stews, soups, stir-fries, and even desserts. Carrot cake from Switzerland or *gajar ka halwa* from India are few of the desserts that use carrots. Carrots are available in red, orange, and dark purple colour. They are available all year round, are rich source of vitamin A and are usually considered to be good for the eyes.

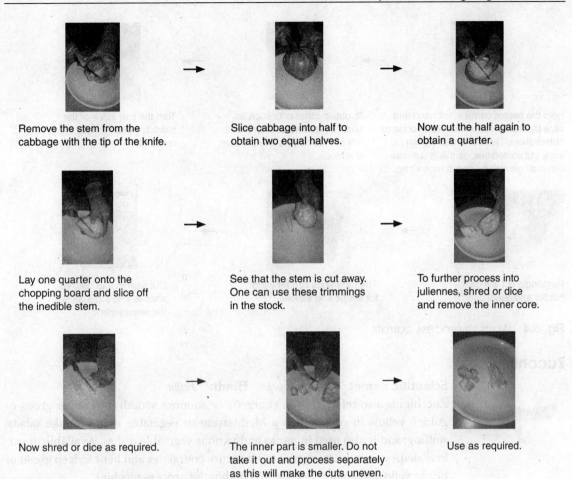

Remove the stem from the cabbage with the tip of the knife.

Slice cabbage into half to obtain two equal halves.

Now cut the half again to obtain a quarter.

Lay one quarter onto the chopping board and slice off the inedible stem.

See that the stem is cut away. One can use these trimmings in the stock.

To further process into juliennes, shred or dice and remove the inner core.

Now shred or dice as required.

The inner part is smaller. Do not take it out and process separately as this will make the cuts uneven.

Use as required.

Fig. 6.2   Ways of processing cabbage

How to Select

While selecting carrots ensure that the colour is bright, they are free from any blemishes, and are firm to touch. A limp carrot will be old and will disintegrate faster in cooking. See Figs 6.3 and 6.4 to know how to peel and process carrots respectively.

How to Store

Store cabbage in 10 to 12°C in the refrigerator.

Hold the carrot in left hand with the root end facing away from you. Place the peeler on the edge of the root.

Drag down the peeler to peel the carrot all the way up to the tip.

Repeat this procedure to obtain a peeled carrot.

Fig. 6.3   How to peel a carrot

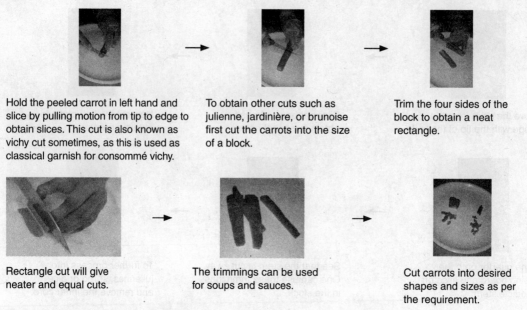

Hold the peeled carrot in left hand and slice by pulling motion from tip to edge to obtain slices. This cut is also known as vichy cut sometimes, as this is used as classical garnish for consommé vichy.

To obtain other cuts such as julienne, jardinière, or brunoise first cut the carrots into the size of a block.

Trim the four sides of the block to obtain a neat rectangle.

Rectangle cut will give neater and equal cuts.

The trimmings can be used for soups and sauces.

Cut carrots into desired shapes and sizes as per the requirement.

Fig. 6.4 Ways to process carrots

Zucchini

Scientific name: *Cucurbita pepo* **Hindi:** *Petha*

Zucchini is also referred to as courgette or summer squash. It is either green or golden yellow in colour. It is a Mediterranean vegetable used to make salads, stuffing, and is also used in pastas and various vegetable dishes. Available in various shapes, usually zucchinis are immature courgettes and hence, deep green or bright yellow in colour. See Fig. 6.5 on how to process zucchini.

How to Select

Zucchini should be firm to touch and the skin should be shiny and glossy to touch.

How to Store

Store zucchini in refrigerator between 4 and 5°C.

Garlic

CHEF'S TIP

One can chop garlic with cold butter. It helps to chop up the garlic finer and keeps it congealed together while chopping. This type of chopped garlic can be used in sauces or in garlic breads also.

Scientific name: *Allium sativum*
Hindi: *Lahsun*

Garlic is used almost all around the world and many cuisines around the world are incomplete without garlic. Europeans use it minced and it is extensively used in Italian and French cuisines. Indians use it for curries and other Asians use it for stir-fries. Koreans eat it raw with barbecued meats. Garlic pod is often consumed on empty stomach to reduce cholesterol in the body and probably this is the reason why the famous garlic raita, called *burhani raita*, is served with ghee-laced-biryani. See Fig. 6.6 to know how to process garlic.

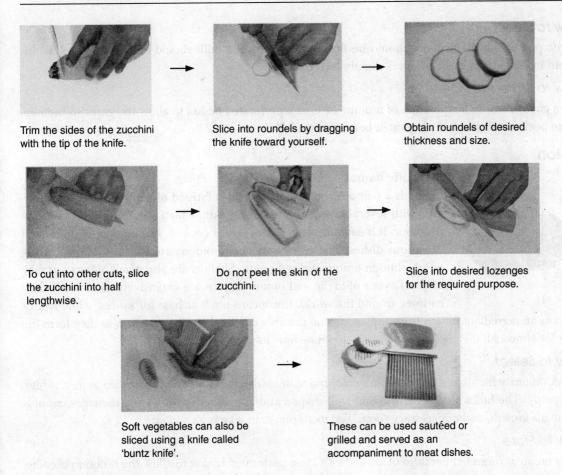

Trim the sides of the zucchini with the tip of the knife.

Slice into roundels by dragging the knife toward yourself.

Obtain roundels of desired thickness and size.

To cut into other cuts, slice the zucchini into half lengthwise.

Do not peel the skin of the zucchini.

Slice into desired lozenges for the required purpose.

Soft vegetables can also be sliced using a knife called 'buntz knife'.

These can be used sautéed or grilled and served as an accompaniment to meat dishes.

Fig. 6.5 How to process zucchini

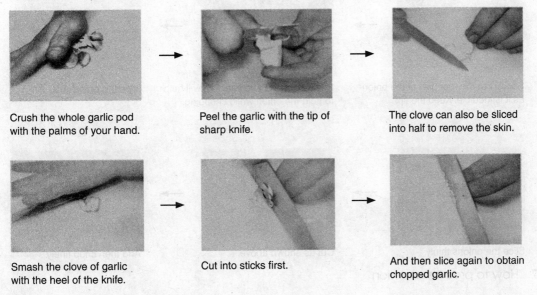

Crush the whole garlic pod with the palms of your hand.

Peel the garlic with the tip of sharp knife.

The clove can also be sliced into half to remove the skin.

Smash the clove of garlic with the heel of the knife.

Cut into sticks first.

And then slice again to obtain chopped garlic.

Fig. 6.6 How to process garlic

How to Select

Garlic pod should be dry and should be firm and plump. The bulb should not be sprouted as this would indicate fresh garlic which is slightly less in flavour.

How to Store

Store garlic at room temperature of around 24°C in a perforated basket to allow the garlic to breathe. Once peeled it should be refrigerated between 4 and 5°C.

Onion

Scientific name: *Allium cepa* **Hindi:** *Pyaaz*

Onion is a bulbous root vegetable that is formed of white fleshy leaves covered with several layers of thin papery skin, which are red, yellow, or white in colour. It is eaten fresh or dried, raw or cooked. It is used as an ingredient in various dishes. The other varieties of onions are white, scallions (spring onion), button onion, and red onion. Shallots are also small onions but they have a flavour of garlic and onion. Onions are extensively used in different cuisines around the world. Europeans use it as base for sauces, while Asians use it as an ingredient for stir-fries. Indian cuisine possibly cannot do without onions as they form the base for almost all the gravies. See Fig. 6.7 to know how to process an onion.

How to Select

Select onions which have dry and pink-coloured skin, with internal leaves which are tender, white, and pulpy. The bulbs should be firm and well shaped and large sized, without any damages, moulds or fungus growth, and unnecessary roots. Ten to 12 pieces in a kg is a good size.

How to Store

Store onion at room temperature of around 24°C in a perforated basket to allow the onion to breathe.

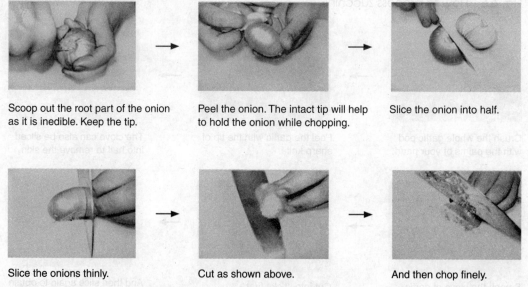

Scoop out the root part of the onion as it is inedible. Keep the tip.

Peel the onion. The intact tip will help to hold the onion while chopping.

Slice the onion into half.

Slice the onions thinly.

Cut as shown above.

And then chop finely.

Fig. 6.7 How to process an onion

Spinach

CHEF'S TIP

Addition of cooking soda while boiling spinach will darken its colour; but spinach will loose its nutritional value.

Always cook spinach uncovered. Refer to pigments mentioned earlier in the chapter.

An important fact to be noted is that the yield of spinach is only 30 per cent. This means that 1 kg of spinach after boiling will yield only 300 g.

This testing is known as 'yield test' and it is important to do yield tests for every vegetable so that the costing of the dish could be done accurately to generate profits.

Scientific name: *Spinacia oleracea*
Hindi: *Palak*

Spinach is a green leafy vegetable and is available in many species. It can be eaten raw; but it is usually cooked. It forms a wonderful accompaniment with meats and is also used extensively in India as a vegetable preparation. Small baby spinach, also known as lamb's lettuce, is always eaten as a salad; whereas larger leaves can be used in stir-fries or simple sauteed with garlic to obtain wilted spinach.

Spinach is packed with nutrients and is a rich source of vitamin C. Spinach can be used in soups as well.

How to Select

While selecting spinach leaves, look for the ones with tender and narrow stems. These would indicate that the spinach is younger. Older spinach leaf is slightly bitter and has a strong flavour.

How to Store

Store spinach in a perforated basket covered with a moist muslin in a refrigerator between 4 and 5°C. The spinach can also be boiled and pasted and this way it can be kept frozen for many days.

Avocado

CHEF'S TIP

Avocadoes turn black very easily when they come in contact with air. So apply lemon juice over the sliced avocadoes or dip them in water flavoured with lemon juice to avoid blackening.

Scientific name: *Persea americana*
Hindi: *Makhanphal*

Also known as 'alligator pear', it is native to Mexico and southern parts of California. It is not found ripe in the market as it never ripens on trees. It has a very buttery taste and, hence, also called 'butter fruit'. An avocado will be bright green in colour and a dark purplish skin would denote that the avocado is ripe. It has a large seed inside which is inedible.

How to Select

6.2

Select avocadoes with green colour. The flesh of the vegetables should be firm. The skin should be resilient to pressure of the thumb. Five to six pieces in a kg is a good size.

How to Store

Store green avocadoes wrapped in newspaper in an ambient room temperature of 22°C. Once ripened they can be refrigerated between 10 and 12°C.

Cucumber

Scientific name: *Cucumis sativus* **Hindi:** *Kheera*

Cucumber is usually eaten raw in salads or dips. It is rarely cooked as it tends to get bitter when cooked. It is also served chilled as an appetizer called crudites with certain dips.

How to Select

It should be straight and should have an even long shape. It should be green in colour, hard to touch, smooth and should have smooth and shiny surface. When broken inside, it should be light and not like a net. Usually four to five pieces in one kg is a good size.

How to Store

Store refrigerated between 5 and 6°C.

Colocasia

Scientific name: *Colocasia esculenta* **Hindi:** *Arbi*

Colocasia is a tuber of a flowering plant that is commonly used in Indian cooking. It is widely used in all parts of India and in some places, such as Gujarat, the leaves of this plant are used to make deep-fried snacks called patra. This plant grows in warm climates.

How to Select

It should be fresh, crispy, and firm. The outer surface should be dry with no excessive stringiness or woodiness. Shape should be uniformly cylindrical and there should be no extra growth.

How to Store

Store at room temperature of 24°C. Do not pile up in a basket; keep spaced out for circulation of air.

Aubergine

Scientific name: *Solanum melongena* **Hindi:** *Baingan*

Aubergine is also commonly known as eggplant or brinjal. This vegetable is believed to have originated in tropical Asia and is used in cooking in almost all the parts of the world, especially in the Middle East. It is available in various shapes, sizes, and colours such as white, purple, green, and even orange. The pea-sized ones, known as pea aubergines, are extensively used in Thai cooking. Aubergine can be cooked by various methods such as frying, grilling, roasting, broiling, etc. Its flavour is enhanced when it is subjected to smoking.

How to Select

It should be uniformly dark purple in colour with a soft sheen. The body should be heavy and pulpy and the skin should be free from blemishes/worms. Holes in a brinjal would indicate worms. Three to four pieces per kg is a good size.

How to Store

Store refrigerated between 5 and 6°C.

Snake Gourd

Scientific name: *Trichosanthes cucumeriana*   **Hindi:**   *Chichinda*

This vegetable is so called because of its shape that resembles a snake. It is from the gourd family and extensively used in Indian dishes, especially in east and south India.

How to Select

Select snake gourds that are pale green in colour with a diameter of 5 to 6 cm. They should be tender and snap crisply when bent.

How to Store

Store refrigerated between 5 and 6°C.

Bell Pepper/Capsicum

Scientific name: *Capsicum annum*   **Hindi:**   *Simla mirch*

When the bell pepper is green it is referred to as capsicum, but when it is coloured it is referred to as bell pepper. It is a vegetable that is extensively used in Mediterranean cooking. It can be grilled, pureed, or sauteed to be served as a vegetable. It can also be hollowed and stuffed and served as a vegetable dish.

How to Select

Choose capsicum that is bright in colour with a waxy sheen. It should have a bell shape and should be crisp with an even colour. Avoid capsicum with any discolouration or soft patches. Six to eight pieces per kg is a good size.

How to Store

Store refrigerated between 5 and 6°C.

Round Gourd

Scientific name: *Praecitrullus fistulosus*   **Hindi:**   *Tinda*

Round gourd belongs to the family of gourd and is used in stews and braises. It is commonly used in Indian cuisine, especially in north India. It can be cut or prepared whole.

How to Select

Select round gourd that is light green with a firm, crisp skin. It should have bright white pulp with small tender seeds. Eight to ten pieces in a kg is a good size.

How to Store

Store refrigerated between 5 and 6°C.

Asparagus

Scientific name: *Asparagus officinalis*   **Hindi:**   *Shatwaar*

Asparagus is a vegetable that is often used in fine dining. It is a shoot which is very rich in fibre. Asparagus is a rich source of vitamin B which is water soluble. It is, therefore, a good idea to poach the asparagus in oil to retain its vitamin content. It can be used as a salad, and also accompaniment to meats in the main course. This vegetable can be served hot or cold and hence, the usage is unlimited.

How to Select

Choose the ones that are bright green in colour with plump, sharp tips which are not bruised. Asparagus should snap when bent. Thick stem will ensure better yield.

6.3

How to Store

Store refrigerated between 5 and 6°C. The shelf life is increased if the bunches are stored upright with base dipped in water, just like flowers.

Artichokes

Scientific name: *Cynara ficifolia* **Hindi:** *Hathi chak*

There are two types of artichokes—Jerusalem artichoke and globe artichoke. The one seen in the photograph is a globe artichoke. Artichokes are often boiled whole in lemon water and the scales are plucked. The base of a scale is eaten as an appetizer. It is also peeled and the internal part, known as the 'heart of the artichoke', is taken out and used as a vegetable accompaniment to meats and other dishes.

How to Select

Choose the ones with minimum weight of 100 to 125 g each. An artichoke should be bright green in colour with firm leaves, tinting purple at the tip. The leaves should be closed and the vegetable should be compact.

6.4

How to Store

Store refrigerated between 5 and 6°C.

Radish

Scientific name: *Raphanus sativus* **Hindi:** *Mooli*

Radish is available in various shapes and sizes. In India the radish from Kashmir is large, white and round, whereas in other parts it is long and white, thick at one end and tapering towards the end. Radish can be cooked, but it is eaten raw as a salad or grated to be used as an accompaniment. In Japan it is known as daikon and is used extensively in cooking.

How to Select

Choose the ones which are 8 to 10 inches in length with stub tips. It should be white in colour and without any grit. Leaves should be green, crisp, and pungent in taste. The thicker the radish, the better is the taste and flavour.

How to Store

Store refrigerated between 5 and 6°C.

Tomato

Scientific name: *Lycopersican esculentum* **Hindi:** *Tamatar*

Scientifically, tomato is a fruit, but is extensively used as a vegetable. It is a versatile vegetable that can be eaten raw as salad or even cooked to make soups or vegetable accompaniments. It can also be used as a colouring agent or souring agent in Indian food and forms base of many types of gravy. It is a vegetable that is used almost in all parts of the world, especially in Spain,

Mexico, and Italy. Tomatoes are available in various colours and shapes. The small cherry-shaped tomatoes are called cherry tomatoes and are used in salads. Green tomatoes are used in Italian and Mexican dishes.

How to Select

Select tomatoes with firm and even shape. The skin should be smooth and without any bruises, cuts, or patches. The colour should be dark orange to bright red without any traces of yellow or green. The flesh should be firm. The skin should be resilient to pressure of the thumb. Eight to 10 pieces per kg (medium sized) and 12 to 15 pieces per kg (small sized) is a good size as per the receiving standards.

How to Store

Store refrigerated between 5 and 6°C.

Potato

Scientific name: *Solanum tuberosum* **Hindi:** *Aloo*

Potato is the most widely cultivated crop in the world and there is hardly any country in the world that does not use potato in one form or the other. This versatile tuber vegetable is available in many colours and sizes and each potato has different usage. The potatoes are usually of two main types—waxy and mealy. The mealy potatoes are starchier and, hence, good for mashed potatoes.

How to Select

Select potatoes which have uniform shape and are dry and heavy for their weight. Surface should be firm and smooth with no deeply set eyes or spots. There should be no decay or discolouration. There should be no germinations on the potato.

How to Store

Store at room temperature of 24°C. Do not pile up in a basket and keep spaced out for circulation of air. Never keep potatoes in a refrigerator as they will turn black.

Beetroot

Scientific name: *Beta vulgaris* **Hindi:** *Chukandar*

Beetroot is a deep red coloured root vegetable, which is a rich source of iron. It is also used as a dyeing agent in the textile industry and can be eaten raw as a salad. Beetroot soup is one of the classical soups and is also the national soup of Poland. Beets are available in two colours–dark burgundy red and golden beetroot. Both the types of beet have a similar kind of mud coloured skin; but the flesh inside a golden beetroot is pleasing yellow in colour.

How to Select

Select beetroot which is dark crimson in colour with a smooth tender surface and is not too fibrous. It should be free of stalks or leaves when received as these parts are inedible and will only increase the weight. The skin should be free of cracks and scars and should be firm and even shaped. Six to eight pieces in a kg is a good size.

How to Store

Store refrigerated between 10 and 12°C.

French Beans

Scientific name: *Phaseolus vulgaris* **Hindi:** *Phalli*

These are from the pods family of vegetables and they yield pulses when dried. Though there are many types of beans available, such as haricot beans, wax beans, French beans, etc. they all have same taste and texture when raw.

How to Select

Select French beans that are light green in colour, long, and crisp. They should snap when bent.

How to Store

Store refrigerated between 5 and 6°C.

Mushroom

Scientific name: *Agaricus bisporus* **Hindi:** *Khumb*

Mushroom is a fungus which is widely eaten around the world. There are many species and varieties of mushrooms available but one has to be careful in selecting mushrooms for consumption, as some of them are poisonous. Morels from Kashmir, porcini from Italy, and enoki from Japan are few of the most expensive mushrooms available.

How to Select

Select mushrooms with white, fresh, unbleached stems that are cut short. They should be devoid of scars or other cultivating effects. The underside of the cap has to be firm and round. This receiving standard is for button mushrooms, as these are the most common ones used in the daily kitchens.

How to Store

Store refrigerated between 5 and 6°C and keep them in an open perforated container to allow air circulation.

Turnip

Scientific name: *Brassica rapa* **Hindi:** *Shalgam*

Turnip is a root vegetable that is often paired with lamb. In France it is called navet and is used to prepare a stew known as navarin. Turnips are also used for making pickles in India. They are used with meats in Kashmir, India.

How to Select

Select turnips which are firm and well formed, and are creamy white in colour with a bluish purple tip. The skin should be smooth and tender. Five to six pieces in a kg is a good size.

How to Store

Store refrigerated between 5 and 6°C.

Ginger

Scientific name: *Zingiber officinale* **Hindi:** *Adrak*

Ginger is a root that is not used as a vegetable but is generally used as a flavouring spice. Ginger is native to South-East Asia and is widely used in many forms. In Indian cooking ginger paste is used as flavouring and thickening agent. In Japanese cooking it is pickled and served with sushi.

How to Select

Select the ones with large-sized bulbs. The ginger should be received fresh and this can be checked by digging the nail of the thumb into it to feel the juice. It should be free of grit or slime. The skin should be thin and should snap at the joint with a crack. The bulb should be fairly even in size and around three-fourth to an inch in diameter. It should not be fibrous as it indicates that it is an old ginger.

How to Store

Store at room temperature of 24°C. Do not pile up in a basket and keep spaced out for circulation of air.

Drumstick

Scientific name: *Moringa oleifera* **Hindi:** *Sajna phalli*

Drumstick is a vegetable that is used in south India and east India. It is stringed and cut into 1 inch sticks and added to stews. It is fibrous and often chewed and seldom swallowed. It is an important ingredient in a lentil curry or sambhar from south India. The leaves of this tree are also used for making a vegetable preparation in south India.

How to Select

Select drumsticks that are tender with no strings. They should have a minimum diameter of 1 cm and should be 12 to 15 inches in length.

How to Store

Store refrigerated between 5 and 6°C.

Okra/Lady's Finger

Scientific name: *Abelmoschus esculentus* **Hindi:** *Bhindi*

Scientifically, okra is a pod vegetable where the entire pod is eaten. In India it is often cooked in oil or roasted, as cooking it as stew makes it slimy. Okra is also an important commodity used in gumbo soup from Caribbean islands. Okra is considered to be very good for skin and in many South-East Asian countries it is often eaten raw as salad.

How to Select

Selec okra which is at least 3 to 4 inches long, bright green in colour, crisp, and tender. It should snap when bent and the seeds should be bright white, small, and soft. Smaller lady's fingers of 2 inch length are preferred in Indian cooking where they are filled with spices and prepared whole.

How to Store

Store refrigerated between 5 and 6°C.

Edamame

Scientific name: *Glycine max* **Hindi:** *Soya phalli*

Edamame literally translates to peas on stem in Japanese. It is categorized under super foods or power foods and is commonly served steamed or boiled. It is sprinkled with rock salt and the whole pod is put in the mouth and pulled out in such a manner that only the beans inside the pod remain in one's mouth.

How to Select

Edamame is handpicked by a farmer to avoid damage to the plant. It is typically harvested before it fully ripens, so that the seed inside is sweet. It is often sold in a frozen state and it must be light green in colour without any blemishes or cuts.

How to Store

Store frozen between −18 and −22°C.

Romanesco/Roman Cauliflower

Scientific name: *Brassica Oleracea*   **Hindi:** NA

This vegetable was first cultivated in Italy in the 16th century. Its unique appearance has made it a popular choice for fine dining and exotic preparations. It is also known as Roman cauliflower as it contains florets, but each floret has a symmetrical cone shaped design that makes the vegetable look very attractive.

How to Select

When selecting Romanesco, choose those that are bright in colour and crisp to touch. Sometimes, the vegetable naturally grows into a deep purple colour, therefore in such cases, choose the ones with deep purple colour. The stem of the vegetable should not be limp and make sure that the florets are not infested with insects.

How to Store

Store refrigerated between 5 and 6°C.

Fiddle Head Fern

Scientific name: *Pteridium aquilinum*   **Hindi:** *Kasrod*

Fiddle head ferns are commonly used in many countries around the world but primarily they are used in Asia. In Asian countries, they are stir-fried along with meats and other vegetables. However, in most parts of India they are used in pickles. In certain states such as Assam, fiddle head ferns are a delicacy.

How to Select

Select those fiddle head ferns which are bright green in colour and do not contain an extra stem at the bottom, as they are hard and chewy. Usually, spring is a good time for getting fresh fiddle head ferns.

How to Store

Store refrigerated between 5 and 6°C.

Topinambour

Scientific name: *Helianthus tuberosus*   **Hindi:** NA

Topinambour, also known as Jerusalem artichoke, sun-choke, or earth apple, is a species of sunflower where the edible tumour is used as a root vegetable. It is popularly consumed in northern and southern parts of America as well as Europe. It is often used in stews or is simply mashed and served along with meats.

How to Select

Do not go for larger topinambour as it is more fibrous and lacks taste. Select similar looking sizes for cooking as it will give you a consistent product.

How to Store

Wash the topinambour and dry it well. Store it at an ambient room temperature in a dry atmosphere.

Celeriac

Scientific name: *Apium graveolens var. rapaceum* **Hindi:** NA

Celeriac belongs to the turmeric family and is obtained from a typical type of celery of which the edible root is consumed. This is a very popular vegetable in South America and the Mediterranean basin. There are many ways of consuming celeriac—it can be shaved thinly and eaten in salads or it can be roasted and even mashed to accompany seafood and meats.

How to Select

A good celeriac should be young and fresh. This can be determined by putting the sharp edge of a knife on it and seeing the juices come out. Young and fresh celeriacs have much more defined earthy flavours as compared to the older ones.

How to Store

Store celeriac in refrigerated conditions for up to 3 months as it can be stored very well in cold conditions.

Yam/Elephant's Foot

Scientific name: *Dioscoreaceae* **Hindi:** *Jimikand*

Yam is a tumour that is obtained from certain varieties of lilies and grasses. It is used in various cuisines around the world, but most commonly in Asia, Africa, and America. Yams can be cooked by using various cooking methods, they can be stewed, boiled, fried, or even roasted or baked. Yams have a hard woody skin and that is the reason that they are called elephant's foot. In India, yams are commonly eaten as a vegetable preparation, especially in Gujarat and southern parts of India. It is important that the yam is cooked before consuming as uncooked yam can give a pricking sensation in the throat.

How to Select

Select the yam that is even and round in shape and is free from blemishes and open cuts. It should be free of soil and its skin should be hard.

How to Store

The yam is better stored at an ambient room temperature.

White Asparagus

Scientific name: *Asparagus officinalis* **Hindi:** *Safed Shatwaar*

White asparagus is not a variety of asparagus; instead it is just asparagus that is grown in the absence of sunlight. The farmers cover the asparagus shoots with a can or any other object to block the sunlight. Such asparagus finally grows to be white in colour. The white asparagus also has a slightly mellowed down taste in comparison to green asparagus and it is used in fine dining restaurants and is served as an accompaniment to meats and seafoods.

How to Select

The white asparagus should not be more than 6–7 inches in length as the stalks could be woody and fibrous. Also, keep in mind to select the asparagus which are at least 1 cm thick as these shoots are very tender and can easily break when cooking methods are applied.

How to Store

The white asparagus is best stored in a perforated container with a damp cloth on top. Another way to store asparagus is to make it stand in a container filled with water in such a way that the base of asparagus remains submerged in water.

Water Chestnut

Scientific name: *Eleocharis dulcis*   **Hindi:**   *Singhara*

Unlike the name suggests, the water chestnut is not a nut but it is an aquatic plant that grows in muddy and marshy water lands. Though it is consumed in many parts of the world, typically it is eaten in Asian countries such as India, China, and Japan. It can be eaten raw, boiled, or roasted. It can also be grounded into flour to make water chestnut cake, which is popularly eaten as a dim sum in China. In India, the chestnut flour is typically eaten during the period of fasting called navrataras.

How to Select

Though usually the water chestnuts used in the hotels are often procured in cans, but whilst selecting fresh ones the outer covering should be bright green and shiny.

How to Store

Store in a perforated container under refrigerated conditions covered with a damp cloth.

Fennel

Scientific name: *Foeniculum vulgare*   **Hindi:**   *Saunf*

Fennel comes from a flowering plant, but we only use the root of the fennel plant which is often known as the fennel bulb. It has a unique and sweet flavour that pairs up very well with seafood. It is a commonly used vegetable in Mediterranean cuisine. It can be eaten raw or braised along with meats and other vegetables to create delectable dishes.

How to Select

Select fennel bulbs that are closed and intact. The appearance should be off white in colour with a light shade. It should be young and juicy and free from blemishes and dirt.

How to Store

Store in refrigerated conditions between 4 and 6°C.

Kohl Rabi

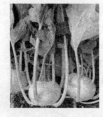

Scientific name: *Brassica oleracea gongylodes*   **Hindi:**   *Ganth gobi*

Kohl rabi is also known as knol khol or German turnip, as it is popularly consumed all across Germany. In India, it is used in Kashmiri cuisine to prepare many vegetarian dishes. It should always be peeled before cooking. Kohl rabi can be eaten raw or braised along with meats to yield delicious stews. Even its leaves are edible and can be used in various preparations. In many countries, kohl rabi is also eaten pickled.

How to Select

Select kohl rabi that is tender and has bright green leaves with tender stems. The skin of kohl rabi should not be very thick.

How to Store

Store in refrigerated conditions between 4 and 6°C.

Brussels Sprouts

Scientific name: *Brassica oleracea gemmifera*   **Hindi:**   NA

Brussels sprouts actually look like tiny cabbages which typically have a diameter of one inch. Its leaves are tender but contain a hard woody inner coat which needs attention whilst cooking. Brussels sprouts are processed for cooking by peeling off the top layer and making an incision with the sharp tip of a knife on a base and then by cooking it by boiling or steaming. One must take care not to overcook this vegetable as it can lose its texture and flavour.

How to Select

Select Brussels sprouts which are tight and compact and have bright green outer covering. They should be free from any holes which indicate infestation by pests.

How to Store

Store in refrigerated conditions between 4 and 6°C.

Unique Mushrooms

Mushroom is a fungus that has been used in cooking since many years. Though it mostly grows in the wild, now there are also farmers who are cultivating mushrooms for human consumption. It's an art to select the right mushroom for human consumption as this genus contains many poisonous species that can harm human health. Hence in this chapter, we will focus on the categories of edible mushrooms. There are hundreds of edible mushrooms, however we will be discussing only the most common ones used in kitchens around the world (Refer to Table 6.5).

Table 6.5   Common edible mushrooms

Name of the Mushroom	Description
Enoki Take	These are long, thin white mushrooms used in East Asian cuisine. These mushrooms are available fresh or canned, the fresh mushroom being preferred. They are traditionally used for soups, but can also be used for salads.
Shimeji	These mushrooms are extensively used in Japanese cooking. Shimeji is very high on umami content and this mushroom can never be served raw due to its typical bitter taste. However, this bitter taste mellows down or disappears upon cooking.

Contd

Table 6.5 (Contd)

Name of the Mushroom	Description
Shitake **CHEF'S TIP:** It is important to cook shitake mushroom thoroughly; raw or undercooked shitake can cause rashes all over the body after 48 hours of consumption, which can last up to 10 days.	Shitake mushroom is a delicacy and is used in almost all Oriental cuisines especially of China and Japan. Shitake is the Japanese name which literally means tree mushroom. Since this mushroom is cultivated on dead tree logs, hence the name. In China these mushrooms are called *Xianggu,* which literally means fragrant mushroom.
Cloud ear	Also known as black fungus, this fungus is so called because of its ear like shape and rubber like appearance. This fungus grows on dead wood, has a dark brown colour and is somewhat translucent. These are perishable and sold dried. When they dry up, they become black in colour. They are rehydrated in water and used in stir fries and soups.
Porcini	Also known as *Ceps,* these mushrooms hold a high stature in Italian cooking. It has a nutty and meaty aroma, hence is commonly used in soups and sauces and even pastas and risottos to give them a unique flavour. It can be used in both fresh as well as dried forms. Dried porcini is often grounded to a powder and used for sprinkling on pastas for added flavour.
Portobello	Portobello is a common mushroom eaten around Europe and parts of America. These are found in two colour variants–brown and white. Usually when this mushroom is smaller in size, it is known by many names and one of the most common names is *button mushroom.* However when the same mushroom is of a very large size, then it is called Portobello.
Morels	Morels are one of the most prized fungus after truffles. They are expensive as they are rarely cultivated commercially and are normally found in the wild. In India, morels are known as *guchhi* and are one of the delicacies for vegetarians. The best morel in India comes from Kashmir. Morels are highly perishable in nature and hence, are always sold in dry form. One has to be very careful whilst cooking morels, as they may contain a lot of mud and mush, therefore should be thoroughly washed before using.
Truffles	Truffles are an extremely sought-after variety of mushrooms which grow underground. They grow in the wild and are never reared. Dogs and pigs are used to hunt these mushrooms. They are also known as diamonds of the kitchen and are commonly used in Italian and French cooking. Truffles are of two types—white and black. The black truffles are also known as summer truffles and the most famous varieties come from Perigord in France. The white truffle is also known as winter truffle and the most famous one comes from Alba in Italy.

CONCLUSION

This chapter covers in detail one of the most common ingredients used in cooking—vegetables. Each vegetable is unique and many of them are very specific to certain countries and regions. For example, drumstick usage is extensively found in south and east of India and it is not used much in any other parts of India. Many vegetables travelled all over the world with invaders and travellers and have ever since been a part of certain cultures and societies. Tomatoes were brought to India from Italy and Spain, but today we cannot think of many Indian dishes without the usage of tomatoes. In this chapter we learnt about the importance of classification of vegetables. It is important to discuss and know about various kinds of these classifications as each vegetable from a family has its own usage and ways of processing.

We also discussed the various kinds of pigments present in the vegetables that are responsible for the wide range of colours present in this mysterious commodity. We came to know about various kinds of acids and alkalis that have an effect upon these pigments and how chefs use this to their best advantage to create an array of vegetable dishes.

We also discussed about the effect of heat on the vegetables and how it affects their colour, texture, flavour, and nutritional quality; with this knowledge we can control these characteristic changes by altering the heat and the time duration for which it can be applied. We also discussed about various cuts of vegetables and also discussed about some Indian cuts of vegetables used in Indian dishes. We then discussed a range of vegetables, their selection criteria and receiving standards and also learnt how we can store these vegetables to maintain their quality and freshness.

In this chapter we also discussed some exotic vegetables and their uses. This chapter lists down various types of commonly used mushrooms in the world today, including truffles and the regions in the world that are the largest producers of truffles.

KEY TERMS

Al dente An Italian term used to denote commodity which has been cooked to a degree where it still is slightly firm to eat.

Alligator pear Another name of avocado.

Anthocyanin A pigment present in red vegetables such as red cabbage.

Anthoxanthin Pigment present in some white vegetables.

Asparagus A kind of shoot vegetable.

Betalains The pigment present in red beetroots.

Braising Method of cooking which uses moist and dry heat.

Brassica Vegetable belonging to cabbage family.

Broad bean Kind of bean used in Western cooking.

Brunoise Vegetable cut into small dices of 2 mm × 2 mm × 2 mm.

Brussels sprouts A vegetable that resembles a baby cabbage.

Buntz Kind of knife with a wavy blade.

Burhani raita Curd preparation flavoured with garlic paste.

Caramelization Browning of sugar on application of heat.

Carotene A pigment present in yellow and orange vegetables.

Celeriac A root vegetable that is used for making mash or is used in salads.

Celery A kind of stem vegetable often used in flavouring soups and stocks.

Cellulose Complex fibrous arrangement present in the stems of vegetables which provide roughage to the human body.

Chateau Cut of vegetable, usually potatoes, turned to resemble a barrel shape.

Chiffonade Leafy vegetables cut into shreds.

Chlorophyll A pigment present in green vegetables.

Cloud ear Also known as black fungus, it is a kind of mushroom eaten in Chinese and Thai cuisines.

Coagulation Hardening of proteins on application of heat.

Colocasia A kind of vegetable from the tuber family.

Dolmas Lebanese preparation of stuffed vegetables.

Edamame The green soy beans in a pod often served steamed in Japanese restaurants.

Enoki take Long and thin white coloured mushrooms used in Japanese cooking.

Fiddle head fern A vegetable that is usually coiled up in appearance. It is also known as *kasrod* in Hindi.

Flavones A pigment present in white coloured vegetables.

Fluting Mushrooms turned with a knife.

Fungi A family which includes variety of mushrooms.

Gelatinization Swelling of starch on absorption of water.

Globe artichoke Vegetable with green scales around it.

Gourds Kind of vegetable available in a shape by which it gets its name.

Greens Term used to denote green leafy vegetables.

Jardinière Vegetable cut into baton of 3 mm × 3 mm × 18 mm.

Jerusalem artichoke A kind of vegetable from the tuber family.

Julienne Vegetable cut into strips of 1 mm × 1 mm × 25 mm.

Kohl rabi A root vegetable that is also known as *ganth gobhi* in Hindi, it is a popular vegetable used in Kashmir.

Lozenge Vegetable cut on a slant angle, sometimes also referred to as diamond cut.

Macaroni A type of pasta.

Macedoine Vegetable cut into large dices of 5 mm × 5 mm × 5 mm.

Matignon Evenly cut root vegetables used as a mirepoix but served along with a dish.

Mirepoix Roughly cut vegetables used for flavouring.

Morels Also known as *guchi* in Hindi, it is one of the most expensive mushrooms used worldwide.

Paring Peeling the skin of a round fruit, such as an apple, using a paring knife.

Parisienne Vegetable scooped with a round scooper.

Paysanne Vegetable cut into geometrical shape of 1 mm.

Porcini An Italian Mushroom used in Italian dishes.

Portabella An Italian mushroom which has a large head.

Portobello A very large mushroom, which is also known as button mushroom when it is young.

Radicchio A kind of lettuce which has slight purplish reddish tinge.

Rhubarb A vegetable stem used in desserts.

Romanesco A vegetable similar to broccoli except that each floret has a unique conical and floral structure.

Savoury Food item which is salted.

Shallots A type of onion that has a flavour of onion and garlic.

Shimeji A type of mushroom used in Japanese cooking.

Shitake A type of mushroom available in China. It grows on Shii trees.

Shocking A process where the cooking of a commodity is arrested by immersing it in cold ice water.

Squash A kind of vegetable available in various shapes and sizes.

Texture The mouthfeel of any food, for example, crunchy, crisp, soft, etc.

Topinambour A European vegetable that is a tuber and is also known as Jerusalem artichoke.

Tourne Vegetable turned to a barrel shape.

Truffle A kind of fungus that is obtained from under the ground and is one of the most expensive ingredients used in European cooking.

Watercress Green leafy vegetables which have very small leaves, usually eaten in salads.

Water chestnut Also known as *singhara* in Hindi, it is a fruit of an aquatic plant, often eaten as a snack in India or as a vegetable in South-east Asian countries.

Yam Also known as elephant foot, it is a root vegetable often used in Indian cooking. It is known by various names in India such as *suran* and *jimikand*.

Yield test Usable weight of a commodity after processing.

CONCEPT REVIEW QUESTIONS

1. How are the vegetables classified?
2. Name two water soluble vitamins and two fat soluble vitamins present in vegetables.
3. Name two fruit vegetables.
4. What do you mean by the term greens?
5. Which mushroom is extensively used in Chinese cooking.
6. Name two mushrooms used in Italian cooking.
7. Name two tuber vegetables.
8. What are stem vegetables the richest source of?
9. Name the pigment present in potatoes. What are the effects of acid and alkali on it?
10. How does anthocyanin react to acid and alkali?
11. Why are green vegetables not covered while cooking?
12. Why does spinach turn colour in iron pot?

13. Why does cauliflower turn yellow when boiled in aluminium container?
14. How do carbohydrates react to heat?
15. What happens to proteins when excessive heat is applied?
16. List four effects of heat on vegetables.
17. What do you understand by the term *al dente*?
18. Is it advisable to use just enough water to cook vegetables? Why?
19. List the various cuts of vegetables.
20. What is the size of brunoise?
21. If you cut jardiniere into equal cubes what do you get?
22. What is the difference between mirepoix and matignon?
23. How would you remove bitterness from bitter gourd?
24. What is the way to process drumsticks for better cooking?
25. How would you boil the lotus root to ensure it stays white?
26. Where does broccoli get its name from?
27. How would you select a cabbage?
28. Why should onions be dry when selected?
29. What would you do to avoid blackening of an avocado?
30. How are Brussels sprouts stored and selected?
31. Write a short note on water chestnut and fennel.
31. Write a short note on 5 unique mushrooms.
32. What are morels and mention their uses?

PROJECT WORK

1. Make a chart and list all the vegetables around you. Now mark the cuts that you can make out of the vegetables as shown in the chart below.

Cuts / Vegetables	Julienne	Paysanne	Brunoise	Macedoine	Lozenge	Jardiniére	Slicing	Turning
Carrots								
Ginger								

2. In a group of three to four, visit the vegetable market for a survey and list all the vegetables that you see there. Classify them in families and record your observations based on seasonal availability and the price range.
3. By now you would have understood the reaction of acid, alkali, and metals on the pigments of the vegetables. Do this practically and record your observations. Also use metals that are not listed in this chapter and critique the results
4. In a group of three to four, choose a family of vegetable and do yield tests for the same. To do this, process it from raw to calculate usable weight. Then boil the vegetable and record the cooking loss. Record your observations and analyse in a group with everybody else's findings.

CHAPTER 7

CLASSIFICATION OF FRUITS AND THEIR USES IN COOKING

Learning Objectives

After reading this chapter, you should be able to:
- identify various types of fruits used in the kitchen and classify them into various categories
- know the way a fruit is processed
- figure out the selection and storage criteria of a fruit
- list the advantages of using fruits in cooking
- list the popular berries that are used for making desserts
- understand the usage of berries in savoury cooking

INTRODUCTION

In the previous chapter we discussed about vegetables. In the coming chapters we shall discuss about various food commodities so that we have a clear understanding of the commodities, before we start applying cooking principles to them to create delectable cuisines. In this chapter we shall talk about fruits, their classification, and usage. In the last chapter we saw that there are certain vegetables which are scientifically categorized into fruits; but here we shall focus on the real fruits which are eaten as a part of dessert.

Fruits are rich source of vitamins and minerals and give the necessary nutrition to our body. Fruits are generally eaten raw, but many of them can be cooked to serve as accompaniments, sauces, or compotes. Some fruits are rich in 'pectin', an enzyme that helps in setting of jams and marmalades. Fruits have been paired with food since time immemorial. Apple with pork, orange with duck, etc. are few of the classical preparations from the Western world. There are many ways to classify fruits; they can be classified as per their texture and flavour or as per their content of flesh. We shall see in this chapter, the criteria of selection of fruits and also the ways of storing them.

CLASSIFICATION OF FRUITS

In botanical terms, a fruit is the ripened ovary of a flowering plant. A fruit usually contains the seeds of the plant which bears it. A number of fruits are also termed as vegetables by some, as there is no single terminology that can accurately fit the variety that can be found amongst plant fruit. Fruits can be broadly segregated on their composition, nutritional value, shape, and flavour. They could be sweet or sour, depending on the variety. Some are fleshy, some are dry, and some are heavy. Fruits get dispersed by falling and rolling away from the parent plant. Grains and nuts also form part of the species, as they are ripened ovaries too. Fruits act as vital means of pollination for plants and hence, help in effective dispersion of pollen for reproduction. It is to be noted that as per culinary terminology, a fruit is defined as a product that has a sweet taste. Fruits generally contain water soluble vitamins such as vitamins B and C. They also have generous amount of antioxidants present in them. It is for this reason that fruits discolour once cut, as the exposure to oxygen instigates chemical reaction with the antioxidants. Examples of such fruits are apple, guava, mango, grapes, etc. Some fruits, known as 'pseudo carp', are accessory fruits, for example, figs. They are not the ripened ovary, but are attached to the plant embryo. Certain fruits are produced without the fusion of the ovary and the embryo. They are artificially produced by a method called 'parthenocarpy', in which pollination is omitted. These fruits are seedless. Plants which are non-fruit producing are known as 'acarpous'. Fruits are widely used in various ways such as cooked, raw, canned, pureed and squashed.

On Basis of Texture and Flavour

Fruits are classified on the basis of their appearance and another way of segregating them is on the basis of their texture and flavour. Broadly they will be classified into:

- Soft fruits   Papaya, banana, melons, etc.
- Stone fruits   Peaches, nectarines, mango, etc.
- Apple and pear family   All apples and pears
- Citrus fruits   Oranges, sweet lime, pineapple, etc.

On Basis of Appearance and Flesh Content

Fruits can also be classified on the basis of appearance and flesh content.

Fleshy Fruits

These are fruits which have supple flesh around the seed. They can be subdivided again into those formed from a single flower and those formed from a group of flowers. Those formed from a single flower are classified as berry, drupe, aggregation of drupes, pomme, and hesperidium. Fruits which grow from a group of flowers and generate only a single seed are sorosis (for example, mulberry), synconium (for example, fig), and coenocarpium (for example, pineapple).

These are further subdivided into the following:

Berry   They are single fleshy fruits without stone, and they have a lot of seeds. For example, banana, kiwi, tomato, passion fruit, and pepper.

Drupe   They are single fleshy fruits with hard stones and single seeds. For example, cherry, apricot, plum, and peach.

Aggregation of Drupes They are fleshy fruits, which have a collection of drupes formed out of the single flower, containing seed in each drupe. For example, raspberry, loganberry, and blackberry.

Pomme These fleshy fruits have a thin skin, and are not formed from the ovary but from another part of the plant. Also known as accessory fruits, they contain seeds in a chamber around the centre. For example, apple, pear, and quince.

Hesperidium It is a single fleshy fruit, which is a berry with a tough aromatic rind. All the fruits belonging to the citrus family lie in this category. For example, orange, lemon, grapefruit, and kumquat.

Dry Fruits

They are divided into dehiscent (in these fruits seeds are contained in a seedpod of some sort which opens to release the seeds) and indehiscent fruits (they do not have seedpod which opens). Dry dehiscent fruits are follicle, legume, silique, and capsule, whereas dry indehiscent fruits are achene, nut, samara, and caryopsis.

Why Eat Fruits?

Fruits have a very positive effect on the brain; the most important substance that fruits contain are water (80 per cent) and natural sugars. Natural sugars help to stimulate the brain so we can think faster and recall information more quickly. Fruits do not contain any amount of bad cholesterol, therefore they are amongst the healthiest foods available. Fruits are rich in fibre and vitamins such as vitamin C. A number of medical practitioners recommend eating specific fruits for specific health problems because of the natural nutrients present in them.

Let us discuss the properties of a few fruits and their usefulness to us.

APPLE

French: *Pomme* **Hindi:** *Saeb*

It is a sweet, astringent fruit. When we refer to fruits or vegetables being astringent, it means the dry, puckering taste caused by tannins found in many fruits. Consumption of apples tends to relieve diarrhoea. The skin of apple is a very good source of vitamin A and is rich in calcium. The vitamin content is five times more in the skin than the flesh and gradually decreases towards the centre of the fruit. The seeds of an apple can cause food poisoning and should never be consumed in large quantities. Apple in India comes from Kashmir and Himachal Pradesh and is usually a fruit of winter season. There are many varieties of apples found in India such as delicious, American, golden delicious, maharaja, and crimson. Small plum size apples are known as crab apples.

How to Select

An apple should be firm to touch and should be free of scars and blemishes. The redness of the apple does not indicate the sweetness of the apple as certain apples are astringent naturally. A shiny skin depicts a fresh apple.

How to Store

Apples should be stored refrigerated at around 4 to 5°C, in a perforated basket so that there is lots of room for it to breathe. It can be stored fresh up to 10 to 12 days, after which the apple's skin will start to wilt.

BANANA

French: *Banana* **Hindi:** *Kela*

It is a sweet and astringent fruit. The ripe fruit relieves diarrhoea, whereas the unripe fruit relieves constipation. It is a high caloric content food with less content of water. The fresh fruit taken along with milk is almost a complete balanced diet. Bananas are often confused with 'plantains'. Plantains are very starchy and are used in savoury dishes in African and Moroccan food. Many varieties of bananas are available in India ranging from small bananas to large red bananas found in Kerala. Banana flower is also cooked as a vegetable and its leaves are used as plates for eating in south India and in some parts of West Bengal. Raw bananas are used as bindings for kebabs and overripe bananas are used for banana bread.

How to Select

Bananas should be firm to touch with an even golden colour. Few black blemishes are natural; but it should not be pulpy when pressed. The dark colour of the skin indicates the ripeness of the banana.

How to Store

Bananas should be stored at room temperature with controlled humidity levels. Never store bananas in a refrigerator as the skin would turn black and look unappealing.

CHERRY

French: *Cerise* **Hindi:** *Glass*

It is a sweet, sour, and astringent fruit. It cleanses the system and is good for dental problems. It is useful in curing insomnia or lack of sleep. Cherries are summer fruits and are grown in Kashmir and Himachal Pradesh in India. There are many varieties of cherries available around the world. Morello, Californian, and maraschino are the most famous varieties. Morello is the traditional cherry used in blackforest gateaux from Germany, whereas maraschino is widely used for production of cherry liquor. Californian cherries are used for consumption as fruits as these are very sweet and black in colour.

How to Select

Cherries should be firm to touch with an even red to maroon colour. The stem attached to the cherry determines the freshness of the same. The bright green colour of the stem indicates freshness. Pulpy cherries are prone to insects.

How to Store

Cherries should be stored refrigerated at around 4 to 5°C in a perforated basket to allow circulation of air. Cherries are perishable and should be used within three to four days of purchase.

FIG

French: *Fig* **Hindi:** *Anjeer*

It is a sweet, cooling, nourishing, and heavy fruit. It is a good laxative. It is a good source of calcium, iron, and potassium and has antibacterial properties. Figs are eaten fresh as well as dried. Fresh figs are highly perishable and form a perfect accompaniment with cold meats. Stewed figs in red wine are also used to accompany roasted or grilled meats. Dried figs are available in the market, pressed into a disc and sewn on a jute string.

How to Select

Fresh figs should be light green with purplish streaks. They should be firm yet soft to touch. A hard fig would denote a raw fig.

How to Store

Figs are very delicate. So they should be stored individually like eggs, refrigerated at around 4 to 5°C in a perforated basket to allow circulation of air. Do not pile figs on top of each other as they would spoil. Figs should be consumed within a day or two of purchase.

GRAPE

French: *Raisin* **Hindi:** *Angoor*

Grape is a sweet, cooling, strengthening, and nutritious fruit. It relieves thirst and burning sensations in the stomach and is believed to be a good cure for fevers. A laxative fruit rich in vitamin B, it helps stop bleeding in gums and removes bitter tastes from the mouth. Grapes are cultivated around the world and are the main ingredient in production of wine. They are available in various colours ranging from green, red to black. Large Californian grapes are widely used on buffets as the large plum-sized grapes look very attractive on the displays. Dried grapes are known as raisins.

How to Select

A bunch of grapes should be compact and heavy for its size. Grapes should be firm to touch and should have a bright colour. As grapes are always received in bunches, check for loose grapes as the shelf life of these grapes will be very less and check for insects inside the cluster.

How to Store

Grapes should be stored refrigerated at around 4 to 5°C in a perforated basket to allow circulation of air. Grapes are perishable and should be used within three to four days of purchase.

GOOSEBERRY

French: *Physalis* **Hindi:** *Ras Bhari*

It is a sweet and astringent fruit and resembles cherry tomatoes. It is sour and pungent when unripe but tastes very aromatic when ripe. It is harvested when its colour changes to a deep orange. It is also known as husk tomato in Mexico. It is useful in curing diarrhoea and bleeding piles. It is widely used in curing of diabetes as well. Also known as cape gooseberry, it is commonly used to make fruit compotes and sauces. Ripe gooseberries are also used as petit fours by dipping them in chocolate.

How to Select

Gooseberries should always be purchased with the parchment that covers them. The dried parchment is the indication that the fruit is ripe. They should be bright yellow to orange in colour and should be soft yet firm to touch. Check for holes in the fruit as there could be insects.

How to Store

Gooseberries should be stored refrigerated at around 4 to 5°C in a perforated basket to allow circulation of air. They can also be kept at room temperature if they are slightly unripe, but then the shelf life will be only three to four days.

ORANGE

French: *Orange* **Hindi:** *Santra*

Orange belongs to the family of citrus fruits and is rich in vitamins A, B, and C and is a good source of calcium. The juice of this thirst-quenching fruit is a popular drink. Oranges are available in various colours ranging from yellow to deep orange and have varied peel thickness. The colour and the thick peel depend upon the weather conditions during the growing season of the fruit. Oranges grown in colder climates have brighter skin. Warm climate results in green coloured skin, even though the fruit might taste very sweet. This versatile fruit can be used in many ways—in milk shakes, jam, jelly, or marmalades. Even the rind and zest of this fruit is added to various breads and pastries for a refreshing tangy flavour.

How to Select

Oranges should be firm and heavy for their size. Thick skin oranges are usually sweet and are often referred to as 'table oranges', while thin skin oranges are juicy and hence, called juice oranges. Juice oranges might not be as sweet as the table oranges but have good yield of juice. The skin of the orange should be smooth and shiny and there should be an even colour on the fruit.

How to Store

Oranges store well at room temperatures, but it is advisable to store them refrigerated at around 4 to 5°C in a perforated basket to allow circulation of air. Do not pile too many on top of each other.

PEACH

French: *Peche* **Hindi:** *Aadoo*

It is a heavy, sweet, astringent, and very tasty fruit. It is a laxative fruit and rich in vitamin A and potassium. It is strengthening and easily digested. This sweet and sour tasting fruit is cooling in nature and hence, abundantly available in summers. Peach is widely used in production of wine, liquor, and brandy. It is eaten as fruit and also canned for various uses. It is a slightly firm fruit so it is often poached and eaten for breakfast. Canned yellow peaches served with raspberry compote is a classical dessert called 'peach melba'.

How to Select

Peaches should be firm to touch with an even orange colour. The fruit when pressed with a thumb should feel juicy and fleshy. They should be free from blemishes and should not have packaging dents.

How to Store

Peaches should be stored in a refrigerator at around 4 to 5°C in a perforated basket to allow circulation of air.

WATERMELON

French: *Watermelone* **Hindi:** *Tarbooz*

This large melon has water content up to 95 per cent and hence, the name water-melon. The most popular fruit to beat the summers, it is eaten sliced or just cut into fruit salad or even relished as juice. It is commonly offered as a welcome drink in most of the hotels as it quenches thirst and is of a pleasing colour. It has a green rind with red or pink flesh and can be round, oval, or oblong in shape. There are certain watermelons which are hybrid and are seedless as well.

> **CHEF'S TIP**
> To check for the ripeness of watermelon, inspect the side on which it is rested. The yellow colour of the side indicates ripe and sweet melon and the whitish part indicates raw melon.

How to Select

A watermelon of 3 to 4 kg size is usually sweet. It should have a smooth, even, and dark green skin. It should be firm and heavy for its size and the surface should be resilient to strong pressure. While selecting a watermelon, one should check for a hollow sound in the watermelon. The hollow sound indicates a sweet melon.

How to Store

Watermelon can be stored at room temperature for around two weeks otherwise it can be stored refrigerated at around 4 to 5°C in a basket.

APRICOT

French: *Abricot* **Hindi:** *Khumani, Khoobani*

It is a sweet and sour fruit. Apricots relieve thirst and are a very good remedy for cough. The fresh fruit is rich in calcium, vitamin A, and sugar. It is a good source of vitamin B complex and vitamin C. Fresh apricots are eaten raw or even stewed for compotes. Dry apricots are used in Indian cooking in Kashmiri and Hyderabadi cuisine.

How to Select

Apricots should be soft, yet firm to touch and should have an even orange colour. They should be free of blemishes and any holes.

How to Store

Fresh apricots should be stored refrigerated at around 4 to 5°C in a perforated basket to allow circulation of air.

CRANBERRY

French: *Canneberges*

It is an astringent fruit rich in vitamin C. It is diuretic and beneficial for kidney and bladder ailments. This sharp tasting red colour berry is very sour and is always cooked with sugar before making it palatable. Cranberry jelly is the classical accompaniment to roast Turkey served during Christmas. Cranberries grow on low trailing vines. It is believed they got their name from the long necked pink blossoms which resembled the neck of the crane. From 'craneberries' they became 'cranberries' over a period of time.

How to Select

It is difficult to find fresh cranberries in India but canned cranberries are always available.

How to Store

Cans should be stored at room temperature. Once opened, empty the contents into a bowl and keep refrigerated at around 4 to 5°C.

DATE

French: *Date* **Hindi:** *Khajoor*

It is a sweet, heavy, aphrodisiac, and cool fruit. It helps to build the body and reduces alcohol intoxication. It is usually consumed raw or with milk. It is a good source of calcium, iron, and sugars. It was one of the earliest fruits available to the mankind and was grown in Mesopotamia, now in Iraq, more than 5000 years ago. It is sometimes called the 'candy that grows on trees'. Dates were consumed on camel caravans across the deserts to get energy and to keep the body cool.

How to Select

It is difficult to find fresh dates in India and hence, they are always available dried and packed. Buy dates that are intact and free from insects.

How to Store

Store dates at room temperature and once opened, empty the contents into a bowl and keep refrigerated at around 4 to 5°C.

GRAPEFRUIT

French: *Pamplemousse* **Hindi:** *Chakotra*

It is a hot and sour fruit. It is an excellent appetizer and promotes salivary and gastric digestion. It is a body strengthening fruit. It is available with white flesh and pink flesh. It is usually eaten for breakfast and classically served with castor sugar and cinnamon.

How to Select

Grapefruit should be firm and heavy for its size. The thick skin indicates sweet fruit. The skin of the grapefruit should be smooth and shiny and there should be an even colour on the fruit.

How to Store

Grapefruits store well at room temperature, but it is advisable to store them refrigerated at around 4 to 5°C in a perforated basket to allow circulation of air. Do not pile too many on top of each other as they will get mouldy if kept for too long in the fridge.

GUAVA

French: *Goyave* **Hindi:** *Amrood*

It is a sweet and a cooling fruit. It is heavy and hard to digest. Eaten without the rind, it may lead to constipation. However, the unripe fruit may be useful in diarrhoea. Decoction of the leaves, when gargled, relieves swollen gums and mouth ulcers. It is useful in gout also.

Guava is a rich source of pectin and hence, it is used for making jams and preserves. This fruit is native to South America and is also widely cultivated in the USA. There are many varieties of guava available some have whitish flesh while some varieties have deep pink flesh.

How to Select

Guavas should be soft yet firm to touch with an even yellow colour. Green colour denotes that the guavas are unripe. They should be free of blemishes and any marks.

How to Store
Guavas should be stored refrigerated at around 4 to 5°C in a perforated basket to allow circulation of air.

MANGO

French: *Mangue* **Hindi:** *Aam*

It is a cool and sweet fruit. Unripe mango is sour, astringent, and pungent. It is also known as the 'king of fruits'. The ripe fruit is a good laxative, increases the urine flow and cools the blood.

The mango originates from South-East Asia, where it has been growing for more than 4000 years. Mango is considered to be a tropical fruit as it grows best in that climate. In India raw mango is used for making pickles, whereas ripe ones are used for milk shakes, ice creams and eaten as it is.

How to Select
Mangoes should be soft yet firm to touch with an even yellow colour. Some varieties of mangoes have green skin also. They should be free of blemishes and any packaging marks.

How to Store
Mangoes should be stored refrigerated at around 4 to 5°C in a perforated basket to allow circulation of air. Do not pile too many mangoes in one basket as it is a very soft fleshy fruit and might get damaged.

PINEAPPLE

French: *Ananas* **Hindi:** *Ananas*

Pineapple is a herbaceous, perennial plant with large pointed leaves. Fresh pineapple is good uncooked and is often used as a topping for a cheesecake or a tart and in fresh fruit salad. In India there are two kinds of pineapples received in the hotel—rani pineapple and normal pineapple. Rani is a superior variety and is slender and longer than the normal pineapple. These can be used in fruit platters. But if the pineapple needs to be cooked into compote, then it is advisable to use the normal pineapple.

How to Select
The skin should be yellow in colour and the surface should have a clean and waxy shine. The fruit should be heavy for its size. The eyes should be fully developed and hard. There should be no internal browning or decay, broken skin, bruises, moulds, or tender spots. The fruit should be light yellow in colour, sweet, and juicy and not fibrous. Each pineapple should weigh between 2 to 2.5 kg. It should have its distinct aroma.

How to Store
Pineapple should be stored in a refrigerator at around 4 to 5°C in a perforated basket to allow circulation of air. Do not pile too many pineapples in one basket as they are very soft fleshy fruits and might get damaged. Store them upright with the stem facing up.

PAPAYA

French: *Paw paw* **Hindi:** *Papita*

It is widely cultivated in all the tropical countries. Papaya is a large fruit with yellowish rind and reddish yellow flesh with a large central cavity with many seeds. Papain, an enzyme found in papaya, is used as a tenderizer. There are three kinds of papaya that are received in hotels—raw, normal, and disco. Raw papaya is used for making salads and relishes. Disco papaya is slender and longer than normal papaya and has dark orange coloured flesh. Papaya is usually used as boats for morning breakfast.

How to Select

Select papaya that is even shaped and more oval and slender in shape. The skin should be yellow in colour with traces of orange. When shaken, the seeds should rattle and the apex of the fruit should easily dent when pressure is applied by hand. There should be no internal browning or decay, broken skin, bruises, moulds, or tender spots. The pulp should be deep reddish orange in colour and deep yellow in case of normal papaya, with a sweet taste. Each papaya disco should weigh 750 g to 1.25 kg and normal papaya between 1.250 to 1.750 kg. It should have its distinct aroma.

How to Store

Papaya should be stored refrigerated at around 4 to 5°C in a perforated basket to allow circulation of air. Do not pile too many papayas in one basket as they are a very soft fleshy fruit and might get damaged. If the papaya appears to be raw, then wrap and store it in a newspaper for a day or two.

MELON

French: *Melone* **Hindi:** *Kharbooza*

Melons are available around the world and there are many varieties of the same. Melon is a great thirst quencher and has always been offered to the travellers in Egypt. Though melons are related to the gourd family, they have a sweeter flesh and hence, used in desserts. Melons combine well with ham and cold meats and thus are mostly served as appetizers also. There are different varieties of melons. These are as follows:

Cantaloupe (known as *charentias* in French) A cantaloupe has an orange flesh, with distinctive rough surface rind with green stripes marking the wedges of this fruit.

Muskmelon   It is beige coloured and has a netted skin and orange flesh.

Honeydew   This melon is the sweetest of all varieties of melons and has green coloured flesh. It has a pale green to yellow skin with netted structure.

Casaba or Japanese melon   It is oval with yellow to orange skin and has a white flesh. It is also called *sarda* in Hindi.

How to Select

Select melons with a weight between 400 and 800 g. They should be sweet smelling with a fairly resilient, even skin. There should be no cracks or blemishes and pulp should be soft and juicy.

How to Store

Melons should be stored refrigerated at around 4 to 5°C in a perforated basket to allow circulation of air.

POMEGRANATE

French: *Pomegranate* **Hindi:** *Anaar*

The fruit originated in the Middle East, and the Moors brought pomegranate to Spain. Pomegranates are round, reddish gold, and about 2 to 5 inches in diameter with juicy, crunchy kernels inside. The pomegranate is held in a cream coloured bitter membrane that is inedible. The dried seeds are also used as a souring agent in Indian cuisine and this fruit is also very popular in the Middle East. In India there are two kinds of pomegranates—kandhari and rani. Kandhari has a golden skin with whitish to light pink coloured seeds, whereas rani has burgundy red skin with deep red coloured seeds. The seeds are always separated from the yellow pith that encircles the seeds as they impart a bitter taste.

How to Select

Skin should be bright red, firm, and relatively thick. It should be spongy, puffy, wrinkled, or have marks and spots. There should be no traces of decay. The seeds should be bright red in colour and sweet to taste. Each pomegranate should weigh at least 300 to 350 g.

How to Store

Pomegranates should be stored refrigerated at around 4 to 5°C in a perforated basket to allow circulation of air.

SAPODILLA/MUD APPLE

Hindi: *Chikoo*

Mud apple or chikoo is also known as sapodilla and is a very popular fruit in India. It has a dark mud coloured skin as well as flesh. It is a very sweet fruit which has long black seeds in it. It can be used in desserts and is usually eaten raw as fruit.

How to select

The skin of the fruit should be mud brown in colour, thin and firm and not spongy, puffy, or wrinkled and there should be no blue or white moulds or traces of decay. It should feel heavy for its size. The weight should be 100 to 125 g per piece. The flesh should be brown, sweet, juicy and not fibrous. It should have its distinct aroma.

How to Store

Mud apple should be stored refrigerated at around 4 to 5°C in a perforated basket to allow circulation of air. Do not pile too many in one basket as they are very soft fleshy fruits and might get damaged.

DRAGON FRUIT

French: *Poire de chardon*

This fruit resembles the fire flames of a dragon. This fruit is native to Vietnam and it is believed that the French brought it to Vietnam over 100 years ago. The fruit is very pretty and has vibrant colours such as purple, magenta, yellow, and blue. Inside this pretty fruit, a soft white fleshy ball of the size of an orange is concealed, with black seeds resembling black poppy seeds. This fruit has a sweet and sour taste and is used for garnishing cakes and pastries or sliced on fruit platters.

How to Select

The skin of this fruit should have a shine and the tips of the protruding tentacles should be soft and pliable. The fruit should feel firm to touch and should be able to depress when pressed slightly.

How to Store

Dragon fruit should be stored refrigerated at around 4 to 5°C in a perforated basket to allow circulation of air.

KIWIFRUIT

French: *Kiwifruit*

Kiwifruit is native to China and is a very pretty fruit when cut open. The dark green coloured flesh with symmetrical arrangement of black seeds and a pretty pattern is concealed under a thin paper like brown skin. They are served peeled as the outer flesh is inedible. One has to be very careful in selecting a kiwifruit, as an unripe one will be sour and can sometimes cause an itchy feeling in the throat and an overripe one will be pulpy and not appetizing. Kiwifruit is used mostly as a decorating fruit and also served on fruit platters to enhance the colour.

How to Select

The skin should be mud brown in colour and paper thin. The fruit should feel firm, yet the flesh should press when pressure is applied on the apex of the fruit. There should not be any wet marks on the fruit as this indicates that it is pulpy and overripe.

How to Store

Kiwifruit should be stored refrigerated at around 4 to 5°C in a perforated basket to allow circulation of air. Do not pile too many of them in one basket as they are very soft fleshy fruits and might get damaged.

A study conducted by Dr Paul Lachance of Rutgers University, New Brunswick, New Jessey, USA, evaluated the nutritional value of fruits to determine which fruit provides the highest nutrition.[1] The analysis determined the nutrient density of the 27 most commonly consumed fruits. The study found kiwifruit to be the most nutrient-dense of all fruits, followed by papaya, mango, and orange. Kiwifruit, orange, and papaya are fruits most appropriate for weight control. Kiwifruit has the highest level of vitamin C, almost twice that of an orange and magnesium. It is important for cardiovascular health.

It is good to know the nutritional value of various fruits as they can be mixed and matched according to one's deficiencies.

BERRIES AND THEIR USAGE IN KITCHEN

Berries are the most common and popular fruit used by pastry chefs all over the world. However, the usage of berries in modern day cooking is not restricted to just desserts, they are often used in savoury dishes as well since they are a rich source of phytochemicals that help fight cancer.

Technically a berry is a fleshy fruit that develops from the ovary of a flower. Many of the berries contain the seeds inside such as blackberry and raspberry. However, certain berries like strawberry have the seeds on the top of the skin. In bakery and pastry berries are used for making sauces known as *Coulis* but apart from the sauces, they can be cooked to prepare jams, fillings for tarts and pies or can simply be used in unprocessed state as a garnish for desserts. The varied colours and textures of berries make it one of the most prominent garnishes for desserts.

Although berries are seasonal in nature, but one can find berries in any season as there are many varieties that grow in summer and a few also grow in winters. Berries need a typical kind of cold climate and soil conditions and therefore they are not grown in all parts of the world. Earlier berries used to be consumed locally, but with advent of technology, they are now traded all over the world and thus sometimes can be an expensive ingredient.

---

[1] These findings were published in the *Journal of the American College of Nutrition*, October 1997 issue.

Many types of berries are found all across the world, but in Table 7.1 we will discuss some of the most common ones that are used in kitchens worldwide.

Table 7.1 Common types of berries used in kitchens

Berry	Description
Strawberry	As per its botanical definition, a strawberry is not a berry, as the berry must contain seeds inside the flesh, but in case of strawberries, the seeds are located on the outer surface of the fruit. Strawberries grew in wild in parts of North America, but were fist planted by the French in the region of Brittany a few hundred years ago. Strawberries are a typical summer fruit in the European continent, but in India this fruit is available from November till March. It best grows in colder climates of North India. Strawberries are best eaten with sweetened whipped cream and they pair up well with chocolates as well. The heart shaped appearance of strawberries make them ideal for desserts served during Valentine's day all across the world. Strawberries can also be stewed or made into delicious jams and compotes. The green part of the strawberry needs to be removed before consumption and the process of removing this part is known as *hulling*.
Raspberry	Raspberries are fleshy fruits that grow on thorny shrubs. They usually grow in winter climate and are available in many colours, the most common ones being red. Black and golden raspberries are unique and are popularly grown in areas around Canada and Scotland. Raspberries are highly perishable and can be used as they are for garnishes or can be cooked with sugar to make jellies, jams, compotes, or a coulis. Raspberries are a rich source of vitamin C and provide folic acid in abundance.
Blueberry	As the name suggests these berries are dark blue in colour. The berries are green when raw and as they ripen they develop a deep blue to a blackish colour with a whitish powdery coating on it known as bloom. Blueberries are available in various sizes and can range from 5 mm diameter to a 15 mm diameter. They can be used fresh or can be cooked with sugar to make jams, compotes, and coulis. Blueberries are a rich source of vitamins and dietary fibre. They are often dehydrated to be used in cookies, cakes, and cereals.
Blackberry	Blackberry is from the same genus as raspberry but is larger in size and when plucked from the plant, it has a green stem attached to the fruit; unlike raspberries, which when pulled from the stem left on the plant leave a hollow centre in the fruit. Mexico is the largest producer of blackberries. They are often used fresh as garnishes in tarts and pies, but can also be stewed with sugar to make jams and compotes.
Mulberry	Mulberry is a fruit of one of the species of deciduous trees and grows in wild temperate regions across the world, especially in Southeast Asia. It is commonly known as *shahtoot* in Hindi and is available in three colours—red, black, and white. In India it is eaten as a fruit but in other parts of the world, mulberries are also used for preparing jams and sherbets.

Contd

Table 7.1 (Contd)

Berry	Description
Elderberry	Elderberries are highly perishable fruits. They have often been used as supplements for medical ailments such as flu and cold. Due to their highly perishable nature, they are rarely sold as a fruit but are often converted into syrups, which are drunk as a beverage.
Dewberry	Dewberry belongs to the same family as blackberry but the fruit is usually of a purplish black colour when ripe. In many countries, they are also known as groundberries and are used as a popular alternative to raspberries and can be used for making jams and jellies.
Currants	Currants are small berries that come from the gooseberry family and are available in many colours, the most popular being red, white, and black. The shiny small berries grow as a cluster on a single stem. The entire stem is used as a garnish in pastry products. Red currants are also used for making jellies and jams, whereas the black variant is usually dried and used as a fruit in Christmas and wedding cakes. The white currants, also known as golden currants, are the sweetest of all varieties and are highly perishable.

FRUITS IN COOKING

The use of fruits in cooking dates back hundreds of years. While fruits are most commonly used in desserts, they can also form part of savoury dishes. Fruits are also found in foods such as cookies, muffins, yoghurt, ice cream, and cakes. Not only are fruits used in preparing a variety of dishes, but they also help keep certain foods fresh and help preserve their colour. There are various ways to cook fruit, and certain considerations should be made beforehand as fruit tends to be delicate, and can disintegrate easily.

Generally, boiling is too harsh a method for most fruits; gentle simmering preserves the texture and shape of fruits. When cooking soft and stone fruits, simply warm them by placing them in a pan of boiled water. Poaching fruit is a similar method, and is a common way to cook fruits such as pears. Bring the water to a simmer and then gently lower the fruit into the pan with a spoon. Immediately reduce the heat so the liquid is barely bubbling, and cook until the fruit is tender. You could also stew fruit, where the saucepan is covered and the fruit is cooked in just enough liquid to cover it. This method helps keep it moist. Fruit poaches well in most liquids, from plain water to dense sugar syrup to wine-poached pears in red wine with ginger sauce.

It is important to make sure that while stewing a fruit, the water ratio is not too high or else all the nutrients will get expelled in the water and the fruit will be left only with fibrous part. Use just enough moisture to cook the fruit. As with cooking vegetables, it is important to retain the crunch while cooking fruits. Barbecuing and grilling fruit leads to very sweet, strong flavours; this is due to the intense heat which caramelizes the sugars. Grilled or barbecued fruits make fantastic desserts, side dishes or appetizers. Fruits such as apples and pears can be grilled and are easy to prepare. This is because they are able to hold their shape and texture while cooking. Softer fruits such as peaches, nectarines, plums, and mangos may become soft if overcooked. Once cut, the fruit should be soaked in water to maximize the amount of liquid inside. This prevents the fruit from drying out on the grill. Adding 1tsp of lemon juice to the water will help the fruit preserve its colour. It is also a good idea to try grilling bananas, tangerines, and pineapples in their skin. In general, leaving the skin or peel on the fruit helps it maintain its structural integrity as it cooks. Sauces of fruits are used to marinate the fruits before grilling them and they can also be served with the fruits after grilling. Among the simplest is may be a puree of the same fruits which have been grilled, augmented with a little lemon juice, and drizzled over the fruit. An even simpler approach is to avoid using a sauce and sprinkling the fruit with brown sugar, cinnamon, or ginger. This method of cooking adds a delicious and exotic feel to barbecues and is a new angle to eating fruit.

Many nutrients in fruits are found just beneath the skin, so it is worth eating the peel if it is edible. Fruit purees and sauces make appetizing additions to desserts and even savoury dishes. For purees and sauces, place the fruit in a saucepan over gentle heat so that it cooks evenly. Stir constantly and then either mash the fruit with a fork and sieve or blend in a blender. For coulis, the fruit can often just be mashed or blended in a food processor and then sieved before serving.

Fruit can be made into jams, jellies, pickles, and chutneys, or can be bottled whole. The most suitable method of preserving depends on the type of fruit and its quality and ripeness. Underripe fruit is fine for chutneys, jams, and jellies but overripe fruit is only good for making chutney and should not be used for making jam.

Drying is also a good way of preserving fruits and it intensifies their flavours. Most fruits can be dried effectively.

CONCLUSION

We read in the chapter about different kinds of fruits used in everyday cooking in hotels. If you research, you will find numerous kinds of fruits that exist in the world. Some are very specific to certain regions and not commonly available in many countries. In the last 5 years India has started seeing such fruits which were not available earlier. Some fruits require certain kinds of tropical climate best suited for their growth and hence we have different kinds of fruits growing all over the world. In India itself, not all fruits grow in every region and thus it is established that climate and soil play an important role in the cultivation of fruits. In this chapter we discussed various classifications of fruits and came to know about various types such as stone fruits and fleshy fruits. We discussed about various other fruits commonly used in hotels and also mentioned about various salient features of selection of fruits for a particular usage. We also read about various ways in which fruits can be stored to maintain their freshness and quality.

In this chapter we have also discussed a range of commonly used berries in the modern kitchen of today all across the world. Berries being a rich source of phytochemicals are one of the crucial ingredients that are used for prevention of cancer.

Acarpous   Plants that do not produce fruits.

Ananas   French and Hindi names for pineapple.

Cerise   French for cherries.

Citrus fruits   Fruits which are tangy and sour in flavour such as orange, sweet lime, and pineapple.

Coulis   Cooked puree of fruit usually used as a sauce.

Crab apple   Type of apple resembling the size of a plum.

Canneberges   French for cranberries.

Drupe   Same as stone fruits.

Goyave   French for guava.

Hulling   Removal of the green leafy covering on top of strawberry.

Juice orange   Oranges with thin peel used for extraction of juice.

Mangue   French for mango.

Maraschino   Type of cherry used in making cherry brandy.

Morello   Type of cherry used in making of black forest gateaux.

Muffins   Dry cakes usually eaten for breakfast.

Nectarine   A type of stone fruit.

Pamplemousse   French for grapefruit.

Parthenocarpy   A method of artificially creating fruits that are seedless.

Peach melba   A classical dessert made with stewed peaches and raspberry sauce.

Peche   French for peach.

Petit fours   Small fancy sweet tit bits served with coffee after a French meal.

Physalis   French for gooseberries.

Phytochemicals   A compound present in fruits and vegetables especially berries, which helps in prevention of cancer in human beings.

Plantains   Starchy fruit that resembles bananas and is used for making savoury mashes.

Poaching   Method of cooking commodity in simmering liquid

Pomme   French for apple.

Stone fruits   Fruits which have seed resembling almonds, for example, peaches, nectarines, and plums.

Table oranges   Sweet oranges with thick skin used for garnishes.

Tangerine   A type of orange which is blood red in colour, usually available in China.

CONCEPT REVIEW QUESTIONS

1. How are fruits classified?
2. What kind of vitamins do fruits have?
3. Which fruit is the richest source of calcium?
4. What do you understand by accessory fruits?
5. What is parthenocarpy?
6. List three stone fruits.
7. List three citrus fruits.
8. List three berries.
9. List five benefits of eating fruits.
10. What is the percentage of water found in fruits?
11. List three varieties of apples.
12. How should apples be selected?
13. List three varieties of bananas. How are they different from plantains?
14. What are the medicinal values of banana?
15. Why should not we store bananas in the refrigerator?
16. List the medicinal values of cherries.
17. Where is morello cherry used?
18. How should you store figs?
19. Describe a gooseberry.
20. What are the selection criteria for oranges?
21. What is the difference between table orange and juice orange?
22. What is peach melba?
23. What are the selection criteria for watermelon?
24. What are cranberries and how are they used?
25. List ways in which mangoes can be used.
26. What method of cooking is usually applied to fruits?
27. List at least five points to be kept in mind while cooking fruits.
28. How do fruits react to grilling?
29. Which fruit has the highest amount of pectin?

30. What is a coulis?
31. Why should berries be part of regular diet?
32. What is the difference between blue berries and black berries?
33. Why is strawberry not classified as a berry in botanical terms?
34. Which berry is commonly used in drinks?

PROJECT WORK

1. In a group of four to five, do a market survey of fruit markets (both local and imported) and make the following observations.

    (a) The location and the reasons for its location
    (b) The modus operandi amongst the wholesalers, retailers, and vendors
    (c) States and their regions supplying the fruits
    (d) The range of fruits and their quality observed
    (e) List the seasonal fruits noticed
    (f) Hygiene and sanitation of the market (its implications to our food production cycle)
    (g) Prices of any five seasonal and non-seasonal commodities

2. Make a chart of seasonal fruits and recommend appropriate fruit baskets for the room amenity and do the costing of the same.
3. Carry out the yield tests of various fruits and record your observations.
4. Research at least three classical preparations of fruits such as apples, mangoes, raspberries, peaches, apricots, and cherries.

CHAPTER 8

STOCKS

INTRODUCTION

Stocks are the flavourful liquids that form the backbone of Western cookery. That is the reason they are termed as *fond* in French, which translates to 'foundation' or 'base'. Stocks are used in almost all the cuisines around the world including South–East Asia. Liquid is the most essential ingredient used for any cooking. Over the years cooks realized that during cooking the nutrients leach into the water, causing loss of flavours as well as essential minerals and vitamins. The need thus arose to use that liquid in one way or the other to retain the nutritional aspects of food. Some ancient cooks used this flavourful liquid as broth and called it soup or simply transformed it into a reduced smooth finished liquid and called it sauce. Stocks give the entire body to a dish and stocks around the world have their own distinctive tastes and flavours. For example, Indian cooking uses a stock called *yakhni* for its biryanis and other dishes, whereas stock flavoured with seaweed is used in Japanese cooking. In this chapter we will discuss various types of stocks and their usage in the gastronomic world.

STOCKS

Stock can be simply defined as a liquid which has been simmered for a long time in order to extract flavours from the ingredients used.

Any type of liquid can be used to start a stock. In almost all the cases, water is the liquid medium used, but then certain stocks can be made using a combination of milk and water as well. However, milk and various other liquid mediums are strictly regional. In classical Western cooking, water is the medium used to start a stock.

Stock is usually made from bones and vegetables. The types of bones used for a stock would depend upon the final usage of the stock. To prepare chicken stock one would use chicken bones only and similarly, to get lamb stock one has to use lamb bones and so on. However, in hotels, four major stocks such as chicken, lamb, beef, and fish are prepared. Fish is a delicate commodity, so fish stock is only simmered for 20 minutes. Beef and lamb stocks can be simmered up to six to eight hours depending upon the usage of the final product. Chicken stock requires three to four hours of simmering to extract the flavours.

Bones are porous in nature and hence, prolonged amount of cooking at lower temperatures is the best method of cooking stocks. Vegetables are again soft in nature so prolonged cooking would make the vegetables mushy and bitter as well. It is not advisable to simmer vegetable stocks beyond 45 minutes.

Stocks are simmered for a long time in order to extract the flavour from the bones and vegetables used. Boiling would not let the flavours leach into the liquid as the pores of the bones would seal in high temperatures and hence, it will hamper the secretion of juices into the liquid. Boiling would also make the stock 'cloudy'. The bones and vegetables contain impurities, which when simmered rise up to the surface and can be skimmed off. This results in a clear stock, which would give a nice visual appeal to the dish. In case of boiling of stocks these impurities would roll boil into the liquid, thereby creating a stock which is not clear and hence, termed as cloudy.

This is the most important part of the stock as this will give a definite characteristic to it. A chicken stock simmered with Indian spices would be used only in Indian cooking, chicken stock with Thai flavourings would be used only for Thai dishes. Similarly, Western stocks would use herbs and other vegetables depending upon the type of stock. The bunch of herbs used to flavour Western stocks is termed as 'bouquet garni'. Certain spices such as peppercorns, bay leaf, and dried thyme are used to spice up the stocks and are known as *sachet d'epices*, literally meaning 'bag of spices'. A stock should be flavoured enough to allow it for easy identification but the flavours should not be overpowering to mask the real flavour of the stock.

Important Terms

Let us discuss a few terms that we come across, when discussing about stock.

Broth   Known as bouillon in French, it is a result of poaching meat or vegetable in a liquid medium. So in basic terminology, broth would be meat and vegetable boiled in water with seasonings and flavourings.

Court bouillon   Water boiled with seasonings and flavourings with an acidic medium, such as wine or lemon juice, is called court bouillon. It is usually used to poach oily fish. Oily fish is very delicate and the acid in the court bouillon will help firm up the fish in cooking process.

Neutral stock   It is usually referred to a veal stock which is white in colour and has a neutral flavour, so it could be used in various dishes. In India due to religious taboo, chicken stock is used as a neutral stock.

Fumet   Fish stock is called *fumet* in French. Many people confuse it with 'glaze', which means a reduced fish stock.

Remouillage   It is also known as 'rewetting' or the second stock. When the stock is made and strained, more water is added to the remaining bones and vegetables and the stock is simmered for around an hour. This is comparatively less flavourful than the original stock; but it can be used for poaching of commodities, etc.

CLASSIFICATION OF STOCKS

The bones and vegetables are simmered to extract colour, flavour, aroma, and body of the resulting stock. The stocks are basically classified according to their colours. These are discussed below.

White Stock

Both white and brown stocks are made from bones and vegetables, however, the process followed for each one is slightly different. Many people think that white stocks are made from white meat and brown stocks from red; but this is not true. In case of white stock the bones are blanched[1] to get rid of the impurities. Vegetables such as leeks, onions, celery, and turnips are used to flavour the stock as red coloured vegetables such as carrots, etc. will change the colour of the stock. No tomato product is used for white stocks.

> **CHEF'S TIP**
> Start the blanching of bones in cold water. Plunging of bones in hot water will seal the pores and all the proteins will coagulate making the stock cloudy in the end.

Brown Stock

In case of brown stocks, the bones and vegetables are roasted or caramelized. Tomato paste is used and it is also sauteed to get a deep brown colour. The colour of the brown stock should be amber colour. The shin bones of beef have the best flavours and hence, the most preferred for brown beef stocks.

> **CHEF'S TIP**
> Season stocks at last, as the stock tends to get salty during the reduction. Also season the stocks moderately as they will be used to prepare other dishes.
> Use whole peppercorns instead of crushed powder as it turns to be bitter.

STOCKS AND THEIR USES

Stocks are the base for any Western cooking. Most commonly, it is used to make soups and sauces; but the usage is not limited to just this. White stocks are used in preparations of white sauces and clear soups, while brown stocks are used in brown sauces, red meat stews, and braised dishes. Stocks can also be used to prepare certain rice dishes such as paella and biryani. Refer to Table 8.1 for ratio of standard flavouring and spices used in stocks.

Table 8.1 Ratio of flavourings and spices used in stock

Standard Mirepoix	Sachet D'epices	Bouquet Garni
	For 4 litres	For 4 litres
50% Onions/* leeks	1 Sprig thyme	1 Sprig thyme
25% Carrots/* turnips	1 Bay leaf	1 Bay leaf
25% Celery/* celery root	3–4 Parsley stems	1 by 6" Section celery
*Use for white mirepoix	1 tsp Peppercorns optional	1 by 6" Section leek
	1 Garlic clove	Optional: 1 by 6" Section carrot

[1] Refer to Chapter 17.

PREPARATION OF STOCKS

○ FISH STOCK

French: Fumet
5 kg Fish bones
100 g Butter
500 g White mirepoix (refer to Table 8.1)
1 litre White wine
4 litres Water
1 *Sachet d'epices* (refer to Table 8.1)
Yields 4 litres

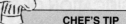

> **CHEF'S TIP**
> To make shellfish stock, use shells of lobsters, prawns, or crabs.

Method

1. Clean and wash the fish bones to remove any dirt.
2. Cut the bones into smaller pieces and sauté with mirepoix.
3. Add wine and cook for two minutes.
4. Add cold water and sachet of spices and let the stock simmer for 20 minutes.
5. Remove from the fire and strain. Discard the mirepoix and bones and cool immediately for further usage.

Usage

1. It is used for making various types of fish soups.
2. It is used for making fish stews such as bouillabaisse.[2]
3. It is used for making sauces, such as fish *velouté*, lemon butter sauce, etc. (refer to Chapter 9), to accompany fish dishes.
4. It is reduced to a glaze and used as aspic[3] on cold poached fish.

Tips for good fish stock

1. Use good quality bones that are free from any foul smell. Heads of fish result in good flavoured stock.
2. Use bones which are free from any meat, as the bones with meat will make the stock cloudy.
3. Sauté the fish bones with mirepoix in very little amount of butter as this will bring out the flavour. Excess butter will result in an oily stock.
4. Do not use oily fish to make fish stock, as the oily fish are too strong in flavour and also it will result in an oily stock.
5. Use cold water for making soup and do not boil the stock. Always simmer the stocks for maximum flavor
6. Do not cook the fish stock for more than 20 minutes, as it will make the stock bitter and too fishy in flavour.

Storage

1. Strain the stock and cool it immediately on an ice bath. If kept at the room temperature, it will give rise to microorganisms to breed as the temperature would be in danger zone (refer to Chapter 1).
2. Store in a clean container preferably of steel, as the other metals would react to cause food poisoning.
3. Label and mark with the date of production for the FIFO concept (refer to Chapter 3).
4. Store at 4 to 5°C for not more than two days. However, for longer storage, set the stock in ice cube trays and once frozen, it can be vacuum packed and stored in the freezer at −18°C for one month.

---

[2] Classical fish stew from France, flavoured with saffron and fennel.
[3] Stock reduced to a gelatin consistency to glaze cooked cold meats for enhancing the look.

◯ WHITE CHICKEN STOCK

French: Fond de volaille blanc
4 kg Chicken bones
500 g White mirepoix (refer to Table 8.1)
6 litres Water
1 *Sachet d'epices* (refer to Table 8.1)
Yields 4 litres

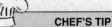

CHEF'S TIP
To make other white stocks, just replace the chicken bones with the bones of the particular animal.

Method

1. Clean and wash the chicken bones to remove any dirt and blood.
2. Cut the bones into smaller pieces, put in cold water and slowly bring to a boil. Discard the water as it will have impurities such as coagulated blood, and refill the pot with fresh cold water and bring to a boil.
3. Lower the heat and simmer the stock for three to four hours.
4. Add the mirepoix only in the last one hour remaining and the sachet 30 minutes before the finishing time.
5. Skim the stock from time to time, as the impurities will rise to the top.
6. Remove from the fire and strain and cool immediately for further usage.
7. Strain the stock and cool immediately.
8. Refill the pot with more water and boil for one hour to get a remouillage.

Usage

1. It is used for making various types of chicken soups. Chicken stock is also used as neutral stock in India as most of the people do not consume veal due to religious reasons.
2. It is used for making stews, fricassées, and ragouts.
3. It is used for making sauces such as chicken *velouté* and jus[4].
4. It is reduced to a glaze and used as aspic on cold poached chicken.

Tips for good chicken stock

1. Use good quality bones that are free from any foul smell. Neck and carcass of chicken result in a good stock.
2. Use bones which are free from any meat, as the bones with meat will make the stock cloudy.
3. Roast the chicken bones with mirepoix to get a brown stock.
4. Skim the scum that keeps rising to the top, as these are impurities and if mixed with the stock will result in a cloudy stock.
5. Blanch the bones and then make the stock.
6. Always simmer the stock for maximum flavour and add mirepoix only during the last one hour and sachet during the last 30 minutes.

Storage

Same as for fish stock

---

[4] To make chicken *jus,* brown the mirepoix and the chicken bones.

○ BROWN BEEF STOCK

French: Estouffade
4 kg Beef shin bones
500 g Mirepoix (refer to Table 8.1)
6 litres Water
1 *Sachet d'epices* (refer to Table 8.1)
50 g Tomato paste
Yields 4 litres

> **CHEF'S TIP**
> To make other brown stocks, just replace the beef bones with the bones of the particular animal.

Method

1. Clean and wash the shin bones to remove any dirt and blood. Crack the bones for better flavour.
2. Roast the bones in the oven till they are brown. Put in cold water and slowly bring to a boil. Discard the water as it will have impurities such as coagulated blood and refill the pot with fresh cold water and bring to a boil. Add tomato paste and simmer the stock for 8 to 10 hours.
3. Add the roasted mirepoix only in the last one hour remaining and the sachet 30 minutes before the finishing time.
4. Skim the stock from time to time, as the impurities will rise to the top.
5. Remove from the fire, strain, and cool immediately for further usage.
6. Strain the stock and cool immediately.
7. Refill the pot with more water and boil for one hour to get a remouillage.

Usage

1. It is used for making various types of soups such as consommé.
2. It is used for making stews, and ragouts and many braised preparations.
3. It is used for making sauces such as *bordelaise* and *espagnole* (see Chapter 10).
4. It is reduced to a glaze and used as aspic on cold poached beef.

Tips for good beef stock

1. Use good quality bones that are free from any foul smell. Shin bones give the maximum flavour and result in a good stock.
2. Use bones which are free from any meat, as the bones with meat will make the stock cloudy. Crack the bones for enhanced flavour.
3. Roast the shin bones till an even brown colour is obtained to get a brown stock.
4. Skim the scum that keeps rising to the top, as these are impurities and if mixed with the stock will result in a cloudy stock.
5. Blanch the bones and then make the stock.
6. Always simmer the stock for maximum flavour and add mirepoix only during last one hour and sachet during the last 30 minutes.

Storage

Same as for fish stock.

○ VEGETABLE STOCK

French: Fond de légumes
2 kg Vegetables
4.5 litres Water
1 *Sachet d'epices* (refer to Table 8.1)
Yields 4 litres

Method

1. Clean and wash the vegetables to remove any dirt. Cut the vegetables into smaller pieces to extract the maximum flavour.
2. Boil vegetables in cold water and simmer for an hour. Add the sachet 30 minutes before the finishing time.
3. Skim the stock from time to time, as the impurities will rise to the top.
4. Remove from the fire, strain and cool immediately for further usage.

Usage

1. It is used for making various types of vegetable soups such as minestrone from Italy.
2. It is used for making stews, ragouts, and many braised preparations.
3. Vegetable stock cannot be made into a glaze as it does not contain gelatine; but reduced vegetable stocks result in essence, which are very flavourful and can be used to enhance the flavour of soups and stews.

Tips for good vegetable stock

1. Vegetables stocks are simple to make and do not require too much of preparation time. It is suggested to make the vegetable stock for its intended use. For example, to make minestrone one would use certain vegetables. After processing the vegetables, the leftover trimmings can be utilized to make the stock for minestrone as it would impart the body and real flavour to the soup. Similarly, for a mushroom soup, utilize the trimmings of mushrooms to prepare the vegetable stock.
2. Skim the scum that keeps rising to the top, as these are impurities and if mixed with the stock will result in a cloudy stock.

Storage

1. Same as above stocks.
2. Strain the stock and cool it immediately on an ice bath. If kept at the room temperature it will give rise to microorganisms to breed as the temperature would be in danger zone (refer to Chapter 1).

○ WHITE CHICKEN STOCK FOR CHINESE DISHES

4 kg Chicken bones
200 g White cabbage
100 g Onions
50 g Sliced ginger
100 g Celery
50 g Coriander roots
6 litres Water
Yields 4 litres

CHEF'S TIP
To make chicken stock for Thai dishes, add lemon grass and kaffir lime leaf.

Method

1. Clean and wash the chicken bones to remove any dirt and blood.
2. Cut the bones into smaller pieces, put in cold water and slowly bring to a boil. Discard the water as it will have impurities such as coagulated blood and refill the pot with fresh cold water, add vegetables and bring to a boil.
3. Lower the heat and simmer the stock for two to three hours.
4. Skim the stock from time to time, as the impurities will rise to the top.
5. Remove from the fire, strain, and cool immediately for further usage.
6. Strain the stock and cool immediately.
7. Refill the pot with more water and boil for one hour to get a remouillage.

Usage

It is used for making various types of Chinese soups, stir-fries, and sauces. It is also used as neutral stock for any meat preparation.

Tips for good chicken stock

1. Use good quality bones that are free from any foul smell. Neck and carcass of chicken result in a good stock.
2. Use bones which are free from any meat, as the bones with meat will make the stock cloudy.
3. Skim the scum that keeps rising to the top, as these are impurities and if mixed with the stock will result in a cloudy stock.
4. Blanch the bones and then make the stock.
5. Always simmer the stock for maximum flavour. Add lemon grass and kaffir lime leaves for Thai stocks.

Storage

Same as for fish stock.

○ JAPANESE DASHI STOCK

4 litres Water
15 cm x 6 cm piece Kombu[5]
100 g Bonito[6] flakes
Yields 4 litres

Method

1. Heat water with kombu and bonito flakes and bring to a boil.
2. Simmer for two to three minutes.
3. Strain and reserve for further usage.

Usage

1. It is used for making various types of Japanese soups such as *udon* noodle soup.
2. Dashi stock is also used for poaching certain vegetables and meats.

Tips for good dashi stock

Do not over boil the dashi stock as it will become bitter.

---

[5] A type of seaweed, also called kelp.
[6] A type of tuna used for flavouring. Bonito flakes are dried flakes of this fish.

Storage

Japanese dishes use everything fresh. So make dashi stock for immediate use and do not store for more than two hours.

○ POACHING LIQUOR

French: Court Bouillon

500 g Onions
300 g Carrots
5 litres Water
200 ml White wine vinegar
5 *Sachet d'epices* (refer to Table 8.1)
Yields 4 litres

Method

Put all the ingredients in a clean pot and bring to a boil.

Usage

Court bouillon is used as poaching liquor for fish, especially for the oily fish, as the acid in court bouillon helps to firm up the flesh of oily fish and it prevents the fish from disintegrating.

Storage

Court bouillon should be made fresh. However, the same court bouillon can be used many times while preparing a dish.

○ INDIAN YAKHNI STOCK

4 kg Lamb shin bones
500 g White mirepoix (refer to Table 8.1)
6 litres Water
4 bay Leaves
20 nos Green cardamom
5 nos Black cardamom
5 nos of 2 inch pieces Cinnamon
2 tbsp of Peppercorn
Yields 4 litres

> **CHEF'S TIP**
> To make chicken *yakhni*, just replace the lamb bones with chicken bones.

Method

1. Clean and wash the lamb bones to remove any dirt and blood.
2. Cut the bones into smaller pieces, put in cold water and slowly bring to a boil. Discard the water as it will have impurities such as coagulated blood, and refill the pot with fresh cold water and bring to a boil.
3. Lower the heat and simmer the stock for three to four hours.
4. Add the mirepoix and the spices only in the last one hour remaining before the finishing time.
5. Skim the stock from time to time, as the impurities will rise to the top.
6. Remove from the fire, strain, and cool immediately for further usage.
7. Refill the pot with more water and boil for one hour to get a remouillage.

Usage

1. It is used for making various types of Indian soups called *shorbas*.
2. It is used in preparation of curries and biryanis.

Tips for good yakhni stock

1. Use good quality bones that are free from any foul smell. Shin bones of lamb result in a good stock.
2. Use bones which are free from any meat, as the bones with meat on will make the stock cloudy.
3. Skim the scum that keeps rising to the top, as these are impurities and if mixed with the stock will result in a cloudy stock.
4. Blanch the bones and then make the stock.
5. Always simmer the stock for maximum flavour and add mirepoix and spices during last one hour.

Storage

Same as for fish stock.

CONCLUSION

The first stocks were simple broths that were the byproducts of poached meats and fish dishes. Before the method of preparing stocks was refined, cooks often braised or roasted meats with a thick slice of ham or veal to give the sauce extra body.

Stocks are the beginning of many sauces. One can make one basic stock, divide it into two or three portions, and flavour each portion with different herbs or vegetables to make three different stocks or sauces.

Since stocks are often more intensely flavoured than the foods they accompany, one must prepare them with the best available ingredients.

Traditional recipes for stocks are divided into white and brown. White stocks are prepared by blanching meat and bones in water, whereas brown stocks are prepared by first browning the bones in the oven, which result in a deep amber colour and a richer flavour. Brown stock can be used as the base for reduction of sauces and stews. One can use white stocks in place of water in many savoury recipes, or as a base for soups and sauces.

The most common white stocks encountered in the kitchen are veal stock, chicken stock, and fish stock. However, in India chicken stock is used as neutral stock.

KEY TERMS

Biryani   Preparation made with rice, spices, and meat/vegetables.

Bouquet garni   French for bunch of herbs used to flavour Western stocks.

Broth   Flavouring liquid obtained by simmering meats and vegetables together. Also known as bouillon in French.

Cloudy   Stock which is not clear. In a clear stock one can see the bottom of the pot in which it is stored.

Consommé   A crystal clear classical French soup, in which the stock is clarified further with minced meat and eggs.

Court bouillon   Water boiled with flavourings and vinegar. It is used as poaching liquor for fish.

Dashi   Japanese stock.

Espagnole   French for brown sauce.

Estouffade   French for brown stock.

Fond   French for stocks, literally translated to foundation, as this forms the foundation of Western cooking.

Fricassée   A white coloured meat stew thickened with cream and egg.

Fumet   French for fish stock.

Kaffir lime   A type of lime used in Thai cooking.

Lemon grass   A herb from Thailand which has a peculiar flavour of citrus lemon.

Minestrone   Classical soup from Italy.

Paella   A rice preparation from Spain.

Ragout   Brown stew of meats and vegetables.

Remouillage Rewetted stock. A quick stock made by adding more water to the strained bones and vegetable to obtain a slightly less flavoured stock.

Sachet d'epices French for a bag of flavouring spices used to flavour the stocks.

Shin bones The bones from the leg of an animal.

Shorba Indian for soup.

Skimming Removing the scum floating on top of stock.

Stocks Flavourful liquid obtained by simmering bones and vegetables for a prolonged time.

Udon noodles Type of Japanese noodles made with buckwheat flour.

Velouté A smooth sauce of a satin finish made by cooking stock with butter, flour, and cream.

Vollaile French for poultry.

Yakhni Stock from India, usually made from lamb and Indian spices.

CONCEPT REVIEW QUESTIONS

1. What do you understand by the term 'stock'?
2. Why is it called 'fond' in French?
3. List few benefits of using stocks.
4. How are the stocks classified?
5. What is the difference between white and brown stocks?
6. List few dos and don'ts of a good stock.
7. What is the difference between a stock and a broth?
8. List the cooking times for at least five stocks.
9. Why should bones be blanched before proceeding to make stocks?
10. Why should one start to make the stocks with cold water?
11. Why should the mirepoix and sachet d' *epices* be added in the last only?
12. Why is skimming necessary?
13. Why would a stock turn cloudy?
14. What do you understand by the word 'neutral stock'?
15. What is a fumet and how do we make it?
16. List few uses of chicken stock.
17. List the storage criteria of common stocks.
18. What do you mean by 'remouillage' and what is the importance?
19. What is the difference between glaze and essence?
20. How is white mirepoix different from the regular mirepoix?
21. Why should the stocks be seasoned moderately?
22. List the ingredients of sachet d' epices.
23. What is a bouquet garni?
24. What is a court bouillon and where is it used?
25. List the salient features of a good fish stock and precautions to be kept in mind while preparing it.
26. List the storage criteria for various stocks.
27. Describe a *yakhni* stock.
28. How is Chinese stock different from dashi stock?
29. How would you prepare estouffade?
30. How are vegetable stocks peculiar in nature?

PROJECT WORK

1. In a group of three to four, prepare various types of stocks, using various commodities such as chicken, duck, lamb, beef, etc. Palate taste the stocks and understand the difference in flavours and colours. Record your finding and share the information amongst the group.
2. In a group of three to four, research about few dishes from at least 10 countries which use stocks. Try those recipes following the methods listed in the chapter and prepare stocks. Taste it amongst the group and make a record of your findings in the following chart.

	Taste	Flavour	Colour	Aroma	Body	Usage
Quail						
Duck						
Wild boar						

CHAPTER 9

SOUPS

INTRODUCTION

In the previous chapter we read about stocks and their versatile uses. One of the most common uses of stocks is to prepare soups. Soups are the second course on the French classical menu and are referred to as 'potage'. We read in the previous chapter that stocks are flavoured liquids obtained by simmering bones and vegetables for a long time. Soups, on the other hand, are dishes that can be served as an appropriate meal. Soups will have textures, colours, and flavours depending upon the type of soup made, while stocks would just be white or brown and usually always a liquid with no texture.

A soup is a flavourful and nutritious liquid food served at the beginning of a meal or a snack. Traditionally in France, *soupé* was a slice of bread on which the contents of a cooking pot (potage) was poured. Soup was designated as unstrained vegetable meat or fish soups garnished with bread, pasta, or rice.

However, it is the good stock which gives the body or strength to the soup. After an appetizer, soup is the first meal in the real sense. It is important to take utmost care in its preparation, as it will create the first impression on the mind of the guest. Soups have many forms—some soups are thin and served as broths, while some are served as clear soups such as consommé. Certain soups are thick and creamy and they could be just pureed or thickened with flour and milk, for example, cream soups. Certain shellfish soups are thickened with rice and pureed such as bisques and so on. It is very important to focus on the texture of a particular soup, as textures are peculiar to many soups. A clear soup will be served crystal clear and the thick creamy soups should have a smooth velvety finish. Characteristic of one soup might differ from another, but few most important points are to be kept in mind while preparing soups. Each soup should reflect its own identity. The flavour of the main ingredient used should remain prominent. Consommé should be clear and not cloudy, and broth should contain even cuts of meats and vegetables, so that it is pleasing to the eyes. The presentation of a soup is also very important as we know that this would create the first impression on the guests.

CLASSIFICATION OF SOUPS

Soups are broadly classified into two types—thick soups and thin soups, which are further classified into various categories. This is done based on the texture of the soups. However, there are certain soups that are neither thin nor thick and so sometimes certain soups are also classified into a category called 'international soups'. These soups would essentially be the national soups of different countries. 'Mulligatawny' from India and 'minestrone' from Italy are two examples of national soups. Figure 9.1 shows the classification of soups.

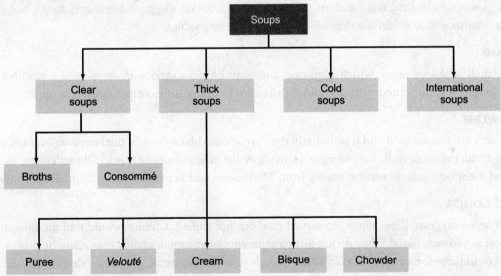

Fig. 9.1 Classification of soups

Broth

A broth is a stock-based soup, which is not thickened. It is served unpassed and garnished with chopped herbs, vegetables, or meats. For example, minestrone, scotch broth, etc. Clear soups are thin like liquid; they never coat the back of the spoon. It is important to have full-bodied thin soups; otherwise they would taste like water. In the previous chapter on stocks, we understood the difference between a stock and a broth or bouillon as commonly referred to in French. While serving broths as soups, one should add reduced stock or glaze to give body to the soup.

Consommé

A consommé is a clear soup which is clarified with egg whites. It is an old saying that if one can read the date on the dime thrown in four litres of consommé, then it is a good consommé. It is named after the garnish used in the soup.[1]

Puree

Puree soups are thick soups made by cooking and then pureeing vegetables or ingredients used in the soup. For example, lentil soup, potato soup, etc. One can roast the vegetables to give a better flavour.

Velouté

A *velouté* is a thick soup, which is thickened with a blond roux, passed and finished with a liaison[2]. They may be vegetable or chicken stock based, for example, *velouté* of chicken. A roux is made by cooking equal amounts of flour and butter over a medium heat. The degree of cooking of the flour gives its name to the roux. A blonde roux is obtained by cooking roux until it turns blonde in colour. The texture of *velouté* is smooth and velvet like.

Cream

A cream soup is a passed thick soup. It may be vegetable based or even meat based; but most commonly vegetables are used to prepare cream soups. In classical recipes it is thickened with béchamel (refer to Chapter 10) and finished with cream. Examples include cream of tomato, cream of mushroom, cream of chicken, etc. Modern trends, however, avoid usage of béchamel sauce because of health reasons and to retain the delicate flavours of the vegetables.

Bisque

It is a shellfish-based soup, which is passed and may be garnished with dices of the seafood used. Traditionally it is thickened with rice and finished with cream, for example, lobster bisque.

Chowder

Chowders are not strained and traditionally they are seafood-based soups thickened with potatoes and finished with cream or milk, for example, clam chowder, seafood chowder, etc. Chowders are from the US and the most classical version comes from Manhattan and hence the name Manhattan chowder.

Cold Soups

As the name suggests, these soups are served cold but not chilled. Chilling would dull the flavours and the soup would taste bland. They do not form a separate classification, as they may again be thin or thick, passed or unpassed. Examples of cold soups include jellied consommé, gazpacho, vichyssoise, etc.

International Soups

Even though we have read about various kinds of soups in this chapter, there are still many other soups from all across the world, the popularity of which over a period of time has given them an identity of being the national soup of the country of their origin. For example, green turtle soup from England, French onion soup from France, and mulligatawny from India.

---

[1] Cuts of vegetables such as brunoise, jardinière, etc. will give name to the consommé (refer to Chapter 6).
[2] A mix of cream and egg yolks used to thicken soups and sauces.

A few countries have more than one national soup due to their diverse tastes and demographics. Many countries like India never had a concept of soup, but during the colonial rule the British army always wanted to start their meal with a soup. Therefore, the royal cooks of the Deccan plateau concocted a soup like dish inspired from the *rasam*. These cooks altered the spice levels and pureed the lentils to give birth to a new soup, which became one of the most popular soups of India known as mulligatawny.

Let us discuss the most famous national soups of different countries from all across the world in Table 9.1.

Table 9.1 Famous soups from around the world

Name	Description
Mulligatawny	This word came from two Tamil words—mullugu and thaneer. The former means black pepper and the later refers to water. The name was probably inspired from pepper rasam or pepper water. The Indian cooks working in the colonial kitchen, prepared a pureed soup from yellow lentils and coconut and flavoured it with curry leaves. The refinement of this soup happened over a period of time and a few Anglo Indian recipes also call for green apples. The soup is traditionally served with boiled rice, boiled chicken, and lemon wedge.
Minestrone	Italy has numerous kinds of soups, as it is considered to be the gastronomical destination by many connoisseurs. There are mostly two versions of minestrone made in Italy—the one that comes from northern Italy is called ligurian minestrone and does not contain tomatoes and the one from southern Italy has tomatoes as well as courgettes. The soup is flavoured with pesto and contains vegetables such as carrots, potatoes, leeks, cabbage, and barlotti beans.
Gazpacho	Gazpacho originates from Andalusia in Spain and is also popularly known as Andalusian Gazpacho. It's a chilled soup and is prepared by blending tomatoes, cucumbers, peppers, one day old bread, and olive oil. The soup is served along with small dices of the vegetables used to give it a texture. In modern trends, chefs have started to make many variations of this soup. Gazpacho made with green tomatoes and watermelon gazpacho are a few examples of these variations.
French onion soup	One of the many popular soups from France, it is a classical popularized in 1800 in the United States of America. Traditionally, the French onion soup is made from beef broth and caramelized onions. The sliced white onions are sautéed in butter over low heat until they caramelize and then a full bodied beef broth is added to make this delectable soup. The onions when caramelized are often deglazed with brandy or Sherry. Usually, this soup is served with a large piece of toasted bread that is trimmed to the size of the bowl and is around 1 inch thick. The bread is covered with grated gruyere cheese and gratinated until the cheese melts. This piece of bread is often called 'croute' or 'diablotin'. Gruyere is a Swiss cheese and is used commonly due to its easy availability; however traditionally French cheese known as 'comte' is used for the same.

Contd

Table 9.1   (Contd)

Name	Description
Cock a leekie	This is the national soup of Scotland and is believed to have originated in France in the 15th century where it was known as chicken and onion soup, until it went to Scotland, where the onions were replaced with leeks. It is fairly a simple broth of chicken stock, leeks sautéed in butter and pieces of chicken seasoned with black pepper and herbs such as parsley and chopped chervil. The name cock a leekie was coined around 18th century, where the cock refers to the chicken and leekie is for leeks.
Polish borscht	This is the national soup of Poland and is believed to have originated in the Ukrainian region of Eurasia. Borscht is made with beetroots combined with meats on the bone that are rich in marrow. After boiling for many hours, the bones are discarded and the meat is added back to the soup along with vegetables such as carrots, cabbage, parsnips and even potatoes. Traditionally, this soup tastes sour as it is flavoured with the juice of fermented beetroots, however now one may add vinegar to give it the same taste. Polish borscht can be served in many ways such as hot or cold, like a broth, or pureed to a smooth consistency.
Green turtle soup	This soup can be considered the national soup of England as it was popularized in London in the 17th century, where turtles were imported from the coast of Atlantic ocean near Brazil and West Indies; they were popularly known as 'sea tortoise'. The difference between a tortoise and a turtle is their habitat. The tortoise lives on land whereas the turtles live in sea or ocean water. Green turtle soup is made from a variety of turtle that is called green turtle; however now any kind of turtle is used for making this soup. It is made by boiling the turtle along with vegetables such as asparagus and herbs like thyme and parsley. This soup is traditionally served as a broth where the boiled turtle meat is cut into small chunks and served in the soup.
Leberknoedelsuppe	The word *leberknoedel* means a dumpling made from the liver and the word *suppe* is soup in Germany and Austria. This is the national soup of Germany and traditionally the liver from beef or pork is used for making the dumplings. Minced liver is mixed with herbs, seasoning, bread soaked in milk and eggs to bind these dumplings, which are then poached in flavourful beef broth. Certain chefs also deep fry these dumplings to increase the shelf life of the soup as freshly poached dumplings need to be consumed as soon as the soup is prepared.
Avgolemono	As the name suggests this Greek soup is made with a mixture of lemon and eggs. Though Avgolemono can also be used as a sauce by itself, in case of it being a soup, a flavourful broth of chicken that is cooked with starchy ingredients such as potatoes, tapioca, rice or pasta is emulsified with a mixture of lemon and eggs whisked together. The preparation of this soup requires skill as it is vulnerable to curdling due to the presence of eggs.
Manhattan clam chowder	This is the national soup of United States of America. This soup is usually made with clams and vegetables such as potatoes, onions, and celery. The soup is made in two variants, one with milk and cream that is popularly made in Rhode Island and the other one has tomatoes, is more reddish in colour and is known as Manhattan clam chowder.

Contd

Table 9.1 (Contd)

Name	Description
Tom yam	This is the national soup of Thailand and is prepared with many kinds of meats and seafood. The name of the soup is suffixed with the name of the meat used, for example tom yam *kai* is made with chicken and tom yam *koong* is made with prawns. Tom yam literally means spicy water. This is a broth that is prepared by boiling meat or seafood with chicken stock and is flavoured with chilli paste, tamarind pulp, and herbs such as kafir lime, lemongrass, and galangal.
Miso soup	Popularly known as *miso shiru*, this national soup of Japan has of late shot to popularity owing to its health benefits and simple preparation. This soup is fairly simple to prepare and here the miso paste is added to the dashi stock (refer to chapter 8 on stocks). More ingredients such as seaweed like *wakame* and silken tofu can be added to the soup. Miso paste is made by fermenting the soy beans with addition of salt and a special edible fungus variety known as *koji*.
Pho	This is the national soup of Vietnam and like miso shiru, this soup has become very popular amongst tourists as a full bowl of pho can turn into a full meal. Considered to be a poor man's meal and often sold as street food in Vietnam, this soup has found its respectable place on hotel menus. It is a broth made with chicken or beef or seafood and is boiled with noodles and flavourful stock that has herbs like cilantro and scallions.
Caldo verde	This is the national soup of Portugal and the name comes from the word green broth. The green typically used in this soup is collard greens along with vegetables such as onions and garlic and it is prepared with generous addition of olive oil. This soup was traditionally vegetarian but new trends of adding meats such as pork and ham has also started.

PREPARATION OF SOUPS

○ MINESTRONE

Country: *Italy*
Yields 10 portions

50 ml Olive oil
100 g Carrot
100 g Zucchini
100 g Potatoes
100 g Onions
50 g Leeks
50 g Broccoli
50 g Cauliflower

Classification: Broth
Do not use extra virgin[3]
Cut into 1 cm dices
Cut into 1 cm dices
Cut into 1 cm dices
Cut into 1 cm dices
Cut into 1 cm dices
Cut into small florets
Cut into small florets

---

[3] Extra virgin olive oil is not used in cooking. It has a very low smoking point and hence, the oil disintegrates in high heat, making it unsuitable for consumption.

200 g Tomatoes	Cut into 1 cm dices
100 g Green peas	Skinned and split
10 g Basil	Shredded
2 litre Vegetable stock	Made from the trimmings of above vegetables
Salt, pepper	To taste

Method

CHEF'S TIP

Season stocks at last, as the stock tends to get salty during the reduction. Also season the stocks moderately as they will be used to prepare other dishes.
Use whole peppercorns instead of crushed powder as it turns bitter after reducing for long time.

1. Wash all the vegetables and process all the ingredients as listed above.
2. Heat olive oil and add chopped garlic. Sauté without colouring and add diced potatoes and carrots. Cover with a butter paper and let the vegetables stew at a medium heat for five to seven minutes.
3. Now add onions, leeks, and 50 ml of stock and stew again covered with butter paper.
4. Add zucchini and tomatoes and stew again in the similar way.
5. Add the leftover vegetable stock and add cauliflower and broccoli. Bring to a boil and simmer for five to seven minutes. Check the seasoning.
6. Add split peas and shredded basil. Serve piping hot.

Service

1. This soup is served in a soup cup garnished with basil sprig and grated parmesan cheese can be served along with it on request.
2. This is a light broth so portion size could be 200 ml of the soup with less stock and more vegetables.

Chef's Notes

1. The choice of vegetables listed above can vary from region to region, so use the stock of the vegetables that you intend to use for this soup.
2. The soup can also have some pasta in it, as this is a wholesome broth usually eaten as a meal with bread. But if you use potatoes then you can avoid other starch. Similarly, instead of peas, barlotti beans or cannelloni beans are also very traditional.
3. Add the vegetables in a sequence. The ones that take longer to cook shall be put in first and so on.

○ CONSOMMÉ

Country: France

Yields 10 portions	Classification: Clear soups
2.5 litre Chicken stock	Cold but not chilled
300 g Chicken mince	Minced without fat
225 g Mirepoix	Cut into brunoise
300 g Egg white	
1 pc *Sachet d'epices*	Refer to Table 8.1
150 g Tomatoes	Chopped
½ pc Onion	Browned on a pan
Salt, pepper	To taste

Method

1. Mix chicken mince with mirepoix, egg whites, and tomatoes.
2. Add the above mix to cold chicken stock and put into a clean heavy bottom stock pot.

3. Put on a medium heat and keep stirring the liquid until the chicken mince starts to form a raft.
4. When the raft becomes a little firm, make a cross in the centre with a knife so that you can see the liquid simmering below the raft.
5. After around 30 minutes place the browned onion inverted in the place where you can see the liquid. The browned onion gives the desired amber colour to the soup.
6. When the raft has hardened and the soup is clear, take it off the fire and carefully strain through a muslin cloth or a cheese cloth to get a clear liquid called consommé.

Service

1. Consommés are served in soup cups garnished with decorative vegetables, custards, or even meats. The name of the consommé comes from the garnish used, for example, jardinière, brunoise and so on.
2. Consommé is a hearty clear soup, so 180 ml per portion should be served.

Chef's Notes

1. To make a double consommé, reduce the consommé to half on a medium flame.
2. Many variations of consommés are there in the world. Italians call it *stracciatella*, when they mix beaten egg, cream, and parmesan cheese into it.
3. Jellied consommés are double consommé mixed with little amount of gelatine (7 to 8 g/litre) and chilled. Since this will be a heavy and rich soup only 140 ml should be served as a portion.
4. The acid from the tomatoes helps the proteins to coagulate and hence, the raft is formed removing all the impurities from the stock and leaving a crystal clear soup.
5. Modern variations of soups, such as vegetable consommé, tomato consommé, etc., are made in a slightly different manner. However, the easiest method is to puree the vegetable and hang it in a cheese cloth to drain out all the juices. Collect these juices and reduce in a pan to get clear consommé. Not all vegetables can be used to make consommés as they do not taste good once the juices are boiled; but the tomato consommé tastes very good when boiled with basil.

○ CREAM OF MUSHROOM

Country: France
Yields 10 portions   Classification: Clear soups
50 g Butter
100 g Onion   Chopped
30 g Garlic   Chopped
500 g Button mushrooms   Washed, cleaned, and sliced
60 ml White wine   Cooking wine
1½ litre Vegetable stock   Preferably made with mush-
                            rooms
1 pc Bouquet garni   Refer to Table 8.1

1 pc Sachet *d'epices*   Refer to Table 8.1
200 ml Cream   Cooking cream
150 ml Béchamel sauce   Refer to Chapter 10
Salt, pepper   To taste

Method

1. Heat butter and sauté onion and garlic till translucent.
2. Add sliced mushrooms and deglaze with white wine.

3. Add sliced mushrooms and sauté well. Add vegetable stock and let it simmer with bouquet garni and sachet of spices, until mushrooms are soft.
4. Puree the soup and strain through a chinois. Put it to boil again and add béchamel sauce and cook till the soup gets a thickened texture.
5. Remove from heat and add cream stirring to get a creamy and smooth texture.
6. Serve piping hot garnished with a sprig of thyme.

Service

1. Creamy soups are served with fried croutons or anything crunchy, to contrast the texture of the creamy soup.
2. Creamy soups are heavy, so serve only 150 ml per portion.

Chef's Notes

1. Roasting the soup provides much better flavour to the soup. Roast the mushrooms with onion, garlic, and olive oil and then proceed normally.
2. Other vegetable cream soups can be made in a similar way, for example, pumpkin, celeriac, asparagus, potato, onions, carrots, spinach, assorted mushrooms, etc.

○ PUREE OF PUMPKIN

Country: England
Yields 10 portions
100 g Butter
250 g Mirepoix
750 g Pumpkin
1 no Sachet *d'epices*
1200 ml Vegetable stock
Thyme
Salt, pepper

Classification: Puree
Or use half oil + butter
Refer to Table 8.1
Cut into slices
Refer to Table 8.1

Sprigs for garnish
To taste

Method

1. Heat butter and sauté mirepoix.
2. Add sliced pumpkin and sauté well. Add vegetable stock and let it simmer with sachet of spices, until pumpkin is soft.
3. Puree the soup and strain through a chinois. Put it to boil again and cook until the desired creamy consistency is obtained.
4. Serve piping hot, garnished with a sprig of thyme.

Service

1. Puree soups are served with fried croutons or anything crunchy to contrast the texture of the creamy soup.
2. Puree soups are heavy, so serve only 150 ml per portion.

Chef's Notes

1. Roast the vegetable to get a better flavour.
2. Pass the soup through a cheese cloth to get the satin-smooth finish.
3. All vegetables can be used to make puree soups.
4. Deglazing with white wine will give additional flavour.
5. While making puree soups with white vegetables, such as celeriac and cauliflower, use only white mire-poix (refer to Table 8.1).

○ LOBSTER BISQUE

	Country: France	
	Yields 10 portions	Classification: Bisques
	150 g Butter	
	1 kg Lobster shells	Cracked
	2 nos Lobsters	For garnish
	300 g Mirepoix	Refer to Table 8.1
	100 ml White wine	Dry white wine
	150 ml Brandy	Classically cognac
	50 g Tomato paste	
	2 litres Fish stock	Preferably with shellfish
1 no Bouquet garni	Refer to Table 8.1	
1 no Sachet *d'epices*	Refer to Table 8.1	
1 no Sachet *d'epices*	Refer to Table 8.1	
50 g White rice	Raw rice	
200 ml Cream	Cooking cream	
Salt, pepper	To taste	

Method

1. Clean and wash the lobster shells and sprinkle with some butter and roast in the oven.
2. Heat remaining butter and sauté the mirepoix along with roasted lobster shells.
3. Deglaze with white wine and crush the shells in the pan with a heavy ladle or a mallet.
4. Add tomato paste and cook. Add brandy and flambé the shell fish.
5. Add stock, bouquet garni, sachet and cover with a lid and let the soup simmer for at least 20 minutes on a medium flame.
6. When done, strain the soup through a conical strainer and put back on fire.
7. Add raw rice and bring to a boil again.
8. Liquidize the soup and pass it through a cheese cloth to get the smooth satin finished soup.
9. Return to boil, check the seasoning and add cream. Do not boil after adding cream.
10. Serve piping hot garnished with pan-fried lobster meat.

Service

1. Modern style of serving this soup would be to place the pan-fried lobster in the centre of the soup plate and the steward should pour the soup from a tea pot or small jug in front of the guest.
2. Bisque is a heavy soup and hence, serve approximately 140 ml as the portion size.

Chef's Notes

1. Many variations of bisques can be made, for example prawns, shrimps, crabs, etc.
2. Use the desired shells to create the bisque so that maximum flavour is obtained.
3. Traditionally rice is used in thickening of this soup; however, modern versions just thicken it with cream.

○ GAZPACHO

Country: Spain

Yields 10 portions	Classification: Cold soups
100 ml Olive oil	Extra virgin
250 g Cucumber	Diced and not peeled
250 g Red bell pepper	Diced
250 g Tomatoes	Diced
50 g Tomato paste	Proprietary
50 ml White wine vinegar	
50 g Garlic	Cloves
100 g Onions	Diced
1 litre Tomato juice	Strained

10 ml Tabasco sauce	Proprietary
200 g White bread	Preferably one day old
Salt, pepper	To taste

Method

1. Marinate all the vegetables, except olive oil and leave them to stand in a cold place for three to four hours or preferably overnight.
2. Blend in a mixer and serve chilled garnished with drizzle of olive oil.

Service

1. Serve chilled with assortment of neatly diced vegetables used in the gazpacho.
2. Croutons are served with chilled gazpacho.

Chef's Notes

1. Modern variations of gazpacho include watermelon and vodka gazpacho, and chilled orange gazpacho. However, the most classical remains the tomato gazpacho.
2. Use ice to blend the soup to get extra chilled effect.
3. If the soup is too thin, it can be thickened by adding more bread. If it is too thick, thin down with tomato juice.

○ CHICKEN VELOUTÉ

Country: France

Yields 10 portions	Classification: *Velouté*
120 g Butter	
250 g Mirepoix	White mirepoix
80 g Flour	Refined
2 litre Chicken stock	Refer to Chapter 8
1 no Sachet *d'epices*	Refer to Table 8.1
120 ml Cream	Whisk cream and egg yolk
2 nos Egg yolks	It is called liaison
Salt, papper	To taste

Method

1. Melt the butter on a medium fire and add white mirepoix.
2. Season lightly and cook without colouring for few minutes.
3. Add flour and cook to form a blond roux.

4. Add chicken stock and keep stirring to avoid lumps.
5. Add sachet and cook on fire for around 15 minutes.
6. Pass the soup through a cheese cloth to get a smooth velvety texture.
7. Put back on fire and bring to boil.
8. Add liaison of cream and egg and do not boil.
9. Serve piping hot.

Service

1. *Velouté* soups are served with fried croutons or anything crunchy to contrast the texture of the creamy soup.
2. These soups are heavy, so serve only 150 ml per portion.

Chef's Notes

1. *Velouté* in French means 'velvet' so the consistency of the soup is very important.
2. Prepare the following *velouté* in the same way—fish *velouté*, chicken *velouté*.
3. Never boil the soup after adding liaison as it would curdle. If it curdles, strain the soup again through a cheese cloth and thicken again with a fresh liaison.
4. Modern style of this service is again done in the same way as that of bisque.

○ CLAM CHOWDER

Country: USA	
Yields 10 portions	Classification: Chowder
50 g Butter	
200 g Bacon	Use bacon with fat
400 g White mirepoix	Refer to Table 8.1
500 g Clams	Cleaned well
2 litre Fish stock	
1 no Sachet *d'epices*	Refer to Table 8.1
200 g Tomato concasse	Refer to Chapter 10
10 g Parsley	Chopped for garnish
Salt, pepper	To taste

Method

1. Heat butter and sweat the bacon and white mirepoix without colouring.
2. Add clams and sauté. Some chefs add white wine to deglaze for a better taste, but it is optional.
3. Add fish stock and simmer the soup with sachet.
4. Add concasse of tomatoes and season the soup.
5. Add fresh breadcrumbs to thicken the soup. Sometimes *beurre mani*é is also added to the simmering soup. *Beurre mani*é is equal amounts of butter and raw flour kneaded together. It is used as a thickening agent.
6. Add cream to the soup and serve hot garnished with chopped parsley.

Service

Serve 180 to 200 ml of portion.

Chef's Notes

1. Chowders are originally from the USA. The most famous one is from Manhattan and it is also called Manhattan chowder.
2. Some chefs believe that cabbage is the most important ingredient in chowder; while others are of an opinion that chowder must always be thickened with potato and milk. The above recipe is how chowders are presently made in most of the hotels.
3. Chowders are always made with shellfish.

MAKING OF GOOD SOUP

The following points should be kept in mind while preparing soup.

1. Always use good quality stock to make soups.
2. Clear soup should be very clear and cream soups should be smooth and velvety. This can be obtained by liquidizing the soup in the blender and passing it through a cheese cloth.
3. The flavour of the main ingredient should stand out in the soup. Extra flavouring, such as mirepoix, should not overpower the flavour of the main ingredient used.
4. For white soups use only white mirepoix and vice versa.
5. The thick soup should never coat the back of the spoon, it should be thick and creamy yet not so thick.
6. Keep skimming the soup and use as less fat and butter as possible, because a soup is considered to be a healthy option and one does not want to see a layer of fat floating on the top. However, in certain broths few drops of oil are added to enrich them. This is known as the eye of the soup.
7. Season the soups moderately, so that the flavour of the main ingredient is enhanced and then guests can always use the cruet set.
8. Use whole peppercorns rather than using crushed ones, as the oil in the crushed peppercorns would make the soup taste bitter. Whole peppercorns can always be strained off.
9. Check the seasoning in the cold soups before serving, as the seasoning goes down once the soup is chilled. The coldness numbs the taste buds and hence, we cannot get the right seasoning.
10. Use appropriate garnishes for the soups. Use swirl of cream on a cream-based soup and not on a broth or consommé.
11. Broth-based soups are robust and full bodied, so use potatoes, dried pastas, and chunks of meat to create them. While preparing broths always add vegetables in order of their cooking. Overcooking of vegetables can make the soup bitter. Pastas and rice may cloud the clear soups, so cook them separately and add to the soup.
12. Use strong stock for making consommé. Modern trends are to serve double consommé. This means that after clarifying the consommé it is further simmered until it is reduced to half.
13. While making seafood soups, use variety of fish for a melange of flavours, with an exception to bisque which would use only one type of shellfish.
14. Do not use very oily fish, such as kingfish or mackerel, to make seafood soups as the oil from the fish lends a very fishy flavour to the soup.
15. For puree soups, use starchy vegetables as they puree well and give a good texture and flavour to the soup.
16. It is not necessary that the pureed soups have to be smooth and velvety. One can roughly puree the soups to have a grainy texture such as corn soups.
17. Add creams and liaisons only in the end and do not boil the soup thereafter, otherwise curdling might take place.
18. Always cook the flour before adding to a soup as a thickening agent, otherwise it will taste raw.
19. While storing the soups in the refrigerator, keep them unfinished. At the time of guest order, heat and add the seasoning and cream or butter as desired.
20. Never store soup beyond two days, as the quality will deteriorate.

MODERN TRENDS OF PRESENTING SOUPS

The service of the soup has always been given a lot of importance as it is the first hot meal after the appetizer and chefs must ensure that the guests are impressed. Even in the olden days, care was taken to present soup in a particular serveware to enhance the look and feel of the soup. A clear soup, such as consommé, is served in white crystal clear porcelain, while robust broths are served in large stone-glazed pottery bowls and so on.

Chefs should serve hot soup piping hot and cold soup cold and not chilled, as the flavours lighten down when the soup is too cold. Chefs should ensure that the dish chosen is hot or cold to pour the soup, as this will directly impact the temperature of the soup.

There are many new types of crockery available to present the soups. Deep plates with large shoulders and small space in between, are used to present soups in a fine dining restaurant.

Teacups are widely used to serve soups. Consommé is also referred to as beef tea and hence, many times served in a tea cup or a consommé cup which has two handles so that the guest can drink the soup from the cup rather than have it with a spoon. Large rectangular plates are used to serve a hot soup in a small cup accompanied with a chilled sorbet on the side.

All the soups in the Tiffins restaurant at The Oberoi, Mumbai are served in a pre-garnished soup plate which is put in front of the guest and the steward pours the soup from a fancy sauce boat or a small porcelain jug (refer to Fig. 9.2).

Thus the soup needs to be given due attention and detail, as it brings in good profit margins to the menu mix.

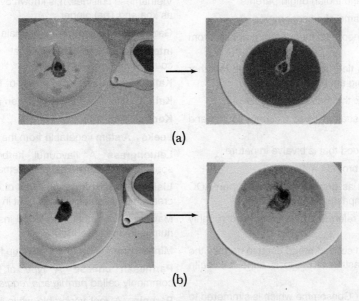

(a)

(b)

Fig. 9.2 Serving of soups
(a) Cream of tomato soup presented in a modern style
(b) Cream of watercress served in a modern style

CONCLUSION

Soup or a *potage* as called in French is one of the most important courses on the French Menu. It acts as an appetiser course rather than a meal and hence the portion sizes are decided accordingly. Heavy soups are served in smaller quantities in comparison to lighter soups such as broths and consommés. Soups are found on every menu today, be it a fine dining establishment or fast food joints, soups are always featured, as most of the people opt for it considering that it is a healthy option, and no doubt it is. A soup is the extraction of flavours blended into a full bodied stock so it would be right to say that stocks form the backbone of soups. The character of a soup depends upon the quality of stock used; the temperature always plays a very important part in service of the soups. As this would be one of the first courses, we need to ensure that the soup makes a positive impact on the guest. Hot soups are more than welcome in all seasons, particularly in winters and on the other hand, cold soups are hot favourites in hot climate. Always warm the soup cups or plates before pouring the soups in them to serve them piping hot.

In this chapter we also discussed about national soups of many popular countries such as France, Germany, and England. We have also discussed a few popular oriental soups such as miso shiru, tom yam, and pho that of late have been popular with travellers around the world. In the last chapter we read about stocks and got to know how to use them in a soup. In the next chapter we will discuss about sauces and you will realize that in many cases a sauce is just like a soup but is served accompanying a dish in much less quantity than a soup.

KEY TERMS

Anglo Indian   A community during the British Raj in India born of British and Indian origin parents.

Beef tea   Another name of a consommé.

Bisques   Velvety smooth textured soup prepared from shellfish.

Beurre manié   Raw flour and butter kneaded together and used as thickening agent.

Chervil   French herb that has a delicate flavour.

Chinois   A conical strainer used to strain liquids and stocks.

Clams   Kind of seafood that is bivalve in nature.

Chowder   Shellfish broth from the USA.

Comte   Also known as gruyere of France, is an AOC cheese made from unpasteurized cow's milk.

Concasse   Usually referred to blanched and roughly cut tomatoes.

Crouton   Deep-fried cube of bread used to garnish the soups. Nowadays toasting of bread is done for health reasons.

Double consommé   Consommé which is simmered to make it concentrated.

Diablotin   A large piece of bread topped with grated cheese and gratinated until it melts.

Full bodied   A term used for describing stocks which have been cooked for a long period of time and thereby have intense flavours.

Galangal   A variety of ginger used in Thai and Vietnamese cuisines. It is known by various names such as kha and Thai ginger.

Gazpacho   National soup of Spain which is served cold.

International soups   National soups of different countries.

Kafir lime   Leaves of a variety of Thai lime plant.

Kai   Thai for chicken, also written as gai.

Koong   Thai for prawns.

Leeks   A stem vegetable from the onion family.

Lemongrass   A flavourful herb/vegetable used in soups and stocks in Thai and Vietnamese dishes.

Liaison   A mixture of one part of egg to three parts of cream, used as thickening agent in soups and sauces.

Marrow   The substance found inside the bone that is nutritious and flavourful.

Minestrone   A broth-based soup from Italy.

Parmesan cheese   A type of cheese from Italy, commonly called *parmigiano reggiano*.

Parsnip   A root vegetable white in colour and rich in starch content.

Potage   French for soups.

Rasam   Thin and clear lentil water tempered with south Indian spices and souring agents such as tamarind, lemon, or tomatoes.

Roux Equal amount of flour and butter cooked to various degrees of colour.

Scotch broth Broth of chicken and leeks from Scotland.

Silken tofu A soft and smooth bean curd that is made from soy milk.

Sorbet Frozen fruit juice or liquid which is usually tangy and is served after the main course to refresh the taste in the mouth.

Stew To cook in its own juices.

Stracciatella Soup from Italy, where a consommé is garnished with a mixture of egg cheese and cream.

Tabasco A proprietary sauce made from chillies.

Vichyssoise Pureed potato and leek soup, often served chilled.

Wakame A kind of seaweed used in Japanese cooking.

CONCEPT REVIEW QUESTIONS

1. How would you define a soup?
2. What is the difference between soups and stocks?
3. How would you classify soups?
4. What is the difference between cream soup and puree soup?
5. List the difference between bisque and chowder.
6. What are national soups?
7. What is the national soup of India?
8. Which is the most famous chowder?
9. Define the process of making a consommé.
10. What are the salient features of a cream soup?
11. At what temperature should cold soups be served?
12. What are unpassed soups?
13. Describe how to prepare a gazpacho.
14. What is the national soup of England?
15. Why are vegetables roasted to prepare soups?
16. How are consommés served?
17. What is a double consommé and how would you make it?
18. What are jellied consommés?
19. How would one clean mushroom?
20. List at least 10 features of a good soup.
21. What care should be taken while reheating the soups?
22. What is a liaison and when is it added?
23. Why do we serve croutons with cream-based soups?
24. What is the thickening agent used in bisque?
25. What is used to thicken the chowders?
26. How would one decide on the quantity of soup to be served?
27. How is gazpacho served?
28. What care should be taken while preparing a *velouté* soup?
29. What are the modern trends followed in service of soups?
30. What kind of crockery and cutlery is used to present soups?
31. How did mulligatawny become the national soup of India?
32. Describe the national soups of any five countries.
33. Describe two kinds of minestrone.
34. Name two soups that can be served cold.
35. Describe how a French onion soup is served traditionally.
36. What do the words cock a leekie mean?
37. How does the Polish borscht get its unique sour taste?
38. What does the word Avgolemono mean?
39. Differentiate between clam chowder and Manhattan clam chowder.
40. Name three oriental flavourings.

PROJECT WORK

1. In a group of three to four, research the various types of platters that can be used for presenting soups in a modern way. Record your observations and share with a group.
2. Divide yourself into seven to eight groups of three to four students in each group. Each group prepares a particular soup with various ingredients. For example, group one makes puree soups of at least six different ingredients ranging from vegetables, grains, and pulses and group two prepares the bisque with at least five different shellfish and so on. Compare the taste and texture of each soup and make necessary observations and share the data amongst each other.
3. Visit various hotels and make observations of the soups served and what kind of accompaniments are served along with them.

CHAPTER 10

SAUCES

Learning Objectives

After reading this chapter, you should be able to:
- identify various types of sauces used in the kitchens and classify them
- understand various kinds of thickening agents used in sauce making
- know the preparation of various mother sauces and their derivatives
- know various types of common proprietary sauces used in the modern kitchens
- know various kinds of contemporary sauces used in modern kitchens
- understand how to make a good quality sauce

INTRODUCTION

In the previous two chapters we read about stocks and soups. Sauces are derived from good flavoured stocks and they are in a way linked to soups. Some soups can be used as accompanying sauces if served in the right portion size.

Sauce is essentially a moist or a liquid component, which is served along with the dish to add contrasting and complementary flavours. Apart from adding moistness to the dish, it also adds to the texture of a particular dish and enhances the taste. Sauces also add a visual interest to the entire dish. Some sauces have been so commonly used with a particular dish, that it has now become a tradition to serve that particular meat or vegetable with that particular sauce. Roast pork which is served with apple sauce and roast turkey which is accompanied by cranberry sauce are some of the classical examples of such dishes.

A sauce is to be served with any kind of food whether it is savoury, sweet, hot, or cold. Very sweet desserts, such as meringue, are served with a chunky tropical fruit salsa to cut down on the sweetness of the meringue and if we look around we shall find many such examples within our own cuisine. A dish without a sauce would seem like an unfinished product. Even snacks, such as samosas and dim sums, are served with respective sauces to complete the experience. Sauces are used for various purposes; however, a sauce must never change the main flavour of the food, rather it should compliment the entire dish to create a fine dining appearance. In this chapter we shall discuss the various classical sauces, proprietary sauces, and also some of the contemporary sauces used in modern cooking. Awareness of healthy eating styles has reformed the way the sauces are made and served.

USES OF SAUCES

Sauces are used for various purposes. Let us discuss each one separately.

Flavour

Sauces add flavour to the dish and at times, some liquids in which foods are cooked, are processed and served as accompanying sauces, or sometimes contrasting flavoured sauces are served to bring about the character of the dish. A succulent, well-roasted chicken is served with its own roast gravy but a duck is served classically with an orange sauce. Apple pie is served with a vanilla sauce, while traditional plum pudding prepared during Christmas is served with warm brandy sauce. The judgement of serving a particular dish with a particular sauce is usually age-old, tried, tested, and perfected with combinations being handed over from generations to generations. These are called classical combinations. Modern chefs contrast their flavours with totally different ingredients. Incorporation of fruits, berries, etc. have now formed an integral part of fine dining cuisine and it would not be uncommon to see many dishes paired up with contrasting fruit-based sauces. The Italian restaurant Travertino at The Oberoi, New Delhi serves roasted quail with Californian grapes and pan-seared pink tuna with fresh morello cherries.

Moisture

The most important reason to serve a sauce is to provide moistness to the food. A meat stew is a meal in itself and contains meat, starch such as potatoes, etc. in a sauce made from the same liquid in which it was cooked. Hence, it can be eaten as it is or with a bread; but a roast leg of lamb cannot be eaten on its own as it would be dry and will definitely require an accompanying sauce to go along with it. In India people normally eat rotis and rice with gravies and curries and these are also examples of sauces.

Visual Appeal

Sauces are also used to provide a contrasting colour onto a plate, so that the overall appearance of the dish is enhanced and it looks like a work of art. Care should be taken while using sauces to offer contrasting colours. The character of the main food item should not be compromised because of the contrasting colours that a chef wants to provide to his/her dish.

Texture

This is one of the most important reasons as to why a sauce is served along with a dish. Sauces add texture to the food and enhance the overall experience. Traditionally, a crispfried texture of the fish finger is enhanced by providing a creamy tartare sauce or even a well-cooked juicy Indian kebab is served with a smooth paste of mint and coriander, to offer the contrast in texture. Crunchy poached prawns served since time immemorial with a cocktail sauce has given rise to a classical starter called 'prawn cocktail'.

Nutritional Factor

The very need of providing the sauces in the first place was to use the liquid that has leached out during the cooking process. While roasting a plump chicken, wonderful natural juices of the chicken ooze out from the flesh and get collected in the roasting pan. These juices contain all the flavours and nutrition that the chicken has to offer and if not served along with the dish, the goodness of the dish will be lost. The juices are collected in the pan and further addition of reduced chicken stock creates a sauce called *jus roti* or roast gravy. This holds true for all the roasted meats. Some of the stews are prepared in the sauces and the dishes are served whole.

THICKENING AGENTS

Before we talk about preparation of various sauces, it is important to understand the various thickening agents used.

Roux

Refined wheat flour cooked with the same quantity of clarified butter is referred to as roux. This word probably comes from the French word *rouge* which means red. The colour of the roux depends on the degree to which it is cooked and the usage of each roux is also defined.

White Roux

It emits an aroma of baking bread and is used for making white sauce and thickening for the cream soups.

Blond Roux

It emits the flavour of toasted nuts and is used to make *velouté* sauce and can also be used in certain cream soups.

CHEF'S TIP
When combining the liquid and the roux, make sure they are at different temperatures. Add hot roux to a cold liquid and vice versa to avoid lump formation in the sauce.

Brown Roux

It emits a deeply roasted aroma and is used to prepare brown sauces. The darker the roux, the stronger is the flavour. It is used for red meats such as beef and lamb. White and blond roux have the same thickening power. Roughly 500 g of roux would thicken 4 litres of liquid and on the other hand brown roux would thicken only 2 litres; because overcooking weakens the thickening power of flour.

Roux can be made in bulk and can be stored in a cool place for a long period of time.

Slurry

A mixture of cornstarch, potato flour or arrowroot, and water is referred to as slurry. It is generally used in Chinese and other Asian sauces and is added to give a shine to the sauce. One should be careful in adding the slurry to the hot liquid as it instantly thickens when added to the boiling liquid. One tbsp of cornstarch would thicken 1 cup of liquid. In case of arrowroot and potato starch, only ½ tbsp per cup of liquid would be required as these are stronger than cornstarch.

Beurre Manié

Equal amounts of butter and flour are kneaded together to form a paste, which can be added to boiling liquids to thicken them. Since *beurre* is not a cooked product, we must ensure that the liquid boils for a considerable time to get rid of the raw flavour of the flour.

Liaison

Usually one part of egg yolk and three parts of cream are whisked together and are used to thicken the sauces. The purpose of the liaison is not only to thicken but to also enrich the sauce. One has to be very careful while adding the liaison, as it should be added into a hot liquid but never boiled again, as the egg will curdle.

Blood

It is rarely used these days because of health reasons; but blood was a very common thickening agent used in the olden times. Dishes that use blood as thickening agents are called 'jugged', for example, jugged hare.

Butter

Butter is also used in many sauces to provide the thickness to the sauce. Cold butter when whisked into a hot sauce gives a shine and thickness. This is also known as 'mounting of sauce' or *monter au beurre* in French.

Vegetable or Fruit Purees

Starch from fruit and vegetable purees used in a dish provides the thickening to the dish and hence, require no other thickening agents.

COMPONENTS OF A SAUCE

Various components (see Table 10.1) form a sauce, such as liquids, seasoning agents, thickening agents, and flavouring agents.

Table 10.1   Various types of components used in a sauce

Sauce	Liquid	Thickening Agents	Seasoning Agents	Flavouring Agents
Béchamel	Milk	Roux	Salt and pepper	Onion, cloves, bayleaf, and nutmeg
Velouté	Chicken stock	Blond roux and liason	Salt and pepper	Tarragon herb, mirepoix, *sachet d'epices*
Espagnole	Brown beef stock	Brown roux	Salt and pepper	Bouquet garni, tomatoes, and red wine
Tomato	Brown stock	Brown roux	Salt and pepper	Tomato, mirepoix
Hollandaise	Clarified butter	Egg yolk	Salt, pepper, cayenne pepper	Lime juice, tarragon

MOTHER SAUCES

It was originally the French who gave nomenclature to the sauces and classified them into various categories. These people defined the structure of 'mother sauces'. These are the main basic sauces, from which all the other sauces are derived. Addition of various ingredients to these basic sauces creates an array of sauces which accompany various dishes, giving them their identity. Let us discuss these mother sauces.

The following are the six basic mother sauces.

Béchamel (White Sauce)

The sauce consists of milk and is thickened with white roux containing equal parts of flour and butter. Marquis Louis de Béchamel (1603–1703), a seventeenth century financier who held the honorary post of chief steward of King Louis XIV's (1643–1715) household, is also said to have invented béchamel sauce when trying to come up with a way of eating dried cod. There are no historical records to verify that he was a gourmet, a cook, or the inventor of béchamel sauce.

Velouté

It literally means velvet. It is a very light blond coloured sauce, made from chicken, fish, or veal stock thickened with a blond roux. In the last chapter we also saw a soup by this name. The procedure is same and that is the reason it was mentioned in the beginning of this chapter that soups and sauces are interlinked to each other in a certain way. *Velouté* is specifically designed to accompany certain dishes and their recipes indicate a specific stock.

Espagnole (Brown Sauce)

Espagnole, meaning 'Spanish' in French was the original brown sauce and still is one of the glories of the French kitchen. The name dates to the eighteenth century and it is believed that the finest ham and tomato, essential ingredients of *espagnole*, were said to have come from Spain. Thickening rich brown stock with brown roux makes the brown sauce.

Tomato Sauce

The traditional French tomato sauce is thickened with a butter roux. However, tomato sauce is more commonly associated with Italian cuisine, and in particular as an accompaniment for pasta. The traditional French recipe included pork, tomato concasse, tomato puree, vegetables, and seasonings that are thickened with roux. The other versions, however, do not contain roux and are based on tomato chunks and puree.

Hollandaise Sauce (Dutch Sauce)

A warm emulsified sauce, it is based on egg yolks and clarified butter. Hollandaise is a French word meaning 'Dutch style'. The sauce is named so as in earlier times 'a la hollandaise' indicated a dish served with melted butter—a reflection of the importance of butter in Dutch cookery. It is thickened with the emulsification of a warm sabayon of egg yolk with the melted clarified butter.

This sauce is a versatile sauce and is served as a topping on a dish and gratinated to give colour, for example, poached fish is served with gratinated hollandaise and also the famous 'egg Benedict' popularly relished in breakfast has hollandaise as the main ingredient. This sauce is served in less quantity as it is a heavy sauce and hence, served as topping rather than a dipping sauce.

Mayonnaise Sauce

The invention of the sauce or the name is possibly derived from three different sources—'mahon', 'manier' (meaning to stir), and 'moyeu' (referred to the middle of the egg). Mayonnaise is a cold emulsified sauce based on egg yolks. If it is not handled carefully, it will separate giving a curdled appearance. It constitutes oil, egg yolk, and vinegar or lemon juice. The key is to keep the ingredients at room temperature and not chilled. A cold bowl, or eggs taken straight from the refrigerator, will prevent the mayonnaise from thickening. This sauce is always served cold and thus forms a base for most of the salads and sandwich fillings. Mayonnaise is generally served as the dipping sauce with snacks.

PREPARATION OF MOTHER SAUCES

BÉCHAMEL

Country: France
Yields 10 portions

100 g Butter	Clarified
100 g Flour	Refined flour
1 litre Milk	
1 pc Pique/cloute	Onion studded with cloves and bay leaf
Nutmeg	Pinch
Salt, pepper	To taste

Method

1. Boil milk with cloute. Then strain and cool the milk.
2. Make roux with flour and butter and cook it to a sandy texture. Add cold milk and keep on stirring with a wooden spoon to ensure no lumps are formed.
3. Cook till the sauce thickens. This would roughly take 20 to 30 minutes on a medium flame.
4. Spice it up with a pinch of nutmeg and use accordingly.

Service

1. This sauce forms the base of many sauces and can be used on its own as well.
2. An essential sauce used in the thickening of the cream soups (refer to Chapter 9).

Chef's Notes

1. The sauce is smeared with butter before storing as this would prevent the formation of skin on top of the sauce.
2. In case while making the sauce, lumps are formed, blend the sauce, strain, and cook again to thicken it.
3. The cloute is studded with cloves for easy removal from the sauce. Use six to eight numbers of cloves on one piece of onion.
4. Bring milk to a boil with cloute and leave it to stand at room temperature to let the flavour infuse. You could strain the milk after a couple of hours and use it for making béchamel.

VELOUTÉ

Country: France
Yields 10 portions

100 g Butter	Clarified
100 g Flour	Refined flour
1 litre Stock	Chicken, fish, veal, etc.
Sachet *d'epices*	Refer to Table 8.1
Salt, pepper	To taste

Method

1. Heat butter, add flour, and cook to a blond roux.
2. Add cold chicken stock (if making chicken *velouté*) stirring continuously to avoid lumps.
3. Add sachet and cook for 15 minutes on a medium flame.
4. Use as required.

Service

1. This sauce forms the base of many sauces and can be used on its own as well.
2. Apart from béchamel, *velouté* can also be used in the thickening of the cream soups (refer to Chapter 9).

Chef's Notes

1. If lumps are formed while making the sauce, blend the sauce, strain, and cook again to thicken it.
2. *Velouté* can also be enriched by adding a liaison.

ESPAGNOLE

Country: France
Yields 10 portions

100 g Butter	Clarified	50 ml Red wine	
120 g Flour	Refined flour	50 g Tomato paste	Proprietary
1 litre Brown stock	Beef stock	150 g Mirepoix	Refer to Table 8.1
Sachet d'epices	Refer to Table 8.1	Salt, pepper	To taste
20 ml Oil			

Method

1. Heat oil and brown the mirepoix on a medium flame.
2. Add tomato paste and cook further. Deglaze with red wine and add beef stock.
3. Add *sachet d'epices* and simmer the sauce for 30 minutes.
4. Heat butter, add flour, and cook to a brown roux and cool it.
5. Add the roux to the hot liquid and keep stirring to avoid lumps.
6. Cook on a medium flame for 10 minutes and use as required.

Service

This sauce forms the base of many sauces and can be used on its own as well.

Chef's Notes

1. In case while making the sauce lumps are formed, blend the sauce, strain and cook again to thicken it.
2. Due to health reasons, adding of the brown roux is omitted now-a-days. Instead the stock is reduced till it thickens naturally. Such kind of sauce is also known as 'jus' and is more flavourful than the original brown sauce.

TOMATO SAUCE

Country: Italy
Yields 10 portions

70 ml Olive oil	
100 g Onions	Chopped
20 g Garlic	Chopped
1 kg Tomatoes	Skinned, deseeded, and chopped
80 g Tomato paste	Proprietary
50 ml White wine	
10 g Basil	Torn
Salt, pepper	To taste

Method

1. Heat olive oil and sauté onions and garlic.
2. Add tomato paste and cook for 10 minutes on a medium flame.
3. Add fresh chopped tomatoes and white wine and cook covered for 30 to 40 minutes on a medium flame or a hot plate.
4. Season and add torn basil in the last 10 minutes of cooking.
5. Store and use as required.

Service

1. This sauce forms the base of many sauces and can be used on its own as well.
2. Apart from usage in pastas, this sauce can be pureed and used in many dishes.

Chef's Notes

1. The above recipe of tomato sauce is very commonly used in modern kitchens. Classical tomato sauce from France is a combination of brown sauce and tomatoes and seldom used nowadays.
2. Apart from basil, oregano also teams up very well with tomato sauce.

HOLLANDAISE

Country: France
10.5 Yields 10 portions

500 g Butter	Clarified
4 no Egg yolks	Separated from whites
15 ml Vinegar	White wine vinegar
Tarragon	Few leaves
10 nos Peppercorns	Crushed
15 ml Water	
Salt, pepper	To taste

Method

1. In a pan reduce vinegar, tarragon, and crushed peppercorns to half. Take the pan off the fire. Add a tablespoon of water and strain this liquid.
2. Combine the reduced liquid with the egg yolks and whisk on a double boiler[1] to a ribbon stage.
3. Add melted clarified butter in a thin stream and continue whisking until a thick sauce is formed.
4. Season the sauce and hold it warm and use as required.

Service

1. This sauce forms the base of many sauces and can be used on its own as well.
2. To taste, hollandaise is usually served gratinated.

Chef's Notes

1. This sauce is very tricky to make as it can curdle easily if care is not taken.
2. If the sauce curdles, add a spoon full of hot water into the emulsion and whisk again. If this still does not fix the problem then whisk one egg yolk with a little warm water on a double boiler and add the curdled sauce in a thin stream to form an emulsion.

---

[1] Round bowl kept in a pot of hot water on fire. The bowl is heated with the hot water underneath.

3. Do not hold this sauce for more than an hour, as it is made of egg and can become contaminated if not stored properly. Preferably make at the time of service.

MAYONAISE

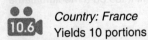

Country: France
Yields 10 portions

4 Egg yolks
400 ml Oil — Salad oil preferably
10 g Mustard paste — Dijon mustad[2]
10 ml Vinegar — White wine vinegar
Salt, pepper — To taste

Method

1. Combine egg yolks, salt, pepper, and mustard paste in a round bowl.
2. Whisk to a ribbon stage and add oil continuously in a thin stream, whisking all the time.
3. If the sauce becomes too thick, then add more vinegar to thin it down.
4. Whisk till a stabilized emulsion is formed and it stands in firm peaks.
5. Use as required.

Service

1. This sauce forms the base of many sauces and can be used on its own as well.
2. It is always served cold and is usually served as a dipping sauce for snacks.
3. Avoid serving mayonnaise in a silver dish, as the egg in this sauce can tarnish the silver dish.

Chef's Notes

1. In the event of curdling of the sauce, add vinegar to rectify. If it still does not bind, add 1 tsp of boiling water and whisk. If still curdled, start with fresh egg yolk and add the curdled sauce in a thin stream to form an emulsion.
2. Use all ingredients at room temperature to get a better mayonnaise.
3. Modern kitchens use ready-made mayonnaise which is mostly eggless. Eggless mayonnaise is made by blending oil with soya lecithin. Most of the food poisoning often happens through mayonnaise as it uses raw eggs and oil to form an emulsion. So using a proprietary mayonnaise is more advisable.

DERIVATIVES OF MOTHER SAUCES AND THEIR USES

Béchamel

It is also known as white sauce. Its derivatives are shown in Table 10.2.

Table 10.2   Derivatives of béchamel sauce

Name of Sauce	Ingredients	Uses
Mornay (cheese sauce)	Béchamel + grated gruyere cheese + liaison	Used in poultry, fish, vegetables, and pasta
Crème (cream sauce)	Béchamel + fresh cream	Used for eggs, poultry, fish, and vegetables
Soubise (onion sauce)	Béchamel cooked with chopped onions and strained + fresh cream	To accompany eggs, veal, and lamb

---

[2] Mustard paste from France

Contd

Table 10.2   (Contd)

Name of Sauce	Ingredients	Uses
Oignons (onion sauce)	Béchamel + chopped onions + fresh cream	To accompany eggs, veal, and lamb
Indienne (curry sauce)	Béchamel + cooked curry powder + onions + chopped tomatoes	To accompany egg, fish, and vegetables

Velouté

It is made from chicken stock and blond roux. Its derivatives are shown in Table 10.3.

Table 10.3   Derivatives of velouté sauce

Name of Sauce	Ingredients	Uses
Allemande (mushroom sauce)	*Velouté* + chopped mushrooms + liaison	Used in veal, vegetables, and poultry
Supreme (strained mushroom)	*Velouté* simmered with chopped mushrooms and strained + liaison	Used for poultry and game birds
Aurore (tomato sauce)	*Velouté* + cooked and strained tomato puree	To accompany eggs, fish, veal, and poultry
Cardinal (seafood sauce)	*Velouté* + lobster butter + fresh cream	To accompany white fish and lobster

Espagnole

It is a brown coloured sauce made from beef stock and tomatoes. Its derivatives are shown in Table 10.4.

Table 10.4   Derivatives of espagnole sauce

Name of Sauce	Ingredients	Uses
Demi-glaze (*jus*)	Equal parts of brown sauce + brown stock and reduced to half	Accompanies red meats and also forms base of other sauces
Bigararde (orange sauce)	Demi-glaze + reduced red wine and orange juice + red currant jelly	Used for duck
Diable (devil sauce)	Demi-glaze + reduced white wine and vinegar + cayenne pepper	To accompany roasted and grilled meats
Madeira (wine sauce)	*Espagnole* + madeira wine	To accompany offal, beef filet, veal, and ham

Hollandaise

It is an emulsion of eggs and butter. Its derivatives are shown in Table 10.5.

Table 10.5 Derivatives of hollandaise sauce

Name of Sauce	Ingredients	Uses
Maltaise (orange sauce)	Hollandaise + juice of blood oranges + blanched julienne of orange zest	To accompany vegetables
Mousseline (cream sauce)	Hollandaise + whipped double cream	Used for fish, chicken, sweetbreads, and vegetables
Moutarde (mustard sauce)	Hollandaise + dijon mustard	To accompany eggs and fish
Béarnaise	Hollandaise + egg yolk + clarified butter + white wine vinegar	To accompany steaks, fish, and eggs
Choron (tomato sauce)	Béarnaise + cooked tomato puree	To accompany eggs and fish

Mayonnaise

It is a cold sauce often used as a spread. Its derivatives are shown in Table 10.6.

Table 10.6 Derivatives of mayonnaise sauce

Name of Sauce	Ingredients	Uses
Aioli (garlic sauce)	Mayonnaise + pounded garlic cloves	Used in fish soups, eggs, vegetables
Chantilly (cream sauce)	Mayonnaise + stiffly beaten whipped cream	Used for vegetable salads
Soubise (onion sauce)	Béchamel cooked with chopped onions and strained + fresh cream	To accompany eggs, veal, and lamb
Verte (Green sauce)	Mayonnaise + puree of blanched spinach, watercress and parsley	To accompany fish, eggs, and vegetables
Cocktail	Mayonnaise + chopped shallots cooked in white wine + tomato ketchup + chopped tarragon + Brandy + cayenne pepper	To accompany seafood
Thousand island	Tartare sauce + ketchup	Served with vegetables
Tartare	Mayonnaise + chopped gherkins + chopped capers + chopped herbs	Served as a dipping sauce with fried fish

PROPRIETARY SAUCES

These sauces are industrially made and have been used in the kitchens since time immemorial. Chinese soya sauce was manufactured in the sixth century BC and is widely used till today. Many of these sauces have been family secret recipes that have been handed from generations to generations and maintain the same quality and at times, even the packaging. Let us discuss few of these common proprietary sauces used in kitchens.

Soya Sauce

It is a basic condiment from South–East Asia and Japan (it is called *shoyu* in Japan and *jiang yong* in China). Soya sauce was first used to preserve food for the winter months, though it is now used as a common flavouring agent in kitchens from east to west.

The sauce is made from soya beans, wheat, water, and salt. Other ingredients can also be used, the most common one being anchovies fish paste. Dark soya sauce is usually used for cooking and a light soy sauce for seasoning.

Soya sauce has the same nutritional value as meat extract and improves with age. In Japanese cooking it is used mainly to season grilled kebabs, tofu, cold vegetable, fish salads, fritters, and sashimi. In China soya sauce is mainly used in marinades and stewed dishes, while in Indonesia it is mainly a table condiment.

Few famous sauces from some countries are as follows:

Kikkoman Soya Sauce (Japan)   It is a regular soya sauce with rich flavour.

Ketjap Manis (Indonesia)   This is a thick, sweet soya sauce also used as a table condiment.

Toyo Mansi (Philippines)   This is a light coloured soya sauce flavoured with a native fruit which is very similar to a lemon.

Making soya sauce is usually left to manufacturers, as it involves fermenting cooked soya beans and wheat that have been salted and injected with an aspergillus mould. The taste depends on the proportion of soya beans to wheat. Good quality soy is aged from six months to two years so that the sauce matures and develops additional flavour. When the sauce is strained from the vats, fairly light soy is produced. The residue may be pressed to extract a thicker liquid, usually called dark soy.

Artificial soya sauces are made with hydrolysed soy protein, artificially coloured and flavoured with caramel and glucose.

Worcestershire Sauce

It is also known as LP sauce. There is a story that two pharmacists by the name of Lea and Perrins formulated this sauce at an order given by one retired army officer, who had savoured this recipe in India. Not impressed with the result, he returned the sauce and it was stored away in a store and was forgotten about. When discovered sometime later, the sauce had matured and was pronounced excellent. The manufacturers then perfected the recipe and developed it for commercial preparation. It soon became popular as a table condiment. Traditional Worcestershire sauce is thin, dark brown, and pungent, with visible sediment. It is soya and vinegar based but also contains an assortment of exotic ingredients, the proportions and precise details of which remain the manufacturer's secret. After the invention of Worcestershire sauce, many other sauces were derived from it by adding few ingredients. The very famous A-1 steak sauce is a beef steak sauce made by combining LP sauce with vinegar, soya, tamarind, sugar, and a variety of spices.

HP Sauce

Also called brown sauce, it is another variety of bottled sauce that is commercially prepared. It is pre-pared in the USA. London club sauce is same as HP sauce, but contains molasses in addition to the other ingredients. London club sauce is made in the UK and exported worldwide. Brown sauces are excellent accompaniments to red meats, providing a sharp contrasting taste.Lighter versions of some brown sauces, made with white wine are available for poultry, light meats, fish, and white sauces.

Barbecue Sauce

Commercial barbecue sauces originated in the USA. Commercial ones are quite similar to homemade ones, but they usually have a higher concentration of salt, sugar, and vinegar to help preserve them. Barbecue sauces may be used in marination, or simply brushed onto the meat while it is cooking. This sauce can also be used as a dipping sauce.

Ketchup

The name is derived from Malaysian *ketjap*, brine in which fish is usually pickled. In South–East Asia, ketchups made from fish and shellfish are still very common; but in the Western world, a ketchup or 'catsup' is a tomato sauce which has been the generic name for any tomato sauce made with vinegar. All ketchups include salt and spices, and often vinegar and sugar. The taste of commercial tomato ketchup is often preferred to homemade tomato ketchup. Commercial ketchup is like a smooth tomato sauce and it is highly seasoned and these days various flavours, such as chilli garlic, Oriental chilli sauce, and sweet and sour tomato ketchups, are available in markets.

Tabasco Sauce

The chilli or hot red pepper is the principal ingredient of commercially prepared hot sauces and relishes. This is an American chilli sauce made in Louisiana, but named after a state in Mexico which boasts about the spiciest chillies in the world. History states that it was made by an American citizen in 1868, when one of his friends brought him chillies from Mexico while he was away on Mexican–American campaigns. Many of the chilli sauces are based on a thick puree of chillies, while others are thinner—the result of long fermentation. Both types are used as table condiment and may also be added to sauces, soups, and stews during cooking. For tabasco, fresh ground chillies are salted and left to mature for up to three years. The liquid is then extracted, mixed with distilled vinegar, and bottled.

Chilli Sauce

It is usually a fairly thick hot sauce as distinct from the thin tabasco sauce. Chilli sauce is prepared from pulped peppers, flavoured with garlic and vinegar, and thickened with cornstarch. Chilli sauce is usually eaten as a condiment with Chinese dishes. This sauce is not to be confused with homemade chilly paste. Commercially made chilly sauces are convenient to use and also they have long to indefinite shelf life if stored in a cool, dry place.

CONTEMPORARY SAUCES

Over the last decade or so our eating patterns have changed dramatically. Our knowledge of the world cuisine has increased, with many of us travelling and eating out more often than ever before. Today's chefs are responding to demands for lighter sauces that are simpler, less rich, and more easily prepared than those of the past, whether they are for everyday use or for special occasions.

With the above objectives in mind, chefs have come up with modern-day sauces which complement, highlight, and enhance the flavour of the dish they accompany, whether it is egg, fish, poultry, vegetable, salad, or dessert. Let us discuss few of these famous sauces used in modern kitchens.

Pesto Sauce

It is a popular sauce from Italy. Pesto means a paste and is popularly made with basil herbs. Northern Italy uses basil for pesto and southern Italy uses tomatoes for pesto.

Method

In a food processor, blend 150 g fresh basil, 60 g fresh flat parsley, 45 g pine nuts, 4 cloves of garlic, 10 g sea salt, 300 ml olive oil, 5 g fresh milled black pepper, and 25 g parmesan cheese to a smooth paste.

Chimichurri

It is from Argentina. It is made in the same way as pesto. It pairs up well with steaks, and is considered to be steak sauce.

Method

Blend together 75 g flat parsley leaves, 2 crushed garlic cloves, 2 small seeded and finely chopped red chillies, 2 tbsp wine vinegar, 1 tsp chopped oregano, and 100 ml olive oil. Process everything until a smooth sauce is made.

Harissa

A violently hot red pepper sauce from North Africa, it is extensively used in Moroccan cooking and is a household staple. Harissa is used as a table condiment, and is an essential accompaniment to couscous.

Method

Grind to a paste 250 g red chilly paste, 150 g red bell pepper puree, 30 g garlic, 2 g oregano, 2 g marjoram, 1 g thyme, 100 ml olive oil, 2 g coriander powder, and 1 g roasted cumin powder and store in a cool palce.

Salsa Di Noci

In Liguria, a region of northwest Italy, this sauce is pounded with addition of walnut and is traditionally served with a pasta called *pansotti*, a pasta stuffed with local wild herbs. It also goes well with other pastas.

Method

To make 300 ml take 2 cloves of crushed garlic, 125 g of walnut blanched and skinned, 3 tbsp of fresh breadcrumbs, 4 tbsp of olive oil, 25 g parmesan cheese freshly grated, 4 tbsp of sour cream. Mix everything and grind it to rough puree.

Romesco Sauce

This spicy red sauce from Catalonia is wonderful with grilled fish and meats. To prepare a milder sauce chilli seeds can be removed.

Method

To make 300 ml take 2 small dried red chillies, 150 ml olive oil, 3 garlic cloves (crushed), 25 g whole blanced hazelnut (roughly chopped), 2 sliced white bread crust, removed and diced, tomato juice, 2 tbsp white wine vinegar, pinch of smoked paprika. Soak the chillies in boiling water, drain, deseed, and chop. Heat oil and add garlic, nuts. Fry till golden brown. Brown the bread cubes in a pan. Warm the tomato juice. Add the nut, garlic, bread cubes, wine vinegar and blend. Season with smoked paprika.

Almond Tarator

This Turkish sauce goes well with grilled and deep-fried foods.

Method

Take 3 slices of white bread and remove the crust and soak in 4 tbsp of milk and leave for 30 minutes. Puree the soaked bread and milk with 175 g of blanched almond, 3 garlic cloves, 120 ml olive oil, and juice of one lemon in a blender. Season and use as required.

Beurre Blanc

An emulsified sauce made with citric juices, white wine, clarified butter, and cream. This French sauce is served with fish and shellfish. There are many ways of making this sauce, but the most common one followed in modern hotels is given below.

Method

Sauté 20 g chopped onion in 50 g clarified butter and deglaze with 50 ml white wine. Add 500 ml fish stock and reduce to a glaze. Add 300 g double cream and cook till reduced to half. Add 10 ml lemon juice only while finishing and serve with the seafood. You can also add butter for an extra gloss and shine.

MAKING OF A GOOD SAUCE

Following are the points to be kept in mind while making a good sauce.

1. Always use good quality stock to make sauces. The sauce for chicken should be made only with chicken stock and so on.
2. The sauce should be smooth and glossy unless specified for a purpose. If the creamy sauce is lumpy, it could mean that the ratio of flour and butter is incorrect. Adding hot liquid to the roux quickly and not stirring continuously could also result in lumpy sauce.
3. Never allow a skin to form on top of a sauce as this will result in lumps, when mixed later. To avoid the formation of the skin, dot the finished sauce with butter or cover it with a greased paper.
4. Cook the sauce for a long time, this would intensify the flavours in the sauce and it would also provide a gloss to the sauce. To make the sauce glossy, pass it through a muslin cloth or a cheese cloth.
5. Always season the sauce in the end, as the reduction of the sauce will result in a salty sauce.
6. Skim the sauce whenever necessary; a greasy stock should never be used for making sauce as this will result in a greasy sauce.
7. Always use correct utensils to make sauces. Cooking white sauce in an aluminium container with a steel whisk will make the sauce turn greyish black and citric sauces made in copper vessels will also get discoloured.
8. The flavour of the main ingredient should stand out in the sauce. Extra flavouring, such as mirepoix, should not overpower the flavour of the main ingredient used.
9. For white sauces use only white mirepoix.
10. Check the seasoning in the cold sauces before serving, as the seasoning goes down once the sauce is chilled. The coldness numbs the taste buds and hence, we cannot get the right seasoning.
11. Add creams and liaisons only in the end and do not boil the sauce thereafter, otherwise curdling might take place.

12. Always cook the flour before adding to a sauce as a thickening agent, otherwise it will taste raw.

13. While storing the sauces in the refrigerator, keep them unfinished. At the time of guest order, heat and add the seasoning and cream or butter as desired.

14. Never store the sauce beyond two days, as the quality will deteriorate.

MODERN TRENDS OF MAKING SAUCES

While in olden times the sauces were creamy and thickened with flour and starch, the modern trend is to avoid using very creamy and flour-based sauces. This change has come over with globalization of the world. The world has shrunk and with people travelling all over the world, they are more aware of sauces used in other parts of the world, such as light Italian sauces, Moroccan sauces such as harissa, or teriyaki sauce from Japan and so on. It is very common to see sauces from various continents served with food of different places. Let us see the various trends in modern sauce making.

1. Stocks are naturally thickened by reducing till a glossy sauce is obtained. Thickeners, such as roux and arrowroot, are not used these days in Western cooking because of health reasons. Flour is a simple carbohydrate and is not as healthy as wholewheat flour.

2. Combination of flavours around the world such as pan-fried fish served with a teriyaki sauce or crispy fried prawns served with spicy tomato salsa is very common on most of the restaurant menus.

3. Olive oil based sauce mixed with ingredients, such as sun-dried tomatoes, capers, olives, and almonds, is a great sauce with grilled fish steaks.

4. Fruit-based sauces, such as fruit salsas, are very common these days. India Jones restaurant of Trident hotel, Nariman Point, Mumbai serves Thai style grilled prawns with raw mango chutney flavoured with curry leaves.

5. Herb-based sauces, such as rouille and pesto, are very commonly served with meats and vegetables.

6. Vegetables are widely being used as sauces with meats and vegetable dishes. Pumpkin, asparagus, and celeriac are few of the vegetables that are roasted and pureed to make cream sauces to be served with various dishes.

CONCLUSION

Sauces are flavourful liquids served with food to add moistness, nutrition, and flavour to the dish. A dish served without a sauce is like a body without the soul. Sauces are served in less quantity so as to enhance the main dish and not to overpower it. Sauces have been served with food since time immemorial, be it Asian, Western, or Mediterranean cuisine. The French were the first to classify the Western sauces into various categories and termed the basic sauces 'mother sauces'. Addition of few ingredients to these mother sauces gave other sauces which are called 'derivative sauces'. Each sauce has been named after the person who invented it, such as Robert sauce, Mornay, etc., while the other sauces have been given coined names for the purpose they have been used for. For example, 'chasseur sauce', which means 'hunter's style', is a sauce that uses forest mushrooms in a brown sauce and so on.

In the French classical brigade, it is very often seen that there is a dedicated person called 'saucier' who prepares all the sauces. In modern kitchens, we still have sauciers attached to commissary kitchens and they are responsible for preparing soups and sauces for the entire hotel.

Apart from boiled, there are also sauces which are emulsification of oil or butter with liquids, for example, mayonnaise and hollandaise.

Then there are proprietary sauces, which are ready made and available in the market off the shelf. Many of such sauces are classical and secret recipes have been handed over from generations to generations, such as LP sauce and soya sauces.

Contemporary sauces are also called 'New World sauces', as these have been recently discovered and it is a very common practice to use such sauces with different kinds of meats and vegetables. Most of the sauces prepared these days are healthy and use less cream and butter.

KEY TERMS

Clarified butter Butter left on a low flame to separate solids and then strained to get a clear butter.

Deglaze To add wine or liquid to hot pan to dissolve the sediments to form a sauce.

Dijon Mustard paste from France.

Egg Benedict English muffins topped with ham, poached egg and gratinated with hollandaise. Usually served in breakfast or brunch.

Jugged Classical dish of hare, where blood was used as thickening agent.

Jus roti See roast gravy.

Jus Also called demi-glaze. It is a reduced brown stock till it is thick and glossy.

Kikoman soya Brand of soya sauce from Japan.

Meringue Dessert made by whipping egg whites with sugar.

Monter au beurre Addition of cold butter cubes blended into hot sauce to give gloss and smoothness to a sauce.

Morello Variety of cherries from France.

Mother sauce Basic sauces from which other sauces are derived.

Pansotti A square sheet of pasta filled and folded to form a triangle.

Pique/cloute Onions studded with cloves and bay leaves used for flavouring white sauce.

Roast gravy Liquid left in the roasting pan after a meat is roasted.

Salsa Spanish for sauce.

Sashimi Japanese preparation where a raw meat is served.

Slurry Mixture of cornflour and water used for thickening.

Tarragon A kind of herb.

Vanilla sauce Sweet sauce made by boiling milk, sugar, vanilla pods, and egg yolks.

CONCEPT REVIEW QUESTIONS

1. How would you define a sauce?
2. What is the difference between soups and sauces?
3. How would you classify sauces?
4. List the basic mother sauces.
5. List five uses of sauces.
6. What are different types of thickening agents and their uses?
7. Describe a roux and its various cooking stages.
8. What do you understand by the term 'deglaze' and what is its importance?
9. What is slurry and where is it used?
10. What do you understand by the term 'jugged'?
11. In which ways is butter used for making sauces?
12. What are the various components of a sauce?
13. What is the difference between *espagnole* and demi-glaze?
14. What are proprietary sauces? List at least five of them.
15. What is a pique/cloute and where is it used?
16. What do you mean by emulsified sauces?
17. List at least two emulsified sauces and four derivatives of each.
18. Describe the process of making a hollandaise.
19. What will you do to rectify a curdled mayonnaise?
20. What is the difference between sauce *oignons* and sauce *soubise*?

21. List at least three derivatives of mayonnaise and hollandaise.
22. List two soya sauces from different countries.
23. What is a Worcestershire sauce and what is its other name?
24. What is an HP sauce and what are its uses?
25. What is the origin of barbecue sauce?
26. What is the difference between tomato sauce and tomato ketchup?
27. What is a tabasco sauce and what are its uses?
28. What do you understand by contemporary sauces? List a few.
29. What is the difference between chimichurri and harissa?
30. List five salient features of a good sauce.

PROJECT WORK

---

1. In a group of three to four, prepare mother sauces and at least three derivatives of each. Compare notes while tasting each.
2. Divide yourself into seven to eight groups of three to four students in each. Each group should prepare various types of contemporary sauces. Compare the taste and texture of each sauce and make necessary observations and share the data amongst each other.
3. Research about various types of proprietary sauces available in the market. List their prices and uses.

CHAPTER 11

SALADS

Learning Objectives

After reading this chapter, you should be able to:
- identify various types of salads and be able to classify them into various categories
- be acquainted with various classical salads and dressings
- have an understanding of emerging trends with regards to salad preparation and presentation
- prepare basic salads and dressings
- familiarize yourself with various kinds of micro greens and edible flowers used in modern salads

INTRODUCTION

In simple words, a salad is a composition of ingredients that can be raw, cooked, or cold, usually served with a dressing and eaten as an appetizer or as a main course. The word 'salad' originated from the Latin word *sal*, meaning salt. The ancient Romans believed that adding salt to green leafy vegetables would reduce bitterness, thus making it easier on the palate to consume. The earliest salads were a mix of greens (sometimes pickled) that were tossed together with some salt and served on the table. In the late 17th and 18th centuries the use of lettuce came into being, which became the most popular ingredient in salads. It was most commonly used as a base or a foundation on which other vegetables and meats were placed. Eventually to increase the moisture content in the salads the use of dressings came into existence. The first dressing was a simple olive oil and vinegar emulsion that was used to add flavour and moisture to the tossed greens; but with the emerging trends and world tastes coming together, chefs are being more creative in giving new dimensions to the humble dish of salad.

COMPOSITION OF A SALAD

Salads have come a long way since ancient Roman times, but even till date a salad will always comprise the following:
- Base
- Body
- Dressing
- Garnish

Base

The salad is built up on the base. It also helps in collecting the excessive dressing that has been used in the salad. Common bases include iceberg cups, chiffonnade of lettuce, to the more contemporary bases such as noodles, avocado halves, and pineapples. The modern trends are, however, drifting away from using the traditional lettuce.

Body

It is the most important part of the salad and is the focal point in any salad which is placed on top of the base. It is the most substantial part which can comprise various ingredients, thus giving the salad its name. The body may include a variety of ingredients such as fish in a tuna salad or apples in a waldorf salad.

Dressing

Dressing is used to moisten and flavour the salad. It also helps reduce the excessive bitterness that some salad leaves may have. It can be served by the side or tossed with the salad. The dressing should be added towards the end or as close to the service time as possible, if not, the greens in the salad will go limp. Dressings can vary from the classical French dressing which comprises vinegar and oil (three parts to one part vinegar) or mayonnaise-based dressings to more contemporary dressings such as blue cheese dressing, fruit-based dressings, and balsamic dressings.

Garnish

Garnish is of prime importance in any salad; it is the focal point of the salad and gives the salad its distinctive nature. It provides the salad with colour, contrast, and elevation. Common garnishes are chopped walnuts, crostini, fresh herb sprigs, fresh sprouted seed leaves, etc.

TYPES OF SALAD

Based on the components of salads they can further be classified as shown in Fig. 11.1.

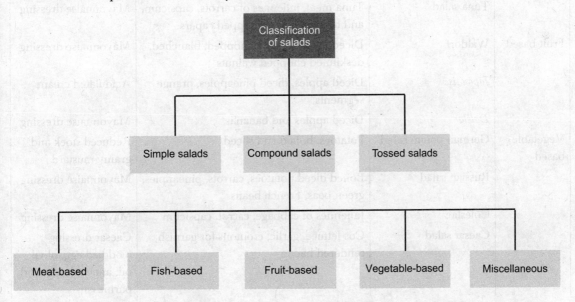

Fig. 11.1 Classification of salads

Simple Salads

These salads comprise only one primary ingredient, which makes up the body, with one or two other ingredients which make up the garnish. These salads are usually tossed with a dressing and one must use ingredients which are grown fresh during the season.

Common examples of such salads are beetroot salad that uses boiled diced beetroot with a vinaigrette dressing or a classical tomato salad where roundels of tomatoes are used, and served with a dressing and garnished with chopped parsley.

Compound Salads

Compound salads are very elaborate in nature and can vary depending on the skill and the imagination of the chef. Compound salads comprise of more than one ingredient unlike simple salads. Compound salads can be classified as follows depending on the ingredient used in the body.

- Fish based
- Fruit based
- Meat based
- Vegetable based
- Miscellaneous salads

They can be understood with the aid of Table 11.1.

Table 11.1   Types of salads

Base	Name of the Salad	Ingredients	Dressing
Fish-based	*Nicoise* salad	Blanched and strung French beans, boiled quartered eggs, quartered tomatoes, kalamata olives, anchovy, vinaigrette dressing, capers, and boiled potatoes	Vinaigrette dressing
	Tuna salad	Tuna meat, juliennes of carrots, capsicum, and tomatoes, and chopped capers	Mayonnaise dressing
Fruit-based	Waldorf	Diced apples, celery chopped, blanched, deskinned chopped walnuts	Mayonnaise dressing
	Japonaise	Diced apples, diced pineapples, orange segments	Acidulated cream
	Dalila	Diced apples and bananas	Mayonnaise dressing
Vegetable-based	German potato salad	Potatoes, boiled and sliced	Reduced stock and grainy mustard
	Russian salad	Boiled diced potatoes, carrots, pineapples, green peas, French beans	Mayonnaise dressing
	Coleslaw	Juliennes of cabbage, carrot, capsicum	Mayonnaise dressing
	Caesar salad	Cos lettuce, garlic, croutons for garnish, rendered bacon	Caesar dressing (coddled egg, olive oil, anchovies, grated parmesan)

Contd

Table 11.1 (Contd)

Base	Name of the Salad	Ingredients	Dressing
Meat-based	*Carmen Hongroise*	Chicken, rice, red pimentos and green peas Bacon juliennes, lettuce, potatoes, and paprika	Mustard dressing Vinaigrette
Miscella-neous	*Indienne salad*	Boiled rice, asparagus tips, juliennes of sweet pimentos, dices of apples	Curry cream dressing
	Fattoush	Chiffonnades of iceberg lettuce, tomatoes, green peppers	Vinaigrette dressing

Tossed Salads

This is a type of salad where varieties of cooked vegetables, mushrooms, truffles, meats, fish, or shell-fish are tossed together with a dressing. Many of the compound salads are also tossed salads, but they are classical and hence, form a different category. All the other creative salads, which are tossed together, will fall in the tossed category.

VARIOUS TYPES OF LETTUCE USED IN SALADS

Lettuce is one of the ingredients that make up the base of a salad. Earlier it used to grow wild but nowadays due to its extensive use for culinary purposes, it is cultivated on a commercial basis. It were the ancient Greeks and Romans who first started consuming salad leaves. These days, it has become more of a fashion to munch on these leaves in a well-dressed salad.

Due to the increase in demand for these leaves, they are available in all supermarkets and even with local vegetable vendors. They are a rich source of vitamins and minerals. It is very essential that the lettuce leaves be thoroughly washed before being used; it is advisable to use a chlorine solution of 50 ppm while washing the leaves, before using them in a salad. Let us see various kinds of lettuce used in salads in Table 11.2.

Table 11.2 Varieties of lettuce used in salads

Name of Lettuce	Picture	Description	Quality Factors
Iceberg		One of the most popular lettuce varieties, it gives a very crisp bite. It is found very commonly in all supermarkets.	The leaves should be tightly packed, and there should be no blackening of leaves.
Radicchio		The leaves have a deep red to pink colour with white ribbing. This lettuce is grown in the dark, thus giving colour to the leaves. It can be eaten raw in a salad, grilled with olive oil, or stuffed.	The leaves should be tightly packed, and should feel heavy in weight, since air is trapped between the leaves, which may have insects.

Contd

Table 11.2 (Contd)

Name of Lettuce	Picture	Description	Quality Factors
Lollo rosso		This lettuce has a mild flavour with tender leaves. The edge of the leaves are ruffled. It makes a good addition to a tossed salad and adds a lot of colour to a salad.	The leaves should be washed thoroughly, since the edges have dust and dirt.
Romaine/cos		It is used in the famous Caesar salad. It has a very good flavour and a crunchy texture. Most common variety is the green Romaine; red is also available.	The leaves should be bright green in colour, with no blemishes.
Red sangria		Thick leaves which are rose red in colour. The heart is pale yellow in colour. It gives a good colour combination in the salad.	The leaves should be washed thoroughly before being used in a salad.
Bib lettuce		This is also called butter head lettuce and has a loose and delicate flavour; it is not commonly available and hence it is very expensive.	Should be stored upright, so as to not damage the leaves.
Rocket lettuce		Rocket leaves also called rucola or arugula have a very peppery and bitter flavour; they are best suited in a tossed green salad and can be eaten raw. This lettuce can also be sautéed or braised and served along with meats as a vegetable accompaniment.	The leaves should be bright green in colour and there should be no holes or any black marks on the leaves, leaves should be washed thoroughly.
Mizuna or Spider Mustard		The leaves are bright green in colour and have a serrated edge. They are tender and have a slightly spicy to mild flavour. They are a great addition to a tossed green salad.	The leaves should have no blemishes and should be bright green in colour.
Dandelions		These leaves are used in a salad to give it a sharp to bitter flavour. The younger leaves can be eaten raw in a salad, while the older leaves are usually blanched since they have a pronounced bitter flavour.	
Frisée		These are curly lettuces, with long tender leaves which are attached to a whitish stem. The tip of the leaves have a serrated edge.	The leaves should be bright green with no yellowing. They should be stored in a paper bag after being washed and drained.

Contd

Table 11.2 (Contd)

Name of Lettuce	Picture	Description	Quality Factors
Kale		Kale is often used in smoothies and is considered to be one of the healthiest lettuces used in salads.	The leaves should be crisp and without any blemishes. Leaves that are smaller in size should be selected as the older leaves can be bitter.
Tat soi		Also known as Chinese flat cabbage, it is a tender lettuce green that resembles a spoon. It is very popular in the US and is of Chinese and Japanese origin.	Leaves that are tender should be selected. The lettuce should have a tender stem that should be white in colour.
Komatsuna		Also known as Japanese mustard spinach, it is one of the most popular salad leaves used in modern menus across the world.	Leaves that are tender, small, and crisp to touch should be selected.
Mesclun mix		Mesclun mix is a mix of three or more baby green lettuces that is colourful.	The heart of the lettuce that is crunchier and of delicate flavour should be selected.
Mache		Also known as lamb's lettuce and corn salad, it is a very popular lettuce as it is small in size and has a soft texture and a nutty flavour.	Leaves that are small, smooth and soft but not wilted should be selected.
Sorrel		The name of this lettuce comes from the French word for *sour* as this lettuce has a typical sour taste and hence, is very popular in salads.	The leaves should be crisp and without any blemishes. Leaves that are smaller in size should be selected as the older leaves can be bitter.
Watercress		This is an aquatic and also a semi aquatic plant that has been consumed by humans for many centuries. Rich in vitamin C and phytochemicals, this lettuce is often used as a garnish owing to its appearance.	Leaves that are smaller and closed like a bunch should be selected. The leaves should be crisp and blemish free.

MICROGREENS

The small baby shoots of any edible plant are known as cress or microgreens. They can be used as a garnish for a salad or even for a main course. Microgreens not only add nutritional value and taste to the dish, but they also add colour and interesting texture to the whole dish. In the last 3-4 years, the use of microgreens has surged in fine dining plating all across the world. A few unique salads can be made completely out of microgreens, but one has to look at the cost aspect as such salads can be expensive due to the cost of the microgreens. Let us discuss some of the most popular cresses used in salads in the following table.

Table 11.3 Common cresses used in salads

Name Of Microgreens	Picture	Description	Quality Factors
Pea shoots		The young tender leaves and stems of pea plant have unique creeper like stems with large flat leaves. They are commonly used as sprinklers on top of salads to give a unique appearance to the salad.	Tender and bright green shoots which are free from dirt and blemishes should be selected. The leaves should be crisp to touch and should have a shine.
Alfalfa		Alfalfa is also commonly known by its other name –*Lucerne.* This is a plant that is commonly used for cattle feeding, but off late the seeds of this plant when germinated yield very thin hair like strands that add flavour and crunch to a dish.	Bright white coloured stems with green tips should be selected. The ones that have thick stems should not be selected as they can be chewy to eat.
Beet greens		The young tender leaves and stems of beetroot plant have unique purplish coloured stems and bright green leaves with red veins. These microgreens look very nice and pleasant as a garnish and have a peppery and nutty flavour.	Tender and bright green leaves that have distinctive red veins and are free from dirt and blemishes should be selected.
Mustard greens		The young tender leaves and stems of yellow or black mustard seeds are a very popular cress used in salads and other dishes. The unique bitter and peppery taste of mustard cress adds a unique taste to the whole dish.	Tender and bright green shoots which are free from dirt and blemishes should be selected. The leaves should be crisp to touch and should have a shine.
Red amaranth		The young tender leaves and stems of the red amaranth plant have a unique colour combination. A bunch of these microgreens can have varying shades from green to red, making them a versatile ingredient for a salad.	Tender and crisp leaves which are free from dirt and blemishes should be selected.

Contd

Table 11.3   (Contd)

Name Of Microgreens	Picture	Description	Quality Factors
Baby basil		The young tender leaves and stems of the sweet basil plant have a unique shape and taste with bright green and shiny leaves. They are commonly used as sprinklers on top of salads to give a unique appearance to the salad.	Tender and bright green leaves that are free from dirt and blemishes should be selected. The leaves should be crisp to touch and should have a shine.
Arugula		The young tender leaves and stems of rocket lettuce have a unique nutty and peppery taste. The unique arrow like appearance of these leaves makes them stand out in a dish and hence, they are very popular as garnish or even as main ingredient in a salad.	Tender and bright green leaves that have distinctive arrow like shape should be selected. The arugula microgreen should be crisp to touch.

EDIBLE FLOWERS

As the name suggests, these are flowers of plants that can be eaten. By far this is the most visually appealing ingredient in any chef's basket. The usage of flowers in food is not new. Arabs have been using rose and orange blossoms in their dishes since a very long time. The crisp fried flowers of zucchini and pumpkin are used in Italian, Japanese and even Indian cuisine. *Gulkand*, a sweet preserve made from rose and honey has been used like a medicine in India for over thousand years.

Flowers have distinctive perfume and flavour and chefs must carefully use them to enhance a dish. Edible flowers can be used in almost all the dishes, particularly salads and desserts. In the last one year there has been a ban on usage of flowers in some countries, owing to food poisoning cases. It is important to understand the species of a flower before putting it on the plate. It is must be ensured that the flowers being used are free from insects as it is a common problem in edible flowers. The flowers are highly perishable in nature and must be used as fresh as possible. The fresh flower must be rinsed in cool water and dried between soft paper tissues so that the tender petals do not get blemishes. Generally as a thumb rule, all edible plants like broccoli, cauliflower, mustard, lettuce plants, yield flowers before they bear a fruit that contains its seeds. Such flowers are commonly used in kitchens.

Table 11.4 shows some of the most popular and common edible flowers used in salads.

Table 11.4   Common edible flowers used in salads

Name Of Edible Flower	Picture	Description	Quality Factors
Nasturtium		Native to Peru, the entire plant can be used in cooking. The leaves that have a mild peppery taste are commonly used as salad greens, whereas the unique red to orange coloured flowers are used as a garnish.	Small flowers that have crisp leaves and tender stems should be selected.

Contd

Table 11.4 (Contd)

Name Of Edible Flower	Picture	Description	Quality Factors
Borage		A popular plant from Middle East, it is now used in many Mediterranean countries as well. The unique shape and the distinctive blue shades of this flower make it a unique garnish for many salads and desserts as it has a unique cucumber like flavour.	Flowers that are small and free from insects should be selected. The whitish stem from the base should be removed as it can be bitter.
Pansies		Commonly known as Viola, they come in various colours such as deep purple, lilac, bright yellow, and orange. An assortment of these flowers add colour to the dish and they have a unique nutty flavour as well.	Small flowers that have smooth and clear petals free from blemishes should be selected.
Calendula		Also known as the poor man's saffron or English Marigold, these flowers are popularly used in food due to their medicinal value. The calendula flowers are bright yellow in colour and have a peppery taste.	Small flowers with bright coloured petals should be selected. The leaves should be crisp and smooth.
Rose		This flower is used in many dishes all across the world especially Middle East and Iran. Fresh rose can be used as a whole when it is in the bud form, alternatively the petals of roses are used as garnish.	Tender rose petals that are free from blemishes should be used.
Daisies		The bright white coloured flowers with yellow centre, make these flowers ideal for green salads, where they stand out in a dish.	Flowers that are small should be selected as older flowers may have a bitter taste.
Lavender		A very distinctive perfumed flower, it is commonly found and used in Mediterranean cuisine. Native to North Africa, this flower has been used as a cure for many ailments. In the culinary world it is commonly used in desserts due to its pleasant smell.	Flowers that are small should be selected as they have a more intense smell.

Contd

Table 11.4 (Contd)

Name Of Edible Flower	Picture	Description	Quality Factors
Courgette flowers		The baby flowers of pumpkin, yellow and green zucchini or any squash can be used as decoration for salads. Their unique velvety texture and a nutty taste add colour, nutrition, and taste to the dish.	Small flowers that are soft and velvety to touch should be selected. The flower should be free from dirt, insects, and blemishes.
Chrysanthemum		It is an ornamental flower with a unique arrangement of petals in a curly style that give it a distinctive shape and appearance. This flower also has a peppery taste and is available in many colours.	Flowers that are small should be selected and they should be cleaned as insects might be trapped in their curly petals.

SALAD DRESSING

The dressing is used in addition to the other ingredients. The salient features of salad dressing are as follows:

- Dressing heightens the flavour of the salad making it palatable and improves the appearance and hence, it is a very important part of the salad.
- Dressing is usually in a liquid or semi-liquid form.
- The dressing is made keeping in mind the ingredients used in the salad. The dressing improves the food value of the salad.
- The dressing should complement the ingredients being used.
- The modern concept of using flavoured oils and reducing the use of fruit juice and extracts in salads is a trend that is fast catching up.

Dressings can be classified as follows:

CHEF'S TIP

Clean all the lettuce and soak the leaves in cold water for five minutes. This will make them very crisp. Drain well and store in a perforated basket, covered with wet cheese cloth until further usage.

Oil-based Dressings

These dressings are temporary emulsions consisting olive oil and vinegar, commonly called vinaigrette. The oil and vinegar emulsion is the most common of all dressings used in salads. It is important to whisk the dressing so as to prevent the emulsion from separating. It is very important to use good quality olive oil and vinegar in order to get good vinaigrette. The dressing can be made in bulk and kept for later use.

Given below is a list of oil-based dressings, which are used in varying proportions.

- American   one part oil   +   one part vinegar
- English   one part oil   +   two part vinegar
- French   three parts oil +   one part vinegar
- Italian   four parts oil +   one part vinegar

However, the oil and vinegar/water separate, with the oil rising to the surface, on standing. Such dressings therefore need to be whisked just before application. Oil-based dressings are best held/stored at near room temperature. The guidelines given below should be followed when making, holding, and applying vinaigrette products.

- For production, choose good quality ingredients, namely oil, vinegar, fresh herbs, and flavourings.
- Prepare and mix dressings well in advance allowing the flavours to fully infuse with each other.
- Hold/store prepared oil-based dressings at near room temperature for best results. Holding in the fridge will cause the emulsion to separate.
- Always re-emulsify the dressings (by agitation or whisking) just before use to ensure the dressing is thoroughly blended together. Oil-based dressings are best stored in squeeze bottles, as it is easy to blend the emulsion by vigorously shaking the same and dispensing it also becomes easy.
- Ensure the ingredients being flavoured/moistened have been properly prepared (for example, drained/dried) before the dressing is applied, otherwise the dressing will not coat the salad. Instead it would separate and form a watery layer beneath the salad, which could be very unappetizing.
- Do not use excessive amounts of dressing but allow the dressing to lightly coat the ingredients. The tossing technique is an ideal way of achieving this requirement.

Fresh Cream-based Dressings

These kinds of dressings are prepared using fresh cream, which is whipped to a coating consistency. The cream is used to moisten the salad and add to the food value. Cream-based dressings have to be stored under refrigeration and have a very short shelf life; hence, they should be made on a daily basis. Acidulated cream is commonly used in salad dressings, which consists of three parts of thin cream and one part of lemon juice or vinegar, salt, and pepper. Acidulated cream is often confused with 'sour cream'. To make sour cream, one has to warm up the cream and add yoghurt culture to it and let it stand in the warm place to set into a curd.

Mayonnaise-based Dressings

Mayonnaise is one of the most popular salad dressings used. It is highly versatile and goes well with fruit, fish, and vegetable-based salads. It should always be stored under refrigerated conditions and should be made in small quantities. Mayonnaise can also be flavoured with the addition of ingredients such as sun-dried tomato mayonnaise, basil mayonnaise, and mustard mayonnaise. Although mayonnaise is a relatively stable emulsion, conditions may arise which will cause prepared mayonnaise to break or separate into an unacceptable product of curdled appearance. To prevent this from happening do the following.

- Add a little warm water to the curdled mix and whisk again.
- If the above does not work, then start with another egg yolk and whisk in the curdled emulsion in a thin stream.

- Store and hold mayonnaise in cool conditions (warm or very cold conditions may result in separation).
- Store in suitable sealed containers to reduce skinning and tainting from other strong flavoured products.

Natural Yoghurt Dressings

People are very conscious of their health these days and hence, yoghurt-based dressings are very common on the menus today. Sometimes yoghurt is used as it is and often it is used to lighten and modify mayonnaise or cream-based dressings. It provides a much lighter product with a piquant taste that is so desirable in salads. Yoghurt used in the dressings should be of thick consistency; if it is not, then it is advisable to drain the yoghurt for some time in a cheese cloth.

EMERGING TRENDS IN SALAD MAKING

Let us discuss the emerging trends in salad making.

Buffet Layouts and Presentation

If we ever glance through the old cook books of the 1980s we would see large platters of salads, garnished with sliced oranges, pineapples, etc. and placed on large lettuce leaves to give lots of colour and variety.

Modern concepts are drifting away from bulky presentations and emphasis is laid on showcasing various kinds of fancy and sleek platters and servewares. Most of the hotels today have dedicated buffet counters, where a salad is served individually on platters with an assortment of dressings to choose from. This allows good portion control and freshness; also the salad does not look all messed up after a few guests have helped themselves from the buffet. Laying out salads on a flat ware or even a hollow ware in a height formation pattern is very modern and fashionable. Salads are even served in broad martini glasses and individual spoons. Companies around the world are designing new plates and platters to help chefs showcase their skills in a modern way.

Healthy Approach

Gone are the days of tossing salads in mayonnaise and other creamy dressings. One can hardly find Russian salads and waldorf on buffet tables and menus of five star hotels. Chefs are laying lots of emphasis on creating lighter salads and using dressings such as olive oil, soya-based creams and fruit and vegetable reductions.

Deconstructed Approach

Let us understand the concepts of a deconstructed salad. Caesar salad consists of cos/romaine lettuce tossed with Caesar dressings, parmesan cheese, bread croutons, bacon crisps, and anchovies. Modern approach would be to serve all these separately and not pre-tossed. This allows the guests to choose how they want to have their salad and also allows them to be creative.

New Approach to Classical Salads

Going with the cliché 'old is gold', many guests would still love to savour the old classical salads, because they have been tried and tested and have been always seen on the buffets during special events and festivities. Modern trends are using the same old recipes and techniques to create the same salads; but are using modern servewares and platters for presenting the same. Presenting traditional coleslaw wrapped up in sliced avocadoes gives an entire new dimension to this humble salad.

Flavour Profiles

With the shrinking of the world and awareness, melange of flavours and combinations has given rise to dishes commonly known as 'fusion cooking'. One has to be very careful in mixing of flavours as it does not take too much effort to turn this very concept into 'confusion cooking'. The mixing of ingredients in a balanced and harmonious manner can produce good results. Combining honey-glazed chicken drumsticks with teriyaki sauce from Japan is quite a hit and so is combination of lemon grass, kaffir lime, chillies, and Western style grilled chicken.

Live Salad Station

Many ingredients are kept separate in different bowls with assortment of dressings and garnishes and is mixed or tossed live in front of the guest. The guest could also choose to mix and match his/her own choices and create a combination for himself/herself. This makes the guest happy as he/she feels involved in the whole process and there is no better way to please a guest.

SALIENT FEATURES OF PREPARING GOOD SALADS

The possible permutations and combinations of ingredients used in the preparation of salads are almost unlimited and this can give wide scope to the person entrusted with their preparation. It does, however, require an understanding of the compatibility of ingredients for colour, balance, and presentation. Thus attention should be given to careful decoration and garnishing of all salads; a good colour balance is essential for mixed or composed salads using a number of different ingredients. Care should be given to the following points.

1. All raw salad ingredients should be fresh and of impeccable quality. If required to be held or stored for a time, they should be kept in a covered container in a refrigerator especially reserved for the purpose. On no account should they be stored close to raw or cooked meat/fish.

2. All leaf salad vegetables should be carefully trimmed of all discoloured or damaged leaves and roots, then washed in cold water, drained, and dried thoroughly. They should not be left to soak in water. Watercress may be held standing in little cold water.

3. Large salad leaves are best if carefully torn into manageable-sized pieces instead of cutting with a knife.

4. Use the core part of the lettuce often referred to as 'hearts of the lettuce' wherever possible, as this part is very crisp and offers a natural crunch to the salad.

5. The cutting of vegetables, either raw or cooked, should be carried out as evenly and neatly as possible. This is essential for good presentation. If cut into julienne, they should not be more than 3 cm in length. If they are too long they can be difficult to manipulate on the plate and awkward to eat.

6. Some items, such as avocado, pears, globe artichokes, and some fruits such as banana, tend to discolour quickly when cut. This can be prevented by preparing them at the last minute and using a lemon-based dressing. Alternatively, they may be sprinkled first with a little lemon juice.

7. Attention should be paid to the optimum period of time required for the marination or maceration of some types of salad. Mixing some ingredients with a dressing strongly flavoured with vinegar or

lemon juice will quickly destroy any inherent crispness. For example, a salad of pimento, fennel, or celeriac will keep its crispness for 30 minutes or so but will become limp and lifeless if held overnight.

8. Where a number of items are used in the composition of a salad, some thought should be given to the balance of flavours and the possible duplication of vegetables elsewhere in the menu.

9. As a general rule, salads comprising raw green salad leaves should be dressed and mixed at the last possible minute and where practical, in front of the customer. In some cases the customer can determine the ingredients and their proportions used in the preparation of the dressing.

10. Most salads can be suitably dressed and presented in glass, china, or wooden bowls. The use of crescent-shaped china still has its adherents as far as the formal dinner is concerned; they can be very useful as a side dish where the amount of salad is small and is served as an adjunct to a main course. Certain types of main dish salad consisting of bouquets of already dressed ingredients are best presented on larger shallow plates.

11. The use of garlic as a flavouring ingredient for salads should be treated with care because of its pungent and all-pervading aroma and taste. For those who like it and are prepared for its consequences, garlic may be added, either chopped or pressed, to the salad or to the dressing. A more subtle and less overpowering way is to rub the salad bowl with a cut clove of garlic before adding the actual salad.

CONCLUSION

Salads have always been a very important course in any kind of cuisine. English call them salads; French refer to them as 'salat' and Italians flaunt them as 'Insalata' whereas Spanish speaking people swear by 'ensalada'. Whatever the name, a salad is relished as a starter or even as a main course. It also forms a great accompaniment to roasts and grills if paired well. In the modern cuisine the importance of salads has been far greater as more and more people are looking for healthy options in their menu and hence greater emphasis is laid on the variety of salads offered on the menu. Both classical and contemporary salads are still found on hotel menus. Salads such as Caesar and Greek are still on the most wanted lists of any menu. The presentation and approach might have changed, but the salads taste the same like they tasted a decade ago. However, the inclusion of the new variety of lettuce, microgreens and the usage of edible flowers have taken the presentation of salads to another level. In this chapter we have discussed various kinds of lettuce that have recently emerged from Japan and other countries. Various kinds of baby leaves known as cress are also being used commonly in most hotels. In this chapter we have also discussed about common edible flowers that are used as a primary ingredient or as a garnish in many salads.

KEY TERMS

Acidulated cream Mixture of three parts of lightly whipped cream with one part of lemon juice.

Aquatic The plants that grow in marshy lands and need lots of water to grow.

Balsamic dressing Balsamic is an aged vinegar from Italy. Three parts of olive oil emulsified with one part of balsamic vinegar.

Blue cheese dressing Lightly whipped cream mixed with blue cheese.

Compound salads Combination of more than one ingredient with dressing.

Crostini Buttered slice of long bread flavoured with herbs and toasted until crisp.

Emulsify Homogenous mixture of two insoluble liquids created by agitation or blending.

French dressing Emulsion of three parts of vinegar to one part oil.

Gulkand An Indian preserve made with rose, honey, and nuts.

Maceration Letting the ingredients infuse with the flavourings for considerable amount of time.

Martini glass A conical-shaped glass used for serving a cocktail called martini.

Microgreens Small baby shoots of any edible plant are known as cress or microgreens.

Phytochemicals A compound present in plants that is responsible for prevention of cancer.

Simple salads Salads comprising a single ingredient and tossed with a dressing.

Smoothie A kind of blended beverage that is made by blending fruits and vegetables with yoghurts.

Sour cream Cream set into yoghurt by addition of yoghurt culture.

Tossed salads Tossing ingredients with dressings in a bowl so as to coat the ingredients with dressing.

Waldorf salad Classical salad made with diced apples, chopped celery, walnuts, and mayonnaise.

Watercress Small leafy vegetable, often used in salads and garnishes.

Yoghurt dressing Drained yoghurt blended with herbs and flavourings to be used as a dressing for salads.

CONCEPT REVIEW QUESTIONS

1. How would you define a salad?
2. What is the difference between dressing and vinaigrette?
3. How would you classify salads?
4. Differentiate between a simple salad and a compound salad.
5. Describe the composition of a salad.
6. List various types of dressings.
7. Name few fruit-based salads.
8. What are miscellaneous salads?
9. Describe a waldorf salad.
10. What is a Caesar salad and what are its components?
11. List at least five salad greens and their characteristics.
12. How should salad leaves be treated and stored?
13. Name one of the lettuces that could be cooked.
14. What is the other name of cos lettuce?
15. What is another name of rucola lettuce?
16. What is a mescullun?
17. Differentiate between French and English dressing.
18. What is an Italian dressing?
19. What are cream-based dressings and how should one store them?
20. What is the difference between acidulated cream and sour cream?
21. How would you rectify a curdled mayonnaise?
22. What are the latest trends in salad making?
23. How could one add a healthy touch to a salad?
24. List few salient features of salad making.
25. How would you impart crispness to the salad greens?
26. Define heart of a lettuce.
27. How could you prevent certain vegetables from discolouring?
28. What care should be taken while tossing a green salad?
29. List the creative ways of presenting a salad.
30. How should the garlic be used in salads?
31. What are the various names of rocket leaves?
32. Name two lettuces that can be used for making a smoothie.
33. Define a mesclun mix.
34. List four microgreens and describe them.
35. What are alfalfa sprouts and how are they used?
36. List any five edible flowers that you can use in salads.

PROJECT WORK

1. In a group of three to four, prepare various dressings and at least three derivatives of each. Compare notes while tasting each.
2. Divide yourself into seven to eight groups of three to four students in each group. Each group prepares various types of salads. Compare the taste and texture of each salad and make necessary observations and share the data amongst each other.
3. Visit hotels in your city and list their approach to salad making. List other creative ways in which you would present salads differently.
4. Collect the pictures of some modern plates and platters and think how you would present salads on them.

CHAPTER 12

INTRODUCTION TO MEATS

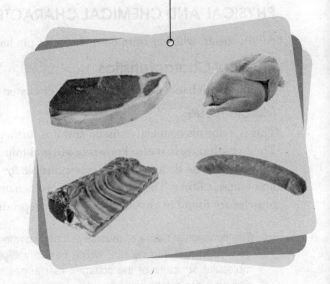

Learning Objectives

After reading this chapter, you should be able to:
- identify various types of meats and classify them
- understand the various cuts of beef, veal, lamb, pork, chicken, and game
- select cuts for classical dishes of each meat
- recognize the importance of selection and storage of each meat
- prepare and process various meats as per standards and specifications
- prepare yield cards and do yield tests for various meats and understand the importance of the same

INTRODUCTION

Many people in the context of meat refer to food which comes from lamb; but in reality, meat is the body tissue of any animal that is eaten as food. This could be meat from chicken, lamb, cow, and even frog legs. In order to classify this broad selection of meats available, the first and foremost approach is to segregate fish from meat. This is done because we would like to refer to meat as a tissue, which is a muscle tissue found in animals. Muscle tissues are involved in helping the animal move its body parts. This chapter discusses various kinds of meats and their usage in cookery. Animals have the same bio-chemistry as those of human beings and they are full of necessary proteins and amino acids that every human requires for his/her basic metabolism. Eating of meats has been prevalent since the time of cavemen and till date domestication of various animals around the world is done for the same reasons. The only disadvantage is that we are exceeding the limits of our protein intake by consuming animal products in a far greater quantity than we actually should. This brings along problems of saturated fats and cholesterol, which subsequently lead to heart disease.

PHYSICAL AND CHEMICAL CHARACTERISTICS OF MEATS

All the meats, whether from cow, sheep, or pig have both physical and chemical characteristics.

Physical Characteristics

The physical characteristics of meat are discussed below.

Muscle Tissue

This is a fibrous connective tissue and it is further divided into skeletal, cardiac, and smooth muscle. The smooth muscle is also known as visceral muscle. This would include all the arteries and the veins in the meat. The skeletal muscle is responsible for most of the muscle weight on a carcass and is made up of muscle fibres. The cardiac muscle, as its name implies, is what forms the animal's heart. Smooth muscles are found in an animal's stomach, reproductive organs, and circulatory system.

> The connective tissue of muscle is mainly composed of collagen and elastin. On heating, collagen gets converted into gelatin, which is easily digestible by the body. As the animal grows in age, molecular structure of the collagen rearranges itself and hence it becomes even more difficult to convert it into gelatin. It is because of this reason that older cuts of meat are difficult to cook till tender.

It is the skeletal muscles that are of the greatest concern to chefs. The muscles are intertwined with fibrous connective tissue which is also known as muscle sheath and fat. Coarse, long muscle fibres yield less tender meat; the thinner, smaller muscle fibres are tenderer. Muscles located along shoulders and legs, which are used for movement, have more connective tissues and are less tender. The muscles in the back are for support and move less and hence are tenderer. Some of the muscles such as tenderloin, which do not receive any exercise are the tenderest cut in an animal and always sold in the market at a very premium price.

Adipose Tissue

This is the tissue where fat is deposited and stored. As the animal ages in life, the concentration of the fat also increases in its body. Initially the fat starts to deposit around the internal organs and the pelvic area but as the animal continues to age, the fat gets deposited externally just beneath the skin. Additional fat now starts to get deposited between the muscles and also within the muscles. This interlacing of fat in the muscle is known as the marbling of meat. Marbling affects the flavour and tenderness of a meat and definitely the well-marbled meats cook to a texture where it is moist and juicy. The juiciness of meats may also be affected by the surface fat on a particular cut of meat. Surface fat protects large roasts and other cuts from drying out, particularly during roasting.

Bone or the Skeletal Tissue

Skeletal tissue consists of the ligaments, tendons, cartilages, and bones of the animal. Bone tissue can yield many nutrients and minerals, when it is used to prepare stock. Ligaments, tendons, and cartilages yield little or no calories.

Chemical Characteristics

The chemical characteristics consist of water, protein, fats, carbohydrates, minerals, and vitamins. Apart from these components, meats also have a pigment called myoglobin. It is this pigment which is responsible for the colour in the meats. Pork and lamb have lesser amount of myoglobin and hence the meat is pink in colour, whereas beef has fairly higher amounts of this pigment leading to dark purplish colour of the meat. Beef turns bright red after cutting because of the reaction of the myoglobin with oxygen.

SELECTING AND GRADING MEAT

Meat grading is one of the very important aspects of meat fabrication and distribution agencies. The meat is priced on the various grades which are governed by very strictly laid out rules by that country of origin. The grading of meat is divided into two types namely quality and yield.

Quality

Quality of the carcass indicates the quality of meat from the carcass. This depends on various factors such as the texture, firmness, and the colour of the lean meat on the carcass. The maturity of the meat and the marbling is also given a substantial consideration.

Yield Grading

It determines the quantity of usable meat that the carcass will yield. Yield grading measures the quantity of saleable meat the carcass will yield as either boneless or semi- boneless retail cuts. This grading also considers internal and external fat.

PROCESSING OF A WHOLE ANIMAL

There are various steps considered before an animal is slaughtered for human consumption. Laws of certain countries, such as the UK, the USA, Australia, and Canada, are very strict when it comes to processing and grading the animals for export, etc. Well, it does not mean that all other countries are not strict; but it is these major countries which do a lot of export of these meats to other countries as well. Let us discuss steps that are followed in the animal processing plants.

Examination

The animals that are brought into the plants for slaughtering are first examined physically for any diseases and other disabilities.

Resting of Animals

The animals are stored in a place for at least 24 hours. This allows them to rest, as there is not much movement allowed. During this resting period they are given plenty of water for the first 12 hours. The animals are then starved for another 12 hours, so that there are no feces left in the intestinal tract.

Cleaning and Sanitizing

The animals are then given a good shower and are scrupulously cleaned and sanitized for the next procedure.

Stunning

This is done to make the animal unconscious. Stunning helps avoid pain to the animal while slaughtering. If the animal is conscious when slaughtered it will quake so much that blood will spread all around. There are different ways in which an animal is stunned. In the olden days, the animal used to be hit hard on the head by heavy hammers. This painful procedure was then replaced by shooting the animal with a captive bolt from a pistol, which was hit on the front head between the eyes. This would render the animal unconscious for a considerable amount of time. This method is still followed in certain places. However, the modern methods of stunning animals with electric tongs is very popular and commonly carried out these days. An electric shock of 70 volts is passed behind the ear of the animal to stun it.

Slaughter

This is done immediately after the stunning; not because the animal will regain consciousness; but it is done because at this time the rate of the heart beat and pressure of blood is very high and this will allow maximum blood to flow out of the animal. The jugular vein[1] is cut to slaughter the animal.

Bleeding

Since blood contains protein it is susceptible to bacterial contaminations. So it is advisable to drain out as much as blood as possible. The animal is hung with the help of hooks with the head facing down so that the entire blood drains out of the animal.

Meat Ageing

Animal carcass must be aged to develop tenderness. Soon after the animal is slaughtered, rigor mortis—a stiffening of muscle tissue—occurs. This condition disappears gradually within a period of seven days in a large beef carcass and around three to four days for a lamb carcass. Cows and pigs are slaughtered very young at the age of two months and five months respectively, hence there is no need to age it. Enzymes present within the muscles work on the connective tissue in the muscles even after slaughter. It is the action of these enzymes that reduces the stiffness caused by rigor mortis. In the process of tenderizing the meat, the enzymes also develop the flavour of the meat. Ageing is very important for chefs who love to serve tender cuts of meats to their guests.

CLASSIFICATION OF MEATS

Before we get into the classification of the meats, let us first understand the nomenclature of these meat products. To a layman, words such as, mutton, lamb, heifer, veal, etc. would seem like something to do with meats, but they would not know what exactly each one would mean. These names of the meats are given to the specific meat that comes from a specific animal. The term beef for instance, refers to the meat from cattle which are over nine months old; all the other cattle which range between three to nine months are classified as calf and the meat from cattle between one month and three months is known as 'veal'. Meat is broadly classified into the following three types.

- **Bovines**   Ox, cow, buffalo, bison, etc.
- **Ovines**   Sheep, lamb, goat, deer, etc.
- **Swines**   Pigs, wild boar, etc.

All animals in these categories differ from each other in size and shape and hence, even taste different from each other. For example, the taste of buffalo from that of cow would be very different and so on.

Meat could be broadly classified into various categories as shown in Table 12.1.

Table 12.1   Categories of meats

Lamb	Beef	Pork	Poultry	Game
Meat from sheep under 12 months of age.	Meat from cow. Meat from young calf is called veal.	Meat from pigs.	Meat from chicken and some other farm raised birds such as ducks, turkey, etc.	Furred wild animals such as deer, rabbits, mountain goats, antelopes, etc. Birds such as quails, wild ducks, pheasants, wild turkey, partridges, etc.

[1] A blood vessel in the neck

CATEGORIES OF MEAT

Let us now understand each meat in detail.

Table 12.2 Classification of lamb

Ram/hogget/ewe	Wether	Kid Lamb	Spring/yearling	Mutton
Male lamb under 1 year is called ram or hogget.	Castrated male lamb is called wether.	Male or female of a sheep that is 30–60 days old.	In the UK and the USA, a lamb between 2–6 months is called spring or yearling.	Lamb above 12 months of age.
Female lamb under 1 year is called ewe.	In India we also call it *khassi*.	In France it is referred to as 'agnelet'.		

Lamb

As lamb comes from a fairly young animal, it is natural that it would not be marbled with fat and hence, it becomes very tricky for chefs to cook the lamb to utmost tenderness and juiciness. Table 12.2 shows the classification of lamb. In the coming chapters we will discuss the methods of cooking in detail and that will give us an indication of not only how a meat should be cooked, but will also help us understand what cut of an animal should be subjected to what method of cooking. Let us understand it with an example. A piece of meat which is very tender and lean, such as tenderloin, can be cooked with minimal amount of heat. Hence, methods such as panfrying, grilling, or even shallow frying, will be the apt methods; but tougher cuts of meat such as the shoulder that undergoes a lot of exercise, needs cooking for a longer period of time. We cannot cook meat at higher temperature for longer duration as it would draw out all the moisture and burn the meat, so longer duration of cooking would entail lower temperatures and cooking methods that use liquid medium to cook, such as boiling,

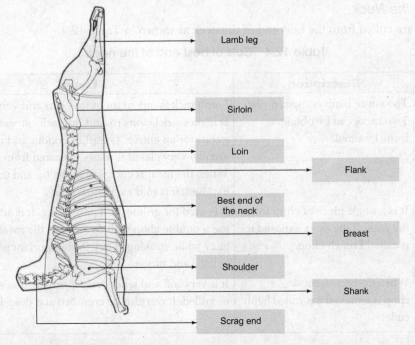

Lamb leg

Sirloin

Loin

Flank

Best end of
the neck

Breast

Shoulder

Shank

Scrag end

Fig. 12.1 Cuts of lamb

stewing, etc. Roasting is another method commonly followed, but then again, not for all the cuts. Since all this while we have been talking about cuts of meat and impacts of cooking methods on them, let us discuss the cuts of lamb in detail (see Fig. 12.1). There is no written standard cuts of meat followed internationally, as each country classifies the meats according to their standards and needs. There are English cuts, French cuts, American cuts, and Australian cuts; but we shall talk about the most commonly followed simple cuts which are internationally used these days.

Shoulder

Various cuts are culled from the shoulder of lambs as shown in Table 12.3.

Table 12.3 Shoulder cuts of lambs

Cut	Description	Usage
Shoulder chops	These come from the shoulder that just begins from the arm. They have a round bone in centre.	Used for grilled steaks, braising and also boiled as they are tough joints of meat and hence moist cooking methods make them juicy. They find good use in Indian curries and also in biryani.
Blade chops	These come from the shoulder towards the neck part.	Used for grilling, boiling, or stewing.
Shoulder roast	This consists of the whole shoulder.	It is usually roasted. It is a very tasty joint but hard to carve as it is very bony. Butchers often debone it, roll it, and sell it as boneless rolled shoulder.
Lamb neck	The neck is the sticking piece from the shoulder.	It is a bony piece, but has lot of collagen present in it. It is often used for stews or is even braised. The meat could be cubed and used for kebabs.

Best End of the Neck

Various cuts are culled from the best end of the neck as shown in Table 12.4.

Table 12.4 Cuts of best end of the neck

Cut	Description	Usage
Lamb rack	The entire lamb consists 8 ribs. Two racks can be obtained from 1 animal.	Lamb rack is one of the prime cuts and very expensive. It makes a delicious roast. One rack can easily serve 3 guests for an entrée. Guard of honour and crown roasts are two very classical roasts prepared from the rack. When the meat is cut away at the tip and the bone is exposed, it is sold as 'French rack'.
Lamb chop	It is a single piece of chop from the rack. If the tip is exposed it is called French chop.	It is used for grilling and pan-frying. It is advisable to use a double chop with two ribs as the meat will remain juicy while cooking. They find important place in kebabs and biryani.
Lamb cult	When the bone from single chop is removed it is called lamb cutlet.	It is very soft and tender and is usually broiled, pan-fried, or grilled. It can also be crumbed and deep-fried.

Loin

Various cuts are culled from the loin of the lamb as shown in Table 12.5.

Table 12.5   Cuts of loin

Cut	Description	Usage
Saddle of lamb	This is the back of the lamb, which hosts one of the few tender and expensive cuts.	The whole saddle can be used for roasting. It could be deboned and rolled to be called rolled loin roast. From that one can also slice off individual deboned rolled loin chops. This cut can be roasted, grilled, or pan-fried. Concealed under the backbone one also finds tenderloin of lamb.
Loin chops	It is one of the tender most cuts of lamb. It is the slice from the saddle of lamb. When both the sides of the saddle are incorporated in the chop it is called double loin chop or English chop.	Since these are very tender cuts, they would be either grilled, pan-fried, or even broiled. When a single loin chop is deboned, rolled and sliced, it is referred to as 'noisette'.
Sirloin chop	The end of the saddle touching the rump where the legs start.	It is not as tender as saddle, because it gets lot of exercise. Used as a sirloin roast, it could also be sliced to obtain sirloin chops. In India it is referred to as *puth* and is used as a cut in biryani.
Loin roast	The saddle of lamb not including sirloin.	As the name suggests, it is used as a roast. If deboned and rolled, it would be same as saddle of lamb cuts.

Lamb Leg

 Various cuts are culled from the leg of the lamb as shown in Table 12.6.

Table 12.6   Cuts of leg of lamb

Cut	Description	Usage
Lamb leg	These are sold single, double, or even boneless.	These are used for roasts. But they can also be cut on bone and used in stews and curries. Most of the Indian hotels prefer to use only leg of lamb to process meat for their curries, kebabs, and biryani. Indian cut called *pasanda* is a thin slice from the leg.
Leg chop	This is a slice of the leg on the bone.	Used for grilling, braising, or stewing. In India it is referred to as *burrah*.
Lamb shanks	This is the lower part of the leg below the knee and above the ankle.	Both the hind legs and the fore legs are used to get shanks. Lamb shanks cook to a moist preparation with lots of gelatinous texture, when braised or stewed. In Hindi it is called *nalli*.
Topside	It is the boneless muscle from the thigh of the leg.	It is one of the tender cuts of lamb and hence it could be broiled, grilled, or poeled. In India it is referred to as *boti*.
Silverside	This is the muscle from the calf of the leg.	It is very tender and can be used for broiling, grilling, and pan-frying. In Indian terminology it is called *kareli*.

Breast

Various cuts are culled from the breast of the lamb as shown in Table 12.7.

Table 12.7   Cuts of breast

Cut	Description	Usage
Lamb ribs	This is the front of the rib cage that protects organs such as the heart.	This also includes 8 ribs from the rib cage and it is very fatty and has lots of collagen. It is best cooked barbequed, grilled, or pan-fried. It could also be deboned and minced. It forms excellent kebabs and in Indian terminology it is referred to as *pasli ka panja* and the mince from it is known as *chikna keema* as it has more quantity of fat. Thin slices from the deboned lamb rib is called *parcha*.
Spare ribs	It is cut from the breast and have cartilage bones on it.	These are used for grilling, broiling, or barbequing.
Riblets	These are individual lamb ribs cut into single pieces.	It can be boiled and deep-fried as snacks. It could be grilled as well.

Mutton/Lamb Specifications—Indian

It is important here to also discuss the specifications of various cuts of lamb received in the hotels. These specifications could depend upon the usage of these cuts in Indian cuisine and the organization. These are shown in Table 12.8.

Table 12.8   Indian cuts of lamb

Cut	Description	Usage
Lamb curry cut	These are 60–70 g pieces of lamb. Curry cut is the combination of boneless and on the bone meat from the leg and shoulder of the lamb.	As the name suggests, this cut is used for lamb curries and stews.
Champ	This comes from the rack of the lamb and each rib is culled out to form a champ.	Champ can be used for kebabs and biryanis. One can also make exotic Indian curries with champs.
Raan	These are the hind legs of the lamb separated from the thigh bone. While receiving lamb legs the following need to be kept in mind. • There should be no flaps of fat or tissue attached to the leg. • The leg should be received without the portion of spine or tail attached. • The colour of the meat should be bright pink. • The meat should not have blue bruises, blemishes, or purple blood clots. • An average weight of 1.8–2.2 kg is good for Indian dishes.	Lamb leg can be used for processing into *boti* and curry cut. The escalope from the leg of the lamb, also called *pasanda*, is used for making kebabs. The whole leg can be used in preparation of Indian roasted lamb leg often served as *raan*.

Contd

Table 12.8 (Contd)

Cut	Description	Usage
Kaleji	This is the liver of the lamb and it should be dark red in colour when received. It should have a shiny smooth exterior and should be free from slime. Always receive liver in a whole piece.	Liver is often paired with kidneys to make a favourite street food called *gurda kaleji*.
Gurda	This is the kidney of the lamb and we should always receive it in brown colour. It should be plump and tender and firm to touch. It should be shiny in appearance.	*Gurda* is often paired with liver to make *gurda kaleji*.
Magaz	*Magaz* is the brain of lamb and is received in whole. It should have pleasant smell and should be free of slime and blood. It should be shiny and plump.	*Magaz* is often used to make curries.
Boti	Boti are the boneless pieces of lamb obtained from the topside of the leg (refer to Table 12.6).	It is used as a cut in curry cut and also used for making curries and kebabs.
Puth	This is the sirloin chop of lamb (refer to Table 12.5).	It is used as a cut for biryani as it is very flavourful.
Burrah	This is the slice of the leg on the bone.	It is used as a curry cut, it can be used in making the popular burrah kebab. It is also used as one of the cuts in birayni.
Nalli	These are lamb shanks (refer to Table 12.6).	It is used for making curries.
Kareli	This is the muscle from the calf of the lamb leg.	It can be cut up and used as a *boti* in curry or biryani.
Pasli ka panja	These are lamb ribs (refer to Table 12.7).	It can be used in making kebabs and also biryanis.

Note: The receiving temperature should be 4°C or less at the time of receiving.

Beef

The largest meat producing countries around the globe are Australia, the USA, Canada, Argentina, and Uruguay. Large numbers of cattle are also found and slaughtered in India; but it ranks amongst the lowest consumers in the world because of religious prohibitions.

The carcass of the beef is huge and hence it is processed into smaller cuts often known as 'retail cuts'. The tenderer cuts come from the less exercised part of the animal such as back loin, flanks, etc. and the tougher cuts are obtained from the leg and the rump. Meat obtained from a young animal is tenderer, compared to meat of the older animal. Beef is said to be the most fortifying and the most nourishing meat amongst all the edible meats (Larousse Gastronomique)[2]. Beef is bright red in colour and is quite firm and elastic to touch. It has a very light and pleasing smell. The meat of the veal on the other hand is pinkish in colour and thus veal at times is classified into the categories of white meats. The marbling of the fat in the beef is very important and this decides the quality of the overall meat. The fat of the beef unlike

---

[2] Encyclopaedia of gastronomy

lamb is slightly more yellowish. The degree of yellow colour of the fat indicates age of the animal. There are many breeds of cows that are reared for milk and meat production. A special breed of cows called *wagyu* is used to produce one of the most expensive beef called *kobe* from Japan. This is one of a kind of meat which is so well marbled that it can also be eaten raw. The fat of beef is called *suet* and in older times it was extensively used in cooking and desserts such as Christmas cakes, mince pies, etc. But the usage of suet is limited only to very classical preparations, as it is saturated fat and hence not very healthy for human consumption.

Just like lamb, the beef cuts are classified as American, French, English, and Australian cuts; but we will discuss the general and the most common terminologies followed in the classification of the beef cuts (Table 12.9).

Indian hotels only use the tenderloin cut of beef as it is one of the tender cuts and is available at much lower cost. We discussed above that due to religious implications beef is not too much in demand.

The cuts shown in Fig. 12.2 are the broader cuts that are also known as wholesale cuts. From these cuts let us now discuss the retail cuts from the beef. The numbers mentioned on Fig. 12.2 are mentioned in the following tables for an easy understanding.

Table 12.9   Classification of beef

Bull/cow	Steer	Heifer	Veal	Yearling bull/cow
Male is called bull. Female is called cow after calving.	Castrated male bull.	Cow which has not calved yet.	Young cattle from 0–3 months.	Bull or cow under 12 months of age.

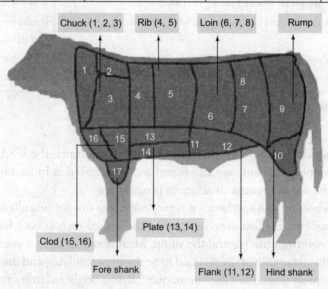

Fig. 12.2 Wholesale cuts of beef

Chuck

Various cuts are culled from the chuck of the beef as shown in Table 12.10.

Table 12.10   Cuts of chuck of beef

Cut	Description	Usage
1 Sticking end or neck piece	This part of the beef is the neck part emerging from the shoulder.	Ideal for slow cooking as this cut of the meat is very fatty and is not tender as some of the prime cuts of beef. The meat from this part could be minced and used for kebabs and forcemeats.
2 Chuck	This is taken from the back of the neck which is the centre between the shoulder of the cow.	The piece of chuck is sold as roasting joint and if cooked in a slow cooking style, this is one of the incomparable roast joints. The ground meat from the chuck is ideal for braising, stewing, or for casseroles.
3 Blade	This cut is from the shoulder of the cow.	This is quite a large piece of meat and the muscle is divided into various cuts so as to get the maximum yields for steaks that can be braised. This also forms an excellent roast joint, as it contains a thick seam of collagen connective tissue, which upon application of moist heat cooks to a wonderful gelatinous texture.

Rib

Various cuts are culled from the rib of the beef as shown in Table 12.11.

Table 12.11   Cuts of rib of beef

Cut	Description	Usage
4 Back rib	The entire rib rack is divided into forerib and the backrib. The backrib is the piece attached to the shoulder.	Used as a roasting joint, classically used as pot roasting or braising.
5 Forerib	From the back of the forerib. Sometimes the bones are cut off to make a square joint which is often used as a roast.	Used for grilling, braising, or roasting. It is also often boned and rolled. This facilitates the easy carving of the meat on buffets. Individual steaks from the boned and rolled rib can be broiled or pan-fried as well.

Loin

Various cuts are culled from the loin of the beef as shown in Table 12.12.

Table 12.12   Cuts of loin of the beef

Cut	Description	Usage
6 Loin	The loin is the most prime cut of the animal. This is divided between loin and sirloin. Some of the famous steaks come from loin of the animal which is mentioned below as 9, 10, and 11.	Since it hosts some of the tender steaks, it would not be a good idea to serve it as a joint for a roast or for that matter deboning and stuffing it would also not be a justice to this prime cut. Methods of cooking such as roasting, grilling, broiling, or pan-frying are apt for this cut as these cuts are lean and thus we should only cook them for short duration of time otherwise they will become dry.

Contd

Table 12.12 (Contd)

Cut	Description	Usage
7 Sirloin	This is the side towards the rump of the animal.	Used for grilling, pan-frying, and roasting. Again being one of the prime cuts of beef, sirloin can be processed into steaks or processed as joint for roasting. It is also served deboned and stuffed. Sirloin can tolerate high temperatures and hence it is ideal for grilling and frying.
8 Tenderloin	This is the tender most muscle available under the loin. Two tenderloins can be obtained from one animal and hence it is very expensive.	The most preferred cut of the animal, as it is ideal for steaks. The entire fillet is divided into various parts which we will discuss in detail in the coming paragraphs. The famous steaks such as chateaubriand come from this muscle.
Rib eye steak	The steak cut from the loin near the rib cage.	One of the tender and expensive cuts used as grilling, pan-frying, and roasting. It is also known by different names such as club steak, delmonico steak, and rib steak.
T-bone steak	This is the centre part of the loin.	It gets its name from the cut on the loin when the T-shaped bone of the spinal cord gets trapped in this cut. When these steaks are cut, the part of the tenderloin muscle becomes the part of this steak and that is the reason they are more expensive.
Porterhoue steak	This is the end of the loin touching the sirloin.	The shape is same as that of the T-bone steak but it is very different from T-steak. Unlike T-bone steak, porterhouse encompasses the heart of the tenderloin called chateaubriand and hence becomes the most expensive steak in this animal.

Rump/Round

Various cuts are culled from the rump of the beef as shown in Table 12.13.

Table 12.13   Cuts of rump of the beef

Cut	Description	Usage
9 Rump	Rump or the round is the backside of the animal where the legs begin.	This is the tougher joint of meat, as it receives lot of exercise and hence does not contain good amount of fat. This cut of meat can be deboned and rolled and served as joint for a roast. The steaks can also be grilled; but one needs to take care while cooking, as it can dry out very fast.
Topside	The part on the top part of the thigh of cow.	Used for grilling, braising, or stewing, can also be roasted as a joint.
Silverside	The lower part of the thigh.	It is a coarser kind of meat and hence is not advisable to use as a roasting joint. The silverside can be ground and used in force-meats and to make kebabs.

Shank

Various cuts are culled from the shank of the beef as shown in Table 12.14.

Table 12.14   Cuts of shank of the beef

Cut	Description	Usage
10 Hind shank	The back legs of the beef below the knee.	It contains a narrow bone called shin bone and since this part of the muscle has loads of connective tissue, it releases lots of juices which reduce to a gelatinous consistency when cooked for longer duration of time. This cut is ideally used to make stocks for soups and sauces. It is also good for braining, boiling, and stewing.
17 Fore shank	The forelegs of the beef.	Used for making stocks for soups and sauces and like hind legs can be used for braising, boiling, and stewing.

Flank

Various cuts are culled from the flank of the beef as shown in Table 12.15.

Table 12.15   Cuts of flank of beef

Cut	Description	Usage
11 Skirt	This is the thin layer of fibrous muscles interlaced with fat that forms the wall around the stomach area.	The tough membranes are cleaned and then the meat is chopped to prepare braises and stews. It can also be ground to prepare kebabs.
12 Flank	It is part of the skirt.	It is a very tender steak hidden under the skirt and is often broiled or pan-fried.

Plate

Various cuts are culled from the plate of the beef as shown in Table 12.16.

Table 12.16   Cuts of plate of beef

Cut	Description	Usage
13 Short plate	This is the middle rib of the rib cage.	Used for grilling and even poaching. It can be deboned and rolled and served as a roast also.
Spare rib	The pair of ribs around the food pipe.	Ideal for barbecues and grilling.
14 Brisket	The lower part attached to the fore shank of the animal.	Classically briskets are used for salted beef, but it can form excellent roasts and grills.

Clod

Various cuts are culled from the clod of the beef as shown in Table 12.17.

Table 12.17   Cuts of clod of beef

Cut	Description	Usage
15 Clod	This is the area below the chuck, above the fore shank.	Used for boiling, stewing, and braising.
16 Thin	This is the extended part of the rib with soft bones and is also referred to as 'leg of mutton' in beef.	Ideal for barbecues and grilling, poaching, and panfrying.

Cuts of Tenderloin

As we saw above, the tenderloin is the tenderest cut of the beef. It is portioned into various cuts that are used for making classical steaks. Figure 12.3 shows the cuts of tenderloin.

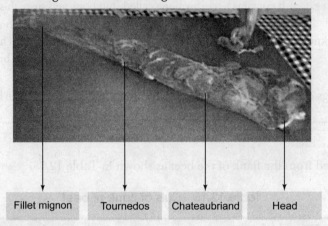

| Fillet mignon | Tournedos | Chateaubriand | Head |

Fig. 12.3   Cuts of tenderloin

Head   This is usually trimmed off and is cubed or grounded for hamburgers and sauces.

Chateaubriand   This is the centre of the fillet. The average weight of the entire tenderloin fillet could weigh between 3.5 kg and 5 kg for each fillet. The chateaubriand comprises 10 to 15 per cent of the weight. On an average it would weigh between 450 to 500 g and this steak is usually served for two people and traditionally would be carved in front of the guest on a trolley called gueridon trolley. It can also be sliced of the weight of 160 to 180 g and served as an à la carte portion.

Tournedos   The next cut from the fillet after the chateaubriand has been taken off. It is usually cut into 60 to 80 g slices and also used to prepare steaks for à la carte.

Fillet Mignon   This is the tail end of the fillet and is sliced into 30 g slices for steaks. This can also be flattened out by hammering for crumbing and frying.

Cooking of Beef

This is the most important part for the chefs when it comes to cooking of beef. Before we do this let us understand a few terminologies.

Larding   Insertion of fat inside a lean meat is called larding.

Barding   Covering of a lean meat with a piece of fat.

Fat plays a very important role in the texture of the meat. When heat is applied to the meat, it is this fat that melts and provides the juiciness to the meat. The muscles which do not receive exercise, such as tenderloin, sirloin, etc. will be having adequate amount of fat and hence the cooked product would be moist and tender. The parts which receive exercise tend to have less fat and such lean cuts of meat need to be larded or barded to cook them to a juicy texture.

The beef is cooked to various degrees of doneness as shown in Table 12.18.

Table 12.18 Cooking stages of the steaks

English	French	Description	Approximate Cooking Time
Very rare	*Au bleu*	Just seared on the hot plate. When cut open, the blood will flow out.	Fifteen to 20 seconds on each side.
Rare	*Saignant*	The meat will still be underdone but blood will be much less.	Two minutes on each side. When the droplets of blood appear on the surface turn the meat again.
Medium	*À point*	The meat is juicy and cooked. When pressed, pink juices come out.	Roughly around 4 minutes on each side. When the juices rise on top and they are pink in colour. When the steak is cut it is also pinkish in colour.
Well done	*Bien cuit*	Meat is firm to touch and the juices are clear.	Seven to eight minutes on each side.

Note: The approximate cooking time refers to a standard steak of 180 g on a moderately heated hot plate.

Selection Criteria of Beef

The quality of beef can be judged from the following points.

- The meat should be firm and bright red.
- It should be well marbled, i.e. it should have a good showing of flecks of white fat.
- It should have a good outside covering of fat, creamy white in colour and of firm texture.
- The bones should be shiny and pinkish with a blue tinge. There should be little or no gristle on steaks.
- Yellowish fat is always a sign that the animal is older or of a dairy breed.

Pork

Pork is referred to as the meat from the domesticated pig. Table 12.19 shows the classification of pig. When we say domesticated, we mean the pig that is specifically reared for the production of meat for human consumption. Pigs feed on garbage and sewage and hence they are the biggest carriers of disease causing germs and insects. Tapeworm is one of the insects associated with pigs and hence animal husbandries that rear pigs for meat always take care while rearing them. In spite of all this it is one of the most widely consumed meats around the world barring the places, where eating pork is a taboo.

Table 12.19 Classification of pig

Hog	Pork	Boar	Suckling Pig
Castrated male pig	Meat from the hogs that are 5 months or older. Older pork meat is used in making processed meats.	Male pig	Baby pig 6–8 weeks old which is still feeding on milk.

Pork meat is eaten in various forms including cooked, smoked, cured, salted, etc. It is one of the most versatile meats that finds a place in the canning industry as well. An English breakfast is probably incomplete if bacon, ham, or sausages are not served with egg preparation. It is one of the meats that can be eaten in breakfast, lunch, and dinner and even during afternoon teas as snacks or filled in sandwiches.

Charcuterie is a French word that forms a category of smoked and cured meats particularly from pig. Pigs have very sensitive noses and this art of theirs helps the man to dig up a fungus called truffle, which holds a very special place on the gastronomic table. Pork is classified as a lean meat. This could be coming as surprise to a layman; but it is true. In spite of being such fatty animals, the fat exists only around the skin and the marbling is not heavily seen in the muscle as in case of beef. Pork fat, often called lard, has been used in cooking and pastry products in the olden times. But since it is saturated fat hence the modern cuisines drift away from using it in the kitchens these days. Traditionally pigs were slaughtered and consumed during autumn. The growing of the pigs in the spring and letting them fatten in the summer yield plump pigs for additional flavour. As during this time apples grow in abundance, it became a tradition to pair the pork with fruits such as apples and apricots. The pairing of pork with such fruits has not yet gone away from the Western table. The biggest consumers of pork are Europe and China with the USA coming third in the list.[3] Figure 12.4 shows the various cuts of pork.

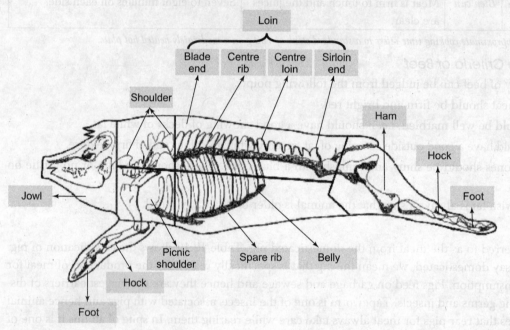

Fig. 12.4   Cuts of pork

Jowl

Various cuts are culled from the jowl of the pork as shown in Table 12.20.

Table 12.20   Cuts of jowl

Cut	Description	Usage
Jowl	This is the chin of the pig.	Ideal for slow cooking as this cut of the meat is very fatty and is not tender as some of the prime cuts of pork. Traditionally pig cheeks are cooked in England and are often referred to as 'bath chaps'.
Jowl bacon	This bacon is prepared from the cheeks of the pig.	It is very popular in the UK and the USA. Jowl bacon is eaten cold.

---

[3] Source: USDA Foreign Agricultural Service, preliminary data for 2006

Shoulder

Various cuts are culled from the shoulder of the pork as shown in Table 12.21.

Table 12.21   Cut of shoulder

Cut	Description	Usage
Shoulder chop	This cut is the section at the top of the foreleg.	Ideal for slow cooking as this cut of the meat is very fatty and is not tender as some of the prime cuts of pork. This makes an excellent roast and it can be sliced into individual chops for grilling. The shoulder can be cubed and used in forcemeats and for stewing. In India pork vindaloo is made using this cut.
Neck end	This is the part of the shoulder closest to the neck.	The fat content in this part makes it ideal for sausage making. It can also be cubed and used for stews.
Picnic shoulder	This is the lower portion of the shoulder attached to the foreleg.	Also known as Boston butt, it is used to prepare famous ham called picnic ham. It can be cubed and made into forcemeat or sausages.

Loin

Various cuts are culled from the loin of the pork as shown in Table 12.22.

Table 12.22   Cut of loin

Cut	Description	Usage
Centre roast/ centre chops	This is the centre of the loin.	The entire centre loin can be roasted, braised, or boiled. Individual steaks sliced from it are called centre chops and are ideal for grilling, pan-frying, or broiling.
Blade roast/ blade chop	This is the part of the loin closest to the shoulder.	It can be cooked in the same way as centre roast.
Loin roast/ loin chop	This is the middle part of the loin.	Same as centre roast and blade roast.
Sirloin roast/ sirloin chop	This is the cut from the end of the loin.	A very moist and fatty cut, it is ideal for roasting. Sirloin chop is also the most preferred cut of meat and can be grilled or pan-fried.
Butterfly chop/ valentines chop/ pork chop	Pork chop comes from the blade or the centre of loin which is the ribs.	An individual chop on bone is called pork chop and when the bone is removed it is called blade chop. When the blade chop is split open to resemble a heart shape, it is called butterfly chop or valentines chop. These chops can be grilled, pan-fried, or broiled. Care should be taken as this part of the meat is very lean and hence can dry up if care is not taken.

Leg

Various cuts are culled from the leg of the pork as shown in Table 12.23.

Table 12.23   Cut of leg

Cut	Description	Usage
Ham	The fresh ham is the hog's hind leg.	The ham is usually cured and smoked and consists of 24 per cent of the total weight of a pig. The legs, if not smoked can be cooked by various methods such as roasting. Honey glazed roasted legs find a special place on the festive buffets. Hams such as Parma ham from Italy, serrano ham from Spain, etc. are the world's popular hams.
Gammon	Also the hind leg of the pig.	The difference from the above is that in case of hams, the leg is removed from the carcass and processed to make a ham. For a gammon the whole animal is cured along with the loin and then it is separated.
Hock	This is the shank portion of the leg. The shank from the front leg is called shoulder hock and the one from the hind is called ham hock.	Hocks are usually cured and smoked and are generally used in flavouring soups and casseroles. The hocks can be cubed up and used in forcemeats or stews.

Belly

Various cuts are culled from the belly of the pork as shown in Table 12.24.

Table 12.24   Cut of belly

Cut	Description	Usage
Spare ribs	These are removed from the belly and are a section that may include the cartilage.	Ideal for barbecues, smoking, and grilling. These are boiled most of the time and then subjected to the mentioned cooking methods.
Pork belly	This is the part located below the loin. It accounts for the total of 16 per cent of the weight of the animal.	The fat content in this part makes it ideal for braising and stewing. Pork belly is cooked on a slow heat for at least five to six hours for the desired tenderness.
Bacon	Since the belly of the pork is very fatty with streaks of lean meat trapped inside, it is smoked to create bacon.	Only the cured and smoked belly will be called bacon. Bacon which is not cured and smoked is known as green bacon and the bacon that comes from the loin of pig is called Canadian bacon.

Selection Criteria of Pork

The pork should be received with skin on. This should be free of hair, should feel moist, and elastic to touch.

- Always receive pork that has light pink flesh which gives a fresh look. This means that the meat should appear moist but not damp or oily looking.
- Select meat where the cut bone appears red in colour. The whiteness of the cut side of the bone indicates the old age of the animal and hence the meat would be less tender.
- When receiving pork always make slashes with the knife on the legs to see if there are any tapeworms.

Poultry

Poultry refers to any domesticated bird that is used for food. It is a term broadly used for birds such as ducks, turkeys, chicken, and even goose. It is believed that Chinese were the first people to raise or farm these birds and then subsequently it was brought to the Western world via Asia, Greece, and Rome. Poultry is a highly nutritious meat as it is classified as a complete protein. It is classified into the category of white meats and it is one of the leanest meats available after the fish. Table 12.25 shows the classification of poultry.

Table 12.25 Classification of poultry

Poussin	Spring Chicken	Broiler	Boiling Fowl	Capon
Chicken weighing 300–400 g.	Chicken weighing 400–450 g.	Chicken weighing between 1–1.5 kg and seven to nine weeks old.	A mature hen of 1 year of age and weighing between 1.5–2 kg.	Castrated male roosters, that are 4–5 months old and weighing between 3–4 kg.

Poultry processing in particular has many interesting terminologies associated with it and it will not be fair if we do not discuss them here. These terminologies would be used as part of a chef's daily work.

Singeing   When the bird is slaughtered the feathers are plucked. In this process some small feathers will be stuck to the skin. Singeing is the process followed to remove such excess feathers present by moving the bird over an open flame. Care should be taken for not burning the skin.

Dressing   This refers to removing of skin off the bird for some preparations. Most of the classical preparations refer to undressed chicken, but since people are now more health conscious it is advisable to serve meats without skin.

12.5

Trussing   This is done in case of roasting. The poultry is neatly tied up with a thread called 'butchers twine' to form a plump shape to look good on the carving table and also since the poultry would be closely tied up there will be less loss of natural juices.

12.6

Spatchcocking   It is the splitting and flattening of the bird, usually done for attractive presentation for baking or grilling. It is termed as *en crapaudine* in French since it looks similar to a flat toad or crapaud.

Basting   It is a term used for meats, especially lean meats. While roasting, the natural juices that have oozed out of the meat are spooned back at regular intervals on to the meat to help it remain moist. Many other methods, such as barding, are common when it comes to cooking of poultry, especially turkey. While roasting the whole turkey, the breasts will cook faster than the legs and hence once the colour is attained on the breast, it is barded with bacon so that it remains moist and does not get overcooked and dry. This is the most commonly eaten meat in India as there are no religious bindings on this meat barring its way of slaughtering.[4]

12.7   **Jointing**   Cutting of bird into pieces is called jointing.

12.8   **Frenching**   When the bone is exposed it is sliced off to expose the bone marrow. This term is called frenching and is done for an aesthetic appeal and also on cooking the protein structure in the bone will coagulate thereby trapping all the moisture inside.

---

[4] Muslims eat halal and Hindus eat *jhatka*

Introduction to Meats   **215**

Types of Poultry

There are many types of poultry birds apart from chicken that are domesticated for their meat and also eggs. These are discussed as follows.

Ducks All ducks are classified into the same family of poultry, but the cooking methods adopted and the taste is strikingly different from a chicken. There are many breeds of ducks available around the world. Some of the famous ones are barbary, peking, nantes, and rouen. A canard is the name given to a duck which is three months old and a duckling is under two months of age.

Turkey A farm-raised bird, it is very popular for its delicate flesh and is always associated with festivities like Thanksgiving Day and Christmas. The size of the bird varies with age but on an average it could weigh between 3 and 11 kg.

Squab It is a 25–30 days old pigeon before it learns to fly. At this age, it would weigh around 600 g and would have reached its maximum limit of growth. Squab meat is very flavourful and is much darker than its other poultry counterparts. It is very tender and retains moisture when cooked. Since the ancient times it has been a dinner entrée for the people from all walks of life.

Goose This is a very interesting species of poultry. It is mostly corn fed and forcibly fed to produce fattened liver called foie gras.

The cuts of the chicken are very simple and not as elaborate as other animals. The most basic cuts of poultry for meat are breast and legs. However, chicken is jointed into various pieces and the cuts are thus named after this. The most common ones are sauté cut or Western cut and curry cut. Figure 12.5 shows the bone joints and the meat in the chicken, because it is very important to understand the bone structure of the chicken as it would help us in jointing it.

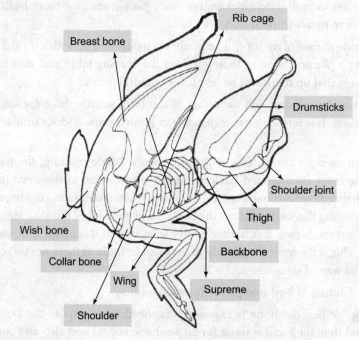

Fig. 12.5 Cuts of meat and bone joints of chicken

Breast

Various cuts are culled from the breast of the poultry as shown in Table 12.26

Table 12.26   Cuts of the breast of chicken

Cut	Description	Usage
Breasts	These can be obtained boneless or with the rib cage bone attached to them.	Chicken is an ideal meat for practically any method of cooking. However, the most preferred ones are grilling, pan-frying, sautéing, or deep-frying.
Supreme	This is the tenderloin of the chicken and is very tender and lean and hence called supreme.	Can be sautéed, grilled, or pan-fried. This is the preferred cut for Oriental satays too.

Legs

Various cuts are culled from the legs of the poultry as shown in Table 12.27.

Table 12.27   Cuts of the leg of chicken

Cut	Description	Usage
Drumsticks	These are the legs of the chicken which are cub shaped.	This can be used whole or even deboned, stuffed, and then cooked. In Indian cooking, it is used in making *tangdi* kebabs.
Thigh	This is the part above the leg joining the hips.	Can be deboned or used whole. Ideal for grilling, barbequing, and deep-frying. In Indian cooking, it is used in making *tikka* kebabs.

Sauté Cut

The classical sauté cut is also known as Western cut. Classically this cut has always been with skin on; but the modern views differ slightly (see Fig. 12.6). There is also another version of this cut which is a little different (see Fig. 12.7). Here the breast is divided first laterally and then the centre piece is divided into two.

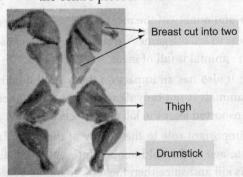

Fig. 12.6 Sauté or Western cut of chicken style 1

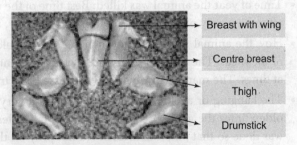

Fig. 12.7 Sauté or Western cut of chicken style 2

Game

Game is a term applied to wild animals that are suitable for human consumption. This consumption of wild animals has been there since the time of cavemen and till date many nomadic tribes living on hunting still feed on this category of meat. Animals ranging from horses to giraffes and zebra are consumed in many parts of the world. Game is again largely categorized into furred and feathered as shown in Table 12.28. Let us discuss about these in detail.

Table 12.28 Classification of game

Furred game	Feathered game
These are animals that have fur on them. Examples are deer, antelopes, bison, etc.	These are wild birds that have feathers on them. Examples are wild turkey, partridge, quails, pheasants, etc.
Furred animals can be large or even small such as hare, rabbit, porcupines, etc.	Game birds can be large such as ostrich, turkey, and small such as quails, partridge, etc.

Large Game

The most common large game meat is venison, which though commonly thought of as deer, is a term that broadly includes the meat from elk, moose, reindeer, caribou, and antelope. Other popular large game animals include buffalo, wild boar, and to a lesser degree, bear. Additionally, there are even rarer varieties eaten around the world such as camel, elephant, kangaroo, zebra, wild sheep, and goats.

Small Game

The most common small game animal is rabbit. Squirrel is also quite popular, followed by beaver, armadillo, and even porcupine. There are many factors that need to be taken into consideration when selecting a game. All the game must be allowed to rest or age after the slaughter to drain away all the blood. The meat of the game animals is strong and very lean as these animals grow and live in the wild and hence get lots of exercise.

The following factors affect the quality of meat of game animals.

- Age: Younger the animal tender the meat.
- Diet: The diet of an animal greatly affects the flavour and quality of the meat.
- Time of year the animal was killed: Best time of the year to kill the game is around the fall, because after the plenty of feeding in autumn and summer, the animal is full of meat.
- How the animal was killed and handled in the field: It also has an impact on the overall quality of the meat. The meat of many otherwise excellent animals is not only damaged but also ruined at times. The manner in which it was dressed and transported after the kill also affects the quality.
- The hanging or ageing of the meat: It plays a very important role in the quality of the meat. In olden times the game used to be hung on tall poles to age in the cold temperature outside. This was done so that other animals could not prey on this kill and since then the term 'hanging game high,' has been associated with game animals.

For maximum tenderness, most game meat should be cooked slowly and not overdone. It can be cooked with moist heat by braising or with dry heat by roasting with an effort to ensure maximum moistness through basting, larding, or barding is imparted to the meat. Following are some of the popular game animals consumed around the world.

Rabbit

The domesticated members of the rabbit family (a rodent relation) have fine-textured flesh that is almost totally white meat. Rabbits are from the category of furred game, which have long ears and long hind legs for running and jumping. They are herbivores with long front teeth for gnawing; some use it to make burrows or tunnels in the ground. Rabbits are smaller and have shorter ears than their relatives—hare. They are plump and less strongly flavoured than their wild counterparts. A mature rabbit averages between 2 and 3 kg, much smaller than its relative, the hare, that usually weighs in between 3 and 4 kg. In India most of the game are farm raised and have a licence to produce meat. This measure is taken to stop the hunting of animals for sheer pleasure and sport. Though hare and rabbit are quite different from each other, both these terms are used synonymously. One of the most famous dishes made with this animal is jugged hare—a classic English preparation that begins with cut pieces of hare that are soaked in a red wine-juniper berry marinade for at least a day. The marinated meat is well browned, then combined in a casserole with vegetables, seasonings, and stock and then baked. When the meat and vegetables are done, the juices are poured off and combined with cream and the reserved hare blood and pulverized liver. The strained sauce is served over the jugged hare and vegetables.

Game Birds

There are various kinds of game birds and some of them are farmed these days making them easily available. Let us discuss a few of them.

Quail

It is a migratory game bird belonging to the partridge family. But when colonists discovered birds that resembled their European version, they called them by the same name. American quail is known by various names depending on the region—bobwhite in the east, partridge in the south, quail in the north, and blue quail in the southwest. Other notable members of this family are California quail, mountain quail, and montezuma quail. They are very social and travel in small groups called coveys. The meat of the American quail is white and delicately flavoured. In general, they should be cooked like other game birds—young birds can be roasted, broiled, or fried and older fowl should be cooked with moist heat. Most of the quail marketed today are raised on game bird farms.

Partridge

The two main varieties of this game bird are the grey partridge and the red-legged partridge, neither of which is a North American native. In various regions of the USA, the name 'partridge' is erroneously applied to other birds including the ruffed grouse, quail, and bobwhite. All of these birds are plump and have white, tender, slightly gamey flesh. They usually weigh around 350–400 g. Partridges can be cooked in a variety of ways including roasting, broiling, and braising. The meat also makes a tasty addition to soups and stews.

Pheasant

The bright colours and long, tapered tail of the cock ring-necked pheasant makes this bird easily identifiable. The ring-necked pheasant is a native of Japan and southern China. They are pretty expensive in India.

YIELD TESTS

Meat is one of the most expensive commodities used in the kitchens of any hotel. Ranging from a meat worth ₹60 a kg to ₹1200 a kg for Australian lamb chops, it becomes important for a chef to take care of the processing and getting accurate yields of the same.

Every hotel does yield testing for each meat and this is done for each cut and specification. Let us take an example of a chicken. To understand the yield from the chicken, we should know for which specified cut are we calculating the yield. Let us understand this more clearly by referring to the sample card of yield test attached below. Let us place the value of 100 per cent to the whole received chicken. The rate of this 100 per cent is ₹60 and thus if the yield is 50 per cent then the rate of that cut would be ₹120 and so on.

The simple formula would be

$$\text{Yield} = \frac{\text{Rate}}{\text{Yield percentage}}$$

In this way we can easily have the yield for each cut of meat that would then help us to arrive at logical and accurate costing.

Yield Card				
Sr. No.				
Name of product:		Broiler chicken		
Grade:		A		
Supplier:		ABC Farms		
Rate:		₹ 60 per kg		
Date of yield test:		1 Dec 2013		
	Qty (kg)	Rate	Amount	Yield (%)
Broiler chicken with skin	1	60	60	100
Broiler chicken without skin	0.85	70.59		85
After removing legs and breasts	0.45	133		45
Boneless chicken	0.35	171.42		35
Chicken tikka	0.3	200		30
Cost per portion				
Selling price				
Cost (%)				
Signature of chef				

Fig. 12.8 Standard yield card format

CONCLUSION

Apart from all the meats listed above, there is yet another category of the meats that is called 'offal'. These have been enjoyed for centuries and also condemned because these include all the other internal organs of the animal such as heart, liver, intestines, brain, and even testicles. Many of these are cooked and eaten directly, whereas others such as intestines are used as cases for sausage stuffing. Apart from these, glands such as thymus and pancreas glands, that are often referred to as sweet breads, are very commonly eaten around the Western world. In north Indian street food, a spicy preparation of kidneys and testicles is very common.

An offal such as ox tongue is boiled and served cold in the UK. Ox tail soup is considered to be the national soup of the UK. Organs such as liver is used to make liver pastes that can be spread on thin crusty breads and served as snacks. The stomach, commonly known as tripe, is used to make a very famous Scottish dish called 'haggis'. The fat around the stomach of a pig is called caul fat and is used for shaping and tying up the meat to make it moist and juicy during cooking.

Meats are a rich source of proteins and on an average a meat contains 20–25 per cent protein. Generally, meat with more marbling has less protein and the best part is that 97 per cent of this protein is easily digestible by the human body. A standard 180 g of meat is enough for a portion and this would provide 85 per cent of the proteins required by the body. The meat is made up of muscle fibre, connective tissue, and fat. Each cut of meat is different in different animals and the different cuts from the same animal react differently to each cooking method and heat applied to them. If a meat is tough and chewy, it is tenderized by chefs. There are various ways of tenderizing the meat. The most common one is to cook the meat on a low temperature for a long duration of time. Some natural foods, such as papaya and pineapple, contain enzymes such as papain and bromenin, which are natural tenderizers and are often used in meat dishes in Indian cooking.

KEY TERMS

Agnelet French word for kid lamb between 30 to 60 days of age.

Bacon Meat cut from the belly of pork and it is always cured and smoked.

Barding Covering the meat with a piece of fat or bacon before cooking.

Basting Pouring the natural juices over the meat that is roasting.

Beef Meat from cow or bull.

Best end The ribs of an animal.

Boti Indian name for large piece of muscle from the thigh of the animal.

Bovines Group of animals comprising of ox, bull, bison, etc.

Burrah Indian name for slice of leg on bone.

Butterfly chop Pork chop without bone and split open.

Canadian bacon Bacon from the loin of the pork.

Canard Duck which is three months old.

Carcass Body of the animal after slaughtering.

Cartilage A soft flexible bony structure as in ear and nose.

Casserole Style of cooking in a closed clay pot.

Caul fat Fat on the lining of the stomach of pigs.

Charcuterie French term for a category that comprises of cured and smoked meats.

Chateaubriand Steak from the centre of the tenderloin of beef.

Chikna keema Indian name for fatty mince obtained from ribs.

Collagen A connective tissue in the muscle of animal that converts to gelatine upon cooking.

Crumbing Covering the commodity with breadcrumbs or similar things before deep-frying.

Delmonico steak Rib eye steak from beef.

Dressing Removing the skin of poultry.

En crapaudine French for flattened chicken after being split from the backbone.

Ewe Female lamb under one year of age.

Foie gras Fattened liver of goose.

French rack The lamb rack where the ribs are trimmed and exposed.

Frenching Slicing off the exposed bone, usually of a poultry.

Game Wild animals commonly eaten by human beings.

Grading   A way of assessing the quality of the meat.

Green bacon   Bacon which is not cured and smoked.

Gueridon   A style of service that is done in front of the guest through a trolley.

Heifer   Cow which has not yet given birth to any calf.

Jointing   Cutting the poultry into pieces.

Jugged   Method of thickening with hare blood.

Kareli   Indian name for calf muscle of meat.

Khassi   Hindi name for castrated male lamb.

Kobe   Special beef from Japan which is highly marbled. It is from a breed called wagyu.

Lard   Fat from the pig.

Larding   Insertion of fat into the piece of meat.

Loin   Saddle of the animal.

Marbling   Interlacing of fat and meat to create a webbed structure.

Meat ageing   Resting of meat in a cold storage to develop tenderness.

Mutton champ   Indian name for lamb chop.

Mutton curry cut   Lamb cut into 50 g piece—boneless or with bone

Mutton   Lamb above 12 months of age.

Myoglobin   Pigment present in the meat that determines its colour.

Nalli   Indian name for lamb shanks.

Noisette   Rolled loin and sliced into thick slices.

Offal   Internal organs of animal that can be consumed safely.

Ovines   Group of animals comprising of sheep, lamb, goat, etc.

Parcha   Indian name for slice of meat from the breast.

Parma ham   Air-dried ham from Parma region of Italy.

Pasanda   Indian name for escalopes derived from the leg.

Pasli ka panja   Indian name for lamb ribs.

Picnic ham   Cured and smoked shoulder with the foreleg attached.

Poeled   A method of cooking where only fat is used as a medium.

Pork   Meat from pig.

Puth   Indian name for sirloin.

Raan   Indian name for leg of an animal, usually lamb.

Ram   Male lamb; also known as hogget.

Retail cuts   Smaller cuts of meat that are individually sold.

Rigor mortis   A natural process where the animal's muscles stiffen up after slaughtering.

Roasts   Large carved out joints of meat that can ideally be roasted.

Scrag end   The area towards the neck.

Serrano ham   Famous ham from Spain.

Shank   The cut below the knee of an animal.

Singeing   Rotating the bird on a spit fire to get rid of feathers.

Sirloin   The end part of the saddle towards the hind legs.

Skirt   The piece of the meat covering the sides of stomach.

Spatchcock   Same as *en crapaudine*.

Squab   Meat of a pigeon.

Steer   Castrated male bull.

Stunning   Making the animal unconscious before slaughtering.

Suet   Fat from the beef.

Sweet breads   Pancreas and the thymus gland.

Swines   Groups of animal comprising of pigs, boars, etc.

Tangdi kebab   Indian name for kebabs made with drumsticks.

Tenderloin   A piece of a tender muscle hidden below the loin.

Tripe   Stomach of an animal.

Truffle   A type of mushroom found under the ground.

Trussing   Tying of a poultry bird to retain shape before roasting.

Veal   Meat from the calf of the cow.

Western cut   Jointing of a chicken into eight pieces.

Wether   Castrated male lamb.

Yield   The amount of usable meat available after processing.

CONCEPT REVIEW QUESTIONS

1. How would you define a meat? What are its compositions in terms of physical and chemical properties?

2. What are the advantages and disadvantages of consuming meats?

3. Define a muscle tissue.

4. Why is the meat of the older animals tough to cook?
5. How is collagen in the meat used to the advantage of cooks?
6. What does a skeletal tissue consist of?
7. What do you understand by smooth muscles?
8. What is myoglobin and how does it affect the characteristics of a meat?
9. How is meat graded and what is its importance?
10. Briefly describe the steps involved in slaughtering of an animal.
11. Why is stunning important?
12. What do you understand by the term rigor mortis?
13. What is ageing of meat and why is it done?
14. How are the meats classified into species?
15. How are meats classified on basis of categories?
16. What is *khassi* lamb?
17. What do you understand by the term mutton?
18. List the cuts of lamb and at least one usage of a cut.
19. What is guard of honour and crown roast and what part does it come from?
20. Which part of the lamb will give us lamb cutlets?
21. List the Indian cuts for lamb.
22. List the receiving specifications of kidneys and lamb fat.
23. What do you understand by the word marbling and how does it affect the quality of meat while cooking?
24. What is the difference between a heifer and a steer?
25. What is *kobe* and why is it so expensive?
26. List the cuts of beef.
27. Which part of the beef would these steaks come from—porterhouse, T-bone steak, chateaubriand, and delmonico?
28. What is the weight of tournedos?
29. Explain the different degrees of doneness in a steak.
30. What is larding and barding and how does it help the texture of the meat while cooking?
31. List the selection criteria of good beef.
32. What is a suckling pig and when is it prepared?
33. What is *charcuterie?*
34. What is the difference between caul, lard, and suet?
35. List the cuts of pork.
36. What is green bacon?
37. Differentiate between hock ham, ham, gammon, and Canadian bacon.
38. Write down the classification of poultry.
39. What is dressing of a chicken?
40. What is basting and why is it so essential?
41. What do you understand by the term trenching?
42. What is trussing and why is it done?
43. How is game classified?
44. What is a yield test and what is its importance?
45. How would you define offal?
46. Write down at least three offal that can be eaten.
47. What are sweet breads?
48. What is the difference between a squab and a canard?
49. Why is spatchcocking also called *en crapaudine?*
50. What is leg of mutton in beef carcass?

PROJECT WORK

1. In a group of three to four, prepare various cuts of lamb, beef, and pork with regard to English, French, and American cuts and compare them with the cuts given in this chapter.
2. Between groups, prepare the recipes pertaining to various meats from the recipes folder in the ORC and critique and evaluate the same.
3. Research and list the various types of steaks that can be procured from beef, lamb, and pork.
4. Compile a list of cold smoked meats, especially from the Italian region and discuss about the different approaches.
5. In a group, research hams from at least five countries and discuss the peculiarities of each.
6. Research on the various game birds and animals eaten around the world and write a short note on each.

CHAPTER 13

INTRODUCTION TO FISH AND SHELLFISH

Learning Objectives

After reading this chapter, you should be able to:

- identify various types of fish and shellfish and be able to classify them
- understand the various cuts of fish and methods of cooking used for the same
- select cuts for classical dishes of fish
- recognize the importance of selection and storage of fish and shellfish
- prepare and process various fish and shellfish as per standards and specifications
- appreciate the usage of famous species of seafood used in modern kitchens worldwide

INTRODUCTION

Fish and shellfish are commonly known as seafood. Though they could be procured from the sea, ocean, river, or lake or they could also be farmed artificially to meet the demands of the customer; they are still known as seafood.

Man has been using seafood in his diet since time immemorial. It is one of the commodities in the kitchen that a chef really feels proud of. Since fish is a delicate meat, one needs to apply delicate cooking methods and also delicate flavours.

In many places around the world, especially in Bengal, India, fish is considered to be vegetarian food, and it is called 'vegetable of the sea'. Similarly, in France it is referred to as *fruit de mer*, which literally translates to 'fruits of the sea'. Whether eaten cooked or raw, the texture of fish and shellfish is always smooth and incredibly versatile.

Fish holds a special place in French classical menu, where it forms a separate course called *poisson*. Fish and shellfish can be used in almost any kind of preparation. The texture of this meat is very different from the other meats discussed in the previous chapter. The physical and the chemical structure of this meat is entirely different from the other meats and that is the reason why it is grouped into a different category altogether. The highly perishable nature of this commodity has led to its preservation, as people did not want to be devoid of the nutrition that it provides. Fish is rich in protein content and some of them are rich in oils, such as Omega-3, which is a preferred choice for people suffering from coronary heart ailments. Doctors advise having fish and remaining off the other meats even when on a diet.

Fish and shellfish are different creatures, but since both come from water we are dealing with them in the same chapter. In case of fish the only difference is in their species and habitat; but shellfish on the other hand differ not only in species but also in their biological composition. Like the name suggests, all the shellfish have shells on them. Some of them live in two shells and are called bivalves such as oysters and clams, while others known as univalves live in a single shell such as abalone. There is another category which wears a kind of armour or a crust and is called crustacean such as lobsters, prawns and crabs. Let us discuss the classification of this category of meat to get a better understanding.

CLASSIFICATION OF FISH

Fish could be broadly classified as shown in Table 13.1.

Table 13.1   Classification of fish

Habitat	Physical Shape	Flesh Type
Salt water—sea and ocean	Round fish	White
Fresh water—rivers, lakes, ponds	Flat fish	Oily

Habitat

Fish can be classified based on its habitat. As we know, fish comes from the sea, ocean, river, or lake, and each of these habitats play a very important role in its taste and texture. As we mentioned earlier, no matter where the fish comes from it is still referred to as seafood. There could be a species of fish that thrives both in sea as well as river. These are then differentiated by their name such as sea sole or river sole. The taste does differ, as sea fish are considered to be healthier because of the salinity in the water. The flavour of such fish is called 'oceanic flavour'.

Freshwater Fish

Some species of fish tolerate and sometimes migrate back and forth between saline waters and fresh waters. Fish that are predominantly inhabitants of fresh water are described as freshwater fish and include bass, perch, pike, smelt, sturgeon, trout, etc.

Freshwater fish are lean because they exercise more—swimming against the current and migrating along the river course. The movement of fish in their breeding season also contributes to their being lean.

Seawater Fish

These are inhabitants of marine water and are adapted to the sea conditions. These fish do not exercise much because the high concentration of salt in the sea water helps them float and they do not have to move against any current. This is the reason why most of these fish are oily and fatty in nature. Oils present in some of them, such as Omega-3, are beneficial for heart patients. The marine fish vary in size, from very small such as anchovies, to very large such as dogfish and sharks. Some other examples of sea fish are codfish (Atlantic cod, Pacific cod, haddock, pollock, silvery pout, Alaskan pollock), cutlass fish, eels, flat fish (flounders, dabs, turbot, plaice, halibut), herrings (Atlantic herring, Pacific herring, sprats, sardines, pilchards), mackerels, mullet, redfish, salmon, tuna (blue fin, albacore, yellow, big eye, skipjack), etc.

Physical Shape

Fish are also classified as per their shape. There are two shapes of fish—round and flat. As the name suggests, flat fish are flat in appearance while round fish are well rounded and plump. There are many ways in which one can differentiate between flat and round fish (Table 13.2).

Table 13.2   Difference between flat and round fish

Flat fish	Round fish
This fish is flat in shape.	This fish is round in shape.
Both its eyes are on one side.	Eyes are on either side of the fish.
It swims flat to the surface of the water with the belly aligned parallel to the water bed.	It swims in a horizontal fashion.
Belly side of the fish is white in colour.	
It yields four fillets.	Both the sides of the fish are of same colour.
	It yields two fillets.
Examples are plaice, turbot, flounder, skate, halibut, sole, etc.	Examples are salmon, trout, snapper, etc.

Flesh Type

Fish are also classified by its flesh type. The flesh of fish is either oily or white. White fish is also referred to as lean fish. All flat fish and many others, such as snapper, cod, etc., are lean fish or white fish. Fish such as herring, mackerel, salmon, sardines, etc. are classified as oily fish. It is very important for chefs to understand these differences in the structure of fish as cooking methods for oily fish and white fish differ. We will discuss this further in the chapter on methods of cooking.

CLASSIFICATION OF SHELLFISH

Shellfish are mainly categorized into crustaceans and molluscs.

Crustaceans

As the name suggests, these have crusts on top or a shell, which act as an armour. Shellfish have a pigment called 'astaxanthin', which on contact with heat turns into a coral red colour, which is much desirable for shellfish. Unlike fish, shellfish do not have any cuts associated with them. They are either cooked in the shell or out of the shell depending upon the end product desired. Few of the common crustaceans are discussed in Table 13.3.

Table 13.3   Common crustaceans

Crustacean	French Name	Type	Processing and Cooking	Other Information
Crabs range from tiny to large varieties. The female crab is called a 'hen'. Crabs are eaten as soft-shell crabs soon after they complete molting or shedding their hard shell.	Crab	Blue crab, brown crab, king crab, snow crab, mud crab, spider crab	1. Since crab has sweet and succulent meat it needs very less time to cook. 2. Mild flavours suit better than creamy sauce. 3. Oriental flavours such as lime juice, coriander, and chilly helps to make good salad. 4. It can be pickled with its flesh removed from shell. 5. Soft shell crabs are deep-fried and tossed with sauces. 6. Crab meat can be used in soups or made into cakes and fried for hot snacks.	The female crab has sweeter meat than the male crab, but they have less amount of meat and hence are smaller in size. Male crabs are considered in cooking as they provide good yield. One should buy crabs alive to ensure freshness and also ensure that they are heavy for their size. Crabs with both claws attached should be purchased, as the claws have the maximum meat.
Crawfish are similar to spiny lobster but do not have claws like them. They have spiny shells and the flesh is in suited for this type of the tail. The head is always used as a decorative piece or used in stocks as these contain the maximum flavours.	Langouste	Rock lobster or spiny lobster, sand lobster	1. The meat can be grilled, sautéed, broiled, and panfried. These can also be stewed in a creamy sauce. 2. Spicy seasoning can produce best results. 3. Oriental flavours are best suited for this type of crustaceans.	These are often mistaken for lobsters. The selection criteria is same as that of crabs.
Crayfish are also used for preparing bisque soup (refer to Chapter 9)	Ecrevisse		They are cooked in the same way as crawfish. Here also only the tail meat and the claws are eaten.	These are also called freshwater lobster.

Contd

Table 13.3 (Contd)

Crustacean	French Name	Type	Processing and Cooking	Other Information
Scampis are also called Norway lobsters or Dublin Bay prawns. It is believed that colder the water in which it lives, better is the taste and the flavour.	Langoustine		1. These are very good for grilling and barbecues. 2. These can be poached in court bouillon and served with melted butter. 3. These are also cooked and served cold on fish platters.	These should be purchased moist and sweet smelling. If these are stale then they smell of chlorine.
Lobster The American lobsters are green in colour and the European lobsters are blue black in colour and have the best flavour and hence are expensive.	Hommard	Maini lobster, bugs lobster, European lobster	1. Cooked with delicate flavours. 2. Boiled in court bouillon and served hot with melted butter or cold with mayonnaise. 3. Lobsters are sautéed with pastas or even stuffed in them. 4. Some very famous preparations such as lobster thermidor, lobster newburg, and bretonne are still very common on menus today.	These should be purchased alive with claws attached. Lobsters should always be proportionately heavy in size. The female lobster has a broader tail compared to that of a male lobster.
Prawns/shrimps are the most popular of all the crustaceans. Technically there is no difference between prawns and shrimps; the only difference is that shrimps are smaller in size. Prawns measuring less than 5 cm are called shrimps.	Crevettes	Cold water prawns, for example, pink shrimps, deep sea shrimp; Mediterranean prawns; warm water prawns, for example, Gulf shrimps, kuruma or Japanese prawns, tiger prawns	1. One must never overcook prawns, as they tend to get tough. 2. These can be poached, grilled, sauteed, deep-fried, or barbecued. 3. These can be used in curries or can be stir-fried. 4. Famous starter called prawn cocktail is served cold.	Prawns should be purchased when they are moist and sweet smelling. If these are stale then they smell of chlorine. Prawns are received in counts such as 10–12, 18–20, 35–40 in 1 kg and shrimps could be in the range of 90–100 in 1 kg. Smaller the prawn better is the taste.

Molluscs

Molluscs are shellfish that have a hard inedible shell. They are classified into three other subgroups—gastropods or univalves, bivalves, and cephalopods.

Gastropods

These molluscs have a single shell. They are also known as univalves. Some of the common gastropods are listed in Table 13.4.

Table 13.4   Common gastropods

Gastropod	French Name	Type	Processing and Cooking	Other Information
Abalones are known as ormer or sea ear. These are large shellfish which are ear-shaped. Sometimes they are called sea cucumber.	Abalone		1. They are usually used in Chinese cooking as appetizers.	
Snails are characterized by a spiral shell. In classical French cuisine these are served as a hot appetizer.	Escargot	Burgundy snail, petit-gris	1. Snails take less time to cook. 2. They are served as an appetizer with vinegar or garlic-flavoured mayonnaise.	
Conches are largely related to whelks and are mostly native to Florida.	Escargot de mer		They are usually beaten by a mallet in order to tenderize them. Conches can easily cause stomach upsets, so they should be boiled in at least two changes of water.	One must buy young conches, which are commonly known as 'Thin lips'. These are usually sold out of the shell.
Whelks are smaller in size compared to conches. They usually have a smaller conical shell and the flesh is more rubber-like.	Escargot de mer		1. These are boiled in sea water or heavily salted water. 2. These can be sautéed after boiling or can also be batter-fried to be served as a hot appetizer.	Same as conches

Bivalves

These shellfish are covered by two shells or valves. Some of the common bivalves are discussed in Table 13.5.

Table 13.5 Common bivalves

Bivalve	French Name	Type	Processing and Cooking	Other Information
Clams are often eaten raw and are a favourite of the Americans. The famous Manhattan clam chowder is considered to be the national soup of the USA.	Poularde	Cherry stone, geoduck, carpet shell, razor clams, long neck, venus	1. Since they are full of sand, they must be washed very well in cold running water. 2. They can be boiled, stewed, or steamed. 3. They can be teamed with pastas or even crumbed and fried and served as hot snacks.	Clams must be purchased tightly shut, as open clams indicate dead clams.
Cockles are similar to clams and sometimes called 'heart clams'. They do not have claws like clams. They have spiny shells and the flesh is in the tail. The head is always used as a decorative piece or used in stocks as these contain the maximum flavour.	Bucards	Dog cockles	Cockles must be cleaned in the same way as clams and the usage of these are same like clams.	They smell sweet when shelled. These should be bought with the shells closed. Cockles with pale flesh are said to taste better than those with dark flesh.
Mussels have sweet tender flesh. They usually have beards which are used as tentacles for locomotion. Mussels must be scrubbed well with a brush to get rid of all the dirt and sand.	Moule		1. Mussels can be eaten raw or cooked. Steaming is the best method of cooking them. 2. Large mussels can also be stuffed and baked and sometimes the meat can be taken out and crumbed and deep-fried.	One should purchase mussels which have as less beard as possible, and are tightly shut.

Contd

Table 13.5 (Contd)

Bivalve	French Name	Type	Processing and Cooking	Other Information
Oysters have a thick greyish green shell. One of the shells is flat and the other is cupshaped. Oysters are eaten raw in the classical French cuisine—an all time favourite and believed to be aphrodisiac.	Huitre	Sydney rock, coffin bay, Tasmanian	1. These are best eaten raw with a little sea salt and squeeze of lemon juice or a dash of tabasco sauce. 2. They can also be poached or steamed and then served with buerre blanc or champagne sauce. 3. Oysters are also crumbed in corn meal and deep-fried.	It is an old saying that oysters should not be purchased when there is no 'R' in a month, for example, May, June, July, and August. It is believed that during such months, they are soft and milky.
Scallops are found in sandy sea beds. One of the two scallop shells is flat and the other one is ridged and curved, which contains the scallop meat.	Coquille St Jacques	King scallop, queen scallop	1. Unlike other bivalves, the scallops should not be alive when cooked. 2. When they are alive they can be open and when tapped a little they close themselves. 3. These also have beards that need to be cleaned before cooking. 4. Scallops require very little amount of cooking and in South America they are even eaten raw drizzled with lemon and olive oil and are called *ceviche.*	One should purchase scallops with their coral and avoid purchasing frozen scallops as they do not have the same taste as the fresh ones.

Cephalopods

Cephalopods means shellfish that have legs over their heads. They are closely related to the snail family but the only difference is that they do not have external shells like the snails; instead they have internal shells that are made from a spongy material. The most common cephalopods are cuttle fish and squids. They have a bulbous head which contains a mouth and two jaws. Tentacles on the head are used for locomotion and they are also covered with suckers. Octopus is an exception as it does not contain any cartilaginous bone but has eight sets of tentacles.

A black coloured fluid is emitted by them to create a smoke screen from the predator. This is commonly known as 'squid ink' and is used for making dyes and medicines. It is also used as a colouring agent in kitchens, especially to colour pasta dough to create black coloured pasta.

Some of the common cephalopods are listed in Table 13.6.

Table 13.6   Common cephalopods

Cephalopod	French Name	Type	Processing and Cooking	Other Information
Squids have tender sac-like body with 10 tentacles. The internal shell is transparent and looks like a piece of glass and is often referred to as 'quill'.	Calamari	Bottle squid, southern squid	1. Squids are very tricky to cook. They should be cooked briefly over a high flame or cooked over low heat for long time or else they will become rubbery. 2. Squids can be cooked in a variety of ways; they can be stuffed and baked or coated with batter or crumbed and deep-fried. It could also be added to pastas and rice or simply cooked with Oriental flavours.	Squids must always be very fresh as they start smelling if they are old. They must be sweet smelling and moist and all the 10 tentacles should be attached. Ink sac must be intact to ensure freshness.
Cuttlefish are similar to squids; but the only difference is that they have a soft bony cartilage which is entirely made of calcium. Their body is broader than squids.			1. The smaller the cuttlefish, the more tender is its flesh. 2. It takes less time to cook on high heat and can be cooked in the same way as squids. 3. Cuttlefish is sautéed in olive oil and garlic or it can be deep-fried.	The receiving standards of cuttlefish are same as that of squids.
Octopus, unlike squids and cuttlefish, do not contain an internal shell or fins. Instead they have 8 equal size tentacles.	Pieuvre		1. Octopus needs slow cooking and is simmered in salted water for a long duration of time. 2. To tenderize the meat it is often pounded with a mallet. 3. Curled up tentacles is an indication that it is tenderized.	While purchasing octopus ensure that all the tentacles are attached. This ensures that they are fresh. They must be sweet smelling and moist. There are two rows of suckers in each tentacle. Inferior quality of octopus will have only 1 row of suckers.

CUTS OF FISH
Before fish are cut, they should be thoroughly cleaned and gutted.

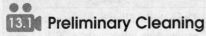 Preliminary Cleaning
This includes the following steps.

1. Clean the fish well and cut off the fins.

2. Scrape off the scales and remove gills.

3. Make a slit in the belly and remove all the internal organs. This is also known as gutting of the fish.

4. Wash them in running water to remove all the blood and any physical impurities.

 Filleting
Filleting is the neat removal of the flesh of the fish from its skeleton to yield sections of fish flesh free from skin and bone. There should not be any flesh remaining on the skeleton of the fish. Flat fish yield four fillets (two from each side) and round fish only yield two fillets (one from each side of the vertebrae).

Filleting a Flat Fish
This includes the following steps.

1. Lay the fish flat on the appropriate chopping board. (The fish must be prepared for filleting, i.e. cleaned, gutted, and descaled.)

2. Cut down the natural centre line of the fish to be boned. Make your initial incision as close to the head as possible to minimize loss of flesh, finish at the tail.

3. Work from the centre, cutting to the left, keeping the knife pressed against the bone to detach the fillet from the bone, lifting the fillet away from the bone as you cut. Repeat the process for the second fillet.

4. When the fillets have been completely removed, turn the fish over and repeat on the underside to yield the four fillets.

Filleting a Round Fish
This includes the following steps.

1. Lay the round fish on the appropriate chopping board (the fish must be prepared for filleting, i.e. cleaned, gutted, and trimmed).

2. Working from the head to tail, cut down the backbone following the vertebrae and over the rib cage. Lift the fillet as you cut so that you have more control over the action of the knife. Detach the fillet at the tail and the head.

3. Turn the fish over and repeat to yield two fillets. The head may be removed before filleting to make it easier to detach the fillets.

Now let us discuss the cuts of fish in Table 13.7.

Table 13.7   Cuts of fish

Cut	Description	Usage
Fillet	It is a boneless cut of fish of entire length from the whole fish.	The whole fillet is further used for processing other cuts of fish. Used for barbecuing, grilling, pan-frying, poaching, steaming, or deep-frying.
Darne	This is the slice of the round fish on the bone. It resembles a steak.	It is an excellent cut for grilling, barbecuing, pan-frying, poaching, and baking. This cut is often used in stews for Indian curries.
Supreme	This cut is often referred to as fillet also. It is the prime cut of fish and is always boneless and without the skin.	This is the most common cut used in many preparations. This cut can be grilled, poached, pan-fried, crumbed, or batter-fried, etc.
Delice	It is not really a cut of fish but is the way cut fish, such as fillet, is neatly folded.	It is used for poaching or steaming. In the Oriental kitchens some flavourful herbs such as scallion, juliennes of ginger, etc. are stuffed before steaming.
Paupiette	This is the thin escalope of a fish that has been flattened.	This cut is often used for stuffing and making roulades. Since the fish is very delicate it is usually rolled in a plastic to retain the shape while cooking. This cut is usually poached.
Plaited	This is also known as en tresse in French. This cut is made by pleating three strips of fish to give a decorative touch.	One can be more creative by pleating strips of different coloured fish such as white fish and pink salmon. This cut of fish is usually poached, crumb-fried, or steamed.
Goujon	This is strips of fish culled from the large fillet and is usually of 8 cm x 1 cm size.	These are usually crumbed and deep-fried and served as hot snacks. This preparation is called fish fingers.
Goujonettes	Same as goujons but half the size of them.	Same as goujons.
Troncon	The concept is same as that of a darne; but this cut is from the flat fish.	Same as darne.
Pave	This is similar to supreme but is always cut in a block or a rectangular slab.	Ideally used for pan-frying or grilling. This cut is most commonly used in fine dining kitchens and it is often served with skin on.
Cravate	It is usually popular only in France. The concept is same like plaited but here the strip of the fish is tied into a knot.	Cooking procedures and methods are similar to plaited.

SOME FAMOUS SPECIES OF FISH

There are more than 30,000 species of fish. There are certain fish that are specific to a country or place and survive in that particular habitat. These are known by the local names where they come from. Some fish are found around the world but with different names. Some of the species of the same fish are of different tastes and even different colour, for example, trout in India is white fleshed but the one found in Yarra River in Australia is pink like a salmon. Let us discuss few of the most common fish used in hotels.

Ahi

This is a type of tuna that can reach up to the weight of 120 kg. The meat of this tuna is pale pink in colour and has a mild flavour. This is generally eaten raw in sushi and sashimi. It is also commonly known as yellow fin tuna, because of the yellow tinge on its tail.

Albacore

It is a highly priced tuna which weighs between 5 to 30 kg. It has high fat content and its the only species of tuna that is white fleshed. It is also canned and its one of the most expensive canned tuna.

Anchovy

This is a small, silvery fish and mostly comes from the Mediterranean and the southern European countries. Anchovies are perishable and that is why they are usually salt cured and canned in oil.

Monkfish

This is a firm-fleshed fish found in salty sea and has a mild to sweet flavour, which is often compared with a lobster. This fish is at times referred to as 'poor man's lobster'. This fish is also known by other names such as angler fish and goose fish.

Bonito

This is the smallest variety of the tuna family and can be moderately to highly fatty. This is one of the strongest flavoured fish amongst the tuna family. This is mostly used in Japanese cooking in dried form to flavour stocks such as dashi and other dishes.

Bacalao

This is a salted cod from the Mediterranean belt, especially Italy and Spain. Processing bacalao for dishes is more of patience than an art, as one has to soak this fish for around 48 hours, constantly changing the water to get rid of most of the salt. This is then only poached or pan-fried. Overcooking of this fish would make it stringy.

Bass

It is found both in seas and rivers. It belongs to the family that consists of grouper, black sea bass, and striped bass. In Australia, bass is also known as *sea perch*, while sea bass is commonly known as *Barramundi*. In Kolkata, the river sea bass is known as *Bhekti*.

Big Eye

This fatty tuna is also called *mebachi* in Japanese cooking, where it is used the most. This tuna has a deep red wine coloured flesh and the belly is often the high-prized cut, which is relished raw in sushi and sashimi.

Black Cod

This salt water fish, which actually is not a true cod, is mildly flavoured with sweet flavour and high fat content. It is often smoked.

Blue Fin Tuna

This highly prized tuna is now difficult to find, as it is the most sought after fish for Japanese sushi and sashimi. This giant fish can grow upto 450 kg. As this fish ages, the meat tends to get stronger in flavour and deep red in colour.

Bombay Duck

This is not a duck; but is a fish found in the Arabian sea along the coast of west India. It can be cooked fresh or is often dried to be used as pickles, and crisp-fried as starters. It is also called *bombill* in Maharashtra, India.

Carp

This fish is usually found in freshwaters and ranges from 1 to 4 kg. This lean white-fleshed fish is often used in the Jewish cuisine.

Catfish

It is a firm-fleshed fish, with low fat content and a mild flavour. Its firm flesh makes it ideal for curries. This fish is called by this name due to the unique shape of its mouth, which is round and has whiskers that resemble those of a cat. Catfish is known by different names in India. In eastern India it is commonly known as *Magur* whereas in north it is called *Singhara* fish. Most of these are found in fresh water but some of the species can also be found in salt water.

Cod

It is very popular in the west, especially the Mediterranean region. It is mainly found in the Pacific and the North Atlantic Ocean. Liver of cod is widely used to prepare a paste called taramasalata which is served as a starter.

Salmon

It is usually found in the Pacific and the Atlantic Ocean. This fish has a very delicate flavour and is of pink to orange hue. It is the most sought after fish on the fine dining table and there can be many species of salmon such as:

Chinook Salmon   It is also known as the king salmon. It can weigh upto 60 kg.

Chum Salmon   It is one of the most delicate flavoured salmon. It has a very light colour and the lowest fat content. It is also called 'dog salmon'.

Coho Salmon   This is commonly known as 'silver salmon' because of its silvery skin. This salmon is high in fat content and has a firmer flesh.

Trout

This fish belongs to the family of salmon and is normally found in fresh water and lakes. There are many species of trout and the most popular one is the rainbow trout, which is also sold under the name *Kullu trout* in India as it is farmed in the Kullu district of Himachal Pradesh. In India trout farming first started in Kashmir. Certain varieties of trout live in sea as well, for example Artic trout and another popular trout from Australia known as the Yarra river trout has pink coloured flesh like salmon.

Sturgeon

The sturgeon is rarely consumed for eating, and is mainly used for its eggs, known as caviar. Caviar is considered a luxury product and is therefore, used in high end hotels. However fishing of sturgeon is now banned as the species is on the verge of extinction and hence has been placed under the category of protected wildlife.

Kingfish

Also commonly known as seer fish or *surmai* in Hindi, it is an ocean fish which is also referred to as the large mackerel fish found in sea and oceans. The oily flesh type of this fish is ideal for making fried fish and is popularly used in a Punjabi dish called *Amritsari machi.*

Mackerel

Mackerel is an oily fish that is rich in Omega 3 fatty acids. The type and the name of the mackerel depends upon the size and the depth at which it lives in the ocean. Mackerels have a shiny skin and are characterized by thin vertical stripes that run across their body.

Sardines

These are small oily fish that are found abundantly in the sea and the ocean. They usually belong to the herring family of fish and are known by this name as they are found in abundance in the sea of Sardinia. Sardines are commonly grilled or pickled and used in Mediterranean cuisine.

Red snapper

It is one of the most popular white fish to be used in hotels worldwide. The origin of this fish is considered to be the Gulf of Mexico and the Indian Ocean. Snappers have a delicate flavour and can pair very well with fresh herbs, chillies, and seasonings. This fish has a red skin and soft white flesh. It can be poached, steamed, grilled, or even pan-fried.

Mahi Mahi

This name comes from the Hawaiian word which means 'very strong'. This fish is found around the region of Gulf of Mexico and Spain and is also known as *Dorado* in Spanish, which means silver colour. Mahi Mahi can grow up to a size of 25–30 kg and has a firm white flesh that makes it ideal for grilling and roasting.

Herring

Normally herrings and sardines belong to the same family and are rich sources of Omega 3 fatty acids; when the fish are small they are known as sardines and when they become large in size, they are known as herrings. These fish are oily in nature and in England, they are often poached in milk and served for breakfast.

Dover sole

It is one of the most prized fish on a chef's table due to its unique flavour and texture of meat. This fish lives in the sea bed of cold waters of the Atlantic and Mediterranean seas and is most suitable for many fish preparations including grilling and deep frying. Dover sole is a flat fish and yields 4 fillets that can be used in various ways.

Turbot

Turbot is a flat fish that thrives in Atlantic, Baltic, and Mediterranean seas and lives in brackish waters. The white and firm flesh of turbot makes it the favoured choice of many chefs due to its retention of white colour even after cooking. This fish can be pan-fried, grilled, and even roasted.

Halibut

It is a flat fish from the flounder family and owes its name to the holy days of Catholics, when it was one of the most popular dishes to be eaten. It is found around the Atlantic coast and is used in many preparations such as pan-fried, grilled, or even fried.

Yellow tail

This is one of the most prized fish in Japan as it lives around the coast of the Pacific Ocean from Japan to Hawaii. It is from the family of fish classified under the name of Japanese Amberjack. The delicate flavour of this fish allows it to be relished both pickled as well as raw in sushi and sashimi. When the yellow tail fish weighs about 3 kg, it is known as *Hamachi* and when the size grows to about 5 kg it is sold under the name of *Buri*.

Pomfret

This is a common fish available in both sea and ocean. Its habitat includes oceans such as Atlantic, Pacific, and Indian and seas such as the Arabian and the Mediterranean. There are many varieties and species of pomfret but the most common ones are white pomfret and Chinese black pomfret. Even though the fish has a flat body, it is still a round fish with delicate and firm flesh. In India it is commonly used for making curries and for preparing tandoori whole pomfret.

SOME CLASSICAL PREPARATIONS OF FISH

Some classical preparations of fish are discussed below.

Rollmop

It is a boned herring fillet, marinated with spiced vinegar, rolled around chopped onions and a piece of gherkin and served with a cocktail stick. Before being rolled up, sometimes the fillets are spread with mustard and sprinkled with capers. A marinade flavoured with juniper berries, cloves, and black peppercorns is poured cold onto the rolled fillet. The rollmops are then left to marinate for five to six days in a cold place and are then served as a cold hors d'oeuvre with parsley and onion rings. They can be bought ready prepared, imported from Denmark or Germany.

Ceviche

It is a dish characteristic of Peruvian cookery, that is based on raw fish marinated in lemon juice and is served with sweet limes, onion rings tomatoes and boiled sweet corn.

Escabeche

The name comes from the word *cabeza* which means a head. It is a spicy cold marinade intended for preserving cooked foods that originated in Spain. It is used chiefly for small cooked fish such as sardines, mackerel, and whiting. The fish are fried or lightly browned; they are then marinated for 24 hours in a cooked spicy marinade.

Nage

It literally translates to swimming. It is an aromatic court bouillon in which crayfish, langouste, small lobsters, or scallops are cooked. They may be served either hot or cold in the cooking stock, which is either seasoned or mixed with fresh cream. Dishes prepared in this way are called *à la nage*.

Sushi

A Japanese dish of cooked seasoned rice, served with a variety of other ingredients, usually raw fish, shrimps, prawns, and vegetables.

Sashimi

A style of serving raw fish. Absolutely fresh fish usually from the sea are used as these fish are healthier and not contaminated. The fillets are thoroughly washed and are thinly sliced and then neatly arranged on the plate. Delicate garnishes are used and the fish is served with wasabi[1] paste, pickled ginger which is called *gari* in Japanese language, and a soya sauce. Wasabi is eaten with sashimi so that it can prevent any food poisoning and the pickled ginger helps to freshen up the mouth after every bite.

SELECTION AND STORAGE OF SEAFOOD

Let us discuss about the selection and storge of seafood.

Selection

There are many different ways of selecting a fish and these do not require any sophisticated tools or deep knowledge. One can easily segregate a good quality fish by carrying out the organoleptic tests, which means using senses such as smell, touch, etc.

Feel the Fish   The fish should be free from slime and should not feel dry to touch. Moist fish indicates the freshness. When the flesh is depressed, it should feel firm and should spring back to its original position. The scales should be firmly attached to the fish.

Smell the Fish   It should have no foul smell. This is probably the easiest way of knowing the difference between a fresh and a stale fish. Since it is highly perishable commodity it will smell very unpleasant if it is spoilt.

Look at the Eyes   The eyes should be bulging and clear. As the fish would age, the eyes would lose the moisture and would sink into the head.

Touch the Fins and the Tail   The fins and the tail should be firm and flexible. Brittle or dry tail is an indication of a spoilt fish.

Check the Gills   The gills should be moist and should have a deep red colour with no trace of grey or brown.

In case of shellfish, try and procure only live shellfish such as lobsters and crabs. Shells of bivalves should be tightly shut and the ones that are open, tap them slightly and if they still don't shut then discard them.

Storage

In ideal situations the fish should be procured as fresh as possible, but in case if it is not possible for various reasons, then one must follow these steps to keep the seafood safe and fresh to eat.

- Rinse the fish in cold water and pat it dry.
- Place the fish on bed of crushed ice and place on the perforated pan to allow for drainage. Cover the fish with additional ice and store in the refrigerator until usage.
- If the fish is not used in the next two days, then take out the fish and repeat the above procedure. While we do that it is important to carry out the organoleptic checks on the fish.
- Clams, mussels, and oysters should not be kept on ice. Store them in mesh bags in the refrigerator. Oysters stored this way would remain alive for at least two to three weeks. Some restaurants have fish tanks where artificial sea water is created to store shellfish such as lobsters and prawns.

---

[1] A pungent root used in Japan. It is antiseptic in nature.

- Do not let the fish to touch the water or to swim in water.
- Do not let the shellfish come in contact with fresh water, otherwise they will get killed.

COMMON COOKING METHODS USED FOR SEAFOOD

One can test for doneness by pressing the flesh with finger. If it is firm then it is cooked, otherwise it would feel soft. One can also insert a knife into the fish, pull back the flesh a little, and take a look. Some of the common methods of cooking used for seafood are as follows:

Grilling

Seafood is just wonderful on the grill. One has to select seafood that can withstand the heat of the grill, and can stand up to the hot iron grills. Use seafood, such as tuna, lobsters, prawns, etc., which are sturdy and do not fall or flake apart. Marinating in oil ensures that the fish does not stick to the grill. Using wood for grilling could also add flavour to the fish.

Broiling

You can broil most seafood even if they are fragile. You can broil them on a bed of vegetables such as chopped onions, celery, peppers, and fresh herbs. You can serve the vegetables right along with the fish.

Sautéing

Sautéing is probably the most common method of cooking seafood. It is a quick way of cooking seafood while maintaining optimum flavour. For best results, when you sauté, cook the side with the skin first (if it is left on).

Cold and Hot Poaching

You can poach fish or shellfish in two ways. Hot poaching cooks the fish in a boiling liquid. You can cold poach seafood by pouring simmering liquid over the fish and allowing it to stand until the fish is cooked. Hot poaching works well when the pieces of fish are large or when the fish is to be served hot. Cold poaching works well with small pieces of fish that do not take long to cook, or when the fish is to be served cold or at room temperature. Remember the more flavourful the poaching liquid is, the more flavourful is the poached fish.

Olive Oil Poaching

Poaching seafood in olive oil is a great way to slowly cook the fish. Cooking seafood in barely warm oil keeps the protein from coagulating. Olive oil poaching lessens the release of natural fat and oil from the fish, which helps to keep it moist. This slow cooking results in a succulent, perfectly cooked piece of fish. This method of cooking is usually used for oily fish rather than lean fish.

Steaming

Steaming fish keeps it extremely moist and flaky. In fact, steaming is a great way to prepare many low-fat dishes and not just delicately flavoured fish. Steaming is very common in many Oriental preparations.

Roasting and Slow Roasting

The only difference between slow roasting and roasting is the temperature. Regular roasting takes place between 160 and 200°C. Slow roasting on the other hand takes place at temperatures as low as 80°C and barely cooking fish would take as long as 20 to 25 minutes. When you slow roast fish the end result is a piece that is so moist that it melts in your mouth and falls apart when you touch it.

Smoking or Curing

A number of popular fish products are smoked or cured or both or they are dried, pickled, or otherwise treated to enhance their taste and prolong their usability. Smoked salmon, pickled herring, smoked oysters, caviar (salted fish roe) belong to this group of fish foods.

Cooking Shellfish

Steamed clams, boiled lobsters—these terms express two things in common to all shellfish cookery—moist heat and low temperatures.

Shellfish are all very lean. Dry heat, high heat, and long cooking times will make them tough and rubbery. Boiling, broiling, steaming, and baking will be the broad spectrum for cooking shellfish.

CONCLUSION

Fish is an aquatic vertebrate that is available to humans in more than 30,000 species. Whatever may be the habitat, whether sea, lake or ocean, these vertebrates are referred to as seafood. There fish and shellfish are classified separately and this classification can be based on their habitat, shape or type of flesh for example oily or lean.

Fish do not have fat like meats; but when we refer to fatty fish, we imply to the oil content present in it. The taste of the fish from the sea or ocean is different from the fish caught in rivers. Cold or warm water also impact the taste of seafood and thus its cost. It is important for chefs to know the shapes of the fish as this would tell them how they should process the fish and what cuts can it yield; also an oily fish will have a different process of treatment even if it is being cooked with the same method as any other fish. For example, a lean fish can be poached in water, milk or any other liquid such as stock, but an oily fish needs to be poached only in court bouillon[2] as the acid in court bouillon will help to firm up the fish by withdrawing oil

---

[2] Refer to chapter on stocks

from the same. Shellfish on the other hand is classified according to the species, for example crustaceans and molluscs. All the meat of the shellfish is concealed within a shell which is an exterior coating that also helps to protect it from its predators. There are some exceptions such as squids and cuttlefish that have a bony structure or a shell inside. We discussed in the chapter the different kinds of shellfish, the various cuts that can be obtained from various fish and the uses of the same. Some of the most common fishes used in our daily operations were discussed and last but not the least, the importance of procuring fresh and live fish and storage of the same in case there is an issue of procurement was also detailed.

In this chapter we have also discussed the many important and common varieties of seafood that are being used in most of the hotels and restaurants in India and abroad. Few species such as trout and its types, sea bass, sardines, mahi mahi, and snapper are discussed in this chapter. Along with the types of seafood, we have also mentioned the methods of cooking that are best suited for different types of seafood.

KEY TERMS

Abalone Univalve from sea, commonly eaten in China.

Ahi Type of tuna that has a light pink coloured meat.

Albacore Type of tuna that has high fat content. Used in canning and is very expensive.

Amritsari machi A deep fried fish preparation from Punjab.

Anchovy A tiny variety of fish that is mostly cured and preserved in oil.

Aphrodisiac Food that activates sexual hormones.

Bacalao Salted cod from the Mediterranean.

Barramundi Australian nomenclature for sea bass.

Bhekti Indian name for sea bass.

Big eye Species of tuna that has a maroon red coloured meat.

Bivalve Shellfish that has two shells.

Bombay duck Species of fish that is found in the Arabian sea near Mumbai in India.

Bonito Species of tuna that has a strong flavour.

Bucards French for cockles.

Buri Yellow tail fish that weighs around 5 kilograms.

Calamari French for squids.

Caviar Roes of sturgeon fish served as appetizers in classical French cuisine.

Cephalopod Literally means foot over heads. This is a category of molluscs.

Ceviche South American pickled fish cured in citric juices.

Chinook salmon Species of salmon also called king salmon.

Chum salmon Species of salmon having lowest amount of fat and is also known as dog salmon.

Clams A type of bivalve, eaten raw or cooked.

Cockles Heart shaped bivalves similar to clams.

Conch A type of univalve that looks like a large snail.

Coquille St Jacques French for scallops.

Cravate Cut of fish where a thick strip of fillet is tied into a knot

Crevettes French for prawns.

Crustacean Shellfish that live in a shell that acts as their armour, such as prawns, lobsters, etc,

Darne Cut of a round fish on the bone that is roughly 1–2 cm thick.

Delice Skinless fillet of a whole fish folded over.

Ecrevisse French for crayfish.

En tresse French for plaited.

Escabeche An art of pickling which originated from Spain and Caribbean islands.

Escargot French for snails.

Fillet Cut of fish which is boneless side of the entire fish that can be with or without skin.

Fruit de mer French word for seafood literally translates to fruits of the sea.

Gari Pickled ginger from Japan.

Gastropods Another name for univalves.

Goujon Cut of fish resembling the size of a finger, also known as fish fingers. It is usually 8 cm long and 1 cm wide.

Goujonettes Half the size of goujon.

Gutting Removal of internal organs from the belly of a fish.

Hamachi Yellow tail fish that weighs around 3 kilograms.

Hen Female crab.

Hommard French for lobster.

Langouste French for crawfish.

Langoustine French for scampi.

Magur Indian word for catfish also known as Singhara in Hindi.

Mallet A square metal hammer used for flattening a piece of meat.

Molluscs These are category of shellfish that have a hard shell in which the muscle grows and lives.

Moule French for mussels.

Mussels A kind of bivalve.

Nage Fish served in a reduced flavourful stock that acts as a sauce.

Oyster Type of bivalve that also yields pearls.

Paupiette It is a slice of fish fillet that is used for stuffing and rolling.

Pave Cut of skinless fish resembling a rectangular block.

Pieuvre French for octopus.

Plaited Three strips of fish interlaced together to form a plait.

Poisson French for fish. It is also the third course in classical French menu.

Poularde French for clams.

Quill Internal shell of squid that resembles a piece of transparent glass.

Rainbow trout Species of trout found in fresh water.

Rollmop Rolled fillet of pickled herring fish served as a starter.

Sashimi Japanese delicacy where a fresh fish is served raw. Usually the meat from the belly is used.

Scampi A variety of crustacean that resembles prawn.

Sea cucumber Same as abalone.

Soft shell crab Tiny crab in which new shell is developing.

Surmai Indian name for kingfish, often used in frying.

Supreme Cut of the fish from the fillet, usually of a weight of 80–100g and without any skin.

Sushi Japanese dish of cooked rice and fish.

Troncon Cut of flat fish on the bone.

Univalve Shellfish that has a single shell such as snail, etc.

Wasabi A pungent root having antiseptic properties, served with raw fish.

Whelks Type of univalve.

CONCEPT REVIEW QUESTIONS

1. How would you define seafood and how is it different from other meats?
2. What are the advantages of consuming seafood over other meats?
3. Why is seafood referred to as *fruits* de mer or fruits of the sea?
4. What are the essential oils found in some fish and how are they useful?
5. How is seafood classified?
6. What is the difference between the classification of fish and shellfish?
7. How is fish from the sea different from the river fish?
8. List few fish from fresh waters and few from salt waters.
9. What is the difference between flat fish and round fish?
10. List at least five flat fish.
11. What is the difference between crustaceans and molluscs?
12. What is the difference between cephalopods and gastropods?
13. Why do shellfish turn to orange colour on cooking?
14. What is a female crab called?
15. How would you differentiate between male and female crab and why is it important?
16. List few features to be kept in mind while buying crabs.
17. What are crawfish and how are they different from lobsters?
18. Define a crayfish. How would one cook them?
19. What are scampis and how are they different from prawns?
20. Write a short note on lobsters, their types, selection criteria, and usage.
21. How would you differentiate between an European lobster and an American lobster?
22. How are prawns received in hotels?
23. Name few common gastropods that are eaten raw.
24. How are snails served?
25. How do you ensure that whelks or conches cook well?
26. How would you select molluscs?
27. What are cockles?
28. How is a mussel processed?
29. What should be the considerations while buying oysters?
30. What are scallops and how should they be cooked?
31. List a few common cephalopods.
32. What is the difference between squids and cuttlefish?
33. List the various cuts of fish.
34. What is the difference between a darne and a troncon?
35. What is the difference between en tresse and 'cravate'?
36. How many fillets would you get from flat fish and round fish?
37. What is a goujon and what are its uses?
38. Give at least three species of tuna and their characteristics.
39. Give at least three species of salmon and their characteristics.
40. Name few classical preparations of seafood.
41. What are the selection criteria of seafood?
42. How should fish be stored?
43. What is unique about storing shellfish and how is it different from the regular fish?
44. List common methods of cooking that would be used for cooking seafood.
45. How would you cook an oily fish compared to a lean meat?
46. In which regions around India would you find trout fish?
47. Caviar is obtained from a particular species of fish. Name the fish.
48. What are the different names of bass fish around the world? Name at least three.
49. Which fish would you choose to make Amritsari mahi and why?
50. Name at least three varieties of fish that are rich in Omega 3 fatty acids.
51. Why is snapper a preferred fish in western cuisine?
52. What does the word mahi mahi signify?
53. Describe the fish yellow tail and its importance in Japanese cuisine.

PROJECT WORK

1. In a group of three to four, prepare various cuts of fish (both flat and round) and compare it with the cuts given in this chapter.
2. In groups, cook the recipes pertaining to various seafood from the recipes folder given on OUP website and present the same for critique and evaluation.
3. Visit a fish market and record your observations regarding various fish available. List them and classify them according to the parameters listed and also write down their local names.
4. In a group, research seafood from at least five countries and discuss the peculiarities of each.
5. Visit few restaurants such as Japanese, Korean, Spanish, and some speciality seafood restaurants and enhance your data bank of seafood recipes.

14

INTRODUCTION TO EGGS

Learning Objectives

After reading this chapter, you should be able to:
- identify various types of eggs and classify them
- understand the structure of eggs
- select eggs for various uses
- recognize the importance of selection and storage of eggs
- appreciate the usage of eggs in cooking to create various delicacies

INTRODUCTION

Science defines egg as a cell from which a living organism takes birth and grows. All animals (including birds) lay eggs, except mammals who give birth to babies. An egg laying animal lays eggs, no matter whether they are fertilized or not. In other words, it does not have to be mated to lay an egg. In order to develop into an embryo, an egg must be fertilized by sperm before it is laid. The fertilized eggs, under favourable conditions hatch into living organisms.

It becomes crucial for humans to select good quality eggs which should be freshly laid, as one does not want to see a structural form of living organism, when an egg is broken. The eggs that we usually get in the market are unfertilized eggs.

There are many varieties of eggs found around the world, but only a few are used for human consumption for various reasons. The eggs can be of fish, poultry, game birds, or even reptiles; but in cooking when we refer to eggs we are always talking about poultry and eggs of birds that are reared for consumption of meat. But then eggs from ducks and even quails have a very special place on the gourmet tables. Eggs can be of various colours and patterns and sizes; the only thing common among eggs is their natural oval shape. A healthy chicken will lay one egg in a day and this will largely depend upon its diet and the time of the day, as sunlight affects the production of eggs. Artificial lights are thus provided in commercial farms to maximize the production of eggs. The size of the egg is largely dependent upon the diet of the hen and also its age. The older hens would lay larger eggs. The colour of the eggs, however, has no bearing upon the colour of the hens; it is probably a fad but many people believe that brown hens lay brown eggs and the white ones lay white eggs. The hens usually cease to lay eggs when they are around three years old. The colour of the yolk however depends largely upon the diet that is fed to the chickens. Feed containing yellower corn will yield yellow yolks as compared to wheat or barley. Sometimes natural products, such as marigold flowers are also added to the feed to darken the colour of the yolks. An egg is a rich source of protein as it has two types of proteins. Egg white contains 'albumen' and yolk contains 'lecithin'. Yolk has fatty compounds and is high in cholesterol, and that is the reason why only egg white is consumed as a healthy option in breakfast compared to a whole egg. Eggs can be put to many uses. Apart from being relished in the breakfast as omelettes, poached or boiled, it can be cooked in curries or even whipped up for delicate desserts and cakes. Eggs can be used for thickening or simply paired with milk to create sauces. Eggs and oil or butter emulsion also form sauces such as mayonnaise and hollandaise. Egg is a versatile commodity and chefs can put it to numerous uses.

STRUCTURE OF AN EGG

Let us discuss the structure of an egg (refer to Fig 14.1) and then we shall discuss different types of eggs and their uses.

Shell

It is the outer covering of the egg and is composed of calcium carbonate. It may be white or brown depending upon the breed of the chicken. The colour of the shell does not affect cooking quality, character, or nutrition.

Yolk

This is the yellow portion of an egg. Colour of yolk varies with the feed of the hen, but does not indicate the nutritional content.

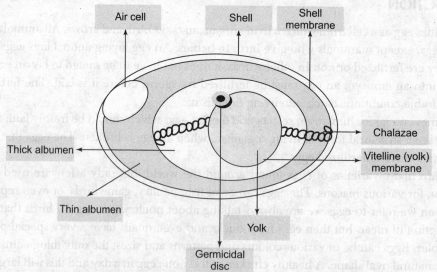

Fig. 14.1 Structure of an egg

Vitelline
It is a clear seal that holds the egg yolk.

Chalazae
These are the twisted cordlike strands of the egg white. They anchor the yolk in the centre of the egg. Prominent chalazae indicate high quality.

Shell Membranes
Two shell membranes, inner and outer membrane, surround the albumen. They form a protective barrier against bacteria. Air cell forms between these membranes.

Air Cell
It is the pocket of air formed at the large end of the egg. This is caused by the contraction of the contents on cooling after the egg is laid. The air cell increases with the age of the egg as there is considerable amount of moisture loss. The eggs are stored with the larger side facing up to keep the yolk in the centre.

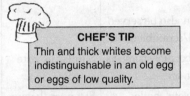

CHEF'S TIP
Thin and thick whites become indistinguishable in an old egg or eggs of low quality.

Thin Albumen
It is nearest to the shell. When the egg is broken there will be a clear demarcation of the thin and thick albumen. As the egg gets older these two albumens tend to mix into one another. This is again a test of good and fresh egg.

Thick Albumen
It stands high and spreads less than the thin white in a high quality egg. It is an excellent source of riboflavin and protein.

CLASSIFICATION OF EGGS

Eggs can be classified into different types as discussed in Table 14.1.

Table 14.1 Classification and types of eggs

Basis of Classification	Source (Bird)	Farm and Feed	Grade	Sizes	
				US	European
Types	Ostrich	Battery farm	AA	Jumbo	Extra large
	Goose	Barn laid	A	Extra large	Large
	Duck	Free range	B	Large	Medium
	Turkey	Organic	C	Medium	Small
Types	Hen	Vegetarian		Small	
	Bantam	Omega-3		Peewee	
	Pheasant	Vitamin enhanced			
	Guinea fowl				
	Quail				

Source of Eggs

Eggs are named after the birds that laid the eggs. Eggs can be distinguished by size and colour of the shell. Various eggs are listed in descending order of their average weights in Table 14.2.

Table 14.2   Source of eggs and their characteristics

Name of Bird	Average Weight (g)	Colour of Shell	Remarks	Taste*
Ostrich	450	Pale yellow	Best used in baked dishes.	Stronger flavour
Goose	200	Chalky white	Best used in baked dishes.	Stronger flavour
Duck	90	Vary from pale blue to white	Higher fat content oilier texture, being rich produces good sponges.	
Turkey	75	Creamy white with light brown specks	Farmed for meat, therefore eggs are reserved for hatching.	Similar in taste
Hen	60	Vary from white to brown	Ideal for many uses, can also be served on its own, fried/boiled/scrambled, etc.	
Bantam	30	White	Can be used as hen's egg, but more numbers are needed as it is smaller in size.	Similar in taste
Pheasant	25	Buff to blue–green	Can be used in many ways, boiled, baked, etc.	Stronger flavour
Guinea fowl	25	Light brown	Ideal for garnishes, salads, even in baked dishes.	Delicate flavour
Quail	18	Pale yellow with dark specks	Light and creamy texture, ideal for garnishes (hard-boiled).	Delicate flavour

* Comparisons made are judged with respect to hen's egg

Farm and Feed

Eggs are named after the farming method employed or the feed provided to the bird to enhance a particular nutrient in the egg. The various eggs falling under this classification are listed in Table 14.3.

Table 14.3   Eggs on the basis of farm and feed

Type	Description	Remark
Battery farm*	These are the most regular commercially produced eggs. In this method, hens are kept in small cages (three to seven per cage) and are fed a high protein diet. This method enables keeping large number of hens in smaller shed, thus there is more yield of eggs at lesser cost.	Regular egg
Barn laid	This method is slightly better for the hens as they are kept indoors in straw covered barns, which are separated into pens and not congested cages. Nutrientwise it is much the regular egg.	Regular egg

Contd

Table 14.3   (Contd)

Type	Description	Remark
Free range	This method of farming involves keeping hens in barns and at the same time, allowing them access to open space with vegetation. The hens get a good amount of exercise, but as the eggs are laid all around the barn and open spaces, egg safety and collection is a challenge, thus increasing cost.	Regular egg
Organic	These are eggs produced from birds, which are fed an organic diet. An organic diet is one which is produced without use of chemicals such as pesticides and fertilizers. These hens are not even fed any hormones or chemicals otherwise used to enhance their growth.	Regular egg free from chemical traces
Vegetarian	These are eggs produced from birds that are fed a vegetarian diet. A vegetarian diet is free from any meat or fish products. Nutrientwise it is much the regular egg.	Regular egg
Omega-3	These are eggs from birds that are fed a diet of canola, linseed, and flaxseed, which are rich vegetarian sources of Omega-3 fatty acids. Omega-3 is beneficial for the heart.	Regular egg with Omega-3
Vitamin enhanced	These are eggs from birds that are fed a diet rich in certain vitamins with an intention of enhancing the particular vitamins' content in the laid egg. Usually birds are fed with a diet of vitamin E, B6, or B12, thus enhancing their content in the egg.	Regular egg with enhanced vitamin content

* Battery farming is the most popular standard farming method employed all over the world.

GRADE

Eggs are named after the quality grades awarded to them based upon set quality factors/parameters. The various eggs falling under this classification are listed in Table 14.4.

Table 14.4   Grading of eggs

Grades*	Quality Factors		
	Shell	Egg White (on breaking)	Yolk (on breaking)
AA	Clean, unbroken	Clear, without spots/clots; thin and thick albumin distinct, with a layer of thick albumin covering yolk	Well centred and rounder in shape.
A	Clean, unbroken	Clear, without spots/clots; thin and thick albumin distinct	Fairly centred and rounder in shape.
B	Clean to slightly stained	Clear, without spots/clots; very less or no distinction between thick and thin albumin	May be off centre and flat in shape.
C	Slightly to moderately stained	Spots or clots may be present; no distinction between thick and thin albumin, egg white is watery and spread out	Off centre and flat in shape. In some cases, yolk even breaks and does not hold shape on breaking shell.

* Food production outlets usually receive AA or A grade eggs for their use.

Sizes

Eggs are named after their sizes, which are assigned to them based upon their weights. The sizing is based upon the weight per dozen of eggs, but is represented as weight of individual egg after taking out mean weight per egg for ease. There are different size nomenclature for classifying eggs on basis of size. Two of the most popular size classification—US and European—are as given in Tables 14.5 and 14.6. In India the European classification is usually followed.

Table 14.5 Classification of eggs by US standards

Egg Size	Weight (g)
Jumbo	73
Extra large	63
Large*	53
Medium	45
Small	40
Peewee	35 and below

* *Large is the most common size.*

Table 14.6 Classification of eggs by European standard

Egg Size	Weight (g)
Extra large	73
Large	63
Medium	53
Small	45 and below

TYPES OF EGGS

We saw the various ways in which eggs are classified. We shall now discuss various types of eggs commonly used in kitchens all over the world. Table 14.7 shows various kinds of eggs.

Table 14.7 Types of eggs

Type	Description	Photograph
Chicken eggs	These are the most commonly eaten eggs around the world. They are available in brown colour and white colour. The brown coloured ones are referred to as *desi eggs* in India.	
Duck eggs	They are darker in colour than chicken eggs and are larger in size too. The duck eggs are stronger in flavour and are always eaten very fresh, as the flavour intensifies with age.	
Goose eggs	Similar in size and colour to duck eggs, goose eggs taste slightly oily as they have more fat content.	
Guinea fowl eggs	They are flecked with brown colour and are boiled between 3–5 minutes and served in salads.	

Contd

Table 14.7 (Contd)

Type	Description	Photograph
Gull eggs	Since sea gulls prey on seafood, their eggs are also fishy in flavour and hence are valued. These are usually boiled for 5 minutes and served cold with celery salt. They are smaller than chicken eggs.	
Ostrich eggs	These weigh around 500 g and are 10 times bigger than chicken eggs. One egg can feed four people and is used in the same way as a chicken egg.	
Partridge eggs	They are tiny in size and cooked medium boiled for 2 minutes.	
Pheasant eggs	These eggs have a natural pinkish hue and are around the size of a quail egg. They can also be used as chicken eggs.	
Quail eggs	They are speckled and slightly brown in colour. They are usually one-third the size of a chicken egg and are usually served cold or set in aspic jelly.	
Plover eggs	They are very similar to quail eggs and are considered to be a delicacy. They are usually served soft boiled.	
Turkey eggs	These are creamy white in colour and speckled with brown colour. At times a turkey egg can be twice the size of a chicken egg. The flavour is same as that of a chicken egg.	
Thousand year old eggs	Also called century eggs, these are a Chinese delicacy — old eggs that are cured for around 100 days. These duck eggs are coated with mixture of lime, salt, tea ashes, and charcoal and buried in the ground to mature. These are usually served shelled, sliced and are also served cold.	

SELECTION OF EGGS

CHEF'S TIP

If eggs of the best quality are desired, medium-sized ones that are uniform in size and colour should be selected. With regard to shape, they should have a comparatively long oval shell, one end of which is blunt and the other, a sharp curve.

Eggs available in the market are graded according to their freshness, cleanliness, size, cracks, and colour. With the exception of their freshness, these points can be readily told from the appearance of the eggs. But in order to determine whether an egg is fresh or not, it is put through a process known as 'candling', by which the interior condition of the egg can be ascertained. This method of determining the freshness of eggs consists of placing a piece of cardboard containing a hole, a little smaller than an egg between the eye and a light, which may be from a lamp or an electric light, and holding the egg in front of the light. The rays of light passing through the egg show the condition of the egg, the size of its air space, the growth of mould, or the spoiling of the egg by any ordinary means.

Another way of judging the quality of eggs is by observing the condition of the surface of the shell. When eggs are freshly laid, the shell is covered with a substance called 'bloom' that gives it a feeling much like that of a thin lime coating deposited in a pan after water boils. This coating disappears gradually as the egg is exposed to air, but as long as it remains, the egg may be considered as fresh and germ-proof. While this way of determining freshness is probably the quickest, it is possible that the quality of some eggs from which the bloom has recently disappeared has not been injured.

One can determine the freshness of an egg by shaking it. When the water inside the shell evaporates, the yolk and white shrink so much that they can be felt moving from side to side when the egg is shaken. More stale the egg, more pronounced the movement becomes. This method should be applied only immediately before the egg is to be used, as the thin membrane between the yolk and the white and the spiral cords that hold up the yolk are liable to be disturbed by the shaking. A test for freshness consisting of placing the eggs in glass containing water will be found effective. A perfectly fresh egg will sink when it is put into the water, but if the egg is three weeks old the broad end will rise slightly from the bottom of the glass. An egg that is three months old will sink into water until only a slight portion of the shell remains exposed; whereas, if the egg is older or stale, it will rise in the water until nearly half of it is exposed.

Check for cracks in the shell which could let bacteria in and contaminate the eggs. They should be less than 21 days old. The yolk should be plump and there should be two layers of white. If the egg is stale the yolk becomes flat ant the distinction between the two layers of white is lost. The final test will be the smell; bad eggs will have an unpleasant odour.

STORAGE OF EGGS

The storage of the eggs could mean two things. First, there is the storage in fridge for daily use and then there is industrial storage that is done at the warehouses for retail markets. The method of industrial storage does not concern chefs as much as the storage factors of the eggs once they are received into the hotels. In the warehouses eggs are kept little above freezing point and the humidity of air and the amount of carbon dioxide in the air are controlled. They will keep about nine months under those conditions. The other methods of long storage of eggs can be achieved by:

Freezing The eggs are washed, sanitized, and then broken into sterilized containers. After combining yolks and whites, they are strained, pasteurized, packed, and quick frozen.

Drying Eggs are broken well mixed and then spray dried at a temperature of approximately 71°C (used by bakers and confectioners).

Let us now talk about the storage of eggs that are received in the hotels.

- Eggs should be used within a month and stored unwashed, with the pointed end down, in the cold part of the refrigerator. Washing the egg makes the shell permeable to smells. So strong smelling foods such as cheese, onion, and fish, should not be stored near the eggs because the egg shells are porous and the egg will absorb strong odours.

- A hardboiled egg will keep for four days if unshelled and two days if shelled. Hard boiled eggs pickled in flavoured vinegar and sterilized will keep for months.

- Once broken, egg yolk will keep for 24 hours and egg white will keep for 6 to 12 hours in a refrigerator.

- A dessert containing raw eggs such as mousse should be eaten within 24 hours. Fresh eggs can be frozen if they are broken into a bowl, beaten and poured into suitable containers.
- These days egg pasteurizers are available in almost all the hotels and it has become a standard to pasteurize the eggs before they arrive in the hotel.
- One must also receive and store eggs in a plastic crate rather than a cardboard box as paper breeds germs and invites pests.

USES OF EGGS

Most of the eggs are enjoyed on their own, served boiled or fried. Eggs also help to make many dishes successful. They add colour and taste to several dishes. We can use eggs in many roles, but these are governed by the major cooking functions performed by eggs.

So, before discussing the uses, it is important to know the three major cooking functions performed by eggs. These are:

- Coagulation
- Leavening
- Emulsification

Coagulation

It is the firming up of the protein on application of heat. We can vary the coagulation to our taste, and have it as soft boiled, hard boiled, fried, or scrambled. When too intense a heat is used, the eggs become over-coagulated. Eggs coagulate at 65°C and continue to thicken till 70°C. This is below the boiling point of water, which is why the eggs should always be cooked over a low fire.

If other ingredients, such as milk and cream are added to the eggs, it may raise the coagulation point. This helps to get a softer scrambled egg as eggs never over coagulate to a hard texture.

Uses of eggs based upon coagulation are given below.

- The process of coagulation thickens custard and sauce.
- Coagulated egg protein helps support cream puffs, cakes, and breads. It binds together foods as in meat loaves and burgers. It also coats foods in the form of egg-based batters.
- To clarify a consomme, a chef beats in egg whites into the soup. The white coagulates, trapping impurities within its strands. This is because, when the egg coagulates, the protein molecules that can be visualized as a network of strands, contract. The network becomes denser because new protein bridges are formed that interlink the molecules tighter.

Leavening

The effectiveness of the leavening depends on the amount of air trapped within the egg. Yolks when beaten transform into thick light yellow foam. Egg yolks, however, hold less air between their molecules as compared to egg whites that have a power of trapping in air bubbles of large size; thus they can be whisked to a very large volume.

A little acid in the form of cream of tartar or a squeeze of lemon juice helps stabilize the foam. When heat is applied, the air in the cells trapped by egg whites expands, increasing the volume to form a light porous structure desirable in a leavened product. However, if over-beaten, the egg whites stretch to a thin film, on application of heat, more expansion breaks this thin film, which means the product rises gloriously at first, but falls back/collapses later.

It is also important to separate the whites and the yolks as even a small trace of yolks will prevent the whites from rising properly.

Use of eggs based upon leavening:

- Eggs are used for making baked goods such as sponges, cakes, etc. Leavening of eggs gives these products a lighter texture that is desirable.

Emulsification

Egg yolk acts as an emulsifying agent, because its protein can wrap itself around tiny globules of oil. Yolk also contains lecithin, which is an emulsifying agent.

Uses of eggs based upon emulsification:

- Oil is added to the yolks to form mayonnaise. The oil is beaten up into tiny droplets, then each drop is caught in an egg yolk protein film, so the droplets remain dispersed in a stable emulsion.

- The emulsifying power of egg yolks also contributes to the crumbly quality of a rich cake.

Apart from these cooking functions, eggs are also used in varying proportions in batters and dough to add colour, flavour, and texture to the products.

COOKING OF EGGS FOR BREAKFAST

Eggs find a special place on the breakfast menu. It is eaten by almost 80 per cent people in the world for breakfast. The egg counters are the busiest in breakfast preparations. Eggs are prepared in various ways but the most common cooking methods used for breakfast are boiled, poached, pan-fried, and also deep-fried.

Boiled

Boiling of the egg can be quite tedious, especially when the guest asks for an egg boiled for a typical time. Some hotels have egg timers to monitor the same. But though they are called boiled, the water should be at a simmer and not at a rolling boil.

The eggs should preferably be at room temperature to prevent the shell from cracking.

Like most other egg preparations which have varying degrees of doneness, boiled eggs are cooked hard, soft, or medium.

Soft boiled—3 to 5 minutes

Medium boiled—7 to 8 minutes

Hard boiled—10 to 15 minutes

The boiling time referred to above is for eggs that are at room temperature and placed in water that is already simmering.

The time spans given are broad due to the different sizes of eggs. For smaller eggs lower the time, and for larger eggs increase the limit. Never overcook a hard-boiled egg as the white would get tough and the egg will darken around the yolk.

Fried Eggs

A fried egg has to be glossy, tender, and moist no matter what the degree of doneness is. Always break the eggs into a bowl, so if there are any blood spots or a broken yolk you can save the egg for some other preparation.

Use a well-seasoned pan, heat it over a low fire till warm, add butter but remember the temperature should be just warm, when you slide in the egg. Deep-fried eggs are traditionally fried in hot oil; but mostly when we refer to fried eggs on the buffet they are pan-fried.

Sunny Side Up   Fried but not turned over or flipped over, the white should be cooked firm and the yolk should be cooked medium.

Over Easy   It is best to make this by placing a sunny side up fried egg under a heat source. The thin film of albumen, which coats the yolk, will cook and form a white layer over the yolk.

Turned Over or Over Well   This is a sunny side up fried egg that is flipped over. Both the yolk and white are to be firm.

Scrambled Eggs   Whites and yellow are beaten together, then fried and broken up while cooking. Scrambled eggs are normally served soft. Some chefs add milk to the beaten egg.

Poached Eggs   It is a simple process, but there are several factors that are critical in achieving a good looking poached egg—one that has a compact, glossy, obviously tender white surrounding an unbroken somewhat thickened yolk. Three major factors that influence the making of good poached egg are:

- The first is the quality and the freshness of the egg. The white will looked ragged and unattractive if the egg is old.
- The second factor is the poaching liquid. This is a liquid with water, salt, and vinegar. Normally a teaspoon of salt and 15 ml vinegar for about 500 ml water should be used. The acid in the liquid helps the protein to coagulate faster and hence retain the shape.
- The third factor is the temperature; it should not be too hot as to create agitation, but hot enough to start cooking the white as soon as the egg hits the water.

Make sure there is enough liquid so that the water temperature does not drop drastically when you introduce the egg. Break the eggs into a bowl and slip it into the side of the dish gently. When done, take them out with a skimmer or a perforated spoon and drain them well. Poached eggs can be served plain on buttered toast or English muffins.

Shirred Eggs

These are eggs which are cooked and served in the same dish, usually a shallow flat-bottomed earthenware dish. The dish is coated with butter and then the eggs are broken into it. Cook it over the fire for a while and then put it into an oven.

Omelettes

Cooking an omelette is difficult to master, as a lot of patience and practice is required. A perfect omelette is fluffy and moist, soft in the centre, yellow in colour with no brown at all. It should be cooked at 60–70°C. It should be oval in shape and one continuous piece. One can add fillings into the beaten mixture or can add it before making the first fold. Some chefs add a little water or milk.

Common pitfalls in making good omelettes are:

- Eggs brown and crisp—too much heat. Cook at 60–70°C.
- Egg white blistered—too much heat or too much fat.
- Odd shaped eggs—eggs not fresh
- Eggs sticking—too much heat, too little fat or porous cooking surface or pan not seasoned well

Seasoning a Fry Pan

A porous cooking surface is a nightmare for an egg cook, and a seasoned or primed surface is necessary. Water destroys a seasoned surface, and cooking a hamburger or anything else in an egg pan causes the same frustration. Egg pans should be used for eggs only. A seasoned surface is one that has been polished at high heat with salt and fat to overcome its porousness.

Following are the steps for seasoning a fry pan.

- Heat the pan on a hot fire.
- Add a few drops of oil and a handful of salt.
- Rub the salt and the oil with a cloth vigorously with pressure into the surface as though you are scouring it with a cleanser.
- Discard the salt and oil mixture and wipe the pan clean with a cloth.
- Repeat until you can cook an egg without it sticking anywhere in the pan.

CONCLUSION

Eggs discussed in the chapter are edible eggs of poultry that include ducks, chicken, and few other farm-raised birds and game birds also used for consumption of meat. Eggs of other animals are known by specific names, for example, eggs of fish are termed as 'roe', etc. There are many factors that influence the egg laying capabilities of birds. One of them is sunlight, and another is climate. Hens lay the greatest quantity and best quality of eggs in April and May. In winter, considerable number of eggs is preserved when they are comparatively cheap and abundant. If stored for too long they are characterized by musty odour and flavour.

Many varieties of eggs are consumed by humans, ranging from hen, to duck, turkey, partridge, and quail. These eggs are different in sizes and colours and also vary in taste. Some taste oilier such as duck eggs. Whenever the quality of an egg is judged, it is always benchmarked against chicken egg as this is the most common egg consumed around the world. The size of an average egg depends upon the breed, age, and feed of the poultry. For this reason eggs are classified according to sizes such as A, AA, and so on. Eggs are also classified upon the basis of source of the bird such as turkey, goose, duck, etc. or sometimes upon the basis of feed and farm such as organic, vegetarian, Omega-3, etc. Eggs are classified differently by the USA and Europe. There is not much difference between the classifications in these categories. In India we usually follow the European classification.

Utmost care should be taken while selecting the eggs, as these are highly perishable if they are not stored properly. Since one can only come to know the real quality of eggs after they are broken, it becomes essential to device some methods where an egg is judged by conducting few tests such as candling, where an egg is viewed against a light. The rays of light passing through the egg, gives an idea about the quality of an egg. The other common test is to float the egg in a bowl of water. A fresh egg would sink to the bottom and the stale egg would float on top.

Eggs find many other uses in the kitchens apart from simply being cooked into a fluffy omelette. Eggs are used in leavening to produce cakes and soufflés and the emulsification properties of eggs find its usage in thickening sauces and dressings. Check the recipes of classical dishes of egg in the online resources provided.

KEY TERMS

Air cell The air pocket at the large end of the shell.

Albumen Protein found in the white of an egg.

Barn laid These eggs are produced like battery farmed eggs; but here the hens are placed in bigger rooms covered with hay.

Battery farm Commercially produced eggs, where the hens are kept in cages and are fed high protein diet.

Bloom The dusty covering on a freshly laid egg, which disappears gradually on exposure to air.

Candling A process where the quality of an egg is judged against light.

Chalazae Fibrous strands that join the yolk and the white together.

Coagulation Firming up of proteins upon application of heat.

Dried eggs Dehydrated egg available in powder form.

Egg pasteurizer A machine used for pasteurizing eggs.

Leavening Incorporating air into dough.

Lecithin Protein found in the yolk of an egg.

Marigold Yellow flower often used as a natural colouring agent in food.

Organic eggs The eggs of hens that have been fed on organic diet.

Peewee An American classification of chicken eggs which are less than 35 g.

Riboflavin A type of vitamin.

Thousand years egg A cured egg for 100 days; a Chinese delicacy.

Vegetarian eggs Eggs produced from the birds that have been fed on vegetarian diet.

Vitelline A clear seal that holds the egg yolk.

CONCEPT REVIEW QUESTIONS

1. Define an egg.
2. What is the difference between a fertilized and an unfertilized egg?
3. Why are there artificial lights in the commercial egg producing farms?
4. What factors influence the size of an egg?
5. How are the eggs classified?
6. What affects the colour of the egg yolk?
7. Draw a structure of an egg and label its various parts.
8. What is the difference between shell membrane and vitelline?
9. What is chalazae and what is its use?
10. How are the eggs classified on the basis of farm and feed?
11. List at least five types of eggs and their characteristics.
12. What is the difference between vegetarian eggs and organic eggs?
13. What are the different grades of egg?
14. What do you understand by the word peewee?
15. How are eggs of duck different from chicken eggs?
16. What is a thousand year old egg?
17. What are the factors to be kept in mind while selecting an egg?
18. What are the various methods used for selection of eggs?
19. How should eggs be stored?
20. Which way should eggs be stored and why?
21. What are the various methods of preservation of eggs?
22. List the various uses of eggs.
23. How does egg help in emulsification process?
24. How are the eggs used in the clarification process of a consommé?
25. Why should eggs be stored away from strong odours?

PROJECT WORK

1. In a group of four, boil 16 eggs at intervals of 1 minute. Record your observations. How does the texture of an egg change with every minute and what happens to the egg if it is over boiled?
2. Do a market survey and study various types of eggs available in the market. Cook the eggs with methods such as boiling, poaching, and pan-frying and study them with regard to textures.
3. Visit hotels and restaurants and make a note of various egg preparations served in breakfast, lunch, or dinner and the various accompaniments served with the same.
4. Make a field visit to commercial egg farms and record your observations on the various procedures followed in the selection and storage of eggs.

CHAPTER 15

SEEDS, NUTS, AND SPICES

Learning Objectives

After reading this chapter, you should be able to:

- understand the different types of seeds, nuts, and spices
- identify various types of spices and nuts and classify them
- know the importance of selection and storage of spices and nuts
- appreciate the usage of these commodities in cooking to create various delicacies

INTRODUCTION TO SEEDS

Science defines seed as a reproductive unit of any plant which produces flowers and is in the form of a grain which is capable of growing into an adult plant if sown. If we closely look around in the kitchen we will find seeds such as spices, nuts, rice, cereals, pulses, and even coffee. Seeds are useful for cooks in the kitchen and can be used in various forms and for various uses. They are classified based on their usage as shown in Table 15.1. In this chapter we will discuss only those seeds that are directly used in cooking as an ingredient and not relate them to seeds which are used in agriculture and farming. We are aware that the same seeds can be processed to produce many other ingredients used in the kitchen such as essence, colours, and oils but we still will limit ourselves to the little seed, which in science is defined as the reproductive unit of a flowering plant. We shall discuss seeds such as nuts and spices. However, we will discuss spices in general and will talk about Indian spices in detail in Chapter 24.

CLASSIFICATION OF SEEDS

There are various ways in which seeds are classified as shown in Table 15.1.

Table 15.1 Classification of seeds

Classification	Description	Example
As flavourings	This category includes all those seeds that are used as herbs and flavourings.	Fennel, caraway, coriander seeds, etc.
As spices and colourings	These are used as spices. They impart natural colouring to a dish.	Cumin, poppy seeds, sesame seeds, annatto seeds, etc.
As food grains	These are seeds of grass plants and are known as cereals.	Wheat, millet, corn, oats, etc.
As pulses	These are the seeds of legumes and are also known as lentils.	Moong, red kidney beans, adzuki, etc.
As nuts	These seeds are found in hard coverings and some grow below the ground as well.	Almonds, walnuts, coconuts, etc.
As seeds for oils	These seeds are rich in oil and can be used for production of oils.	Mustard, linseed, grape seed, sesame, etc.
As beverage	The most popular beverage, coffee, comes from this category. This category is also used in the production of chocolate.	Coffee, cocoa beans, etc.
Miscellaneous	These seeds are those which do not fall in the above categories, but have specific use in the kitchen.	Basil seeds, which swell up on soaking are served with frozen milk dessert called kulfi in India.

SEEDS AS SPICES

Many seeds are used as spices, either by themselves or in combination. The spices could again be divided into categories such as flavouring spices, colouring spices, aromatic spices, thickening spices, etc. and we shall discuss these in detail when we talk about spices in Chapter 24. Table 15.2 shows the common seeds used as spices around the world.

Table 15.2 Common seeds used in cooking

Name/Hindi name	Description	Usage	Storage
Aniseeds/*vilayti saunf*	Aniseeds have a sweet and aromatic bouquet and distinctive liquor flavour. The spiciness of the seeds is similar to that of fennel.	Aniseeds are used in savoury and sweet cooking. They are used to flavour fish soup, sauces, breads, cakes, biscuits, and confectionery.	Should always be bought in small quantity and stored in airtight containers as the volatile oils evaporate over a period of time.

Contd

Table 15.2 (Contd)

Name/Hindi name	Description	Usage	Storage
Dill seed/*sowa*	Dill seeds have a sweet and aromatic bouquet with a hint of caraway in flavour. There are two varieties of dill seeds, one comes from North and South America, and the other comes from China and Japan.	Dill seeds go well with fish dishes. They also go well with egg and potato dishes.	These should always be bought in small quantity and stored in airtight containers.
Caraway seeds/ *shahi jeera*	Caraway seeds are mostly grown in Europe and Holland. These seeds have warm, sweet, and slightly peppery aroma. They are also chewed to sweeten the breath.	Caraway seeds are mostly used in German and Australian cuisine. These feature mostly in cabbage soup, breads, cheese, and liqueur.	These should always be bought in small quantity and stored in airtight containers.
Celery seeds/ *ajmud*	Celery seeds are the seeds of the vegetable celery. These are small and grey brown in colour. They have strong and bitter flavour.	Celery seeds can be added in making bread or cheese crackers. They can be sprinkled over lightly boiled carrot, grilled tomatoes, and salad. These are also complementary to egg and fish dishes.	These should always be bought in small quantity and stored in airtight containers.
Fennel seeds/ *saunf*	Fennel seeds have warm and sweet aroma with a flavour of anise. Fennel is best bought in seed form.	Fennel complements pork, lamb, veal, and also goes well with fish. The crushed seeds can be used for flavouring sauce. Fennel seeds are used in various curry powders. They are also used in savoury dishes, breads, cakes, and pastries.	These should always be bought in small quantity and stored in airtight containers.
Sesame seeds/ *til*	These are available in two varieties—black and white sesame seeds. These seeds have a nutty flavour. The flavour of sesame seeds is improved further by roasting.	These are very popular in the Middle East. They are used in making paste, halwa, and confectionery.	These should always be bought in small quantity and stored in airtight containers.
Pumpkin seeds/ *kaddu key beej*	These seeds are richer in iron than any other seeds and are an excellent source of zinc.	These seeds are widely used in South American cooking, where they are roasted and ground to make sauces.	These should always be bought in small quantity and stored in airtight containers.

Contd

Table 15.2 (Contd)

Name/Hindi name	Description	Usage	Storage
Hemp seeds/ *bhang*	Hemp seeds give a nutty flavour. These are believed to have been first cultivated in China and then spread to other parts of the world.	They can be used in various sweet and savoury dishes. They are also used as a drug in India.	These should always be bought in small quantity and stored in airtight containers.
Sunflower seeds/ *surajmukhi key beej*	Sunflower seeds are pale green tear drop shaped seeds, which have a crunchy texture and an oily taste. These seeds are rich in vitamin E.	Sunflower seeds are used in salads, rice pilaf, and cous-cous and are used in bread dough, muffins, casseroles, and baked dishes.	These should always be bought in small quantity and stored in airtight containers.
Lin seeds/*alsi*	These seeds are golden in colour and are also known as flaxseeds. They are a rich source of poly unsaturated fat, including the essential fatty acid.	Lin seeds can be added to muesli and other breakfast cereals, mixed in bread dough or toasted into savoury salads.	These should always be bought in small quantity and stored in airtight containers.
Poppy seeds/ *khuskhus*	Poppy seeds can be black or white in colour. They give a nutty flavour.	The black variety is used in breads and tea cakes and they are used in Germany and eastern European pastries and tarts. In India these are used to thicken sauces and curries. In Jewish cookery they are pounded to a paste, sweetened with honey and used to coat dumplings and bagels.	These should always be bought in small quantity and stored in airtight containers.
Nigella seeds/ *kalonji*	These seeds have little aroma but when rubbed they give a peppery aroma. These are sometimes used as a substitute to pepper. They also taste like herb similar to oregano or carrot.	Nigella is one of the five spices of Bengali five spices. It is known as *kalonji* in India and is used widely in Indian cooking. These seeds are often sprinkled on naan bread. In the Middle East, these seeds are often sprin-kled over cakes and breads.	These should always be bought in small quantity and stored in airtight containers.
Angelica seeds	Angelica is from the parsley family. The flavour of the seed is like celery.	Angelica is mostly used in cookies, breads, and pastries.	These should always be bought in small quantity and stored in airtight containers.

Contd

Table 15.2 (Contd)

Name/Hindi name	Description	Usage	Storage
Annatto seeds	Annatto seeds are used both as spices and as dye stuff. They are washed and dried separately from the pulp of the seed pod for culinary use. They may be added directly to a cooking liquid or infused in hot water until the desired colour is obtained and then used for stocks or colouring rice.	In some places in the western part of the globe, these seeds are used to colour confectionery, butter, smoked fish, and cheeses such as Cheshire, Leicester, Edam, and Muenster.	Annatto seeds should be kept out of light in an airtight container.
Basil seeds/ *subza* or *tootmalanga*	Basil originated in India and Iran. Italian sweet basil is different from Thai and Indian basil. The seeds of basil have medicinal properties apart from being used in Asian drinks and desserts.	The seeds when soaked swell up to a gelatinous mass with a black spot inside. They are used as an accompaniment to kulfi.	They should always be bought in small quantity and stored in airtight containers.

NUTS

Nuts are one celled fruits with a dry shell, but only hazelnuts, chestnuts, cashews, and acorns fall in this category. Other nuts are edible kernels from which the fruit wall has been removed (almonds, walnuts, coconut, and macadamia nuts); some are seeds (Brazil nuts, pistachios, and pine nuts); or pulses (peanuts). It is interesting to know that all nuts are not seeds, but by and large most of the nuts are seeds. Nuts are highly nutritious and full of flavour. They are rich in vitamins (B complex and E), as well as minerals (magnesium, potassium, iron, calcium, and phosphorous). They contain huge amount of calories. Nuts are rich in monounsaturated and polyunsaturated fats, with the exception of Brazil nuts and coconuts which are high in saturated fat but not cholesterol. In general nuts are large, dry, oily seeds or fruits of some plants. Most nuts which are seeds are from the fruits of a tree, while others such as groundnuts in reality are pulses. Nuts are used in various ways in the kitchen. They can be flaked, ground into powder; and some nuts are also used as butter and oils such as peanut butter and walnut oil. Nuts can be eaten whole and they are a great source of protein, good fats, essential minerals, and vitamins. They can be used as a crust on roasts and can also be used in desserts such as cakes. So it is interesting to know that all nuts are not seeds, but in large most of the nuts are seeds. In Table 15.3, we shall discuss various types of nuts.

Commonly Used Nuts

Nuts are edible seeds of dried stones of fruits. They are extensively used in cooking around the world. Some of the most common nuts used in the kitchens are discussed in Table 15.3.

Table 15.3   Commonly used nuts

Name/Hindi name	Description	Usage	Photograph
Almond/*badam*	It is the most popular common fruit. Actually it is a seed of a plum-like fruit and is commonly grown in California (USA) and Kashmir and Himachal (India). Greeks were the first to grow them in Europe.	• Thickener for sauces, base for biscuits and cakes • Main ingredient in pastry • Making pralines and nougatines • Making nut butters and marzipan • Crunchy topping for tarts, desserts • Making oils, extracts, and liqueurs	
Walnut/*akhrot*	This is native to Asia, Europe, and North America and is the second largest nut consumed after almond. The seed has high oil content and is concealed within a wrinkled woody shell.	• Base for biscuits, ice creams, and cakes • In cooking • In baking—breads, pies, cookies • For adding crunch to salads • Base for sauces—tarator • In making oils	
Pistachio nut/ *pista badam*	It is native to Central Asia and finds usage in garnishes.	• Used in sweetmeats and cookies • Decorating pates and terrines • As topping—cakes • Used in ice creams and sauces • Salted and roasted as snacks	
Hazelnut	The hazelnut looks like a chickpea and is enclosed in a brown shell. It often resembles a heart shape and is mostly grown in Europe and China.	• Adds flavour to all baked goods • Ground hazelnuts are used in cakes and ice creams • In making nut butters and marzipan • To add crunch to salads and meats • Hazelnut paste is used in confectionery to make chocolates	

Contd

Table 15.3 (Contd)

Name/Hindi name	Description	Usage	Photograph
Cashew/*kaju*	In the sixteenth century, the Portuguese brought them to India and today India is the largest producer of cashew. It is also found in the Amazon region. Cashew is the seed of a plant that has irritating oil which if consumed bites the throat and hence the plant is always heated and the seeds are removed carefully.	• Used in savoury curries • Used in sweetmeats • Added to biscuits, cakes, and ice creams • Can be roasted and salted to make snacks • Also fermented to make alcoholic drinks	
Pecan	It is a species of the hickory plant, and is native to Mississippi. Pecan nut kernel resembles walnut, the difference is that the kernel is long and flat and much darker in colour. The shell is also smooth like hazelnut.	• Used in confectionery • Ground into powder and used • Used as a crust on meats	
Coconut/*nariyal*	The word coconut is derived from the Portuguese word cocoa, which means monkey as the 3 dots on the nut resemble a face. It contains water and a fleshy part.	• Grated as topping • Added to biscuits, cakes, and ice creams • Used in savoury curries • Coconut water, especially from green and tender coconuts, is a refreshing drink • Shredded and dried for various uses • Coconut milk is obtained by grating the flesh of the coconut and squeezing the milk	
Peanut/ *moongphali*	It is the seed of a legume that grows below the ground. It is native to South America, but the world's largest producers are India and China.	• Used in savoury curries • Used in sweetmeats and cookies • Eaten as snack, either roasted or fried • Peanut butter is used as it is or in confectionery and bakery	

Contd

Table 15.3 (Contd)

Name/Hindi name	Description	Usage	Photograph
Macadamia	It is the nut of a small evergreen tree, and is very common in Australia. It looks similar to hazelnut the only difference is that it is probably 3 times bigger than a hazelnut.	• Used in stuffings, cakes, and cookies • Eaten as snack • Used in making chocolates and other confections	
Brazil nut	As the name suggests, this nut is native to Brazil and an attempt to cultivate this in any other country has failed. The Brazil nut tree is around 150 ft tall and 6 ft in diameter. The size of the fruit of this tree is 3 kg and contains 15–20 seeds from which the nuts come. The calorific value of one Brazilian nut equals to that of one egg as they are high in oil content and also rich in sulphur and other amino acids.	• Used in stuffings, cakes, and cookies • Eaten as snack	
Candlenut	It is used in Indonesia and Malaysia. This nut resembles macadamia nut but is larger in size. It has a fatty taste and is mostly ground with herbs and vegetables to be made into curry pastes.	• Used in curries and stews	
Pine nut/ *chilgoza*	Also called pignolia, it is a naked seed of different species of pine. Pine seed starting from sowing will take about 75 years to reach commercial levels of production and hence it is expensive.	• Used in sauces and chutneys—such as pesto and tarator • Used for stuffing • Used as a crust • Roasted and eaten as a snack • Used as a garnish	
Chestnut/ *singhara*	It is native to America and is often used in Chinese cooking. It has high carbohydrate content and is advisable to mature it in room temperatures to let the carbohydrates convert into natural sugars.	• Eaten smoked or boiled as snacks • Steamed and often used as vegetables in Chinese stir-fries • Chestnut paste is used in desserts • It is often ground as a flour and eaten during fasts in India	

Handling Nuts

CHEF'S TIP
To make powders, grind the nuts with little amount of icing sugar, as nuts contain oil and will form into a paste rather than a powder.

Nuts can be used in a variety of ways as mentioned in Table 15.3. Generally nuts can be processed in the following ways.

- Nuts are easily chopped if they are warm and moist as most of the nuts contain oil and hence they become more pliable if warmed.
- Roasting or toasting nuts brings out their flavour and once shelled and skinned they can be placed in a medium oven (175°C).
- Nuts can be blanched in hot water for a couple of minutes to remove the skin. This is usually done for pistachio nuts as blanching deepens the colour and also for almonds to make it into a paste for making marzipans, etc.

SELECTION AND STORAGE OF NUTS

Unshelled nuts keep well almost anywhere if they are protected from heat, air, light, and moisture. Shelled nuts on the other hand, do not keep as well, and should be stored tightly covered in a cool dark and dry place. Fresh nuts are available with their shell and are best for long storage. Nuts in their shell should be heavy for their size, and intact with no cracks or holes. The larger the size of the nut, the better will be the taste. Unshelled nuts are sold in bulk and may be roasted or raw, left whole, slivered, sliced, or broken into large pieces. Any nut with visible mould should be avoided. All these forms have specific uses in the kitchen.

Certain factors to be kept in mind while storing nuts are given below.

- As nuts contain substantial amounts of oil and easily turn rancid, exposure to light, moisture, or heat will reduce their shelf life.
- Nuts are best kept in their original packaging, airtight containers, refrigerators or freezers if kept longer than a month.
- For short periods nuts can be stored in an airtight container in a dark cool place.
- Nuts in their shell keep longer than any form of processed nuts. Freezing nuts with their shell on is best.
- Nuts that are soft or stale may be partially restored by roasting in the oven.
- Salted nuts have a freezer storage life of six months as compared to plain ones, which can keep up to 12 months.
- Toasting and roasting reduce the freezer storage life, so delay toasting and roasting till required.
- Nuts and seed should be used directly from the freezer without thawing.

Packaged nuts may be processed with preservatives and dyes, and are often heavily seasoned with salt. Dry roasted nuts usually do not have any additional fat but may be roasted with different types of fat. To ensure storage, nuts dipped in chocolate, spiced nuts, sugared, or yogurt-coated nuts are also available.

CONCLUSION

In this chapter we dealt with seeds that are used as nuts and spices and also discussed various other seeds used in cooking. We shall discuss the seeds used as spices in detail in the section on Indian cooking. We have understood the importance of seeds, in this chapter, and also saw how seeds from various plants are categorized into different categories which could then be categorized further into other categories. For example, spices are seeds that are further classified into colouring, flavouring, spice, etc. We broadly classified the seeds into various categories. We discussed about seeds that are used as spices or condiments depending upon the usage. Some uses of the seeds are discussed in the chapter;

however, the usage of these seeds can be unlimited and can depend upon the creativity of the chefs. Many chefs use variety of roasted seeds and use them to crust pieces of steaks to offer a texture to the product. We also discussed about nuts in this chapter. Nuts comprise a culinary category that includes the edible seeds of fleshy fruits as well as true nuts, or one seeded fruit with a hard layer corresponding to the flesh of culinary fruits. Nuts are less important in the human diet than grains and legumes. They have a high fat content—in the order of 60 per cent by weight—and this limits their usefulness as a protein source by limiting our capacity for them.

KEY TERMS

Annatto A seed often used as colouring agent. Yellow butter is coloured with annatto.

Bagels Ring-shaped bread of Jewish origin.

Bhang Hindi name for hemp seeds.

Casseroles Style of cooking in the clay pots, where the food is served in the same pot.

Couscous Small starchy granules made of wheat used in Moroccan cooking.

Kulfi Frozen and reduced milk-based dessert from India.

Linseed Type of seed used in a cereal.

Marzipan Pliable almond and sugar dough, used in confectionery.

Millet Kind of cereal.

Muesli A combination of certain cereals and nut served as a breakfast cereal.

Muffins Small individual cup cakes served as breakfast rolls.

Nougatine Caramelized sugar and almonds, which can be used roughly crushed.

Praline A paste made of caramelized sugar and almonds used as fillings.

Subza Hindi name for basil seeds.

Sweetmeats A combination of chopped dried fruit used in confectionery.

Tarator A Mediterranean dip of almonds and garlic.

Terrines A cold meat preparation of ground meat made in an earthen mould.

CONCEPT REVIEW QUESTIONS

1. List at least three differences between spices and nuts.
2. What are seeds? List at least five seeds which are not nuts or spices.
3. How are seeds classified?
4. Give few examples of seeds that are used for oils and for beverage.
5. How are basil seeds used in India?
6. Name at least five seeds found in the western parts of the globe, which are used in cooking.
7. List few uses of caraway seeds, dill seeds, and aniseeds.
8. Define a nut and give examples of at least five nuts that are not grown in India.
9. Write down the nutritional breakdown of nuts.
10. Write down how nuts can be used in the kitchens.
11. Write a short note on Brazil nuts.
12. What is the difference between a cashew nut and a candlenut?

13. What are the few uses of hazelnut?
14. How is walnut different from pecan nut?
15. In reality what is a peanut?
16. Which nut is used in Indonesian curries?

17. Which nut is usually used during fasts in India?
18. List few salient features of nuts.
19. How would one remove the skin from the nuts?
20. List at least five precautions while storing nuts.

PROJECT WORK

1. Do a market survey and study various types of seeds, spices, and nuts available in the market. Take photographs and research about their uses and classify them as shown in Table 15.1
2. Visit a dry fruit market in India and research about various species of the same nut and their uses. Record down the observation and make a report.
3. Research about the various types of oils made from nuts and seeds and see which ones are used for what purpose and why.
4. Do research on various types of seeds that could be used as a beverage and list all the beverages made from seeds.

CHAPTER 16

INTRODUCTION TO RICE, CEREALS, AND PULSES

Learning Objectives

After reading this chapter, you should be able to:

- understand the difference between cereals and pulses
- identify various types of rice, cereals, and pulses and be able to classify them
- understand what impacts the cooking of these commodities
- select pulses and rice for various uses
- recognize the importance of selection and storage of rice, cereals, and pulses
- appreciate the usage of these commodities in cooking to create various delicacies
- comprehend the use of local and home grown ingredients such as millets and analyse it with other super foods sold in the global market

INTRODUCTION

In Chapter 15 we read about the classification of seeds. We understood that seeds can be classified in various ways such as flavourings, spices and colourings, food grains, pulses, nuts, oils, beverage, and so on. We discussed seeds, nuts, and spices in general. In this chapter we shall focus on cereals and pulses and one of the cereals, 'rice', will be discussed in detail, as this versatile cereal is eaten around the world and exists in various types and sizes. As described in Chapter 15, cereals are the seeds of grass plants and pulses are the seeds of legumes.

Table 16.1 shows the difference between cereals and pulses.

Table 16.1 Difference between cereals and pulses

Cereals	Pulses
These are the seeds of grass plants such as wheat, rice, barley oats, etc.	These are the edible seeds of certain legumes.
These are usually ground into flours.	These are soaked and boiled to be made into stews, soups, salads, etc.

Contd

Table 16.1 (Contd)

Cereals	Pulses
These are rich sources of complex carbohydrates.	These are rich sources of proteins.
Cereal is usually covered with a husk that is often eaten as bran, which is a rich source of fibre. It is also added to flours to make bran bread or breakfast cereals.	The husk of pulses do not find any usage like those of cereals.
Examples are barley, corn, buckwheat, etc.	Examples are kidney beans, peas, black gram, etc.

PULSES

Pulses are often referred to as lentils and these are fibrous and starchy but high in protein content as compared to cereals. Pulses are the edible seeds of plants belonging to legume family which includes beans, peas, and lentils. These are dried as soon as they are plucked from the plant, so as to retain the flavour, texture, and plumpness. Pulses are an important source of protein and iron. Pulses store well for long periods approximately six months to a year if kept in a dry, airtight container away from light, but it is best to eat them as soon as possible, as they toughen on storage and take longer to cook.

CHEF'S TIP

A general rule for cooking pulses is to add salt to the ones which are de-shelled and no salt has to be added to the ones which have skins or shells, as cooking time increases if salt is added to these types of pulses. Acidic ingredients, such as tomatoes, should be added only after the pulses get cooked because it tends to toughen the skin of the pulses.

CHEF'S TIP

Never serve kidney beans, black eyed beans, or soya beans sprouted as they contain toxic substances and can cause severe food poisoning if consumed raw. One must boil these for at least 10 minutes to get rid of the toxins.

Cooking of Pulses

Majority of the pulses found, with a few exceptions, need to be simmered for a considerable amount of time in a covered pan. When baking them in the oven sufficient amount of water is to be added. Pulses can be cooked by various cooking methods; but the most common ones are boiling, stewing, and sometimes braising. One must have seen some people roasting pulses in sand; well it is a common practice on the streets of India, where in winters it is not unusual to see people munching on them as snacks.

There are certain pulses which have toxins in their skin and can cause food poisoning if consumed raw or semi cooked, for example, kidney beans, soya beans, etc. One should always make sure to at least boil these beans in fresh water for 10 minutes, after it has been soaked in water overnight, before consumption. Since the beans are dried, it is often advisable to soak them for sometime to reconstitute, as this reduces the cooking time quite considerably.

COMMON BEANS

Let us read about some beans throughout the world in Table 16.2. Some of these might not be commonly used in India, but they are quite commonly used in Western cuisine.

Table 16.2 Various types of beans

Beans	Alternative/ Hindi name	Description	Photograph
Haricot beans	Navy beans, Boston beans, pearl haricots, and fagiolis.	Small white beans, which are roughly oval-shaped and flat.	
Flageolet beans	Green haricot beans	They range from creamy white to light green. Flageolets are removed from the pod when tender and just before they mature. The caviar of beans, flageolets are tiny, tender French bush type beans that are very popular in French cooking.	
Pinto beans	French beans, snap beans, bush beans, pole beans, pinto beans	These are smaller, paler version of borlotti, with orange pink skin specked with rust colour fleck. Pinto beans have a beige background strewn with reddish brown splashes of colour. 'Pinto' in Spanish means 'painted'. When cooked, their colour splotches disappear, and they become pink.	
Kidney beans	*Rajma*	These are glossy mahogany red kidney-shaped beans which retain their colour even after cooking. These famous beans come from Jammu and Kashmir in India.	
Cannellone beans		Creamy white Italian beans which are slightly larger and fatter than the white haricot beans. These beans often pair up well with pastas.	
Chickpeas	Garbanzo beans, *Lahori choley, Kabuli chana*	These resemble shelled hazelnut beans. These are used in the Middle East to make hummus and in India to make *pindi chole*.	
Black beans	Turtle beans	Shiny black kidney-shaped beans, descriptively named probably in reference to their shiny, dark, shell-like appearance.	

Contd

Table 16.2 (Contd)

Beans	Alternative/ Hindi name	Description	Photograph
Black eyed beans	Black eyed peas, cow pea, *lobiya, rongi, feijão*	Known as black eyed peas and cow pea in the USA. Commonly used in India, these beans are creamy white in colour, kidney-shaped, and have a black spot.	
Borlotti beans	Fasiola beans, cranberry beans	These are oval beans having red streaked, pinkish brown skin and a bitter sweet flavour.	
Ful medames		These are small Egyptian beans and a member of broad bean family. They have a strong, nutty flavour and a tough light brown skin outside.	
Moong	Mung, green gram, golden gram	These are olive coloured beans. They are commonly sprouted and used in Chinese cooking.	
Aduki beans	Adzuki beans	These are tiny deep red beans with sweet nutty flavour, known as 'king of beans' in Japan.	
Butter beans	Lima beans, Madagascar beans	The term butter bean is used for a large, flat variety of lima bean.	
Broad beans	Fava beans	These beans are a great source of proteins, carbohydrates as well as vitamins. They are wide beans with large flat seeds. Their colour ranges from green to brown when dried.	
Black, red, green, and white soya beans	Soy	These are small oval beans, with colour ranging from streamy yellow to brown and black.	

Contd

Table 16.2   (Contd)

Beans	Alternative/ Hindi name	Description	Photograph
Green and brown lentils	Continental lentils	These are also referred to as continental lentils and are disc-shaped.	
Red split lentils	*Masoor*	These are orange red coloured split lentils.	
Puy lentils		These are tiny dark blue marbled lentils of French origin.	
Yellow lentils	*Arhar dal, toor dal,* pigeon pea	These beans were commonly fed to pigeons. They are commonly used in Indian cooking.	
Yellow and green split peas	*Vatana, matra*	In India these are commonly used in Maharashtra and in street food of north India, where it is served as a curry with yeast leavened bread called *kulcha*.	
Brown speckled rattle snake beans		These are less starchy. They do no require soaking and will cook in less than 1 hour. They are well suited to salads and purees.	
Black grams	*Urad dal*	These are available whole, split, or without husk and all three variations are used for different products. The whole dal is used for making famous Indian *dalmakhni*.	

Contd

Table 16.2 (Contd)

Beans	Alternative/ Hindi name	Description	Photograph
Moth beans	Papillon, haricot, dew bean	These beans are very small and brownish in colour, with a creamy yellow colour interior.	
Garden peas	*Mattar*	Green peas are often eaten fresh, but they are also sold as frozen in the super-markets.	
Bengal gram	*Chana dal*	It is a good source of protein and is often fed to horses. It is found in Bengal and hence the name.	
Appaloosa bean		This bean is multi-coloured and tastes best with chilli peppers, cayenne peppers, and mustard. It also does not need soaking and can be cooked in less than an hour.	
Congo	Guga peas	African origin peas; also found in the Caribbean Islands and India.	
Lablab	*Moth ki dal*	Hard shelled beans found in India, which must be shelled before cooking.	
Black kidney beans	*Kaale ranwah*	Black coloured version of the normal kidney beans found in the Carribean Island and Mexico.	
Black chickpeas	*Kaala chana*	They look and taste very different from the more common white kind, *Kabuli chana*.	

CEREALS

As we have read above, cereals are seeds of grass plants and they comprise edible grains such as wheat, oats, rice, etc. The term 'cereal' is derived from the Roman name *ceres* after the goddess of agriculture. Agriculture first flourished in the Middle East, Central and South American countries and till this date the beans that come from this part of the continent are even unheard of in India, where cereals and pulses form the diet of 95 per cent of homes. Cereals are rich source of complex carbohydrates and hence these are used in various forms and at almost any time of the meal. In breakfast cereals can be had with milk or cooked with milk to be made into porridge. In lunch and dinner they can be had in the form of pastas, flat breads such as Mexican tortillas and Indian rotis. We shall discuss the most commonly used cereals in hotels and then we shall discuss rice in detail as this cereal is consumed as staple food by almost half of the world's population (refer to Table 16.3).

Table 16.3 Various types of cereal

Cereal	Alternative/ Hindi name	Description	Photograph
Pearl barley	Barley/*jowar*	This is the husked and polished berry of barley and is used in soups or salads.	
Buckwheat	*Kuttu*	This cereal is native to China and is used for making noodles. Sobas noodles in Japan are also made of buckwheat. Buckwheat flour is commonly eaten in India during fasts.	
Maize	Corn, *masa, makki*	This cereal is native to Mexico, where maize corn is ground into a flour called *masa harina* or cornmeal. This is not to be confused with corn flour. In India, *makki ki roti* from Punjab is very common. Polenta from Italy is also made of cornmeal. This is one of the flours that is free of gluten and can be easily given to people suffering from celiac[1] disease.	
Oats	*Vilayti jai*	Oats are used in various forms; they can be ground to various degrees of fineness or can be soaked and pressed to be served as rolled oats. They can be cooked with milk or water to be served as porridge. Broken oats are also known as grits and this cereal is commonly eaten for breakfast.	

Contd

---

[1] Gluten allergy

Table 16.3 (Contd)

Cereal	Alternative/ Hindi name	Description	Photograph
Wheat	*Kanak*	Wheat is native to Mediterranean region and the Middle East. Wheat can be used in various forms. Wheat can be ground to wholemeal flour or refined to create the white flour. It can also be cracked and used as *burghul*, as known in the Middle East or *dalia* as known in Hindi.	
Sago	*Sabudana*	It is a processed starch that is extracted from palm trees. This is mostly known to Asians and in India it is often used in making desserts and also used in making papad.	
Bran	*Kanni*	This is the brown outer husk or outer layer of any cereal grain. This is a rich source of fibre and is added to the flours to make high fibre breads.	
Rice	*Chawal*	It is available around the world in different shapes and sizes. It can be boiled or steamed.	
Sorghum	*Jowhar*	One of the most popular millets used in Africa and Asia. In India it is commonly ground into flour and used for making rotis. The nutritional quality of it being rich in calcium and iron makes it ideal for people allergic to gluten as it is also gluten free.	
Foxtail millet	*Kangni*	In India, we always used to consume various kinds of millets, but now as we are getting developed as a nation, these millets are being labelled as pigeon feed. The foxtail millets are available in seed form, rice form and even in the form of semolina. They can be used for making salads or simply cooked like a rice.	

Contd

Table 16.3 (Contd)

Cereal	Alternative/ Hindi name	Description	Photograph
Finger millet	*Mandua/Ragi*	One of the most consumed millets in Southern India that at times it even replaces wheat and rice. It is one of the richest sources of calcium and can be used in breads and even idlis.	
Pearl millet	*Bajra*	Pearl millet or *Bajra* is commonly used in India for making rotis or even health drinks in Rajasthan (*bajrey ki raab*). It is a rich source of iron and magnesium and helps in formation of milk in lactating mothers.	
Barnyard millet	*Jhanghora*	This millet is popularly used in making rotis and can also be used for many other Indian dishes like dosa/idli. It is a rich source of fibre that gives a feeling of fullness, thereby advised by Ayurveda as weight loss grain.	
Kodo millet	*Kodra*	This millet which is rich in calcium, magnesium, and iron is the substitute for rice as it is cooked in the same way as rice. One can also use the cooked grains of kodo millet and combine them with other vegetables to make healthy and nutritious salads.	
Proso millet	*Barri/Chena*	Like all other millets, this millet is popularly known to help in making bones strong due to it being a rich source of calcium and phosphorous. Proso millet can be cooked like any other grain into porridge or can be used in salads as well.	
Quinoa		Originated in Peru, quinoa is not a cereal in the real sense but is a seed of a variety of amaranth plant. It is being discussed here because off late it is considered to be a super food and is the richest source of plant based protein that is used in replacement of many cereals. It can be cooked like rice or powdered to be used in breads.	

Contd

Table 16.3 (Contd)

Cereal	Alternative/ Hindi name	Description	Photograph
Chia		Chia seeds are native to South America and are a rich source of Omega 3 fatty acids. They come from a plant called dessert plant that belongs to the family of mint. Like basil seeds, chia seeds swell up in water and are considered to be one of the super foods that help combat cardiovascular diseases by lowering the blood pressure and triglycerides in the body.	

RICE

The Chinese are believed to be the first to have cultivated rice; but there is no historical evidence to this fact, as historians believe that rice was cultivated around the world simultaneously. Rice is believed to have been cultivated 3500 years ago in Africa and around the same time it was also cultivated in the Middle East and Mediterranean Europe. Rice then spread through to France. It is believed that it were the Spaniards, who brought rice into South America.

Rice needs abundant moisture to grow and hence it grows in places which receive plenty of rainfall in the season. It grows in subtropical climates which are neither too hot nor too cold. In India rice is the staple food for almost 70 per cent of the people. People in north India eat more wheat as compared to rice, because of the excessive production of wheat in this region. Agriculture is the main source of income for people in rural India and most of their land is used to cultivate rice and wheat. India is believed to be the second largest producer of rice in the world after China. The king of rice or basmati is exported to many parts of the world. Apart from basmati, India exports other types of rice to various countries. Since there are so many varieties of rice it becomes necessary to classify them. Most of the times rice is classified on the basis of the size of its grain and the length of the grain usually defines the style of cooking the rice. Rice is not washed for some dishes such as risotto before cooking, giving it a unique character. Rice can be boiled or steamed; soaking of rice aides cooking of rice and makes the rice less sticky as most of the starch is washed away. Table 16.4 shows the classification of this versatile grain.

Table 16.4 Classification of rice

Long Grain Rice	Short Grain Rice	Glutinous Rice	Wild Rice
Brown rice	Pudding rice	Chinese rice	Giant Canadian
White rice	Carolina rice	Thai rice	Red camargue
Basmati	Java rice	Japanese rice	
Brown basmati	Italian rice	Sushi rice	
Surinam rice	Spanish rice	Shinmai rice	
Jasmine rice			
Patna rice			

CLASSIFICATION OF RICE

Long Grain Rice

These are used for savoury dishes and are three times longer and wide as well. After cooking, each grain separates individually. One of the famous long grain rice of India is basmati which literally translates to 'queen of perfume'. The best basmati rice comes from old Karnal district in Haryana which is also called the rice bowl of India. The grain of uncooked basmati can be as long as 7 mm in length and around 2 mm in breadth. The nutty aroma of this rice comes out better when cooked with little amount of *desi ghee*[2]. We will discuss more about basmati, but let us first read more about other long grain rice.

Brown Long Grain Rice

It is the rice with the bran but without the husk. This is equivalent to wholemeal or wholewheat bread. It has got a specific nutty flavour and good texture. Even a shorter grain is available which is mostly used for sweets but is milled to remove the bran. The difference between brown rice and white rice is not only the colour; but also the milling process. A grain of rice has seven layers, when the outer layer and the husk are removed, brown rice emerges. When the rice is further polished and milled, it results in white rice. That is the reason why brown rice is considered to be healthier than white rice.

White Long Grain Rice

This is one of the most popular varieties majorly produced in the USA, India, China, Thailand, and Malaysia. It is fully milled and all the outer covering and bran is removed; this is why it appears to be shiny. This feature is often referred to as polished. This rice goes well with continental dishes and Asian food.

Basmati Rice

This rice is rightly considered to be an essential ingredient for biryani and is often known as the queen of rice. It comes from Punjab and the foothills of Himalayas in India; the soil and climatic conditions of these regions support the growth of basmati rice. In Hindi the word basmati translates to 'queen of perfume'. This rice has a unique flavour and texture. It is a long grain rice which becomes even longer after cooking which results in its great texture. It is mostly used in savoury dishes.

Brown Basmati

It is same as all brown rice with the bran but husk removed. It tastes same as basmati rice with additional features of brown rice.

Surinam Rice

It comes from South America. It is a very thin long grain and is difficult to obtain.

Texmati Rice

This is the aromatic rice that comes from Texas in the USA, hence the name texmati. This grain of rice is a cross between Indian basmati and long grain rice. It definitely has more aroma than long grain rice but less than that of basmati. Texmati is available in white and brown colours. Since the brown rice has more bran, it takes longer time to cook the same.

---

[2] Clarified butter

Thai/Jasmine Rice

This rice has got a milky aroma and fragrance to it which makes it a good combination with spicy Thai and Indonesian dishes. Used for both savoury and sweet dishes, it cooks quickly and salt is never added to it while cooking. In Thailand it is also known as *hom mali.*

Patna Rice

Initially all the rice (long grain) exported to Europe came from Patna, Bihar in India, hence the term Patna rice. It still persists in the USA. However nowadays the term refers to the rice from Bihar region. It is a multipurpose rice which is light, fluffy, and dry, with little firmness to it in the centre. It is suitable for all kinds of Indian dishes as it takes flavourings well.

Short Grain Rice

These are also known as round grain because of their shape and are mostly used to make sweets. Some grains are medium sized and are used in savoury preparations as they give a creamy texture, for example, arborio rice is used for risotto and short grain *Valencia* from Spain is used for paella.

Pudding Rice

This is a type of rice which is plump and white in colour. This rice is suitable for creamy milky puddings as it absorbs lots of liquid and swells up and this in turn results in a smooth and creamy dish.

Carolina Rice

This is a high quality short grain rice from the USA and gets its name from the state where it was first produced. It is used for savoury and sweet dishes and is also widely used as filling for different vegetables.

Java Rice

It is used for making desserts as it absorbs a lot of liquid and is an intermediate size grain. It is also considered to be of high quality.

Italian Rice

Italian rice is classified as *ordinario, semi fino, fino*, and *super fino* according to the grain size. Risotto rice generally comes in category of *fino* and *super fino* varieties.

Arborio   It is one of the most popular known risotto rice and comes from north west Italy. It has got a larger plump grain as compared to finer varieties of risotto rice as mentioned below.

Vialone Nano   It is a plump grain and a firm inner starch which gives risotto a slightly more bite.

Carnaroli   It is a cross of an Italian and a Japanese rice and is known as a premium variety of risotto rice. It has a softer starch outside that dissolves during cooking leaving the inner grain which gives it a firm bite.

Risotto rice absorbs more liquid than other rice and thus more liquid is added at frequent intervals.

Spanish rice

Rice is grown extensively in Spain mostly in the swampy regions of Valencia. It is mostly a short grain rice but longer grain variety is also available that is mostly stirred into soups. Slightly sticky short grain is used for a very famous dish called paella. Rice is graded by the amount of whole grains by weight present.

Categoria Extra (red label)   It is a fine rice with 95 per cent whole grains.

Categoria Uno (green label)   It has 85 per cent whole grains.

Categoria Dos (yellow label)   It has 80 per cent whole grains.

Bahia   It is medium grain rice for paella.

Bomba   It is plump rice for paella like risotto rice which absorbs more liquid.

Glutinous Rice

This type of rice does not have any content of gluten but is called so due to its stickiness after cooking. It is mostly popular in Asian cooking. It is also referred to as sticky or sweet rice.

Chinese Glutinous Rice

It is a generic name for short grain rice, japonica, or *geng* rice. Black, red, and white varieties of this rice are available.

Thai Glutinous Rice

It is available in black (deep blue) and white varieties and is popular for making sweet puddings and desserts.

Japanese Glutinous Rice

This short grain rice is slightly sweet in taste. It becomes sticky on cooking which helps in picking it up with chopsticks. It becomes quite tender after cooking but retains its crunchiness. Uses of this rice are quite flexible; it can be served in breakfast as well as in other meals along with meat. Popular brands are *mochigome* and *koshihikari*.

Other Japanese Rice

Some Japanese rice which are not glutinous are discussed below.

Sushi Rice

It is not glutinous but becomes sticky after cooking which helps in making sushi. Although many varieties are available the most popular are Japanese rose, kokuho rose, and calrose. This rice comes from short grain Japonica rice. If this rice is not sticky, it feels dry.

Shinmai

It is one of the most prestigious Japanese rice which is available in late summer. It is the first rice of the season. Since it is high in moisture content it requires less water during cooking.

Wild Rice

Wild rice is not actually rice but the grasses that grow in marshy areas of North American great lakes. It has got a nutty aroma and is quite expensive and hence used in combination with white long grain rice.

Giant Canadian Wild Rice

It is quite similar to the American variety but it is longer, glossy, and has a superior flavour. It grows on lakes on the western coast of Canada where native people beat these grasses and the grains that fall in their canoes are theirs and the ones that fall down are for the following year's harvest.

Red Camargue Rice

This red rice is a cross-pollination between white rice and wild red rice. It comes from Camargue region of France. It is red in colour and as it is cooked the colour intensifies and even the water gets a deep red colour. *Wehani* is a popular variety of this rice that is grown in California.

Miscellaneous Rice

Various other types of rice are discussed below.

Parboiled Rice

As the name suggests, this rice after being harvested is soaked and then steamed in large steam pressures. This is done at controlled temperatures and controlled moisture so as to discourage the swelling of the grains. The rice is then dried and finally milled. This process changes the nutrition of the rice kernel as well as its colour as the rice becomes light yellow and this colour usually fades away when cooked. Parboiled rice has higher vitamin content as compared to normal milled rice and is widely used in Asia and in south and east of India. As this rice is not sticky when cooked, each grain fluffs up.

Red Rice

This rice comes from the japonica family and is different from camargue rice. Red rice has a shorter and a wider grain. It takes around eight times the amount to cook when compared to normal rice which takes up only one and a half and two times of water. Red rice in India is also known as Kerala rice, as it is very popular in the state of Kerala. It is considered to be very healthy rice and hence is found on the healthy choices on the menu. It is believed to have lower cholesterol. When cooked this rice retains its redness but the grain bursts to show a fluffy white grain with streaks of red around it.

OTHER RICE PRODUCTS

Rice is also processed in various forms of different products as it is a versatile ingredient. These are as follows:

Flaked Rice

It is mostly available in Asian markets and is used as a cereal, and for various other savoury preparations. It is also known as *chiwda* in Hindi and is used for making a popular dish known as *poha*, which is commonly had as a snack or as breakfast.

Ground Rice

It is used mainly for sweets such as rice puddings, biscuits, etc. It is also used for baking purpose as it gives a good sandy texture. It is a good substitute for wheat flour, especially for all those who have allergy from gluten.

Rice Bran

It is mostly used as an animal feed but in Japan it is roasted to make a mash which is fermented to make a base for pickling vegetables known as *nuka-zuke*.

Rice Cakes

In Japan these are known as *mochi* and are made from glutinous rice. These can be fried, grilled, or also boiled to be served as an accompaniment.

Rice Wine

Sake, mirin, and various other rice wines are also made from rice. Vinegars are produced as a result of over fermentation of wines and commonly available rice vinegars are white, red, and black vinegar.

COOKING RICE

Generally rice is boiled or steamed. The key to cook rice is that how much proportion of water to take while cooking it and also the time that should be taken to cook it.

The proportion of liquid depends upon:

- tightness or looseness of the cover or lid;
- desired moistness of the finished product;
- variety, age, and moisture content of the rice used; and
- method of cooking used that is either absorption or drainage method.

Drainage Method

While cooking with drainage method, 4 litres of boiling water should be used for every 500 g of rice. When tender and cooked the rice has to be strained to drain away all the water. Then it should be spread on the table to dry a little. This method would result in producing separate and unstuck grains. This method is used for basmati, brown basmati, red camargue, and brown rice.

Absorption Method

This method is commonly used for making pilaf or buttered herb rice and it is more like braising. Rice is first sautéed in fat and then cooked with liquid preferably in oven. Fat gives flavour and also keeps grains separate. Absorption method is used for basmati, Thai fragrant rice, short grain rice, and glutinous rice. It may also be used for brown rice and American long grain rice. In this method washed rice is cooked in a covered pan with a tight lid until the rice is cooked and is tender. This process can take more than 25 minutes to cook.

While cooking risotto the method used is neither drainage nor absorption. In this method the rice is first sautéed with fat or oil and then stock or liquid is added slowly and then risotto is worked upon and then some more liquid is added until risotto is cooked but yet firm or what is called *al dente* (to the teeth). As the rice leaves all its starch in the pan, the dish gets a lot of creaminess. The normal arborio risotto which is very common takes 18 minutes to cook on a slow flame. Since one does not want any starch to be lost, risotto rice is never washed before cooking.

While cooking sushi rice the rice is washed five to six times or until the water is clear in which the rice is washed. Traditionally the rice was left under running water for a long time so as to wash it and also to avoid damage to the grains. The ratio of liquid to be used while cooking is 1:1 or 1:1.25 depending upon the maturity of the rice. To cook, rice and water has to be put in a pan and boiled uncovered. Once it starts boiling, lower the flame and cover. It should be cooked for approximately 10 to 15 minutes or as suggested on the packet. When rice is cooked it must be spread on a *hangiri* (traditional Japanese vessel made from wood) and fanned continuously to cool down. The fanning while cooling rice results in a shiny grain of rice much desirable for sushi. Check the recipes of classical dishes of rice given in the ORC.

SELECTION OF RICE

Selecting rice is a very tricky thing. One cannot judge the quality of the rice by its colour as it is available in different shapes and sizes. However there is one thing that we must check out—the difference between new rice and old rice. This must be again asked of the supplier as it is difficult to make out the new rice from the old one. One could easily spot small insects called weevils in old rice and they can easily be washed away. Many people prefer rice with weevils, as it is an indication of good old rice; but in hotels one cannot afford to risk to serve customers with any of the infested rice, so industrially cleaned and picked rice is delivered at a premium price.

If the rice is new, then there would be more starch in it and the grains will be sticky when cooked. This might be a good idea for rice puddings such as *kheer* but it will not be used in preparing pilaf or good biryani.

One should purchase rice from reputable suppliers and ensure that they are free of any contaminants such as stones and pebbles.

CONCLUSION

There are various types of seeds used in everyday cooking. We use them as a spice such as mustard seed and cumin, as a grain like wheat, rice and so on and also as a main ingredient like chick peas, lentils and many more. We also consume seeds as nuts like almonds, cashew nuts, etc. Since the variety and usage of seeds is so diverse, it becomes crucial for us to categorize them. In this chapter we have discussed only two seeds namely cereals and pulses and we are going to discuss all the others in the coming chapters.

Cereals are the seeds of the grass plants whilst edible seeds of certain legumes are known as pulses or lentils. Lentils can be stewed, boiled, used in salads or even sprouted; but then we must be careful as all the pulses cannot be sprouted as some contain poisonous compounds that need minimum cooking of 10 minutes. Cereals on the other hand are ground into flour and made into breads or are eaten as an accompaniment such as rice. Cereals are a rich source of complex carbohydrates and fibre. Cereal is had in various forms in breakfast as breakfast cereals. It is important to understand the difference in commodities as it largely impacts the cooking methods that it will be subjected to eventually. Though pulses do not have classifications but various types of pulses have been mentioned in the chapter with regards to their description and also Hindi names. Their photographs are also given (Table 16.2) for easy identification, but it is important to mention at this stage, that many pulses look alike in colour and shape. The same pulses are known by different names in certain parts of the world and so in the Table 16.2 various other names of the pulses are also listed.

Cereals have been discussed with more emphasis on rice as this is the most common ingredient eaten around the world. There are many varieties of rice found around the world and it is believed to be one of the most ancient food grains that was cultivated around 5000 years ago. Rice is mostly a staple for Asians and is eaten with stews and curries. Unlike other cereals and pulses, rice is definitely classified into various categories as there are many types of rice and the usage of these also vary according to their structure and species. Certain rice is used for pilaf and biryani, where it is important to have all the grains separated from each other and hence long grain rice is used. Certain preparations like risotto from Italy need to be starchy and hence Arborio is used. Sushi rice in Japan is different from wild rice which is an aquatic grass from Caribbean Islands and the method of cooking it is also different. The rice is classified on the basis of its grain such as long grain, short grain, glutinous and wild rice. The chapter also discusses various types of millets that are indigenous to our country and are the power house of nutrition and energy. The usage of these grains will also help the farmers in India as it costs very less to grow millets and is more profitable for a farmer. We have also discussed few of the seeds such as chia and quinoa that are considered to be super foods and are being widely used in the world for nutrition and health benefits.

KEY TERMS

Adzuki beans   Red beans from Japan.

Al dente   Italian word that means cooked to the bite or crunchy.

Arborio   Italian short grain rice which is starchy.

Arhar   Hindi name for yellow lentils often called pigeon pea.

Bahia   Medium grained rice used in paella.

Barlotti   Kind of bean which is speckled and resembles red kidney bean. It is often called *chitra rajma* in Hindi.

Basmati   Long grained rice cultivated in India.

Biryani   Indian preparation of long grain basmati rice layered with meats and spices and cooked covered on slow heat.

Bajre ki raab   A drink made from pearl millet and yoghurt tempered with spices.

Bomba Plump rice also used for paella.

Bran Outer covering of grains obtained after milling.

Breakfast cereal Processed cereal usually eaten for breakfast such as rolled oats, corn flakes, wheat flakes, etc.

Burghul Cracked wheat grain commonly used in the Middle East.

Celiac Allergy to gluten.

Cereals Seeds of edible grass plants.

Chiwda Hindi for flat rice flakes.

Dalia Hindi for *burghul*.

Dosa A flat pancake of rice and lentils from Southern part of India.

Fava bean Large flat bean obtained from broad bean and is green in colour.

Ful medames A kind of bean found in the Middle East.

Garbanzos Italian for chickpeas.

Gluten Protein found in some flours.

Glutinous rice Rice that has lots of starch and cooks sticky.

Hom mali Thai name for Jasmine rice.

Hongiri Wooden bowl used for cooling sushi rice.

Idli A steamed preparation of rice and lentils from Southern India.

Jowar Hindi name for barley.

Kheer Rice pudding from India.

Kuttu Hindi name for buckwheat.

Lentils Another name for pulses.

Lima beans Large flat beans which are white in colour.

Masa harina Italian for corn meal.

Masoor Hindi name for red split lentils.

Matra Hindi for dried yellow peas.

Mirin Rice wine from Japan.

Mochigome Famous Japanese brand of Japanese glutinous rice.

Paella Spanish preparation of rice with assortment of meat and seafood.

Parboiled rice Rice soaked and steam cooked prior to milling.

Pilaf Indian preparation of rice with herbs and spices.

Poha Savoury preparation from India using flat rice.

Polenta Ground corn meal used in Italian cooking.

Pulses Edible seeds of certain legumes.

Puy Type of lentils from France; usually braised.

Red rice Kind of wild rice that is red in colour often known as camargue rice. It is a type of pollination between white and wild rice.

Risotto Italian dish prepared with Arborio rice.

Rolled oats Soaked and rolled oat grain, often cooked with milk as a porridge.

Roti A flat Indian bread made from a dough and cooked on griddle.

Sushi rice Rice from Japan that is used for making sushi.

Sushi Japanese delicacy of cooked rice and raw meats, usually seafood.

Texmati Long grain rice from Texas, USA.

Tortilla Flat bread from Mexico. It can be made from corn or flour.

Urad Hindi for black gram lentils.

Vatana Hindi for dried green peas.

Wehani Variety of red rice that grows in California.

Wild rice Aquatic grass that is often cooked as rice.

CONCEPT REVIEW QUESTIONS

1. Define a seed and list the various seeds discussed in this chapter.
2. List at least three differences between a cereal and a pulse.
3. What are pulses? List down at least five Indian pulses.
4. What should be kept in mind while cooking pulses?
5. How would you differentiate between a new and an old pulse?
6. How does the age of the pulse affect its cooking?
7. Why should you refrain from adding acidic ingredients such as tomatoes in the beginning of cooking?
8. What do you understand by the term 'reconstituting the pulses'?
9. Why should we be careful in sprouting lentils?
10. What do you understand by the words *sabut, chilka*, and *dhuli* ?
11. Define a cereal.
12. Name at least five common cereals used in breakfast.

13. What is the cereal used during fasts in India?
14. What is gluten-free flour and how is it important?
15. What is the difference between *burghul* and *dalia*?
16. What is bran and how is it useful?
17. How is rice classified?
18. Name at least four types of long grain rice, short grain rice, and glutinous rice.
19. What is a wild rice?
20. Give three uses of each of the rice from different categories.
21. What is the difference between brown rice and white rice?
22. How is texmati different from basmati?
23. Give two kinds of rice from Japan.
24. What is the other name of Thai rice?
25. What is paella and what kinds of rice are used for making the same?
26. What is so special about risotto rice and what are the types?
27. How is Spanish rice graded?
28. Write three uses of glutinous rice.
29. What is parboiled rice and how is it different from other rice?
30. How is red rice found in India different from red camargue rice?
31. List few by-products of rice.
32. What precautions should be taken while cooking rice?
33. What are the various methods of cooking rice?
34. How would you select rice?
35. What would be the difference in the usage of old and new rice?
36. Name two super foods.
37. Name five millets that can be used in Indian cuisines.
38. What is Kodo millet and what are its uses?

PROJECT WORK

1. In a group of four, cook lentils with at least four different cooking methods and compare the results of the same.
2. Do a market survey and study various types of cereals and pulses available in the market. Take photographs and research about their uses.
3. Visit hotels and restaurants and make a note of commonly served cereals for breakfast.
4. Research about the various traditional rice preparations from India and understand the different kinds of rice used in the same. Write your observations and prepare few of these dishes in groups to compare the results.

CHAPTER 17

METHODS OF COOKING

Learning Objectives

After reading this chapter, you should be able to:
- use various methods of cooking
- understand the importance of a cooking method with regard to taste and textures
- know the utensils and equipment used in various methods of cooking
- be aware of the heat transference methods involved in each method of cooking
- comprehend the temperature ranges of each method of cooking
- know the usage of the cooking methods to create various delicacies

INTRODUCTION

Understanding different cooking methods is important because there is a relationship between cooking methods, the expected result, and the best types of food to use. For example, it is hard to complete a recipe successfully for poached salmon if one does not know what poaching means.

If you understand how a cooking method works, you can choose the best foods to use with that method. For example, if you have snapper fish, you would use a method that enhances the flavour of this delicate fish, such as grilling, steaming, or smoking and not stewing or braising, which would disintegrate the fish.

The methods of cooking are broadly classified in Fig. 17.1.

These days with the advent of technology and new equipment, there are many ways of cooking available, which are not necessarily the methods of cooking; but can be described as modes of cooking. Common examples of these are microwave cooking, induction cooking, and infrared cooking. Infrared and induction cooking were discussed in Chapter 4 and we shall discuss microwave cooking in this chapter as many people these days classify it under the methods of cooking.

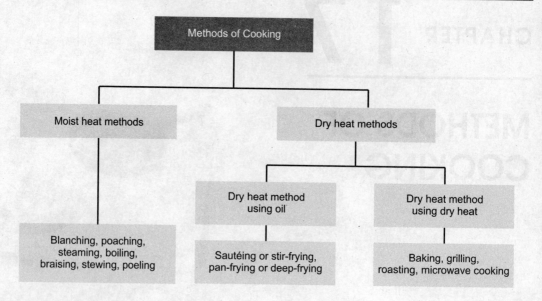

Fig. 17.1 Classification of methods of cooking

The first step towards being a chef is to know the ingredients, ways of processing, selection, and storage and we have substantially covered these in the last few chapters. This chapter on methods of cooking is the second step to actually start cooking the ingredients and hence the knowledge of various methods of cooking becomes crucial. To create a dish sometimes more than two to three cooking methods are used and one can alter these methods to create diversity in the product. For example, some meats are first pan-fried and then roasted. Meat can also be smoked, shredded, and then stir-fried to create a delicacy. Some dishes have been time tested and are referred to as classical dishes and some are created by chefs and categorized into contemporary cuisine.

In this chapter we will discuss the methods of cooking with regard to transference of heat and ranges of temperature, which have to be kept in mind as they are crucial to any cooking method. The important points of each method must be carefully read as they help us in understanding that method better. We shall then discuss food commodities which can be prepared with different cooking methods and the reasons behind the same. We shall also deal with the range of equipment and utensils used for different cooking methods. There are a total of 13 methods of cooking (refer to Fig 17.1) and we will discuss them individually.

BLANCHING

The word blanching comes from the French word *blanc*, which means to whiten. Blanching actually is not a direct method of cooking, it is usually the preliminary process followed to prepare the food which would eventually undergo another method of cooking. The whitening of the food as described above, gastronomically means brightening of the colour of the vegetable and as we read in Chapter 6, blanching is incomplete without 'shocking or refreshing'—a term used when the commodity is brought straight from the boiling liquid and plunged into cold water to arrest the cooking, brighten the colour, and pre-serve the nutrition of that particular ingredient. Blanching is done for various reasons and one of them is brightening of the commodity. In broader aspects, this method of cooking is used for several products for various purposes, which could range from blanching for removing strong flavour from a commodity to even increasing its shelf life. More uses of blanching are discussed in detail in Table 17.1.

The different ways of blanching are discussed below.

Hot Water Blanching

In this method of blanching the food commodity is immersed in the boiling water and cooked for the required time. The process is completed after shocking or refreshing; however, certain starchy commodities such as potatoes, etc. are not refreshed as they have a tendency to absorb moisture.

Cold Water Blanching

In this method of blanching the food is immersed in cold water and then brought to a boil. The food is blanched for the required time and then refreshed depending upon the type of commodity. Usually all the vegetables which grow above the ground are blanched with hot water method and all the vegetables which grow below the ground are blanched with cold water method. For example, blanching potatoes in cold water will make them more absorbent and hence, it will be good to use them for roasting, where they will absorb the oil and the flavours and become crisp.

Oil Blanching

This method of blanching uses oil as a medium to blanch the commodity and is usually done to remove the skin from certain vegetables such as peppers. This method is widely used for blanching potatoes for French fries.

Each method of blanching is used for different types of food commodities and it serves different purposes.

Heat Transference

All the principles of cooking rely upon the heat transference methods. We read about the effect of heat on vegetables in Chapter 6; let us now discuss the methods of transference of heat. Usually there are three main types of heat transference (refer to Fig. 17.2).

Conduction

It occurs in all principles of cookery. We normally talk about metals being good conductors or bad conductors. Conduction means travelling of heat within a solid matter. The heat travels from the heat source to the cooking vessel to the liquid and then to the contents, where it is transferred (conducted)

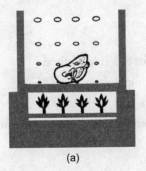
(a)

(b)

(c)

Fig. 17.2 Types of heat transference
  (a) Conduction, (b) Convection, and (c) Radiation

through the food. A simple example of conduction is a pan of water on top of an electric stove. Heat from the stove is conducted through the pan and is transferred to the contents (the water) which, provided with sufficient temperature and time, will boil. Food placed in the pan will conduct heat from the water and again, provided with sufficient time, will cook (Fig. 17.2).

Convection

It is the circulation of heat, which occurs in liquids and gases. Convection with gases (air, steam) works on the simple principle of 'hot air rises'. In liquids (water, stock, etc.) the same principle applies—the liquid at the bottom of the pan is heated and becomes lighter, therefore it rises and is replaced by the heavier and cooler liquid from the surface. The movement between gases and liquids forms a current or circulation of continual motion (as long as heat is applied). These days convection ovens are available in the market where a fan is placed behind a heating element and this fan then 'convects' the heat emitted by the element to the entire space in the oven thereby aiding in cooking food (Fig. 17.2).

Radiation

Radiated heat is transmitted in short waves from the heat source directly onto the surface of the food. The heat is absorbed at the surface and the molecules become agitated causing friction. Heat is then conducted through to the centre of the food (Fig. 17.2).

The main method of heat transference in blanching is conduction. The heat is transferred to the surface of the body through oil or water which in turn gets heat from the cooking vessel which gets heated from the heat source. The heat reaching the surface of the food then gets conducted to inside of the food thereby cooking it to the required degree of doneness. There is also a circulation of the heat within the medium and hence it will not be wrong to say that even convection is taking place.

Temperature Range

To cook food we have to transmit heat through any medium of cooking as discussed in the section on heat transference. Thermometer probes are available to check the internal temperatures of any food commodity and mostly the food is cooked when the internal temperature reaches 70°C, but this would again depend upon the degree to which we want the food commodity to get cooked. The texture of the cooked product will depend upon the time duration of heating it. The following is the general temperature range that is applied to various commodities to cook them.

- Water at a temperature between 65 and 100°C.
- Steam at a temperature between 100 and 120°C.
- Hot air at a temperature reaching 280°C.
- Fat (frying) at a temperature between 130 and 180°C.
- Little fat (pan-frying) at a temperature reaching 180°C.

In case of blanching, the following temperature ranges are used for different methods.

- Food that is blanched in hot water will be plunged into boiling water at 100°C.
- Food that is blanched using cold water will be brought to a boil eventually.
- Food that is blanched in oil will be plunged into oil at a temperature between 130 and 165°C.

The required temperature will be determined by the thickness, texture, and density of the food and the standard required.

Blanching and Its Uses

Blanching is used for various reasons. Even though it is the pre-processing of a food commodity, there is more to do with blanching rather than just cooking of food. Refer to Table 17.1 to have a better understanding of this cooking method.

Table 17.1 Uses of blanching

Food Commodity	Method of Blanching	Reason for Blanching	End Use
Animal bones	Cold water	To whiten the colour and impurities	White stock
Turned potatoes	Cold water	To make them absorbent	Roasted potatoes
French fries	Hot oil	To allow bulk preparation in advance with no spoilage	Fried as required for service
Tomatoes	Hot water	To remove skin	Used in salads sauces, garnishes, etc.
Vegetables growing above ground	Hot water	To par or precook—blanching often retards the action of enzymes, thereby allowing them to freeze for storage	In preparation for freezing or for final cooking later
Meats	Hot water	To close the pores thereby retaining colour and nutrients	Cold meat platters, main courses
Nuts	Hot water	To remove the skin	Puddings, salads and garnishes, etc.
Strong flavoured vegetables	Cold water/hot water, depending upon the vegetable	To remove the strong flavours, for example, from herbs	Salads, main courses, etc.
Bacon and ham	Hot water	To remove excess salt	Breakfast ham and bacon, etc.
Bell peppers	Hot oil	To remove the skin	Salads, soups, main courses, etc.
Leafy vegetables	Hot water	To brighten the colour	Various uses

Equipment Used in Blanching

Various types of equipment are used to effectively blanch the food commodities (Table 17.2).

Table 17.2 Equipment used in blanching

Equipment	Description	Photograph
Stock pot	This is available in various sizes and is used to prepare stocks. Bones are blanched in it.	
Dipping basket	This is used to lower vegetables in the boiling water, and is also for easy removal.	
Western spider	This large perforated spoon is used for removing commodity from liquids.	
Chinese spider	This large perforated spoon made from wires is used for removing food commodities from liquids.	
Steam jacket kettle	This is used for blanching and cooking stocks, etc. in bulk.	
Deep fat fryer	This is used for frying in oil blanching methods.	
Sauce pan	This is used for heating water and making sauces.	

POACHING

Poaching can be defined as a moist cooking process in which food is gently simmered in the liquid, which is brought to, and maintained at a temperature just below boiling point. Poaching is that method of cooking where a food commodity is simmered in a liquid. Liquid simmers whereas the food in it poaches. Poaching is done for shorter duration of time and the simmering of food for longer periods of time will be almost close to the method of cooking called boiling.

Water or flavoured liquids can be used for poaching (Chapter 8). Even milk or wine is used at times for poaching. Poaching takes place at a temperature below the boiling point of water, for example, between 70–96°C approximately. The type of food that is to be cooked determines the temperature. Fruits, such as apples or pears, require a lower cooking temperature because of their delicate texture, especially if they are to be served in their natural shape. Eggs require higher cooking temperature allowing the albumen to coagulate quickly to envelop the egg yolk and therefore to retain an ideal shape.

The main method of heat transference employed in poaching is the same as for blanching. The main reasons for poaching food are given below.

- It is a fast method of cooking food.
- Food is moved as little as possible so it does not break apart.
- Poaching liquid forms the base of many sauces.

Heat Transference

The method of heat transference in poaching is conduction. Heat is transferred to the food through the cooking medium which in turn gets heated from the metal it is in and the metal gets the heat from the heat source. Just like blanching there is a circulation of heat within the cooking medium and hence convection also takes place. If poaching is done in the oven then maybe radiation will also take place.

Poaching in 'bain-marie'—to cook in a water bath—is very commonly used for holding of food. Some food items are cooked in bain-marie and this can be called an extension of poaching as here very gentle foods are poached for longer duration of time with application of gentle heat. This method of cooking is suitable for cooking pates and terrines, egg custards, some baked cheesecakes, etc.

Temperature Range

The temperature range of poaching is usually between 70–96°C. It must be taken care that the temperature of the water never comes to a boil and should always stay under 100°C.

Poaching Points

The following points are to be kept in mind while poaching.

Only prime quality food is to be used as this method of cooking is done for less duration as the liquid keeps drawing out the nutrients from the food commodity. It is essential to use prime quality food because this method of cooking is very gentle and hence not suitable for second quality ingredients.

1. Always keep the temperature below the boiling point, i.e. 100°C.
2. Food may be covered; but it is not necessary unless cooking in oven.
3. Keep liquids to minimum to preserve the nutritional value of food commodity.
4. Cuts of fish should be placed in hot liquid, whereas the whole fish should be placed in cold water and then heated, as the whole fish will lose its shape if immersed in boiling liquid.
5. Food being poached should be of same size to ensure even cooking.
6. The food commodity should be completely immersed in the liquid.
7. Different food commodities require different temperature ranges and this must be strictly followed to get the desired product.
8. Acid medium is required for poaching, as acid hardens the proteins whereas fats and oils soften the proteins.
9. The ratio of water to vinegar should be 10:1 while poaching eggs.

Poaching and Its Uses

The various uses of poaching are shown in Table 17.3.

Table 17.3 Uses of poaching

Food Commodity	Poaching Liquid	Use
Chicken	Chicken stock	It is used to make poached chicken with *velouté* sauce (refer to sauces and its derivatives in Chapter 10).
Cuts of lean fish	Fish stock/milk	It is used to make poached fish with *velouté* sauce (refer to sauces and its derivatives in Chapter 10). Haddock fish is poached in milk and served in breakfast.
Cuts of oily fish	Court bouillon	The acid in the court bouillon firms up the fish.
Eggs	Water and vinegar	It is used to poach eggs for various breakfast preparations and classical preparations such as egg Benedict, etc.
Whole fish	Court bouillon—start from cold liquid	If the whole fish is immersed in hot liquid it will lose its shape as the proteins will coagulate and due to shrinkage of the skin the fish will become crooked.
Fruits	Flavoured sugar syrup depending upon the fruit (cinnamon goes well with apples, etc.)/wine	It is used to poach fruits for using as compotes for breakfast or sauces. Poached pears in red wine are classical Italian desserts and are called *para en vino tinto*.
Meats	Meat stock depending upon the type of meat being used	It is used to poach the prime cuts of meat to be used in cold meat platters, etc. or even used as main course accompanied with the relevant sauce.

Equipment Used in Poaching

Various types of equipment are used to effectively poach food commodities (Table 17.4).

Table 17.4 Equipment used in poaching

Equipment	Description	Photograph
Sauteuse	Pan with slanting sides.	
Perforated spoon	Spoon with holes to remove food from liquid.	
Fish slice	An angular flat spoon with perforations to remove the fish from the liquid.	

Contd

Table 17.4 (Contd)

Equipment	Description	Photograph
Salmon kettle	Large oval pan with perforated inserted tray used for poaching whole fish, especially salmon.	
Gastronome tray	Trays with raised edges, also used for presenting food on the buffet. It is available in various sizes.	
Sur la plat	Round white porcelain dish with handles; used for presenting poached eggs.	

BOILING

Water boils at 100°C and that becomes obvious when we talk about the principle of cooking called boiling. The food gets heated by convection currents in hot water and this includes all the variants such as blanching, steaming, and poaching, therefore there are no Maillard reactions (refer to Chapter 12). Boiling is more relevant to vegetables. They are cooked quickly in rapidly boiling salted water, and for a very short time. This gives a firm and pleasant texture, keeps the colour, and ensures that the vegetable will retain most of its nutrients. Unlike poaching, the amount of water used should be at least four to five times the volume of the vegetables being cooked so that the water can regain its boil quickly after the vegetables are plunged in. The amount of salt used is the reverse of what one might expect. For example, asparagus may take about five to six minutes to cook, so the salt does not have the time to permeate into the vegetable. Therefore the salt content of the water should be high. The water should be at full boil and the pan should be uncovered; volatile acids present in all green vegetables will escape; but if the pan is covered they would find their way back into the water and discolour the vegetables (refer to effect of heat on pigments in Chapter 6). Just like poaching there can be various types of liquids used for boiling such as stocks, milk, court bouillon, etc. Generally, food cooked by this principle is simmered rather than boiled. Simmering is a more gentle process, which takes place just below 100°C therefore ensuring better retention of quality, colour, and texture and reducing the amount of shrinkage. The difference between poaching and simmering is minimal, poaching is normally for short periods whereas simmering is usually for longer periods and will initially be brought to 100°C. Rapid and rolling boil happens at 100°C while simmering at 85 to 96°C.

Heat Transference

The method of heat transference in boiling is similar to that of poaching; i.e. conduction. Heat is transferred to the food through the cooking medium which in turn gets heated from the metal it is in and the metal gets the heat from the heat source. Just like blanching and poaching, there is also a circulation of heat within the cooking medium and hence convection also takes place.

Temperature Range

The food items are rarely boiled unless intended for a particular use. The food is mostly cooked in a simmering temperature ranging from 85 to 96°C. Some foods such as pastas and rice are rolled boiled to cook rapidly.

Boiling Points

The following points are to be kept in mind while boiling.

1. The food item should be completely immersed in the liquid for an even cooking.
2. This method of cooking is usually used for tougher and stronger flavoured meats as long duration of cooking in moist heat tenderizes the fibres thus making the meat soft.
3. Remove any scum regularly or else it will boil back into the food.
4. Keep replacing the evaporated liquid as required.
5. The food item may be covered to return to boiling point rapidly but cover should then be removed.
6. Vegetables grown above the ground to be started in boiling water.
7. Vegetables grown below the ground to be started in cold water.
8. Eggs to be started in boiling water.
9. Temperature of the water to be around 100°C for rolling boil and for simmering for longer duration of time, usually at 96°C.
10. Food being boiled should be of same size to ensure even cooking.

Boiling and Its Uses

Boiling is done for several reasons (Table 17.5).

Table 17.5 Uses of boiling

Food Commodity	Liquid	Use
Vegetables	Water/stock/milk	• To remove strong flavour from certain vegetables. • To gradually soften the fibrous textures of vegetables such as turnips and carrots, etc. • To extract excessive starch from potatoes, etc. • Cooking in milk retains the white colour of certain vegetables such as garlic, lotus root, raw banana, etc. • To reduce the cooking time to minimum.
Meats	Stock/milk	• Boiling is usually used for second cuts of meat, preserved or salted food are boiled to get rid of excess salt and flavours. • Tripe is boiled in milk to avoid discolouration. • To remove strong flavours from leg of mutton.
Pulses and cereals	Stock/water/milk	• Pastas, rice, and pulses are boiled to avoid clogging. • Milk is boiled with rice and sugar to create rice puddings. • Other cereals are also boiled as per usage.
Eggs	Water	• Boiled eggs are very common for breakfast and salads.

Equipment Used in Boiling

Apart from equipment such as stock pots, sauce pans, perforated spoons as discussed earlier, various other types of equipment are used to effectively boil food commodities (Table 17.6).

Table 17.6 Equipment used in boiling

Equipment	Description	Photograph
Brat pans	Large equipment used for bulk cooking.	
Chinois	A conical strainer used for straining liquids.	
Colander	A large perforated bowl used to strain rice.	
Boiling kettle	Same as steam jacket kettle (refer to Chapter 4).	

STEAMING

> **CHEF'S TIP**
> The action of baking powder is retarded by the steam at higher pressure; it is thereby advised to cook dishes which rely upon the aeration by baking powder in high speed convection steamers.

Steaming is one of the most commonly used methods of cooking, especially these days when people are more conscious about their health. In layman terms it is a principle of cooking where the food is placed over boiling water and is covered with a lid to trap the steam. This steam then cooks the food. Technical progress has tremendously influenced this type of cooking. Actually we can choose between cooking pot and inside grill, as used by people earlier, or the new pressure steamer, combo- steamer, or pressure cooker for large quantity of food.

This cooking technique is intimately related to boiling. Food is cooked over water at a rolling boil, at boiling point—when the water molecules are changing to steam or gas—and thus the temperature is slightly higher than the boiling point of water. Steaming cooks the food by surrounding it with hot steam. The steam circulating around the food provides an even, moist environment that allows the food to retain most of its flavours and natural juices. You can add herbs, stock, wine, or beer to enhance the flavour. Foods such as vegetables can be cooked fairly rapidly with this kind of technique. It is most suitable for delicate foods which would easily break up while rolling around in a liquid. This is a very ancient art of cooking and Asians are believed to have used natural objects, such as bamboo pipes, to steam. Till this day traditionally dim sums are steamed in special bamboo baskets called dim sum baskets. This method of cooking is not very popular in the Western world, where it is used to prepare food for patients and people on restricted diets, as there is no yield of liquid that could be turned into an accompanying sauce. Probably this is the reason for using this method for second quality meats; on the contrary the Asians use this for prime quality meats.

Heat Transference

Heat is transferred by the steam that is generated from the boiling water. The steam condenses on the colder food commodity and turns back into liquid. It is important to keep the food on a perforated pan so as to allow the drainage of the liquid back to the boiling medium. The food being cooked on the perforated pan is subjected to convection because the heat from the steam is circulated around it. The heat is also transferred through conduction as the perforated pan gets heated up and conducts the heat through to the commodities kept on it.

Temperature Range

There can be different temperature ranges at which steaming takes place. If we go back to our elementary school science, we read about three states of water namely solid (ice), liquid (water), and gaseous (vapour or steam). To change any of the state of this matter, we need to apply heat. The temperature of the steam will be a little higher than that of the boiling water; but it will depend upon the equipment that a commodity is being cooked in. The temperature of the steam changes with the change in pressure according to the type of steamer used. The temperature range for steamers will be as follows:

With Pressure   High pressure compresses the heat thereby increasing its temperature between 120 and 180°C.

Without Pressure   This involves cooking in a pot with lid on and not allowing much steam to escape. The temperature range will be just above the boiling point and could be in a range of 103 to 105°C.

With Water Steam   The device which uses water steam is a multifunctional machine. It is in fact an improvement of the air convection oven. It works with fluids, hot air, and steam, which can be used separately, simultaneously or one after the other. The temperature could be between 105 and 120°C.

With Dry Steam   The steam produced in the engineering department is utilized in laundry and in kitchens. The steam-operated machines such as steaming kettles or coffee machines use this type of steam. The temperature will range between 100 and 105°C.

Steaming Points

The following points are to be kept in mind while steaming.

1. Usually second quality of meat is used with the exception of chicken and fish.
2. Use foods that will not disintegrate if they are subjected to longer duration of time.
3. Certain foods such as steamed puddings should be covered with foil or plastic to avoid absorption of liquid into the commodity.
4. Perforated trays should be used to allow the drainage of the liquid.
5. Usage of modern high pressure steamers will preserve the nutrition and colour of the vegetables.
6. Be cautious while cooking with steam at higher temperatures as the steam condenses on the skin to give a boiling effect thereby burning the skin twice and causing twice the pain. Always stand back at a distance whilst opening high pressure steamers, to allow the steam to escape first.

Steaming and Its Uses

The uses of steaming for various food commodities are shown in Table 17.7.

Table 17.7 Uses of steaming

Food Commodity	Method of Steaming	Uses
Vegetables	Convection steamers/high pressure steamers	• It is used to retain the colour. • It is used to retain the nutritious value. • It is used for preparing vegetable dishes as accompaniments or it is used as a main dish.
Vegetables	Convection steamers/high pressure steamers	• It is used to retain the colour. • It is used to retain the nutritious value. • It is used for preparing vegetable dishes as accompaniments or it is used as a main dish.
Meats	Atmospheric steamers	• Moist heat at not very high temperatures aids in softening the skeletal muscles into gelatine there by creating a soft succulent meat.
Pulses and cereals	Pressure cooker	• In Indian homes, one cannot do without pressure cookers as it is used to prepare rice, lentils, and many other curries. • It aids in faster cooking of pulses and cereals.
Potatoes	Convection steamer	• The water does not come in contact with the potatoes and thus keeps their nutritional value intact. • It does not discolour the potatoes.
Frozen vegetables	High compression steamer	• It retains the colour and the nutritional value of the frozen vegetables.
Fish	Dim sum basket/convection steamers	• It is used mainly for Chinese style steamed fish. • Delicate fish does not disintegrate with convected steam.

Equipment Used in Steaming

Apart from equipment such as stock pots, sauce pans, perforated spoons, etc. as discussed earlier, various other types of equipment are used to effectively steam food commodities (Table 17.8).

Table 17.8 Equipment used in steaming

Equipment	Description	Photograph
Pressure cooker	A sealed pan with pressure releasing valves used for cooking pulses and cereals.	
Atmospheric steamer	A common steamer used in professional kitchens to cook meats and vegetables.	
High compression steamer	A high speed steamer capable of cooking smaller quantities very quickly.	

Contd

Table 17.8 (Contd)

Equipment	Description	Photograph
Convection steamer	A steamer where steam is convected. Usually present in combination ovens, where it is used to steam fish and vegetables.	
Perforated tray	A gastronome pan which is perforated and is used for steaming commodities so as to allow the drainage of the liquid.	
Dim sum basket	A bamboo basket with a lid used to steam dim sums.	

STEWING

Stewing may be defined as a long slow, gentle, and moist cooking process, during which small, tough, second grade quality meats are made tender and palatable. This is generally true but in real sense stewing refers to cooking a food commodity in its own natural juices. *Etuver*, the French word for stewing, literally translates to cooking in its own juices. It does not necessarily mean that stewing will be for longer duration of time as this would really depend upon the commodity being used. For example, lamb stew will take more time than stewed apples, etc. Stewing of meat takes longer time as one would use tougher cuts of meat such as shanks and shoulder (Chapter 12). Tougher cuts of meat contain high amount of collagen and gentle heating for longer duration of time will break down the connective tissues and convert them into gelatin which is easily digested by the body. As this method breaks down tougher cuts of meat, it is a very economical method of cooking. It is one of the oldest methods, where the meat and vegetables were cooked in cast iron pans over wooden charcoals to create a wholesome meal. It is also one of the most nutritious methods of cooking as all the juices of the commodity are retained in the dish. Though this method was primarily used for meats, now it is not uncommon to see other commodities, such as fruits and vegetables, also being subjected to such methods. This method is widely used in Indian cooking, for example, for making gravies and curries. The advantage of cooking meat by this method is that since it is cooked in an enclosed container, all the juices form a natural sauce which is served along with the dish. The stews are normally served in a bowl along with the sauce in which it was cooked.

Stews are basically classified into the following three categories.

Brown Stew

As the name suggests, this stew is brown in colour and hence, more commonly used for red meats such as beef, lamb, or pork. The meat is either marinated with 'seasoned flour' or flour is sprinkled on top during the cooking process. Let us take the example of lamb stew for better understanding. Heat oil and cook the mirepoix of vegetables until translucent. Add meat and cook over high heat so that the meat is seared from all the sides and acquires a deep brown colour. Add a spoonful of tomato paste to give a deeper red colour and add lamb stock and stir the meat. The liquid mixes with the flour to create a sauce. Add enough liquid to barely cover the meat. Put a lid on top and cook in an oven or on top of a stove at a low to medium heat to let the lamb stew in the sauce. Keep checking the meat at frequent intervals so as to replenish the lost liquid and to stir the meat to avoid it from sticking to the base of the pot. When the meat is done, remove the mirepoix and put the lamb stew back on stove. You could enhance the stew by adding some turned potatoes, mushrooms, and pearl onions and cook till the vegetables are tender. The time of cooking will depend upon the quality and cut of the meat.

There will be lot of things common in most of the stews as described above and these would be—searing the meat, cooking with the stock of the same meat, covering with a lid, and sauce being formed as a part of the dish. The brown stew is commonly known as ragout in France.

White Stew

As the name suggests, this stew is white to blonde in colour and hence, mostly used with white meats such as poultry and veal. Unlike brown stew the flour is not browned and can be just sprinkled on top of the seared chicken and cooked under a radiated heat such as salamander and then liquid should be added to form into a sauce. Another method could be to prepare a sauce such as *velouté* and stew the seared chicken in the sauce. In France white stews are called fricassée.

Miscellaneous Stew

These are those stews that are neither brown nor white. For example, *ratatouille*—a Mediterranean stew of vegetables with tomatoes and basil, *bouillabaisse*—a classical Provencal dish from France, etc. are common stews which are neither brown nor white. Some stews, such as *mouilles mariniere* (stewed mussels), are stewed in white wine and some herbs.

Heat Transference

A stew can be prepared on top of a gas burner or in an oven depending upon the type of commodity being used and hence the temperature ranges for both these methods will be very different from each other. On top of the burner the heat will be transferred mainly through conduction and little convection may also occur as stews are always cooked with lid on. In an oven, apart from conduction and convection, some radiation will also occur. However the radiated heat will not directly affect the stew as the food is covered with a lid, but this heat will conduct through the metal to the food.

Temperature Range

As discussed above, a stew can be cooked in an oven or on top of a stove. It is easier to monitor the temperature on top of the gas burner and it is not so easy to monitor the temperatures when the stew is in the oven. We have also read that stewing is done for longer duration of time and if the temperature is high then the proteins in the meat will become quite firm leaving the meat dry and chewy, so it is important that the temperature is maintained between 80 and 90°C. If the stew is cooked in the oven then the temperature of the oven needs to be at 180°C.

Stewing Points

The following points are to be kept in mind while stewing.

1. It is a very gentle method of cooking.
2. The amount of liquid used is very small, just enough to cover the food commodity.
3. Cheaper and less tender cuts of meat are used as the slow gentle cooking renders them tender. Meat with bones may also be used to extract the gelatine.
4. The liquid is served with the food and forms a sauce with it.

5. The liquid should be gently simmering; if it boils rapidly the proteins will toughen.
6. The cooking pan is covered with a well-fitting lid.
7. The condensation that forms on the lid helps in self basting or in keeping the food moist.
8. Ideal stewing temperature on the flame is approximately 80°C.
9. Ideal stewing temperature inside the oven is 180°C.
10. The meat is always cooked well done as it is a slow and long process, thus all the microorganisms in the food are destroyed.
11. All the nutritive value of the food is preserved.
12. Thickening agents may sometimes not be required as the connective tissues break down into gelatine which gives thickening to the stew.
13. Heat transfer is by conduction and convection. When finished off in the oven radiation also takes place.
14. Stews can be broadly classified as brown stews, white stews, and miscellaneous stews.
15. Liquids used include stocks, wines, water, milk, etc.
16. Since a lot of connective tissues are broken down the food is very easy to digest.

Stewing and Its Uses

Stewing can be used for cooking various food commodities (Table 17.9).

Table 17.9 Uses of stewing

Food Commodity	Type of Stew	Classical Dishes
Neck of lamb	Brown	Navarin d'*agneau*—a lamb and turnip stew
Diced shoulder of lamb	Miscellaneous	Lamb curry
Vegetables	Miscellaneous	*Ratatouille*
Apples	Miscellaneous	*Compotes*
Chicken	White	*Fricassée depoulet aux champignons*—stewed chicken with mushrooms
Diced topside of beef	Brown	*Ragout de boeuf jardinière*—beef stewed with vegetables
Fish	Miscellaneous	*Bouillabaisse*—assorted fish and shellfish stew with saffron

Equipment Used in Stewing

Many kinds of pots and pans with fitting lids can be used for stewing. Traditionally, cast iron pans were used to stew the meats; but bulk cooking can be done in large stock pots, tilting brat pans or even in steaming kettles as discussed in the above cooking principles. Apart from these, several other equipment may be used (Table 17.10).

Table 17.10 Equipment used in stewing

Equipment	Description	Photograph
Forced air convection oven	Oven where the heat is circulated by an internal fan. It is commonly referred to as combi oven as it can cook both by dry heat and steam methods.	

Contd

Table 17.10 (Contd)

Equipment	Description	Photograph
Cocotte	An earthenware container with a lid.	
Crock pot/slow cooker	Electrical cooking appliance with ceramic container that maintains relatively low temperatures ideal for stews.	
Tagine	Typical cooking equipment used in Moroccan cooking, which is a cast iron pan with a conical lid used for stewing.	

BRAISING

Braising is a long, slow, gentle, and moist process during which commodities particularly covered with a liquid are cooked in an enclosed container in an oven. Braising is usually applied to larger, tougher joints of meat, game, poultry, and occasionally fish, which require extended periods of cooking time in order to become tender, palatable, and more easily digestible.

Larger braised joints will require carving when cooked, after a suitable resting period. Alternatively, small pieces of meat (but not diced meat) may also be braised, for example, braised steaks, braised lamb chops, etc.

As in stewing, because of extensive slow cooking time, the collagen (collective tissue) is allowed to convert to gelatine without toughening the protein in the meat. In short, braising is a combination of three methods of cooking namely stewing, steaming, and roasting. Like stewing, braising uses moist heat method where liquid is added to the food commodity that has been seared in hot oil and the food is cooked with a lid on. The only difference is that the liquid is added to cover the food about one half to three quarters only. The food is thus stewed as well as steamed as the large open area allows the steam to be trapped inside the pot and circulate around the food. This technique is good for meats which are not tender, such as pot roast, because slow moist cooking in liquid helps to tenderize the meat. The meat is done when it is falls of the bone. During cooking the meat releases its flavour into the liquid, which can be reduced to make a sauce. Braised food is cooked in the oven. The heat should always be low, ideally around 140°C. Braising takes a long time, but the end result is definitely worth it. As long as there is water in the casserole the temperature will not be more than 100°C, irrespective of the oven temperature. If the oven temperature is too high it will evaporate the water in the meat tissues and make it hard and dry. When the meat is done the lid is removed and the meat is cooked openly to get some colour thereby accomplishing the method of cooking called roasting. Some chefs do not leave the meat open in the oven if they have seared it well in the initial stages of cooking.

Often it is recommended that some flour should be added at the start for binding. The addition of flour at this stage will absorb all the fat, and all this will form an emulsion later to be used as a smooth and glossy sauce. All cooking juices without any kind of thickening will be thin, but it will be delicious. If you want a thick sauce, you must remove all the excess fat in the end and then thicken with cornflour.

The sequence that is usually followed while braising a food commodity is given below.

1. First sear the meat on a hot skillet (to close pores) and add vegetables or tomatoes or deglaze with red wine. Cover one-fourth of the meat with stock, cook in oven taking care of replenishing the liquid to avoid drying of the meat and cook covered in an oven.

CHEF'S TIP
To avoid discolouration of vegetables cover them with buttered parchment paper or wax paper to avoid colour alteration.

2. For vegetables (Belgium endive, fennel, etc.) use a matignon (refer to Chapter 6) or onions, or lard (i.e. for red cabbage). Lay vegetables on top of matignon and cover one third of the piece with stock. Cover and braise in oven.

Braising may be divided into two main categories—brown and white.

Joints or cuts of meat for brown dishes are sealed by frying or flash roasting before being braised and are served with a brown sauce. White dishes are not seared and are made from white meats such as poultry, etc. The commodities may require blanching and would then be cooked in white stock, the accompanying sauce being made from the resulting cooking liquor.

There are certain similarities between braising and stewing; however, there are also several differences (Table 17.11).

Table 17.11 Difference between braising and stewing

Braising	Stewing
Large joints of meat or small cuts of meat are used.	Small cuts or diced meat are used.
The food is half to two-thirds covered in liquid.	The food is totally submerged in liquid.
The sauce is finished after the process of cooking is complete.	The sauce is cooked at the same time as an integral part of the dish.
Braised dishes are cooked in an oven.	Stews may be cooked either on top of a stove or in an oven.

Braised foods are both nutritious and economical as cheaper and tougher cuts are utilized and little of the natural flavours and juices are lost during cooking. The cooking liquor, which contains juices which may be seeped from the joint, is often used as a base for the accompanying sauce.

A combination of herbs, spices, wine, and vegetables are often used as a marinade in which meat or games are marinated before braising. Apart from adding flavour, the acidity in the wine helps to tenderize the joint.

Lean meat can be dry to eat and for this reason topsides of beef and particularly game, for example, venison, is often larded prior to cooking. Larding is the insertion of fat into lean meat by use of a larding needle. Brown braised meats should be sealed prior to braising. The colour from the brown surfaces adds a certain richness, which develops during the process, giving the dish its characteristic colour, aroma and flavour.

The matignon (bed of root vegetables) is used to add flavour to the finished dish and to prevent the joint coming into direct contact with the cooking vessel.

To glaze braised joints simply remove the cover and napper or coat over with the sauce regularly for the final 15 to 20 minutes of the cooking time. This is known as basting and it forms an important part, especially while roasting meat. A glaze will develop and set on the surface of the joint, which will enhance the appearance, and presentation of the finished dish.

Heat Transference

Before the meat is put for braising, it is first seared on a hot pan to seal the pores. During the sealing of a joint, heat is being transferred by conduction. During the process, while the joint is in the oven a combination of heat transfer occurs. Convection takes place as heat circulates within the oven chamber. Conduction also occurs through the cooking vessel (container) to the commodities. Radiation occurs during the later stages when the cover is removed and the food is exposed to the direct heat source. So this method of cooking utilizes all the methods of heat transference.

Temperature Range

The internal temperature of the sauce and the food commodity should be between 80 to 90°C and since braising is done in an oven it will be important to keep the temperature control to 180°C. The process, as described in stewing, should be long and slow.

Braising Points

The following points are to be kept in mind while braising.

1. Use cheaper or tougher joints of meat as this is a moist cooking method which makes the meat tender.
2. Joints may be cooked on a very hot pan, turning evenly to seal the outer covering before braising as it gives a deep colour to the joints.
3. Joints may be marinated. Marination tenderizes and adds flavour to the meat.
4. Joints may be larded (inserted with strips of pork fat).
5. Joints may be glazed to form a nice colour and crust.
6. Always cook covered initially and then uncover and roast the joint for final finishing.

Braising and Its Uses

Braising is used for various food commodities (Table 17.12).

Table 17.12 Uses of braising

Commodity	Method of Braising	Liquid for Braising	Classical Dish
Fennel	White	White chicken stock	*Fenoulis a la crème*
Celery	White	White chicken stock	*Celeries polonaise*
Saddle of lamb	Brown	Demi-glaze	*Selle braises*
Topside of beef	Brown	Demi-glaze	*Boeuf braise*
Duck	Brown	Demi-glaze	*Canard braise aux petit pois*
Pheasant	Brown	*Jus lie*	*Faison braise*
Sweet breads of veal	White	White veal stock	*Ris de veau bonne maman*

Equipment Used in Braising

Just like stewing, many kinds of pots and pans with fitting lids can be used for braising. One should ensure that the pot can fit inside the oven, or one could braise in a roasting pan with lid. Apart from these, some other equipment may also be used (refer to Table 17.13).

Table 17.13 Equipment used in braising

Equipment	Use	Photograph
Frying pan	To sear the meat and to evenly brown it.	
Sauce pan	To prepare the sauce from the strained liquid.	
Oven	To cook the braised meat.	

POELING

Poelet is a French word that does not have any English translation; some people refer to it as pot roasting, which is not a correct translation, as this principle of cooking is unique in itself. This method of cooking utilizes only prime quality foods. It is usually used to prepare first quality meats, poultry, and some game and the process involves cooking of meat in a closed container in melted butter. Since the food will be cooked covered, all the moisture will be retained inside the pot and we can say that the meat will be cooked in its own juices. Butter will soften the proteins and the meat will cook soft and tender. The melted butter can then be strained and reused later.

Just like braising, poeling is a combination of three methods of cooking such as: Stewing—as the meat is being cooked in its own juices; Steaming—as the dish is being cooked in a closed pot with lid on so all the steam generated stays inside thereby aiding in cooking; and Roasting—as in the final stages of poeling the meat is browned under radiated heat to give a glazed look on the meat for presentation and texture.

This method is more suitable for meats that have natural fats present in them. However, game and poultry which are lean need to be barded or larded first before cooking by the method of *poelet*. Just like the method of roasting, the meat here should regularly be basted to help keep it moist and juicy and again like braising or stewing, the remaining liquid and sediments can be used as the foundation for the accompanying sauce. There are some similarities between roasting and poeling; but then there are significant differences (Table 17.14).

Table 17.14 Difference between poeling and roasting

Poeling	Roasting
The meat is not seared before poeling.	The meat is seared before roasting.
The food is cooked with butter as the main liquid.	No liquid medium is used while roasting.
The dish is always covered while cooking.	The roasts are never covered while cooking.
Poeled dishes can be served along with vegetable mirepoix as a stew.	Roasts are generally carved and served with an accompanying sauce.

For any poeled dish, one could use an earthenware dish such as casserole or cocotte as seen in braising above. For bulk cooking one could use a gastronome. First put the mirepoix of vegetables at the bottom. If the mirepoix is intended to be served with the dish then use matignon (as seen in Chapter 6 it is a mirepoix), which is cut into defined shapes so that it is more presentable. Then place the piece of the meat on top of the mirepoix and bard the meat if it is a lean piece of meat. Then pour melted butter generously over it and sprinkle salt and pepper. Now cover the food with lid and cook it in the oven until the meat is done. The final stage of cooking will be done by glazing the meat under radiated heat. During this stage, one must ensure that the mirepoix is browned but not burnt as this would give a very unpleasant taste to the sauce. Now baste the meat with the liquid in the pan as this will help the meat to be juicy and moist. Lift the meat out of the liquid and let it rest. Now strain off as much of butter from the pan taking care to retain all the sediments as this will form the foundation of the sauce. Now deglaze the sediments with wine or spirit as required by the recipe and cook until the sauce begins to thicken. Now add *jus lie* or demi-glaze and cook until a sauce is formed. This will ensure that all the flavours are incorporated into the sauce. Strain the sauce and keep aside for later use. Now the meat can be carved and served with the sauce or just cut and combined with the sauce and served in a casserole.

Heat Transference

The process of poeling is done in an oven for better results. During the process of poeling, heat is being transferred by conduction. During the process, while the meat is in the oven a combination of heat transfer occurs. Convection takes place as heat circulates within the closed container. Conduction also occurs through the cooking vessel (container) to the commodities. Radiation occurs during the later stages when the cover is removed and the food is exposed to the direct heat source. So this method like braising utilizes all the three methods of heat transference.

Temperature Range

The internal temperature of the sauce and the food commodity should be between 80 to 90°C and since poeling is done in an oven it will be important to keep the temperature control to 180°C.

Poeling Points

The following points are to be kept in mind while poeling.

1. Prime quality meat, poultry, and game are used.
2. No liquid is added to the dish. Melted butter is the only liquid medium used.
3. The pot is always covered and cooked in the oven.
4. The food is basted occasionally to keep it moist and juicy.
5. The final finishing of the food commodity is done by glazing in the oven.
6. Accompanying sauce is made from the sediments remaining after draining the melted butter.

Poeling and Its Uses

Poeling can be used for various commodities (Table 17.15).

Table 17.15   Uses of poeling

Commodity	Classical Dish
Fillet of beef	*Fillet de boeuf richelieu*
Cushion of veal	*Noix de veau jardinière*
Loin of pork	*Longe de porc normande*
Saddle of venison	*Rable de venison diable*
Saddle of hare	*Rable de lievre grand veneur*
Pheasant	*Faison careme*
Guinea fowl	*Pintarde souvaroff*
Duck	*Canard aux jus d'orange*
Poussin	*Poussin en cocotte grand mere*

Equipment Used in Poeling

The equipment used in poeling are the same as those used in braising.

ROASTING

Roasting is the most preferred method of cooking used in Western cookery. Roasting also uses prime quality meats and some large roasts which are to be used on the carving board are also referred to as joints. Roasting is done for meat, poultry, game, and certain root vegetables such as carrots, potatoes, celeriac, etc. There are many different types of interpretations of the word roasting around the world and some of these are as follows:

Spit Roasting   It is very popular in Greece and some other Mediterranean countries. Here the meat is marinated and cooked on an open flame.

Tandoor   It is a method of roasting in Indian cooking in which a clay oven is used, where the meat is chargrilled to impart a smoky flavour to the dish.

Barbecue   Barbecue or open fire roasting is again very similar to spit fire; the only difference is that the meat is cooked over the radiated heat from the fuel and not directly cooked in flame.

Pot Roasting   This is a very ancient method of cooking meats in a closed container over a stove or wood charcoals. The dry heat created inside the pot covered with a lid helps to cook the meat.

CHEF'S TIP
The average cooking time followed for most meats is 20 minutes per 500 g and 20 minutes extra. For instance if the roast weighs 2 kg then the approximate time for roasting will be 80 minutes for the meat and 20 minutes there of, so the total time will be 1 hour and 40 minutes.

Rotisserie   This is the modern version of spit roasting, where the meats are skewered onto a metal rod that rotates in front of a heat that is produced either by the electric filament or by gas burners. The Lebanese dish *shawarma* is prepared using a similar technique; the only difference being that the rotisserie is an enclosed cabinet and shawarma griller is an upright equipment where the slices of the meat, usually chicken, are placed and rotated.

In all the above methods of roasting there is one thing in common and that is food cooked by the dry heat surrounding the commodity. Roasting is usually done in ovens with dry convected heat. Roasted food is often basted with the natural juices which flow out of the meat while roasting and therefore it is essential that the meats are roasted over a trivet of bones or mirepoix so that the meat does not touch the liquid. If this happens then the method of cooking will become braising rather than roasting. While roasting meat, one must sear the meat on a cast iron pan or cook in the high heat so that the meat attains a brown colour. Many chefs believe that this procedure will sear the meat and all the juices will be locked inside; but this belief is not true, as right after sealing, even when the meat is kept to rest, the blood can be still be seen on the board. Care should be taken not to burn the mirepoix as the remaining sediments in the pan shall become the foundation of the accompanying sauce. After the meat is roasted, it should not be carved immediately. One should rest the meat for the following reasons.

- When the meat is roasting, the liquids rise up to the surface waiting to gush out. If the meat is sliced there will be loss of juices rendering the meat dry and chewy and hence resting allows the juices to settle back into the meat thereby allowing the juices to be distributed evenly resulting in a juicy roast.

CHEF'S TIP

Hold the tip of the index finger with the tip of thumb and feel the fleshy part on the palm under the base of the thumb. This will give you an indication of the doneness of the rare meat. Now hold the tip of the second finger and press to feel the doneness of medium meat. The tip of the third finger and the feel of palm give you the idea of medium to well done meat and the tip of the last finger and the feel of the palm give you an idea of well done meat.

- The meat fibres relax if held at room temperature and it becomes easy to carve.
- When the meat is rested after roasting, certain juices will come out and then it will be easier to present the meat on the carving board. Otherwise it can be very messy.

The meat is always processed and dressed before roasting and this is done for various purposes such as:

- The wish bone in the poultry is removed so that carving can be done easily.
- The poultry is trussed—tying of the poultry with a thread so that it retains the shape during roasting and also the surface area of the meat is reduced to discourage the evaporation of juices as retention of juices helps this lean meat to remain moist.
- Carving of the shoulder of the lamb becomes very easy if it has been boned and rolled up.

However, the time of roasting will depend upon many factors such as size of the joint, the evenness of the cut of meat, type of oven such as convected oven or a regular oven, the quality, and the grading of the meat. Even the age of the meat plays a very important role in the roasting time and doneness. The following points can be helpful in checking the doneness of meats.

- Appearance to the experienced eye cannot be relied upon every time, so sometimes the colour of the juices which run out from the meat can give us an indication about the doneness of the joints. If the juices are clear then the meat is well done and if there is still some pink blood then it is medium and so on.
- Resistance to pressure is another indicator of the doneness of the meat. Raw meat has fairly soft texture and as it cooks the protein coagulates and loss of moisture firms the meat up.
- A meat thermometer or a probe is the best and accurate method of testing the doneness of meat.

Heat Transference

In this chapter when we refer to roasting, we talk about oven roasting. Again like braising and stewing, roasting is a combination of all three methods of heat transfer namely conduction through the roasting pan to the product; convection, as the roasting is done in convected ovens, where the heat is circulated around the food commodity; and radiation as the rays of heat from the oven surface help in browning of the food commodity.

Temperature Range

Temperature ranges are very difficult to determine in roasting, as this principle is diverse and the temperature would really depend upon the size and type of commodity being roasted. However, if convection ovens are used the temperature could range between 180 and 240°C. It is always advisable to put the meat at a high temperature and when the meat has slightly browned, the temperature can then be lowered to 180–200°C.

The internal range of the cooked meat will depend upon the type of meat and the doneness required. Usually an internal temperature of meat may vary between 60 and 80°C. Meat temperature probes are available in the market to check the doneness of the meat.

Roasting Points

The following points are to be kept in mind while roasting.

1. Suitable for prime quality meat, poultry, game, and fibrous vegetables.
2. No liquid is added to the meat. It is just marinated with little fat, herbs, and flavourings.
3. No cover is used while roasting.
4. Basting is essential to keep the roast juicy and moist.
5. Meat has to be trussed or processed before roasting.
6. Meat has to be rested after roasting.
7. The residual sediments are deglazed with wine and stock and served as sauce with the roasts. These are often known as jus roti or the roast gravy.
8. Roasts are always placed on trivet of bones or mirepoix of vegetables.

Roasting and Its Uses

Roasting can be used for cooking various commodities (Table 17.16).

Table 17.16 Uses of roasting

Commodity	Degree of Cooking	Use
Vegetables	Well done	Used as accompaniment, or roasted vegetable soup or salads
Potatoes	Well done	Accompaniment to the roasts
Loin of venison	Medium to well done	Carving roast for buffet or served as an à la carte
Saddle of hare	Medium to well done	À la carte for dinner service
Top side of leg of lamb	Medium	À la carte for dinner service
Rack of lamb	Medium	To prepare roasts such as crown roast, guard of honour (refer to cuts of meat) or used as an à la carte dish
Chicken	Well done	Roast for buffet or à la carte

Contd

Table 17.16 (Contd)

Commodity	Degree of Cooking	Use
Boned out sirloin	Rare/medium	Roast for buffet or à la carte
Pheasant	Well done	Roast for buffet or à la carte
Guinea fowl	Medium to well done	Roast for buffet or à la carte
Duck	Well done	Roast for buffet or à la carte

Equipment Used in Roasting

Apart from the forced air convection ovens and combi ovens, following are the other equipment used in roasting (Table 17.17).

Table 17.17 Equipment used in roasting

Equipment	Description	Photograph
Roasting tray	A cast iron pan used for roasting meats and bones.	
Bain-marie	A container placed in a hot water container to hold sauces and accompaniments for the roasts.	
Carving board	A wooden or a nylocast board to carve meats in front of guests.	
Carving knife and a fork	A large knife and a fork to carve the meats in front of the guests.	
Microwave oven	An oven that works on microwaves.	
Thermometer probe	A needle with a temperature gauge at the back to record the internal temperature of the meat.	

GRILLING

It is another prime method of cooking, where both the meats and their cuts should be of prime quality. As we progress in the following paragraphs we will discuss the cuts of food commodities which will affect the texture of the food while cooking by this method. Grilling is often confused with broiling. The term broiling refers to food cooked under radiated heat. Broiling would mean different things in different contexts. We can broil or lightly roast the spices on a hot skillet to accentuate the flavours. Traditionally, meats are broiled under radiated heat and this could be best achieved under a salamander. Cooking under salamander is also sometimes used to give colour to a particular product, especially in case of cheese and egg-based sauces and is often referred to as 'gratinating'. All of the above mentioned methods can be grouped into a larger principle called grilling. This principle of cooking relies on dry heat and hence it is important that prime quality of meat is used for a good end product. Usually fish and poultry are cooked by this method. Heat is radiated onto the food and the heat can be applied from top or from bottom or from both the sides depending upon the dish and the equipment used. The length of cooking time cannot be established for this method as it would depend upon many factors such as size of meat and the cooking equipment being used. The degree of cooking will also depend upon the degree of cooking required for meats. For example, cooking of steaks (refer to Chapter 12). Foods thicker than 5 cm are seldom grilled as because by the time the temperature inside the meat reaches the optimum temperature, the outside surface will char and get burnt. Grills are served with variety of sauces and accompaniments such as potatoes and vegetables. Barbecuing can be a type of grilling unless done in spit fire method, where it will be known as roasting.

Heat Transference

Grilling relies upon radiated heat; but as with other methods of cooking the heat is also conducted through to the food. Since the radiated heat is directly upon the food and not indirectly through another receptacle, it is a fairly rapid cooking method.

Temperature Range

Temperatures are often difficult to determine in grilling process as there are many factors that influence the temperature. One of them is the usage of equipment and the other is the fuel used in cooking. Many other factors such as quality of meat and the cut and thickness of meat will also affect the range of temperature that is applied to the food item. The temperature of the radiated heat source may be at times as high as 700°C, but the surface temperature of the food will be between 180 and 200°C. Modern research says that use of lower temperatures yield tender and succulent product.

Grilling Points

The following points are to be kept in mind while grilling.

1. Prime quality food and cuts should be used.

2. The food item should be of regular shape for even cooking.

3. In grilling one must start with high heat and then reduce the heat considerably to cook the product.

4. The meat used should be well marbled and if not should be well larded or barded by fat.

5. Place the food onto well-oiled grill or bars so that the meat does not stick and loose its aesthetic appeal and taste.

6. Regular basting with marinade or oil is required to keep the food moist.

7. Most of the grilled meats are served with flavoured butters often known as compound butters, which melt and form a butter sauce.

Grilling and Its Uses

Grilling is used for cooking various food commodities (Table 17.18).

Table 17.18   Uses of grilling

Commodity	Degree of Cooking	Use
Fish	Well done	Fish is usually marinated and then dusted with flour to prevent it from sticking on the grill. The whole fish can be grilled too in the similar way.
Shellfish	Well done	Mostly shellfish such as prawns, lobsters, scampies, crayfish, etc. are often marinated and grilled.
Fillet of beef	Rare/medium/well done	Can be grilled whole like chateaubriand or grilled like individual steaks.
Lamb chops	Medium to well done	Whole rack or individual lamb cutlets are grilled and served as a main course.
Tenderloin of lamb	Medium	À la carte for dinner service.
Rack of lamb	Medium	To prepare roasts such as crown roast, guard of honour (refer to cuts on meat) or used as an à la carte dish.
Pork chops	Well done	Marinated and grilled pork chops are served as a main course in lunch and dinner.
Sirloin steak	Rare/medium/well done	Grilled for à la carte.
Vegetables	Well done	Grilled and served as accompaniments or can be used as main course.
Small chicken	Well done	Small chicken flattened like a toad (refer to cuts of chicken in Chapter 12) is often marinated and grilled and served as a main course.
Sausages	Well done	Grilled sausages are commonly served as an accompaniment with eggs or can be served a as a meal.

Equipment Used in Grilling

The grilling equipment can vary considerably and each has a different purpose and usage. From large grills used in the traditional kitchens to portable grills which serve as the best equipment for afternoon picnics or camping are versatile and operate on various fuels such as charcoal, gas, etc. and even electricity (Table 17.19).

Table 17.19   Equipment used in grilling

Equipment	Description	Photograph
Bar grill	It is usually gas operated and has lava stones placed on it. Lava stones are available in the market and they absorb heat and the heat is then radiated on to the food. Otherwise the method of cooking will be more of spit fire roasting than grilling.	

Contd

Table 17.19 (Contd)

Equipment	Description	Photograph
Charcoal grill	Wooden charcoal is used as a fuel in this equipment and the food is grilled when the coal has burnt and turned red.	
Ridged griddle plate	A ridged iron plate gives grill marks on the food commodity. Though not a traditional method, it is convenient to use.	
Hot plate	A large metallic plate used to grill burgers and escalopes of meat.	
Contact grill	Often used for grilling sandwiches and for escalopes of chicken, to cook the meats on both the sides at the same time.	
Salamander	Often referred to as a grill, but it is an equipment that radiates heat and is used for gratinating and broiling as well.	

SAUTÉING

Sautéing comes from the French word *sautir* which means to jump. In this method the food is continuously tossed in a pan. In Indian cooking, *bhunana* is closely related to sauté method of cooking and stir-frying of food in other Asian cooking is similar to sauté. Principally, sautéing is in ways similar to pan-frying but there is a little difference in the cooking methods, which we shall see while dealing with frying. Sautéing can be done with raw food, semi processed food, or cooked food. In sauté, the food commodities are cooked in a rapid way, quickly moving the food on a slanted pan known as sautoir or a smaller version called sauteuse. This principle of cooking should use prime ingredients which are first sautéed and then mixed with sauce. This principle should not be confused with stewing. Stews are food commodities cooked in liquids which form a sauce, whereas in sauteing, the food is quickly stir-fried and mixed with appropriate sauce and sometimes mixed together on a medium heat. The sauces of the sautéed dishes are always finished after the meat has been removed from the pan and then served poured on top of the dish. Let us discuss the method of sautéing by taking an example of the preparation of *tournedos* of tenderloin.

Marinate the *tournedos* steaks in salt and pepper and heat some clarified butter in a frying pan. Arrange the steaks onto the frying pan and turn over the side to sear the meat on the other side as well. Remove the steaks from the pan and arrange the steaks on a plate. Strain excess butter from the pan carefully to reserve the sediments. Deglaze the pan with brandy and allow all the flavours to mingle together. Reduce the sauce and pour over the sauteed steaks. In this preparation we saw that the principle of sauteing is closely related to the method of pan-frying, but in few preparations the meat is quickly stir-fried till cooked and the sauce is prepared later on from the residual sediments.

Heat Transference

The principle of sautéing if done on a gas burner uses the conduction method, as the heat is being transferred from the metal pan on to the food.

Temperature Range

The temperature range of the products has to be carefully controlled by the chefs. Too high a temperature will burn the meats and too low a temperature will make the food absorb the fat and thus become soggy. Even cooking is at times ensured by moving the pan from a high heat to moderate heat, etc. Fairly high temperatures are used with care of regulating it at frequent intervals to achieve the best results. The temperature range can vary from 150 to 200°C and this wide range is dependent upon various factors such as type of cut of meat and the required degree of cooking.

Sautéing Points

The following points are to be kept in mind while sautéing.

1. Prime quality of food has to be used.
2. No liquid is added to the food until it is removed from the pan and the residual sediments are deglazed and then formed into a sauce.
3. The food item has to be placed into high fat to seal the food and then cooked on a low heat.
4. Certain food items such as sauté cuts of chicken with bone may be covered and finished cooking in the oven. The procedure of making sauce however will be same as that of the other sautéed products.
5. While cooking chicken, place the legs and the thigh first and add the breast after some time for even cooking of the product. Breasts are tenderer and will cook faster than the legs.

Sautéing and Its Uses

Sautéing can be used for various food commodities (Table 17.20).

Table 17.20   Uses of sautéing

Commodity	Method	Use
Potatoes	Potatoes are to be sliced thin, blanched in salted water, and then sautéed to give a colour to potatoes.	It is used as an accompaniment to meats and poultry as a starch component.
Mushroom	Mushrooms are to be sliced thin, and sauteed directly in oil or butter.	It is used for salads, soups, and accompaniments.

Contd

Table 17.20 (Contd)

Commodity	Method	Use
Vegetables	Boiled and kept as a *mise en place*. It would be sautéed in butter as and when required.	It is used as a vegetable accompaniment, or as a salad.
Meat such as noisette of lamb/cutlets-veal loin, etc.	Sautéed in butter and served with sauce.	It is used as a main course.
Strips of tenderloin fillet	Sautéed quickly to prepare stroganoff.	It is used as à la carte for dinner service.
Supremes of chicken	Sautéed and served with sauce.	It is used as à la carte for dinner service.
Whole chicken	Cut into sauté cut (refer to cuts of chicken).	It is used as a main course.

Equipment Used in Sautéing

Since the food is mostly sautéed over the stove, there are no specialized equipment used apart from normal frying pans and sauce pans. Chinois may be required to strain the sauce. Some other equipment used are shown in Table 17.21.

Table 17.21 Equipment used in sautéing

Equipment	Description	Photograph
Sauteuse	A small pan with slanting sides, for tossing food.	
Wok	It is usually made of cast iron and can be with or without a handle. It is mainly used for stir-frying foods in Asian cooking.	

FRYING

Frying is probably the fastest cooking method used in the kitchen. It also involves cooking of only prime quality food, where emphasis is laid on the cuts, as it directly affects the product. There are also several safety factors which need to be taken into consideration while deep frying. As this method utilizes oil as a medium, the temperature range can go as high as 190°C. When we refer to frying we always refer to deep fat frying though there is another type of frying called shallow-frying. As the name suggests, in case of deep fat frying the food shall be immersed in oil, whereas in case of shallow-frying, the oil will come up to only half or even less than half the thickness of the food commodity. We shall see the difference between these in Table 17.22. We also have to take care of the moisture content in the food, as hot water in the oil causes the oil to splutter and splash thereby creating burn marks on the skin. Always ensure that any moist commodity is well dried or coated with a crust before deep-frying. This would give a great texture to the product and altering the crust can give variety of products. The most common method of coating is called crumbing as *a l'anglaise* in French. The food is dipped in seasoned flour and then in the dredged in egg and finally coated with bread crumbs before deep-frying.

When the method of cooking by fats and oils is discussed, it becomes important to highlight two main factors that influence the quality of oils:

Smoke Point   This is the stage when the oil begins to give a haze or smoke.

Flash Point   This is the stage where the temperature surpasses the smoke point and at this point the oil becomes vulnerable and can catch fire and ignite.

Care must be taken to select the fats and oils which are capable of being heated up to high temperatures; otherwise they will reach the flash point very quickly. That is the reason that butter cannot be used for deep-frying, as when you melt butter, within no time it turns to brown colour, often called *beurre noisette* or brown butter and thereafter it turns black, often called *beurre noir* or black butter. *Beurre noisette* is used as a sauce with fish and pastas. As a guide, oil will be ready to use once it gives off a faint blue haze and we can see the heat rising up. There are many factors that need to be taken care of while frying and we shall discuss them as 'frying points'.

It is interesting to know how food actually gets cooked by the deep-frying method. When the food is immersed into hot oil, the surface of the food gets sealed by the hot oil, thereby preserving all the moisture in it and due to temperature the moisture gets converted into steam which in turn cooks the food. The traditional methods of deep-frying were to fill up a pan with oil and use it as a fryer; but with advancement in technology, sophisticated deep fat fryers are available in the market with dipping baskets, which facilitate the removal of the food thereby allowing maximum drainage of oil. Certain battered foods should not be placed in dipping baskets as the batter tends to stick to the meshed basket thereby discolouring the oil. During frying the food needs to be turned once or twice for even browning. While frying it might be necessary to stir the food slightly so that it does not stick to one another to form one big lump. Table 17.22 shows the difference between shallow-frying and deep-frying.

CHEF'S TIP

If by any chance the oil catches fire, do not ever put water, the oil is lighter than water and will float on top, thereby causing more damage as it will splutter all around. Just cover the apparatus and cut off the oxygen supply, or use a foam type extinguisher. Do not panic and get help immediately if the fire is out of control.

Table 17.22   Difference between shallow-frying and deep-frying

Deep-frying	Shallow-frying
Food item is immersed in oil.	Food item is only immersed in oil half way through.
Does not matter which side is put first on the pan.	The presentation side is put on the pan first.
Cooking medium is oil and high quality fat and not butter and other fats that have low melting point.	Butter can also be a cooking medium apart from oils.
The cooking is purely by conduction.	The cooking is by conduction and convection also.
The food is cooked in frying equipment or wok or *kadhai*.	The food is shallow-fried in a pan or a slightly concave flat plate such *tawa*.

All foods cooked by both deep-frying method and shallow-frying method need to be coated. Traditionally coatings can be of different types; some are given below.

Flour, egg wash, and breadcrumbs	*Paner a l'anglaise*	English style
Milk and flour	*Paner a la francaise*	French style
Flour and batter	*Paner a l'orly*	Orly style
Matzo meal and beaten egg	*Paner a la juivre*	Jewish style

Heat Transference

There could be more than one method of heat transference depending upon the type of frying. Deep fat frying utilizes conduction as the heat is transferred from the heat source, through the cooking medium (oil) to the food. And in shallow-frying the heat is also conducted through the pan to the food. Here the conduction is not only through oil, but is through the metal on which food is placed. Therefore in shallow-frying, some amount of convection is also taking place as the air is circulating around the product.

Temperature Range

The temperatures will vary according to the food being fried and the end usage of the product. For example, samosas will be fried in medium temperature to make them crisp whereas crisp fried spinach will be deep fried in hot oil. Cooking samosas in hot oil will make them brown on the outside but they will remain underdone on the inside, and the spinach in moderate oil will become limp as it will absorb oil. The temperature range may vary from 160 to 190°C and this will depend upon the smoke point of the oil.

Frying Points

The following points are to be kept in mind while frying.

1. Prime quality food has to be used.

2. Fill two-third of the deep fat fryer with oil as sometimes while lowering the food into the oil the oil froths up and can spill over creating a mess.

3. Do not overload the fryer with the product as the temperature of the oil will drop and the product may become limp after absorbing the warm oil.

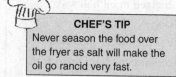

CHEF'S TIP
Never season the food over the fryer as salt will make the oil go rancid very fast.

4. Always dry the ingredients prior to lowering in the oil. If the recipe requires coat the food commodity with coatings.

5. Make sure the food is at the right temperature. Frozen food might lower the temperature of the oil and hence the oil must be at high heat when doing so, certain products such as frozen French fries are fried in the frozen state.

6. Strain the oil regularly to get rid of the little particles that lie at the bottom of the frying equipment.

7. When adding the next batch of food to deep-fry, one must wait for the temperature to come back to the optimum temperature required and this is also known as 'recovery time'.

8. Turn the foods once or twice during the cooking to ensure even browning.

9. In case of shallow-frying, make sure that the side of the product that is intended to be presented to the guests is put on the pan in such a way that the presentation side is down on the pan.

10. The shape, density, and precooking temperature of the food are important factors when considering the time and temperature for frying. To ensure even cooking, items of the same size, density, and type should be cooked together. The time and temperature will also depend upon the water content of the food and how quickly the heat may be conducted through it.

11. Some foods will float to the surface of the oil when they are cooked. However, this is not a sign that can be taken as correct for all foods, as some float almost immediately. Fish in batter may cook before the batter is cooked, in which case it is necessary to hold the fish below the surface with a spider to finish the cooking and browning of the batter. Apple fritters and doughnuts will need to be turned half way through the cooking time, to ensure even cooking on both sides.

12. When dredging the food with flour, make sure to shake off excess flour, otherwise it will burn in oil. Make sure that the food coated with flour is fried instantly otherwise, the flour will absorb moisture from the food and become sticky.

13. Wherever batter is used, ensure that the batter is not too thick, otherwise the food will be very doughy. It is advisable to check a small quantity prior to frying all.

14. While frying, lower the food away from you into the hot oil to avoid accidents.

15. The food always needs to be drained on absorbent paper to soak excess oil, otherwise the food will be greasy.

Frying and Its Uses

Frying can be used for cooking several food commodities (Table 17.23).

Table 17.23   Uses of frying

Commodity	Coating	Frying	Use
Fish	Egg and flour/bread crumbs/poppy seeds/sesame seeds/tempura batter	Deep-frying/shallow-frying	Fish fingers as snacks/fish and chips for main course/battered fried fish for snacks and main course
Shellfish	Chinese batter of flour and cornflour	Deep-frying	Used for stir-fries in Chinese preparations
Escalopes of meats	Breadcrumbs	Shallow-frying	Tenderloin steak is flattened out and coated in flour, egg and bread crumbs to be shallow fried—known as steak Milanese
Vegetables	Batter	Deep-frying	Used as snacks such as pakoras, tempuras, etc.
Kebabs	Poppy seeds/semolina/pressed rice	Shallow-frying	À la carte for dinner service
Supreme of chicken	Breadcrumbs/crushed nuts/seeds	Deep-frying	Chicken nuggets

Contd

Table 17.23 (Contd)

Commodity	Coating	Frying	Use
Rolled breast of chicken with butter	Breadcrumbs	Deep-frying	Chicken a la kiev
Eggs	Batter/bread-crumbs	Deep-frying/ shallow-frying	Used in egg preparations for break-fast and meals
Fruits	Batter	Deep-frying	Fruit fritters used in desserts
Sausages		Shallow-frying	Served as snacks

Equipment Used in Frying

Various types of equipment are used in frying (Table 17.24).

Table 17.24 Equipment used in frying

Equipment	Description	Photograph
Deep fat fryer	It has a thermostat for regulating the temperature.	
Kadhai	Indian wok traditionally used for frying.	
Pressure fryer	Commercial deep fat fryer which cooks with pressure.	
Spsider	A large wired spoon to lift out food from the oil.	
Jhara	A large slotted spoon used in Indian kitchen to remove food from oil. It can be a thin mesh also to filter the oil.	
Fish slice	An angular flat perforated spatula to lift fish out of the oil.	

BAKING

Baking is a cooking principle which is applied only to bakery, confectionery, and patisserie, which would essentially be flour-based products. There are few exceptions as some potatoes are baked with their skin and sometimes pastas are also baked. Oven is integral to cooking, as one always has to use an oven to accomplish baking. This principle of cooking is versatile, as the same ingredients such as flour, sugar, and eggs can be used to prepare breads, cookies, and delicious desserts. We shall discuss the types of these ingredients in detail in the section on flour and bakery.

Baking cooks food by surrounding it with hot, dry air. This is similar to roasting, except that one does not baste the food as one would during roasting. Baking is carried out in a conventional oven where the food is cooked by dry heat and sometimes the moisture available in the food acts as a steam thereby modifying the dry heat of the oven.

Heat Transference

As in roasting and poeling in which also food is cooked in an oven, similar methods of heat transference (conduction and convection) are used in baking. As the convected heat comes in contact with the food, it is conducted through the goods being baked. Radiated heat from the source of equipment will also be absorbed by the food to a great extent. Thus this method of cooking also utilizes all the methods of heat transfer.

Temperature Range

It becomes very challenging to determine the exact temperature of baking as different products require different range of temperatures. Some commodities require high temperature during initial cooking to create what is known as 'oven spring' and then the heat is lowered to cook the product through. Some products like cake sponge sheets get cooked only at high temperature. The temperature range can be between 100 and 250°C. A temperature of 100°C is used mostly to dry out certain products such as meringues.

Baking Points

The following points are to be kept in mind while baking.

1. Food is baked open on trays and seldom covered.
2. It is usually related to the flour-based products.
3. The oven is preheated before the baking goods are placed inside.
4. After baking, the products need a wire rack to cool down without absorbing any moisture, which could result in mould later on.
5. The baking goods need to be fresh and not stored for long.

Baking and Its Uses

Baking is used to cook various commodities (Table 17.25).

Table 17.25   Uses of baking

Commodity	Temperature of Cooking	Use
Bread rolls	180–200°C	Used with the soups and meals
Sponges	200°C	Used for making cakes and pastries
Turn overs	180°C	Used as snacks
Quiches	160°C	Used as snacks or main course

Contd

Table 17.25 (Contd)

Commodity	Temperature of Cooking	Use
Goods made with choux paste	200°C and reduced to 180°C	Used for making pastries, cakes, and desserts
Goods made with puff pastry	180°C	Used for making pastries, cakes, desserts, and some savoury products
Goods made with cookie doughs	180–200°C	Used to prepare cookies, tarts, and flans
Fruits	180°C	Baked apples, pears, quinces, etc. are often baked and used as dessert

Equipment Used in Baking

Various types of equipment used in baking are shown in Table 17.26.

Table 17.26 Equipment used in baking

Equipment	Description	Photograph
Baking trays	Often known as sheet pan, they can be of iron, or teflon coated for non stick.	
Baking moulds	Containers of various shapes for baking bread, muffins, etc.	
Cutters	There are various shapes of cutters for biscuits and cookies.	
Dough mixers	There are various kinds of dough mixers to knead the dough, cream the butter, and whisk the eggs. It has three attachments such as dough hook, flat paddle, and the balloon whisk.	
Ovens	Baking ovens, with steam attachment for better baking.	

MICROWAVE COOKING

Microwave ovens are a fairly recent cookery innovation. There are many inventions such as infrared cooking and induction cooking; but these are really not principles of cooking but equipment through which cooking is conducted.

Microwaves are in fact electromagnetic waves operating at high frequency, which can penetrate food causing molecular disturbance in the water content. The waves cause agitation in the water molecules, which in turn generate heat through friction without any chemical changes to the food.

Specific cooking containers have been designed for use in the microwave ovens. They are perhaps as expensive as glass, china, or porcelain dishes suitable for use in microwaves. Microwaves will not be absorbed by china, glass, or porcelain containers and will pass straight through, in the same fashion as light rays pass through a window.

Plastics that can withstand very high temperatures are also suitable. China plates with gold or silver inlaid bands should be avoided as the metal content will reflect the waves back into the magnetron causing possible damage.

The three main functions of a microwave oven are as follows:

Cooking

Microwave oven is not used for cooking as much as it is used for the other functions listed below. The food in the microwave heats from inside out as the magnetron rays pass through the food. Now a days, microwaves are available with heating filament as well, that allows for the browning of food, which is a feature quite helpful for chefs.

> **CHEF'S TIP**
> Do not use crockery which has gold or silver inlaid bands, as the metal content will reflect the waves back into the magnetron, thus causing damage to the equipment.

Defrosting/Thawing

Microwave is used for defrosting or thawing of food. Frozen meat is usually removed from the freezer and kept in a refrigerator set between a cold temperature of around 4–5°C—this is also known as defrosting or thawing. The meat is kept in the refrigerator for 12 hours or until it is completely defrosted. However, this is a time consuming process and microwave aids in defrosting the commodity in a few minutes. It is possible to defrost only smaller quantities of meat at home in microwave ovens; but in a professional kitchen, where large quantities of meat are used, it is advisable to thaw them in refrigerators.

Reheating

Microwave in professional kitchens are rarely used to cook as hotels cook in bulk and therefore its usage in kitchens is limited to reheating.

Most chefs still prefer traditional methods for prime cookery but acknowledge the benefits of microwave cookery. For example, during very busy meal periods reheating food quickly and efficiently portion by portion eliminating waste and providing good quality food is possible with the help of microwave ovens. Frozen food can be successfully defrosted. Various levels of power and timings can be pre-programmed to help the kitchen team.

The magnetron is a tube inside the microwave oven, which produces the microwaves. It is imperative that microwave ovens are not switched on unless there is something inside. If the waves are bounced back from a metal surface or if there is no food placed to absorb the waves one could cause permanent damage to the magnetron.

Modern microwave ovens have convection fans for even distribution of heat and also heated filaments to achieve the browning effect as in the case of roasts.

Heat Transference

Microwaves would only penetrate to a depth of approximately 35 mm; any further heat is conducted through to the centre of the item being cooked or heated. The depth of penetration depends upon the size, shape, density, uniformity, and arrangement of the food in the oven. Large, dense, and irregular-shaped items will not cook evenly or successfully in a microwave. The microwaves pass though the food thereby agitating the molecules to cause friction that in turn produces heat; therefore heat is transferred by conduction. Also there is some amount of convection happening as the heat generated is circulating around the food. Modern microwaves can also radiate heat to get the desired browning effect.

Temperature Range

Temperature ranges are impossible to determine as no heat is actually generated by the oven, it is generated only within the food. Cooking or heating can be controlled by both adjusting the power input to the oven and by varying the amount of time the food is exposed to microwave power.

Microwave Cooking Points

The following points are to be kept in mind while cooking in microwave oven.

1. Food of uniform shape and size should be used in microwave cooking.

2. It is more efficient in smaller quantities.

3. No metal containers should be used in microwave.

4. Normal microwave ovens do not brown food but there are models available now which have a heating element that can brown the food as well. They are better for reheating than cooking.

Microwave and Its Uses

The uses of microwave ovens in cooking are shown in Table 17.27.

Table 17.27   Uses of microwave

Food Commodity	Function of Microwave	Menu Example
Potatoes in jacket	To cook	Baked jacket potatoes to be served with chives and sour cream
Peas	To reheat	Buttered peas
Frozen minced beef	To defrost/thaw	Bolognese sauce
Precooked chicken pie	To defrost and reheat	Chicken pie

EQUIPMENT USED IN MICROWAVE COOKING

CHEF'S TIP
Never cook a whole egg in microwave as it would explode.

Various types of equipment are used in microwave cooking (Table 17.28).

Table 17.28   Equipment used in microwave

Equipment	Use	Photograph
Microwave oven	To reheat, cook, and defrost food.	
Glass bowls	To cook or reheat food in microwave.	
Plastic containers	Special hard plastic containers which are micro-wave safe are used for reheating food.	

CONCLUSION

The knowledge of equipment, fuels, vegetables, and meats were covered in the last few chapters. By now you would have understood the importance of knowing your commodities and the effect of heat on the textures of food. In this chapter we discussed the methods of cooking and now we know the different methods of cooking used for different commodities to achieve different results.

For your understanding the recipes pertaining to each method of cooking are provided in the ORC and once you are thorough with this chapter you can produce food comparable to any international standards.

We discussed about the thirteen methods of cooking—blanching, poaching, stéaming, boiling, braising, stewing, poeling, sautéing, deep-frying, baking, grilling, roasting, and microwave cooking.

We discussed each method of cooking in detail with regard to its methods of heat transference, temperature ranges used, and various uses of these methods of cooking. We saw how the same commodity can be used for different preparations by using different methods of cooking. For example, chicken can be boiled to make stock and the same chicken if roasted can be used as a main meal. We also discussed about the equipment used in these cooking methods. In the next chapter we shall discuss some more commodities such as flour, fats and oils, sugar, and milk products, which will form a base for our bakery and confectionery section to follow.

KEY TERMS

Bain-marie   Water bath containing hot water used for holding hot food.

Baked cheesecake   A cake made by using cream cheese and is baked in a water bath.

Bar grill   A grill having metal rods.

Barbecue   To roast or grill on charcoals.

Blanc   French for white; used in cooking to denote the brightness of a commodity.

Broiling   Cooking under dry radiated heat.

Choux paste   Cooked paste of flour eggs and butter used to make pastries and desserts.

Coagulate   The stiffening of proteins with application of heat.

Contact grill   A piece of equipment which closes like a book.

Dim sum   Steamed flour bund usually stuffed with assortment of meats and vegetables and served as main meal or snack in China.

Dipping basket   Meshed basket with handle used for lifting food out of oil or hot liquids.

Flash point   Term related to heating of oil to appoint where it catches fire.

French fries   Potato cut into long batons, blanched and deep-fried to accompany sandwiches, fried fish or eaten as snack.

Fricassée   White coloured stew.

Glazing To place the meat under radiated heat to brown the surface and bring shine to it.

Gratinating Grilling a dish usually glazed covered with sauce and cheese.

Joints Large pieces of meats used in carving.

Kadhai Indian wok used for stir-frying and deep-frying.

Matzo meal Ground corn or cornmeal.

Meringues Egg whites and sugar whipped up together to form a whip cream consistency.

Napper To coat a dish with sauce or liquid.

Oven spring The initial rise in a food when put to be baked in oven.

Paner French word for coating the food.

Pate A paste of meat usually poached in water bath and served cold.

Pot roasting Roasting food in a closed container.

Prime quality Best quality and tender cuts of meats.

Probe Thermometer with a sticking needle to check the internal temperature of the meats.

Puff pastry A flaky pastry.

Quiches Small savoury made with assortment of meats and vegetables and baked with egg custard.

Ragout Brown coloured stew.

Refreshing Also known as shocking, it is immersion of boiled commodity into cold water to arrest cooking and to brighten the colour.

Rotisserie Equipment used for roasting.

Salamander Equipment used in the kitchen that radiates heat from top.

Sauce pan Cooking pan with straight sides, usually used to prepare sauces.

Seasoned flour Flour mixed with salt and pepper.

Second quality meat Tougher cuts of meat such as shoulder, rump, etc.

Shawarma Lebanese preparation of meat.

Smoke point Related to heating of oil where it starts to give a smoke.

Spit roasting Cooking over open fire.

Stock pot A large pot with handles to prepare stocks.

Sur la plat Porcelain dish with handles, used for presenting egg dishes.

Tandoor Clay oven from India, which is fired with wood coal.

Tawa Equipment made of iron and used for making rotis, etc.

Tempura Japanese style of preparing fritters.

Terrine A cold meat or vegetable preparation made in earthenware mould.

Tripe Stomach of lamb.

Trivet A rack of bones used as a base for roasting.

Turn overs Savoury made by puff pastry.

Turned potato Cut of potatoes where it is trimmed to a barrel shape.

CONCEPT REVIEW QUESTIONS

1. List the various methods of cooking and list down the classification also.
2. What do you understand by the term blanching?
3. List the steps of blanching.
4. How many types of blanching are there and what is their significance?
5. In case of vegetables how does one decide which method of blanching should be used?
6. What are the three methods of heat transference?
7. Explain convection.
8. List the range of temperatures while blanching a food commodity.
9. List at least five uses of blanching.
10. List at least three poaching liquids.
11. What is the ratio of vinegar to water in case of poaching eggs and why is vinegar added in the water?
12. What is the temperature at which poaching takes place?
13. What is the difference between poaching and simmering?
14. List at least five important points to be kept in mind while poaching.
15. Why is whole fish poached from cold liquid?
16. Why is steaming not a very popular cooking method in the West?
17. List at least four equipment which can be used in steaming.
18. Why should not the products have baking powder in case we intend to steam them?
19. What are the ranges of temperature in steaming?
20. What is the basic difference between stewing and braising?
21. What is the difference between a ragout and a fricassée?

22. List at least five salient features of stewing.
23. Name few classical dishes prepared by stewing method.
24. What is a tagine and how is it different from a cocotte?
25. Define poeling. What is unique about this principle of cooking?
26. What is the difference between poeling and roasting?
27. Name at least five classical dishes prepared by poeling method.
28. How many kinds of roasting are there?
29. How is roasting different from baking?
30. What is a trivet and what are its uses?
31. Define *jus roti* and *jus lie*.
32. Why should the meat be allowed to rest prior to carving?
33. How would you calculate the time for roasting any food commodity?
34. Define grilling and differentiate among grilling, broiling, and gratinating.
35. What are the temperature ranges in case of grilling?
36. Name few types of equipment used in grilling.
37. What is sautéing and what are the salient features of the same?
38. How would you define frying? How many types of frying are there?
39. Define smoke point and flash point.
40. List various classical styles of coating foods before frying.
41. List five safety features to be kept in mind while deep-frying.
42. What are the temperature ranges while baking?
43. What do you understand by the term 'oven in spring'?
44. What is microwave cooking and what is the principle behind its working?
45. What care should be taken while using microwave oven?
46. What are the advantages and disadvantages of using microwave ovens?
47. Why should we be careful in selecting porcelain which will go in the microwave?
48. List the three main functions of microwave cooking.
49. What are the methods of heat transference in the microwave?
50. Mention few salient features of microwave cooking.

PROJECT WORK

1. In groups of three to four, blanch commodities for various purposes as listed in this chapter and compare the results. Also one group should blanch in the wrong way, for example, blanching potatoes in hot water instead of cold water as mentioned in this chapter and compare the difference.
2. Prepare ragouts, fricassée, and miscellaneous stew and compare the results. Make a note of temperatures and time used and the methods of heat transference.
3. Prepare a meal and list the various cooking principles which were used to create a dish. To prepare a meal, sometimes a combination of cooking methods is used.
4. In a group of three to four, roast the various types of meat and record the time taken by the same. Now compare your results with the initial weight of the meat.
5. In a group, boil 30 eggs at an interval of 30 seconds and compare the results. Record your observation and notice how heating time affects the quality of product.

CHAPTER 18

BASIC COMMODITIES USED IN BAKERY AND PASTRY

Learning Objectives

After reading this chapter, you should be able to:
- understand the different types of flours and their uses
- know various types of gluten-free flours and their importance in modern baking
- identify various types of oils and fats and know the characteristics of the same
- recognize the importance of raising agents in bakery and their functions
- understand various types of sugars and sweeteners which affect the quality of baked products
- appreciate the usage of various dairy products in cooking as well as baking and confectionery to produce various types of desserts

INTRODUCTION

In the previous chapters we read about various commodities such as fruits and vegetables, meats and eggs, various kinds of seeds such as cereals, pulses, nuts, and spices. After understanding these commodities we related them to the principles of cooking and learned to create dishes or at least cook the classical dishes in a more structured manner. Similarly, before we go on to create pastry and bakery products, it is important for us to know about the basic commodities which are used in bakery and pastry. Though these commodities are also used extensively in other principles of cooking, we will be discussing them as a foundation to the section on bakery and pastry. Since this is a much specialized field in cooking, hotels have a separate pastry and bakery department which supplies desserts and breads to the entire hotel. Cooking is an art combined with science and this holds 100 per cent true for bakery and pastry. As we read in previous chapters, many guests are allergic to gluten in flour and to serve such guests, the chefs should be aware of gluten-free flours. Similarly, the knowledge of various types of sugars will help chefs to produce pastry goods of different textures and mouth feel. Dairy products which form the base for most of the pastry products are discussed in this chapter.

FLOUR

Flour is obtained when grains and pulses are milled. Milling can be of various degrees to give a particular structure to the product and the usage of each milled product will be different from the other. For example, the milling of wheat grain can produce bulgur, couscous, semolina, and flour. In bakery when we refer to flour, we always mean refined flour, unless specified as wholemeal flour. The shelf life of unmilled grain is shorter than that of milled grain. So care should be taken while storing the flour.

Wheat is among the most extensively cultivated cereal crops in the world; it is a member of grass family and is botanically named *triticum*. It is an annual or biennial crop which grows in the temperate regions of the world. Commercially wheat is classified into three broader sections.

1. *Triticum vulgare*, used for making baker's flour
2. *Triticum durum*, used for making pastas
3. *Triticum compactum*, also known as clubbed wheat, used for making low-gluten flours

STRUCTURE OF WHEAT GRAIN

Wheat grain can be divided into three major parts—the bran, the embryo or the germ, and the endosperm; wherein bran is 12 per cent, germ is 3 per cent, and the endosperm is 85 per cent (Fig. 18.1). The wheat grain consists of six principal layers, the first three together are known as pericarp, which includes epidermis, epicarp, and endocarp. The second three layers are together known as seed coat, which has testa, nuclear layer, and the aleurone layer. The germ has three main components which later develop into wheat plant; these are the plumule which develops into green shoot; the radical which gives rootlet; and the scutellum which is the storehouse of the vitamin content of the grain.

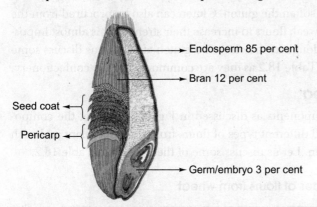

Fig. 18.1 Structure of wheat grain

The endosperm constitutes almost 85 per cent of the total wheat grain and has starch cells, soluble and insoluble protein, oil, moisture, sugar, and minerals. The white flour or the refined flour is milled out of the endosperm and the wholemeal which is healthier option as the wholewheat grain is milled for the same. There are many by-products from wheat and for ease of understanding they are given in Table 18.1.

By-products of Wheat

There are many by-products of wheat, which are used in the kitchen in one form or the other. Let us discuss these in Table 18.1.

Flour is one of the structural ingredients used in pastry and bakery kitchens. There are many different kinds of flours used in the pastry kitchen and each flour has a different role to play in the final outcome of the product. Therefore, it becomes important to choose the right type of flour for the right type of product. You would commonly hear chefs using words like strong flour and weak flour. These merely indicate the amount of gluten present in the flour. There are two types of non-soluble proteins

Table 18.1 By-products of wheat

Product	Description
Wholewheat	Unrefined or minimally processed wholewheat kernels
Cracked wheat	Coarsely crushed, minimally processed wheat kernels
Bulgur	Hulled, cracked hard or soft wheat, parboiled and dried kernels
Semolina	Grounded polished wheat kernels with bran and germ removed
Couscous	Semolina pellets, often parcooked
Farina	Polished medium ground wheat cereals
Bran	Separated outer covering of wheat kernels and flaked or powdered
Germ	Separated embryo of wheat kernels, flakes

in the flour—'glutenin' and 'gliadin'. When the dough is kneaded these two proteins combine to produce gluten in the dough. Without gluten there will be no such thing as raised bread. Gluten provides elasticity to the dough, which in turn traps the air and gas released by yeast and forms a sponge-like texture in the baked breads. The gluten in the flour can be altered by various methods. Manipulating the dough for longer duration of time or adding some acid, such as lemon juice, will strengthen the gluten strands and addition of oils and fats will soften the gluten. Gluten can also be procured from the market as a commercial product and added to weak flours to increase their strength. It is almost impossible to knead corn flour and rice flour into a dough as they have no gluten at all. Let us discuss some of the flours obtained from the wheat kernel in Table 18.2 as they are commonly used in confectionery.

Types of Flours Obtained from Wheat

The wholewheat grain consists of various components as discussed in Fig 18.1. Each of the components are milled in various proportions to yield different types of flours from the same plant and each one has a particular usage in the bakery kitchen. Let us discuss some of these flours in Table 18.2.

Table 18.2 Types of flours from wheat

Name of Flour	Description
Wholemeal flour	Also called *atta* in India, it is the whole milled wheat kernel. The flour is cream to brown in colour as it has the bran grounded with it. It is not advisable to sift the wholewheat flour as most of the bran, an important dietary component, will be lost.
Graham flour	It is usually found in the USA and the milling concept of this flour is very interesting. The wheat kernel is separated into its various components such as endosperm, germ, and bran. The endosperm is ground finely to produce white flour with gluten, whereas germ and bran are ground till coarse. The milled flour is then mixed back to yield graham flour. In case of non availability of this flour one can mix refined flour, bran, and germ in the ratios that they naturally exist in the grain (refer to Fig. 18.1).
Brown flour	It is almost 85 per cent of the grain millet, where some amount of bran has been extracted. It is nutritious as it has high percentage of germ.

Contd

Table 18.2 (Contd)

Name of Flour	Description
Strong flour	It is milled from hard flour, in other words from high-protein flour. The strong flours absorb more water than weak flours, as gluten can absorb twice their own weight of water. This flour is used for products which will have a high rise in the oven such as yeast breads, choux pastry, and puff pastry. Strong flour is also known as baker's flour.
Weak flour	Weak flour is also known as soft flour or cake flour. As the name suggests, this flour has less gluten and hence, it is used for products that need a softer texture such as cookies and cakes and sponges.
All purpose flour	The all purpose flour is a blend of flours and has medium strength. In India, all the refined flour that we get is all purpose flour.
Cake flour	Refer to weak flour.
Pastry flour	It is a very finely ground polished flour of soft wheat kernels, usually enriched and bleached.
Self-raising flour	This flour is usually of medium strength and contains baking powder in a proportion. Since the flour contains moisture, this can react with the baking powder lessening the effect of baking powder and hence, it is not advisable to buy the commercial self raising flour. It is advisable to use 60 g of baking powder to 1 kg flour to make it self raising. This flour is commonly used to make afternoon cookies called scones.

TYPES OF FLOURS

Flours are not only derived from wheat but also from other grains and seeds. It is very important for chefs to have knowledge of such flours as they can make different products with the range of the flours which are healthier. Also, since many people suffer from gluten allergies, it is important for chefs to use products which are gluten free. Many types of grains are available in the market but few of the popular flours derived from them are discussed in Table 18.3.

Table 18.3 Types of flours from various grains

Name of Flour	Description
Rye flour	Rye flour does not have as much gluten as in popular flour and hence, it is sometimes mixed in proportions with flour for the production of breads. Breads which use only rye flour are more dense and chewy. This flour is majorly used in the Russian and Scandinavian breads. Rye flour dough is quite heavy and sticky.
Spelt flour	It is quite popular in European countries such as Germany, France, and Switzerland. It is made from spelt which is a species of wheat. It is a good source of vitamin B.
Rice flour	It is the finely ground polished rice with a similar texture of corn starch, usually used as thickening agent. Rice flour is free of gluten and if the dough has to be made one would have to make it with hot water.
Maize flour	Popular in Mexico, this flour is made from cooked maize corn and then grounded. It is also known as *masa harina*. This flour has also been used in India since time immemorial and a very popular north Indian dish called *makki ki roti* is made from it. This flour is also free from gluten.

Contd

Table 18.3 (Contd)

Name of Flour	Description
Cornflour	It is made by grounding the white heart or the germ of the corn kernel, one of the widely used thickening agents in Chinese cooking. This is also free of gluten and usage of this flour in products gives crispness to the product. It can also be added to strong flour to turn into weak flour. Commercial custard powder is also made with cornflour with colour and flavour added. Cornflour is not flour, but it is actually a starch.
Arrowroot	This flour is finely milled from the arrowroot plant. It has the same properties as cornflour and the uses are very similar. It is widely used for making glazes.
Barley flour	Made from the pearl barley, it has low gluten content with mild flavour.
Buckwheat flour	It has distinctive greyish brown colour with earthy bitter taste. It is used to make classical preparations such as Russian blinis, pancakes, and French galettes. In India it is widely eaten during fasts and is commonly known as *kuttu ka atta*.

GLUTEN-FREE FLOUR

Apart from the flours discussed above, it is also very important for chefs to know about gluten-free flours, as the demand for the same is increasing constantly. Some of the gluten-free flours are discussed in Table 18.4.

Table 18.4 Types of gluten-free flours

Name of Flour	Description
Amaranth	It is a green leafy vegetable related to spinach and beets. Tiny seeds of this plant are often ground into nutritious flour. It is light brown in colour and has a nutty aroma.
Rice flour	Refer to Table 18.3.
Pulse flour	These are the seeds of many edible legumes and can be ground into flours for use in gluten-free breads. Chickpea flour is very commonly used in India and Mediterranean countries.
Maize flour	Refer to Table 18.3.
Buckwheat	Refer to Table 18.3.
Chestnut flour	It is a smooth shelled nut; it is usually roasted and ground into flour. In India it is called *singharey ka atta* and is commonly eaten during fasting.
Barley flour	Made from pearl barley, it has low gluten content with mild flavour.
Cottonseed flour	The seeds are commonly used for making margarines or cooking oils, but these seeds can be ground into flour which is quite nutritious.
Flaxseed flour	It is an ancient seed which has been used in medicines since time immemorial. It is used whole, toasted, or ground into flour. It is believed to be a good cure for diabetic patients and is believed to lower cholesterol levels.
Millet	It can also grow in areas which do not get much rainwater. In India millets are commonly known as *bajra*.
Quinoa flour	This is one of the grains which has the highest amount of protein. It is mainly found in China; but its popularity is catching up with the Western world as well.

Contd

Table 18.4   (Contd)

Name of Flour	Description
Soybean flour	It is high-fat and high-protein flour which has a strong distinctive nutty flavour.
Sunflower seed flour	It can be dried, roasted, and ground into flour. It can be combined with other kinds of gluten-free flour as it has a very nutty flavour.

RAISING AGENTS

Some ingredients play a very vital role in baking. Raising agents are those ingredients which are responsible for the (inside) chemical changes in baking. They are also known as leavening agent. Some of these are available naturally such as yeasts, whereas some of these are produced by chemicals such as baking soda. They are added to batters and doughs to help them rise. The action of moisture, heat, or acidity (or a combination of the three) triggers a reaction with the raising agent to produce carbon dioxide gas, which becomes trapped as it bubbles through the dough. When you put the particular product for cooking or when the dough cooks the bubbles become set in the mixture, as the protein in the flour coagulates upon coming in contact with heat thus giving breads, cakes, scones, etc. a soft sponge-like texture.

CHEF'S TIP
When using dry yeast instead of fresh yeast, use only half the mentioned quantity of fresh yeast as dry yeast is very strong.

Types of Raising Agents

Some of the common raising agents used in bakery are shown in Table 18.5.

Table 18.5   Types of raising agents

Name	Description	Storage
Baking powder	Baking powder is used as a raising agent for a number of doughs and batters such as cakes, scones, puddings, and biscuits. Baking powder is made from a combination of alkaline and acid substances. The composition of baking powder is usually of cream of tartar and bicarbonate of soda which react when they come in contact with moisture and warmth to produce carbon dioxide gas in the form of small bubbles. Baking powder is usually a single acting agent, which means it reacts as soon as it comes into contact with any liquid. Hence, it is extremely important to work quickly once milk or water comes into contact with the dry ingredients so that the resulting carbon dioxide does not get a chance to escape.	One should always store baking powder in airtight containers free from any moisture because slight presence of moisture will start the reaction in it. We can also make our own baking powder by mixing half the quantity of bicarbonate of soda to cream of tartar.

Contd

Table 18.5 (Contd)

Name	Description	Storage
Bicarbonate of soda	It is also known as baking soda or bicarbonate of soda or cooking soda, which is used in a variety of dishes such as biscuits, batters, pudding, etc. As mentioned earlier it can be mixed with cream of tartar to produce baking powder. It usually reacts in the presence of any acidic medium such as sour milk, buttermilk or orange juice, which causes carbon dioxide gas to release causing the desired result in baked goods.	The shelf life of baking soda can easily be around 3 years if stored in a cool, dry place, however, if it gets damp or moist it will loose its effectiveness. The best way of testing baking soda is to take a little powder and add lemon juice to it; it will immediately start to fizz which also indicates that the powder has been stored properly as stale powder will not give you the desired result.
Cream of tartar	Cream of tartar is fine white powder which is extracted from the tartaric acid that crystallizes in wine casks during the fermentation process of grapes. It is also known as potassium salt and has a number of uses. It may be combined with bicarbonate of soda to produce baking powder; it can also be added to increase stability and volume of whisked egg whites when making meringues or folded into cake batters. When boiling potatoes or cauliflowers it can be added to them to reduce any discolouration. Adding a small amount to sugar syrups will prevent them from crystallizing and hence used in sugar work and decorations.	It should be stored in airtight containers, as it has ability to absorb moisture.
Salt of harts horn	It is originally made from the ground antlers of reindeer; this predecessor of modern baking powder is made with ammonium carbonate. It is commonly used in Scandinavian countries to produce light biscuits with a crisp texture such as gingerbread. It has an unpleasant ammonia flavour; hence, it is best used in biscuits which allow this ammonia flavour to dispense off while baking.	It should be stored in an airtight container in a cool, dry place as it has strong flavour, which can be absorbed by other foods.
Yeasts	It is a single cell fungus that feeds on simple sugars to produce carbon dioxide, gas, and alcohol. It is used to ferment fruits, grains, etc. to produce wine, beer, and other spirits. It is also used as a leavening agent to produce a wide range of bakery products. There are two major types of food yeasts which are commonly available, one is non-leavening yeast known as brewer's yeast, while the other is leavening yeast known as baker's yeast.	Baker's yeast is sold in fresh blocks which must be used within a few weeks of purchase, or as dried granules which can be stored up to a year. Fresh leavening yeast may also be sold as starter yeast which is traditionally used to make sour dough or sweet breads known as starter breads.

FATS AND OILS

The scientific term 'lipid' comprises a group of substances which include natural fats and oils. Both these lipids consist of fatty acids and glycerol. The only difference between fats and oils is that oils are liquid at room temperature, whereas fats are solid as they contain saturated fatty acids. The exception to this rule is coconut oil and palm oil, which are solid at room temperature. When talking about fats and oils you will come across many terms such as MUFA (monounsaturated fatty acids) and PUFA (polyunsaturated fatty acids). Saturation means the density of fat, in other words it is the molecular structure of the fat where the carbon atoms are bonded with oxygen and hydrogen. The saturation is increased artificially by adding hydrogen into fat by a process known as saturation or hydrogenation of oil. The oil is converted into margarine by passing hydrogen into it to make it saturated. This is done to stabilize the fats and oils and therefore the shelf life of the product increases as it does not oxidize easily. Fats are naturally saturated with hydrogen whereas some fats are artificially transformed from oils such as margarine; such fats are known as transfats and they are unhealthier as they are the prime cause of cardiac disorders.

Fats and oils do not dissolve in water; but they can be emulsified with water to produce salad dressings and sauces. Fats along with carbohydrates and proteins make up the major components of food. Fats and oils are extracted from animals and seeds such as nuts, spices, etc. Animal fats such as suet, lard, etc. come naturally from animals such as beef and pork, respectively, and products like butter are mostly churned out from cow's milk. The choice of oil or fat used for cooking gives an instant recognition to the origin of the dish. For example, dishes made in olive oil are from the Mediterranean region, and those made in sesame oil are from Asia. Even in India mustard oil is more prevalent in the east, coconut oil in south, and ghee and butter are commonly used in the north.

USAGE OF FATS AND OILS IN COOKING AND BAKING

Fats and oils are the prime ingredients in any dish around the world. In all the methods of cooking discussed in Chapter 17, we saw the importance of oil, especially in cooking methods such as frying. Fats and oils give richness, variety of textures, and smoothness to the foods which otherwise may be too dry to eat. The melting points of oils and fats are very important for chefs, as they decide the usage of the particular lipid in a dish. One cannot deep-fry in butter as it will burn to black when it reaches the temperature of frying.

Fats and oils are used in baking cakes to moisten the batter and improve the keeping qualities of a cake. Fats have the ability to retain a certain amount of air during the preparation stage. During the cooking process, the fat melts and produces steam which in turn aerates and lightens the product. Flavoured oils are key ingredients in salad dressings. Peanut oil, for instance, is agreeably light in a dressing, while olive oil is unmistakably rich and distinctive. All purpose oils, such as corn and sunflower oil, have no strong flavour to impart, but even small quantities of nut oils such as walnut oil or infused oils such as chilli oil, add a distinctive flavour to the dish. Fats and oils can be categorized into animal fats and vegetable fats. Fats are often treated to obtain products for cooking and the two most common terms in cooking are 'rendering the fat' and 'clarifying butter'.

RENDERING THE FAT

Rendering is a process where the fat is melted on heat to separate the skin and non-fatty membrane. This is done over low heat and it is a good idea to mix with water and bring to a boil. Reduce the flame and let most of the water evaporate leaving behind clear fat which can be strained and stored away. The other method is to keep the fat on low heat. All the fat will melt leaving some crisp skin behind, which is known as 'crackling' and this can be then used in salads and garnishes.

CLARIFYING BUTTER

Clarifying butter is a process where the water and the milk solids are removed so that the butter becomes more stable and can be used for cooking without changing its properties. To prepare clarified butter, slowly melt the butter so that the water evaporates and the milk solids settle at the bottom of the pan. The clarified butter is then either spooned out or carefully poured out leaving the milk solids behind. Clarified butter has a higher smoking point because the milk solids are removed, meaning that you can heat up the butter to a higher temperature without burning or smoking.

Let us discuss the types of fat used in bakery and confectionery in Table 18.6 (animal fats) and Table 18.7 (vegetable oils).

Table 18.6   Animal fats used in cooking

Type of Fat	Description
Lard	This is the fat from pork and is commonly used in cooking and baking. With more awareness of health the use of lards and other natural animal fats is restricted to special dishes only. Animal fats are rendered before using. Rendering is a process where the fats are heated on a low heat to remove the non-fatty membrane. The non-rendered fat has more pronounced flavour and it is commonly used to line pates, etc. in *charcuterie.*
Suet	Fat from beef is called suet. Due to its stability it used to be a very common ingredient for short crust pies, etc. but butter has majorly replaced all the natural saturated animal fats.
Margarine	This is an emulsion of water and oil. It mainly contains vegetable oils but at times it may contain a mixture of both animal and vegetable oils. These oils are then saturated by addition of hydrogen which makes it more stable and increases its melting point. The handling of this fat becomes very easy in warmer conditions and it can cream very well to give more structure and volume to the baked product. Margarine is used mainly in pastry work and is hardly used in cooking.
Butter	Dairy butter consists of about 80 per cent fat and 20 per cent water and whey. It is the milk protein in the whey that makes butter spoil quickly and together with milk sugar (lactose), causes it to scorch when overheated. In the West most butter is made from cow's milk, but elsewhere butter is made from the milk of water buffalo, yak, goat, and sheep.

Contd

Table 18.6   (Contd)

Type of Fat	Description
	The quality of butter is affected by the cream used for it, which in turn is influenced by the season and the feed of the animal. Colour varies from very pale to deep yellow, but producers may add colouring to butter, particularly salted butter, so that it looks uniform throughout the year. Sometimes the cream is allowed to ripen or lactic yeast is added to give the butter a pleasant acidity and nutty aroma. For health reasons, most butter is pasteurized, which means that the milk used to make it is sterilized by heating it briefly to destroy any harmful bacteria. In some countries raw butter is also available; it has a better taste, but does not keep well. Butter is also graded according to quality in many countries.
	Unsalted butter is made from fresh cream. It is especially appropriate for delicate pastries, cakes, and icings, where even a pinch of salt would easily stand out. In most European countries, most of the butter sold is unsalted and made from ripened cream, while in the UK and the USA, the reverse is true. However, both salted and unsalted kinds are usually available. Butter may be clarified to separate the fat from the water and milk solids, so that the remaining fat does not scorch or turn bitter. Clarified butter may be heated to a much higher temperature than regular butter and is good for sautéing. Creamed butter can be mixed with various flavourings to form compound butter. Savoury butter is a popular accompaniment to meats, fish, and vegetables; sweet butter includes sugar, and flavourings such as vanilla, grated citrus rind, or liqueur.
Ghee	Ghee is the primary cooking fat in India, and in many Arab countries, where it is called *samna*. It is a type of clarified butter, which is simmered until the moisture evaporates and the butter caramelizes, producing a strong, sweet flavour. It is made from both buffalo's and cow's milk and can be bought in jars or made at home. Flavourings are often added—in India these might include bay leaves, cumin seeds, cloves, fresh ginger, turmeric, peppercorns or chilli peppers; in Arab countries, herbs such as oregano or thyme are often used. Since butter is expensive, ghee is sometimes made with part butter and part margarine.
Poultry fat	Chicken fat also known as *schmaltz* is commonly used in Jewish cuisine. Duck fat and goose fat have also been used for making duck *confit* and in France they are regarded superior to lard and suet.

Table 18.7   Types of vegetable oils used in cooking

Type of Oil	Description
Corn oil	It is made from corn or maize and has a smoke point of 230°C and is most suitable for deep-frying.
Cottonseed oil	Cottonseed oil is commonly used in cooking and it also has a very high smoke point like corn oil.
Sunflower oil	Most of the vegetable oils are made from seeds and sunflower oil is no exception. Vegetable oils can be unrefined or refined. The refined process increases its smoke point thereby making it more suitable for frying. It has a smoke point of 232°C.

Contd

Table 18.7 (Contd)

Type of Oil	Description
Groundnut oil	Also known as peanut oil, it is suitable for deep-frying as it has a high smoke point. People who are allergic to peanuts will not be affected by peanut oil as the proteins have the allergen and not the fats. It has a smoke point of 225°C.
Palm oil	It is the most widely produced oil in the world. It is made from the seeds of the fruit of the palm. It has a reddish hue because of the presence of beta carotenes, which fade away, when the oil is heated. It is very stable oil and hence, commonly used to make margarine. This oil is semi solid at room temperature.
Rapeseed oil	Also known as canola, it is the healthiest of oils used for cooking and the literal translation of canola is Canadian oil low in acid content.
Mustard oil	Mustard oil is made from mustard seeds and is very commonly used in India. This oil is deep yellow in colour and has a very strong flavour. Mustard oil should be allowed to smoke for at least 2–3 minutes to allow the pungent flavours to escape. This oil is commonly used to pickle vegetables.
Olive oil	It is made from olives and is commonly used in all the Mediterranean countries. This oil has a low smoke point of 165°C and hence, not very suitable for deep-frying.
Nut-based oil	Made from various nuts and seeds of nuts. Oils such as hazelnut oil and walnut oil are very strong and only a few drops of them are added to salads or soups and they are never used for cooking. They are very expensive and rich in Omega-3 fatty acids. Nut oils on application of heat may turn bitter and have a very low smoke point.
Vegetable oil	Some vegetables, such as avocado, pumpkin, etc., are commonly used to make oils used to make margarine. Avocado oil is very expensive and used to flavour salads and dressings.

MILK AND DAIRY PRODUCTS

Milk is a major ingredient in our diet—poured over cereals, put in tea and coffee—and it also is a part of many dishes, especially desserts such as ice cream, custard, pancake, rice pudding, etc. It is particularly high in calcium, but it is also fairly good in fat too. Milk is mainly made up of water and has many nutrition contents such as proteins, fats, and carbohydrates. Milk also has a sugar called lactose and this is the reason why milk changes colour when heated for a long time. In cooking, milk is mainly used as a poaching liquor to make white sauce and other sauces. In bakery, it is used as a liquid in place of water to enrich the dough or used as a liquid in making creams and pastes.

CHEF'S TIP
When using acids, thicken the milk with starch. This will avoid the milk from curdling, as starch will hold the proteins together and the milk will not split.

In pastry products milk improves the texture, flavour, nutrition value, and the quality of the product. When milk is used in sauces or as the base for desserts, low heat should be applied to the milk as both the odour and colour of the milk gets adversely affected by the intensity of heat provided. Heated milk forms a layer of protein on the base of the cooking equipment and this coagulated protein can burn

Table 18.8   Composition of milk

Constituents	Percentage Present in Milk
Proteins	11
Fats	3–6
Minerals	1
Vitamins	1
Carbohydrates	2
Water	82

if care is not taken. The burnt milk has a very unpleasant flavour which is never desired in cooking. When the acid medium is added to the milk, it coagulates resulting in cheese, curd, or paneer.

The composition of milk is shown in Table 18.8.

There are many types of milk consumed though mostly it is cow's milk, goat's milk, and sheep's milk. Milk needs to be pasteurized in order to be safe for consumption.

Pasteurization of Milk

Almost all fresh milk is marketed as pasteurized these days, as this is a precaution to guard against food poisoning. To pasteurize milk one has to heat it to a high temperature below boiling point by one of the following two methods.

Flash Method   Milk is brought to 71°C and held there for at least 15 seconds.

Holding Method   In this method the milk is heated to 62°C and held at this temperature for at least 30 minutes.

Types of Milk

It is important to know various kinds of milk used in cooking and especially in pastry the type of milk plays a very important role. The whole milk will give a different product from the milk powder or milk solids and so on. Table 18.9 shows the various types of milk.

Table 18.9   Types of milk

Type of Milk	Description
Whole milk	It can be cow's milk or milk from buffalo, sheep, or even goat. This milk contains at least 3.5 per cent of butterfat, which gives it the wholesome taste.
Homogenized milk	It is whole, pasteurized, and treated so that its fat globules are broken to the extent that there is no separation of fat from the milk. It is a mechanical process which reduces the size of the fat and then mixes them together. This is the best milk to use for tea and coffee as the fat does not separate and float on top.
Skimmed milk	The fat from the whole milk is removed by a centrifugal force. The fat from the milk is sold separately as cream. The skimmed milk has a trace amount of fat present that can be lower than 1 per cent.
Buttermilk	It is a by-product obtained while making butter. When the butter is churned, the whey which is left behind is known as buttermilk. Today buttermilk is made from pasteurized milk with an addition of lactic acid bacteria. This milk can be used for making sorbets and ice creams.
Dehydrated milk	This is whole milk from which the water is removed by either spray drying or by roll drying processes. Milk powder is used in breads or cookie doughs to provide enrichment.

Contd

Table 18.9 (Contd)

Type of Milk	Description
Condensed milk	This is reduced milk, in which sugar and stabilizers are added to produce a thick and viscous creamy liquid. Condensed milk is used to make various cakes and pastries. It is commercially made and sometimes the sealed condensed milk cans are boiled in water for several hours to produce caramelized cream which can be piped on pastries.
Dried milk solids	This is obtained by cooking milk over a slow heat till most of the liquid evaporates and the remaining solid mass is sold as milk solids. They are also known as *khoya* in Hindi and are extensively used in Indian desserts.

CREAM

CHEF'S TIP

Cream should be whipped at a temperature of around 4°C. It will be also helpful to chill the bowl to allow the little dissipation of heat. Whisking of cream over a bowl of ice will help the cream to whip faster and reduce its chances of curdling into butter.

Cream is the butterfat content of whole cow's milk, separated from the water. The principal difference between the various types of cream—single cream, double cream, whipping cream, clotted cream, and soured cream—is the balance between water and butterfat. This will determine their consistency, whether thick or thin. Other differences are in the way they have been made and their time for maturing, which results in different tastes. Some of the common types of creams used in cooking and confectionery are given in Table 18.10.

Table 18.10 Types of creams in cooking

Type of Cream	Description
Single cream	This cream contains not less than 18 per cent butterfat. It cannot be whipped as it contains too little butterfat. This cream is also known as cooking cream as it is commonly used in cooking for making sauces or used in dressings and soups as well. Single cream in pastry is used for custards, etc. which will eventually be baked and cooked.
Double cream	This cream contains not less than 45 per cent butterfat. It cannot be whipped too much as it will turn into butter. It can be used to enrich sauces, but may curdle if boiled along with acid ingredients.
Whipping cream	This cream contains not less than 38 per cent butterfat. It is perfect for whipping. After whipping you will find a difference in texture and a change in volume. Sweetened or unsweetened cream can be used in desserts or can be used as an accompaniment, and is incorporated in mousses to lighten them.
Clotted cream	This cream contains not less than 55 per cent butterfat. It is already very thick so it can be used as it is and not whipped. This cream is served along with afternoon cookies called scones.

Contd

Table 18.10 (Contd)

Type of Cream	Description
Sour cream 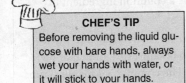	This is a single cream which contains about 20 per cent butterfat, but has a souring culture in it. It is matured through the same process followed for setting curds. This cream can be shaped into *quenelles* and served with soups or salads. It can also be served as a sauce in itself.
Acidulated cream	This is cooking cream with addition of lime juice to make it sour. This cream is used for making dressings for salads.
Imitation cream	As the name suggests, this is not a real cream. The butter fat in the milk is replaced by vegetable oils and that is the reason it is popularly sold as cream with zero cholesterol, as we know that cholesterol is present only in animal fats and not in vegetable fats. This cream is prepared commercially and is commonly used in kitchens because of its stability and keeping quality. The taste is very synthetic and one can combine whipped cream and imitation cream in equal parts to get a stable and better product as compared to the imitation cream. Imitation creams can be frozen and are available in sealed tetra packs.

SWEETENERS

Sweeteners are the soul of all desserts. When we refer to desserts, they have to be sweet. Sugar is one of the most important ingredients used in confectionery and its usage is not only limited to providing sweetness, but it has various other uses such as altering the texture of products, giving colour to the baked goods, etc. Sugar also delays the coagulation of proteins in eggs and promotes aeration in a product, etc. It is available in various forms such as grain sugar, icing sugar, breakfast sugar and this categorization is basically done on the basis of shape and size of the sugar crystals. Apart from these there are still various other types of sweeteners used in cooking, especially confectionery. Honey, corn syrup, treacle, etc. are other very commonly used sweeteners in cooking and the choice would depend upon the kind of texture the final product is aimed at. For example, while making meringue the sugar crystals will not dissolve by the time meringue is completed, and usage of any other sugar apart from icing sugar in butter cream will form small crystals, which will give a grainy finish to the cake. Sugar also lowers the freezing point in ice creams and hence, we must be careful in adding the amount of sweeteners in ice cream—less will make it set like ice and more will hamper the setting of an ice cream.

> **CHEF'S TIP**
> Before removing the liquid glucose with bare hands, always wet your hands with water, or it will stick to your hands.

Let us discuss various types of sweeteners commonly used in baking and cooking in Table 18.11.

Table 18.11 Types of sweeteners used in cooking

Type of Sweetener	Description
Granulated sugar	Sugar crystals usually obtained from sugarcane. This is the regular white sugar which is used in homes. Usage of this sugar will find its place in any preparation which has sufficient liquid to dissolve it. For example, whipping eggs, making sugar syrups, cooking *sabayon* over double boilers, etc.

Contd

Table 18.11 (Contd)

Type of Sweetener	Description
Castor sugar/Breakfast sugar	It is commonly used in breakfast, for tea and coffee. It is a small evenly graded sugar crystal which dissolves quickly and is easier to dissolve in the creaming methods. It is more expensive and hence should not be substituted for grain sugar.
Icing sugar	Granulated sugar is crushed into fine powder and has a small percentage of corn starch added to keep it smooth and free flowing. Icing sugar is used for creaming methods where it would be used as icing for cakes and pastries. Icing sugar can also be sifted on top of dry baked sweet products as a garnish.
Brown sugar	This is a granulated sugar which is available in variety of shades of brown. The darker brown sugar is also known as 'demerara sugar' and darker the colour, more pronounced is the flavour. Brown sugar is the residual sugar obtained during the process of refining sugar. Many people mix the granulated sugar with caramelized black sugar syrup to colour the sugar; but the flavour is not the same.
Golden syrup	It is a thick amber coloured liquid obtained from sugar during the refining process. It is treated with acid to cut down on its sharp taste. It looks similar to honey and is used in making confectionery products.
Corn syrup	It is chemically refined syrup made from corn kernels. This is usually obtained as a clear liquid and the coloured corn syrup is artificially coloured. It is very sweet and contains high amount of fructose. It is a common ingredient in processed foods and beverages.
Maple syrup	It is a sap of maple tree and a natural sweetener. It is boiled down to thick syrup. Pure maple syrup is very expensive, as to obtain 1 litre of maple one has to boil down at least 40 litres of maple sap. The commercial maple syrups are corn syrups with a percentage of maple syrup added to them. It could be added in the range of 2–6 per cent. The percentage of the maple is always mentioned on the bottle and this decides the price of the product.
Treacle	When the sugarcane juice undergoes refining, it undergoes many stages. In the first stage the white sugar or the raw sugar is removed. The remaining sugar syrup is used to make treacle which is stronger than golden syrup but less than molasses.
Honey	Honey is a natural sugar obtained from bee hives. The colour and flavour of honey will vary with its source. Some commercial honey farms allow bees to suck the nectar from only one particular flower to produce the honey of that flavour. One can use honey in most of the baked products but care has to be taken as honey can caramelize even at lower temperatures.
Date sugar	It is obtained from drying and pulverizing dates. It is very sweet and although it does not dissolve very well, it is used in many baked products.
Palm sugar	Palm sugar is traditionally made from the sap of palmyra palm or the date palm. It is extensively used in Asian cooking, especially Thai. Nowadays sugar from sago and coconut is also made and sold as coconut sugar and sago sugar.

Contd

Table 18.11 (Contd)

Type of Sweetener	Description
Jaggery	Jaggery is a product made in India, Africa, and South America. It is produced from sugarcane and is healthy and nutritious as the whole sugarcane juice is cooked with molasses. The colour of the jaggery or *gur*, as commonly known in India, can be light to dark depending upon the degree of cooking.
Molasses	Molasses is the by-product of sugar from sugarcane. There are three stages of refining of sugar and with every stage a residual sweetener is left behind which is known as molasses. As the stages increase, the colour and the flavour of the molasses become more dark and strong.
Invert sugars	These are sucrose-based syrups that are treated with acids or chemicals. The acid breaks the sucrose molecule into glucose and fructose. Since there are now two molecules of sugar it will be sweeter than sucrose. Corn syrup is a type of invert sugar and this property of inverting sugar does not let the sugar to crystallize easily and hence the product stays moist.
Liquid glucose	Liquid glucose is obtained by treating the corn slurry by acid—a process known as hydrolysis. This is chemically made and results in a thick viscous liquid that is used to produce candies by not allowing the sugar to crystallize and also acts as a preservative. Liquid glucose contains the dextrin gum which retards the crystallization of sugar. When added to products, it makes them pliable and hence very commonly used to prepare garnishes and decoration pieces with sugar.
Isomalt	It is a natural sugar substitute and in reality it is a sugar alcohol. It is available in crystalline forms and is used for preparing sugar garnishes as it is more stable than sugar and does not caramelize thereby giving an appearance of thin glass sheets.
Sugar substitutes	These are chemically produced and have no nutrition value at all. Saccharin and cyclamates are best known and more commonly used in food items, especially for people who are diabetic. It is slightly bitter in taste and is used as a sweetener in low calorie or diet soft drinks.

CONCLUSION

In this chapter we discussed about the structure of wheat grain (Fig. 18.1) and saw that wheat grain is made up of 85 per cent endosperm from where the refined flour is milled. The rest of the grain comprises of 13 per cent of bran and 3 per cent germ. The entire wholewheat grain is milled to obtain wholemeal flour which is healthier, as it contains fibres and proteins in the germ.

From one wheat grain we can obtain various kinds of flours and by-products as discussed in Table 18.1. Apart from the by-products of wheat, we discussed various types of flours obtained from wheat grain (Table 18.2) used in cooking, especially to produce variety of breads. Keeping in mind the people who have special dietary requirements, such as allergies to gluten, we have also discussed variety of other flours (Table 18.3), some of which are free of gluten (Table 18.4). We also discussed various types of raising agents which aerate the product chemically. Some of these aerating agents are also natural, such as yeasts, and form the basis of many breads and cookies.

Various types of raising agents, both chemical and natural, are discussed in Table 18.5. Fats and oils which are commonly used in kitchens and in baking were also discussed in Tables 18.6 and 18.7. Oils and fats were discussed in this chapter and we came to know about the various types of uses of the same. All the oils, except for coconut and palm oil, are liquid at room temperature and some oils are modified so that they can be as stable as fats, by a process known as hydrogenation, whereby the fats get saturated.

Most of the animal fats are saturated naturally; but some are artificially saturated and these are called transfats. We also discussed the basic procedures for rendering and clarifying the fats.

Then we dealt with the dairy products, which form the structure or the soul of pastry products. Milk and cream are invariably used in cooking; but in pastry they find a special place with each type of cream used for a specific purpose. We also discussed imitation cream that is made from the vegetable oils. Then we discussed one of the key elements, sweeteners, used in preparing desserts. Sweeteners are available in various forms and the most common ones used in the kitchens are discussed in Table 18.11. In the coming chapter we shall discuss various kinds of bakery techniques and some products made from these.

KEY TERMS

Atta Indian name for wholemeal flour.

Bajra Indian name for millet flour.

Blinis A small thick pancake made from buckwheat flour.

Bulgur Cracked wheat used in the Middle East.

Buttermilk Whey left behind after churning out butter from cream.

Cake flour Weak flour used in cakes.

Canola Oil from rapeseed; the abbreviated form of Canadian oil low in acid content.

Centrifugal force Spinning of the liquids in a machine which allows the two liquids at different densities to separate from one another.

Charcuterie French for section in kitchen which deals with cold meat preparations such as sausages, salamis, etc. It usually comprises of pork products.

Clarifying A procedure usually used for butter, where it is melted to get rid of milk solids, thereby making it stable to cook.

Confit Stewing of duck or goose in pork fat or duck fat to produce soft textured meat.

Cooking cream Cream which can be boiled and hence used in cooking.

Crackling The crisp piece of meat or skin left behind after the meat has been rendered.

Farina It is flour which is ground coarse and is usually of a polished grain.

Gluten A protein obtained when the wheat flour is kneaded with water.

Hydrogenation Saturating the oils with hydrogen atom.

Invert sugar A sugar that has been altered chemically with addition of an acid. This acid does not let the sugar to crystallize easily.

Khoya Indian name for reduced milk solids often sold in cakes.

Kuttu ka atta Indian name for buckwheat flour.

Leavening agents Products used in aerating a baked good. It can be natural like yeasts or chemicals such as baking powder and cooking soda.

Lipid Scientific term for fats and oils.

Makki ki roti Popular north Indian bread made from corn meal.

Margarine Artificially saturated vegetable oils; sometimes it may also contain animal fats.

Meringue Stiffly beaten egg whites with castor sugar to resemble whipped cream. It can be used as a base for desserts or dried in oven and used as garnish.

Milling Grinding of a commodity to yield flours of different textures.

MUFA Monounsaturated fatty acids—they have only one double bond in the fat structure.

Paneer Indian name for cottage cheese.

Pasteurization Heating of milk or dairy products to kill most of the harmful bacteria.

Pate French for pastes usually from meats.

Potassium salt Another name for tartaric acid, which is obtained naturally during the wine fermentation process.

PUFA Polyunsaturated fatty acids—they have more than one double bonds and are healthier.

Quenelles An oval shape provided by moving a creamy substance between two spoons.

Rendering Separating fats from the muscle or skin by applying low heat.

Samna Local name for clarified butter in the Middle East.

Schmaltz German name for chicken fat often used in Jewish cooking.

Scones English preparation of flour, self raising flour, and milk. It is served during afternoon tea.

Singharey ka atta Indian name for water chestnut flour.

Soft flour Another name for weak flour.

Strong flour Flour having high amount of gluten.

Transfats Oils which are saturated artificially.

Weak flour Flour which is low in gluten content.

CONCEPT REVIEW QUESTIONS

1. List the most common ingredients used in baking and confectionery.
2. What are the most common types of wheat available and what is their speciality?
3. Draw and label the structure of wheat.
4. What are the major components of wheat grain?
5. List at least five by-products obtained from the wheat grain.
6. What is gluten and how is it formed?
7. How does gluten present in flour help us to bake products?
8. How can we make the gluten soft or firm?
9. Why cannot we make bread from cornflour alone?
10. List and describe at least five types of flour obtained from wheat.
11. What is graham flour and what is the substitute in case of its non availability?
12. Define self-raising flour. What is the ratio?
13. List at least five types of flour which can be used for making gluten-free breads.
14. Name at least two flours used in India during religious fasts.
15. What is *masa harina* and how is it used in Indian cooking?
16. What is the difference between arrowroot and cornflour?
17. Name three products made from buckwheat flour.
18. What are raising agents and what is their role in cooking?
19. Name at least one natural raising agent and two chemical raising agents.
20. What care should be taken while using dry yeast instead of fresh yeast?
21. Define lipids and give one major difference between oils and fats.
22. Name two oils that set solid at room temperature.
23. What are transfats and why are they unhealthy?
24. Define MUFA and PUFA. What are their roles?
25. How is oil saturated? Define the process.
26. Name at least three animal fats and three vegetable oils and their uses.
27. Name at least three oils that are more suitable for deep-frying.
28. Write a short note on canola oil.
29. Name the constituents of milk.
30. What are the various types of dairy products used in the kitchens and what are the specific uses of each?
31. What is the difference between dehydrated milk and dried milk solids?
32. Name at least three types of creams used in cooking and pastry.
33. Name the various kinds of sweeteners used in confectionery.
34. What is the difference between corn syrup and golden syrup?
35. How would you differentiate between molasses and treacle?
36. What are invert sugars and what are their uses?

PROJECT WORK

1. In groups, make the recipe of bread by just altering the different types of flours such as rye, buckwheat, wholemeal flour, rice flour, maize flour, and barley flour. Compare the results and write down your findings in the chart below.

Flour	Taste	Texture	Crumbs	Weight	Volume

2. In groups, try out the gluten-free recipes as provided in the ORC and compare the results.
3. Make a recipe of scones by altering the raising agents and compare the products with one another. See which has better volume and flavour and why.
4. Make a recipe of pound cake as shown in the ORC by altering the recipe by adding at least five different types of sweeteners available. Compare the results and understand the effects of various sweeteners on the product.

CHAPTER 19

BREAD FABRICATION

Learning Objectives

After reading this chapter, you should be able to:
- understand the different principles of bread making
- know various steps involved in production of breads
- prepare various kinds of breads and shapes thereof
- know about various types of international breads
- recognize the importance of ingredients and processes involved in bread making, thereby understanding the basic faults in bread production
- appreciate the storage of various breads and their usage in kitchens and serving techniques

INTRODUCTION

This chapter will deal with the intricacies of bread making. Bread making can be said to be an art which is carried out scientifically. The ingredients used in making bread should be of good quality, and also the optimum temperatures and humidity levels should be strictly maintained. How each ingredient should be treated while bread making will vary depending upon the quality of the ingredients and the place where it is made. Whether a standardized product with characteristic qualities can be obtained, will be determined by the different treatments given to the same product at different places. In this chapter we shall talk about the commodities used in bread making and also see the role of individual commodities in the process. Once we understand the role played by such ingredients we can modify breads to produce a particular product, with a desirable structure. This will help us to evaluate breads and if a particular bread has not turned out to be as it was supposed to, then we can relate to what went wrong with the recipe. But many times the quality of the bread is affected not by the ingredients alone, but also the process involved in making it. In the previous chapter we saw how mechanical kneading of flour leads to the development of gluten. Thus, if a bread does not have a good structure, just the quality of the flour is not to be blamed, as the problem could be in kneading, baking style, time, and temperature. This chapter will help in understanding the historical significance of breads and relate it to the present technology and methods. This chapter will also discuss the function of various ingredients and how they affect the final product and storage qualities. It will also deal with various bread improvers and the effectiveness of each type.

The origins of the first bread are unknown, but wheat has been cultivated for more than 5000 years. The leavening of the bread must have been purely accidental—the previous day's dough would have been left over and the wonderful fermented smell would have urged the baker to bake it. Till date in many traditional bakeries the concept of sour dough or starter is still prevalent. Some bakeries have secret guarded recipes for sour dough.

UNDERSTANDING BAKING

It is not enough to have a knowledge of commodities, working terminology, and types of aeration when it comes to bread making. One also needs an understanding of baking to ensure the product would not be substandard or spoilt, resulting in a waste of time and materials. Simply setting the oven to the recommended baking temperatures will not always guarantee success. There are many factors that affect the baking temperature and the baking time. They include the following:

Shape and Size of Products Being Baked

Generally speaking, the thicker the product, the longer it takes to bake. The baking temperature must be lower, or else the outer portion will be burnt before the middle is cooked. Thin products will bake more quickly, and so the oven temperature can be higher.

CHEF'S TIP

- Getting to know your oven is an important factor in successful baking.
- Have the accuracy of your thermostat checked regularly.
- Do not open the oven door during the early baking stages and avoid opening it too frequently as some products may collapse. The moisture tends to escape, thereby dropping the pressure and temperature in the oven, which results in collapsing of products.
- Place items to be baked on a tray in an interlocking manner to allow free passage of air through them so that they bake evenly.
- Close the oven door slowly.

Oven Humidity

Oven humidity is critical for some types of baked products. A cake will need a pan of water in the oven to provide extra humidity. The moisture will delay the formation of the crust until full expansion has taken place. The result should be a cake with a flat top and a crust of a pleasing colour. On the other hand, an oven full of cakes will provide sufficient humidity for proper crust formation.

Oven Overloading

Oven loading must be taken into consideration because the temperature of an oven filled with products will fall. The drop in temperature will depend on the type of product being baked and the size and type of oven used. To counter this factor, use a higher initial temperature. While placing items to be baked on a tray, place them in an interlocking manner to allow movement of air between the products being baked.

Density of Products Being Baked

Products containing a high proportion of sugar, fat, eggs, and fruits will need a lower temperature, and will take longer to bake. You may need to line baking trays with several layers of paper insulation to prevent the bottom from being burned.

Type of Oven

Commercial pastry ovens are thermostatically controlled by three switches—one maintains the preset temperature and the other two control the top and bottom heat of the oven. This flexibility allows a chef to bake different items more effectively by having more or less heat in the area required. A convection (fan-forced) oven maintains an even temperature throughout the oven. A standard type of oven has a varying temperature within the chamber, with the top area having a higher temperature. The general rule is that yeast and pastry products are placed towards the top of the oven while cakes are placed towards the centre.

INGREDIENTS USED IN BREAD MAKING

Every ingredient used in the making of bread has a particular role to play in achieving the final, desired product. These ingredients, however, perform well only when certain conditions are met and are highly dependent on each other to perform that particular function to the desired level. For example, yeast performs well in the presence of sugar as well as moisture. Thus it becomes essential to understand the nature of each of these ingredients in detail, how they will affect the final product, and how to manipulate these materials to achieve the desired products.

Flour

It is the main ingredient used in making breads. We discussed various types of flour (wheat, rye, multi grain, etc.) in Chapter 18. Usually strong flours are used in bread making. Wholewheat flours have lesser concentration of gluten as the bran content is more. This causes a weaker structure in the bread. Since the bran particles are slightly abrasive, they cut the gluten fibres resulting in a loaf with a smaller crumb. The presence of the bran particles also allows higher moisture absorption, resulting in a short fermentation time. When the germ is present in the flour there is a higher enzyme activity, as a result of which the gluten develops faster and the breads are made with a shorter fermentation time.

Water

Water is the most commonly used liquid in bread making. It moistens the flour and helps in forming the dough. It also aids in the baking process. Water performs the following three main functions in the bread dough.

1. Helps hydrate and moisten the insoluble proteins.
2. Disperses the yeast amongst the entire dough.
3. Binds the flour and other ingredients into a dough.

It is observed that the water content in the dough greatly affects the rate of fermentation. The speed of fermentation is greater in ferment and dough process as compared to sponge and dough process, which has an increasing level of hydration. As the fermentation time increases it becomes essential to reduce the water content to effect a higher ripening of the dough. The amount of water present will also greatly affect the texture of the final dough obtained. Table 19.1 shows the uses of different types of dough.

Hard water has a higher alkalinity. As yeast works best in an acidic medium, fermentation can be slower in the initial stages if hard water is used. However, as the fermentation proceeds the acids produced will neutralize this alkalinity and then the fermentation will continue at a brisk pace. Also, the alkalinity and the mineral salts will tighten the gluten and thus the dough will be firmer. Very hard

Table 19.1 Uses of different types of dough

Dough	Use
Batter	Waffle, *jalebi*, pancake, cake
Exceptionally soft dough	Brioche, *baba* dough
Soft dough	Bread rolls
Moderately stiff	Crusty bread, malted and rye breads
Very stiff	Fancy breads for decoration

water also has magnesium sulphate which has a retarding action on the yeast. Breads can be made with both hard and soft water, provided the physical adjustments are made. When the dough is kneaded for longer duration of time, the temperature in the dough increases due to friction. This has to be watched carefully as the temperature of the dough should not go beyond 25°C for the yeast to start working. In such cases a baker often uses ice to make the dough. Ice keeps the fermentation activity of yeast at an ideal rate for gluten ripening. The quantity used will vary depending upon the time of kneading of the dough or the friction factor, and the dough temperature required. Ice used must be in the form of flaked ice so that it is evenly distributed in the bread dough and causes an even cooling of the dough. It can be safely said that 5 kg of ice will be equivalent to 4 litres of water.

Yeast

Yeast is a single cell microorganism which causes the leavening in the dough. It converts the natural sugar in the flour into tiny bubbles of carbon dioxide that are trapped in the dough. During baking these bubbles expand to give the texture and lightness to the dough.

CHEF'S TIP
When using dry yeast in the recipe, use half the amount of fresh yeast mentioned, as dry yeast is stronger.

Yeast is available in two forms—dry and compressed. The ideal temperature for yeast to act is 25°C. The primary function of yeast is to change sugar into carbon dioxide so that the dough is aerated.

When dispersed in water with yeast food, the yeast exudes an enzyme that changes sucrose into dextrose, which is then absorbed by the yeast cell. Inside the cell, this is broken down into carbon dioxide and other by-products. Yeast also has enzymes which change protein into simpler compounds which can pass through the yeast cell membrane.

Yeast works best within a temperature range of 25 to 40°C. Above this, fermentation becomes rapid but the yeast gets weaker successively and is finally killed at 70°C. At this temperature, yeast is completely retarded though it is not damaged. Yeast can never dissolve completely in water, though it is just dispersed well into it. One could use a whisk to effectively distribute.

Compressed yeast must be cold to touch and must possess a creamy colour breaking with a clean fracture. If it is light in colour, and is dry, warm, with a pungent odour, it is in poor condition and the quality of bread might not be good. If it is dark brown in colour with a soft sticky consistency and an unpleasant odour, it is unsuitable for use.

Salt

The main function of salt is to control the action of yeast as it slows down the fermentation process. It should be mixed with flour for best results. It also provides flavour to the bread. It also affects the quality of the crumb, crust, and colour of the baked product. So salt mainly performs the following functions:

- imparts flavour;
- gives stability to gluten;
- controls the rate of fermentation;
- retains moisture; and
- affects the crust colour and crumb, due to control on the rate of fermentation.

More salt or less salt will adversely affect the final product as shown in Table 19.2.

Table 19.2 Effect of less and excessive salt

Less Salt	Excessive Salt
Large volume—as there is more breakdown of sugar into carbon dioxide	Tightening of the gluten gives a dense structure
Less crust colour	Dark crust colour as sugars are not broken down
Weak crumb structure	Crumb structure resembles cheese, as not enough gas is produced

Sugar

The main function of sugar is to act as food for yeast. It helps in developing flavour and colour. Sugar is the primary food that the yeast feeds on to produce alcohol and carbon dioxide. With the exception of lactose, yeast can break down all the other sugars present in the dough, either naturally in the flour or as an addition of sugar, mainly sucrose or sometimes, maltose. Flour naturally contains about two and a half to three per cent of sugar in the form of sucrose and maltose. This is enough for the yeast in the initial parts of the fermentation. However, in the final proof when maximum of the sugar is required to be broken down for an optimum rise, the natural sugars are exhausted and the addition of sucrose or maltose is required. Like salt, too much sugar or less sugar will impact the dough texture (refer to Table 19.3).

Table 19.3 Impact of sugar

Less Sugar	Excessive Sugar
Not enough volume	Large volume—as there is more sugar available for breakdown
Less crust colour	High crust colour
Weak crumb structure	Weak structure of bread
Lower yeast activity as yeast works best in 10 per cent sugar solution	Lower yeast activity as yeast works best in 10 per cent sugar solution

Impact of Sugar

Sugar has a solvent effect on gluten and this greatly affects the quality of the crumb in bread loaves. To counteract this, a mineral improver is used and excess salt is used as salt has a stabilizing effect on the gluten.

Sugar has many roles to play in dough. Few of these are as follows:

- Sugar is the primary food for the yeast.
- It helps improve the crust colour.
- Sugar also acts as a preservative and thus behaves as an anti-staling agent.
- Some sugars act as bread improvers.
- Sugar helps the bread to retain moisture, thereby keeping the bread moist.
- Some sugars impart flavours, for example, treacle, honey, and demerara sugar.

Milk

It makes the bread whiter and softer, and provides moisture and a distinct flavour. Milk also has a physical effect on bread in the form of the tightening effect of gluten by the action of 'casein' or the milk protein. However, boiling or pasteurization neutralizes the effect to a great extent. Lactose or milk sugar is the only sugar which cannot be fermented by yeast and hence it remains in the dough right till the end, resulting in a good crust colour. Milk is generally used in powdered and skimmed form and hence the amount of water taken up in the dough is slightly more, though not considerably.

Egg

Eggs are used for richness and to give lightness and colour. Eggs are again rich in protein and hence will tighten the gluten strands, but this effect gets balanced, as the fat in a yolk helps to soften the gluten as well. The use of eggs will yield softer bread. In many types of bread where a hard structure is required like hard rolls, one does not use eggs in the recipe.

Oil/Fat

It is used to provide flavour and softness to the texture. Different kinds of fats are used for different breads such as olive oil for *focacia* (Italian bread). Fats have a physical effect on breads rather than any chemical reaction. Fat being a shortening agent reduces the toughness of the gluten and mellows the final product. Fat also has lubricating effect on the fine gluten strands giving extra volume to the final product. These strands begin to slip over each other and thus affect the final quality. As the amount of fat increases, the fermentation rate decreases. This is because the fat will form a thin layer on the yeast cell membrane hindering the release and the absorption of the materials. Thus yeast quantity is slightly increased.

CHEF'S TIP

When using high amount of fat in a recipe of bread, always use milk, as the milk protein will have a tightening effect on gluten and thus it will offset the shortening effect.

Effects of Fat

The effects of using fat are as follows:

- It increases the nutritious value of the bread.
- It reduces elasticity, softens the crust and the crumb.
- It helps retain moisture in the baked product, thereby keeping it moist.
- It increases volume if used extensively.
- Fats such as butter and lard give flavour to the product.
- If used in large amounts, it retards fermentation.

Bread Improvers

Flour is of variable quality and hence it becomes necessary at times to add bread improvers to the dough to bring the final product to a set standard. Bread improvers may be divided into three main categories. These include:

- those of mineral nature, used by the miller.
- those of organic nature, mainly enriching agents.
- those of the mineral and organic categories which are also foods for yeast.

Mineral improvers are popular because they increase the yield of the bread by necessitating the use of extra water. Some of the mineral improvers have a slight drying effect on the crumb.

PRINCIPLES BEHIND BREAD MAKING

19.1 There are various stages in bread making and these are very crucial as each stage has to be carefully followed to obtain the desired result. We have discussed the role of various ingredients in bread making. Let us now understand how these ingredients are mixed to obtain a loaf of fresh bread.

Collecting the Mise en Place

The most important thing required in any pastry operation is to collect your *mise en place*. This will allow you to do things in a planned manner and the product also will be of the desired quality. Weigh all the ingredients as per the recipe and make sure the ingredients are at the required temperature. If the recipe calls for ice water, then use ice water. Substituting cold water from tap will not give the desired results. Weigh using a digital scale, as accuracy of ingredients is very important in pastry. Sift the flour to remove any impurities. Test the freshness of yeast by checking the following:

- It should have a fresh smell.
- It should be firm and should crumble easily.
- The yeast shown be of 'fawn' colour.
- It should become liquid if creamed with little sugar.
- The temperature of the yeast should be in the range of 5°C.

Select and prepare the bread tins. Always use thick and heavy pans for baking bread as they can withstand the high temperatures of the oven, without getting deformed with the heat. Shapes of bread rely on the mould used. Grease the mould with oil properly, to avoid the baked bread sticking to the mould.

Make sure that the temperature of the oven is at the required degree as the temperature of baking is very crucial and would change with different types of breads.

Mixing of the Ingredients

Mixing of the ingredients involves much more than just mixing everything together to form a dough. There are many methods in which bread can be mixed or kneaded and we shall discuss them individually as they form the basis of bread making. Broadly these mixing methods are classified under three headings such as:

- Straight dough method
- Ferment/sponge method
- Salt delayed method

Straight Dough Method

This is one of the most popular methods used in bread production and as the name suggests, is simple and straight forward. The fermentation time can vary between 30 minutes and 14 hours. The time of fermentation can be controlled by the moisture content, yeast content, and the salt content. Whole-wheat breads are made with one hour of fermentation as the doughs absorb more water, than in white flour. The time of the fermentation can also vary with the type of ingredients. Very strong gluten flour will require a long fermentation time to help in the softening and mellowing of the gluten. However, a wholewheat bread or germ bread will require a shorter fermentation time due to the high enzyme activity in the germ of the wheat grain, and the higher water content in the dough. The shortest method is the 'no time dough method' which calls for a high percentage of yeast (two and a half per cent) and the dough is directly made, scaled, and moulded. This is not a very good method of bread making and must be resorted to only in dire circumstances.

This method is not very efficient as it has certain limitations such as:

- There is not enough time for the gluten to ripen or mellow down and the bread contains only carbon dioxide gas and in reality there has not been enough fermentation in the dough, so the bread lacks flavour.
- The finished product is generally of poor quality and the bread becomes stale quickly because of insufficient gluten ripening.
- The bread structure also will show uneven expansion as the gas is not evenly distributed in the gluten network.
- The bread will lack the characteristic aroma of a well-fermented bread as there is not enough time for the various chemical changes to take place.

Germ breads are made with this method, due to the high enzymic activity that causes the dough to ripen quickly. The dough is made warm preferably to help develop the flavour of the dough quickly. The most commonly used straight dough processes are of one to five hours of bulk fermentation. This is the time from the dough making to the scaling of the dough.

The temperature of the dough increases with time as the fermentation is an exothermic reaction—involving release of some heat energy. Thus one must be careful in mixing the ingredients because if the temperature goes beyond 50°C, the yeast will die. It becomes difficult to control the fermentation process in the long processes. The longer processes are used only when the dough or the gluten is too harsh to be made into bread and the entire gluten can stand long fermentation strains.

As the fermentation time increases, the gluten softens to a larger extent. Thus the water content is also reduced. Along with this, the salt content is increased and the yeast content is lowered. This will lower the fermentation rate and help conserve maximum gassing power in the final stages. The very long process is not widely practised and is replaced by a shorter sponge or a ferment and dough process as described here.

Ferment/Sponge and Dough Process

Breads and buns can be made in two stages to help the fermentation and yet achieve better dough ripening. These are:

1. Ferment and dough process
2. Sponge and dough process

Ferment 'Ferment' is a proportion of water, yeast, yeast food such as sugar, and just enough flour to make a thin batter. The yeast readily disperses in the water and begins assimilating the food dissolved in the water. It begins fermenting immediately and multiplies and is soon active and vigorous. This makes it ready to undertake the harder work of fermenting the dough. Ferment is made and kept until it shows a sign of collapse. This is when it is considered to be at its optimum for bread fermentation. Usually, 30 minutes to one hour for fermentation is sufficient for achieving good results.

Ferment is usually used for doughs when they contain rich ingredients and are high in sugar concentration. Usually the ideal concentration for yeast to work is 10 per cent sugar. Thus, ferment made with this concentration will give the bread a boost. A flying ferment is haphazard guess of water, yeast, sugar and flour which is allowed to stand only till the rest of the ingredients are weighed and the dough is prepared—approximately 10 to 20 minutes. This is done to activate the yeast and many books mention it as creating a well in the centre of flour and breaking up yeast with water and sugar and sprinkling little flour on top. When the bubbles start to appear on top, it is an indication that the yeast is active.

Sponge 'Sponge' can be said to be a stiffer version of the ferment. The rate of fermentation is hence lesser and the sponge is kept for a longer time. It is made by mixing a part of the flour, yeast, sugar, and salt (sometimes not), and some or all the water. The speed of the fermentation is controlled by the amount of yeast added, addition of salt, water content and the temperature of the sponge as well as the holding temperature. When the sponge rises and collapses, the remaining materials are added to make dough which is then given bulk fermentation.

The size of the sponge in relation to the dough will give the name to the process. For example, one-fourth sponge, one-third sponge, half sponge. The sponge quantity will go only up to 75 per cent of the dough content.

The main purpose of the sponge is to help develop a mellow flavour which is the result of the long fermentation. This is done without subjecting all the gluten to the harsh fermentation process and thus staggering the quantity of gluten present in the final product. This prevents a weak structure or a collapse of the bread. In most bakeries a portion of the previous day's dough is added to achieve this effect. The dough thus ripens well at walk in temperatures 5 to 7°C for a long period of time (16 to 18 hours minimum) and gives excellent flavour to the bread. This is also known as sour dough or ferment. In Italian this ferment is known as *biga* and in French it is called *levain*. In India it is known as *khameer*.

One must be careful while kneading the dough. Many dough mixers have two speeds such as slow and high speed to knead the dough. Many bread recipes call for kneading the dough at slow speed for couple of minutes and then increasing the speed to high. Usually bread is kneaded until a film is formed when the dough is stretched. This is also known as windscreen test.

Salt Delayed Bread Making Process

It is an excellent process used initially for harsh gluten flours, but now widely used for all bread making processes as it drastically reduces the fermentation time without giving any change in quality. This process calls for the omission of salt in the first stages of dough making. As was discussed earlier, salt is helpful in controlling the pace of fermentation by the yeast and hence when the salt is omitted in the first stages, the action of the yeast will increase. The gluten will ripen or soften well due to the rapid action of the gases released. The chemical changes that take place in the dough will also become fast and the effect of the acids produced will be visible in a shorter time.

The salt is added later on in the following three ways.

1. By sprinkling the salt over the dough
2. By using some water reserved from the original quantity
3. By using some fat to incorporate the salt

This process is the best method of conditioning dough without using higher yeast contents or an increase in fermentation temperature or time.

Proving

The next step is to let the dough to ferment. 'Proving' means to let the dough rise to at least double in size. This is done to let the yeast break down sugar into alcohol and carbon dioxide. The gases thus released help to ferment the dough and distribute uniformly. The ideal temperature for proving is 32°C. Proving is done in three stages. One is done after kneading, called first proving; the second is done after 'knock back' (refer to the next step given below), called intermediate proving; and the final proving is done after shaping the bread. So we can say that fermentation is done for the following reasons.

- It helps in production of carbon dioxide gas which aerates the dough.
- It helps to condition the dough. This is through the enzymatic action due to reduction of natural sugars for assimilation by yeast.
- It helps to reduce the proteins to simpler nitrogenous compounds for growth and development of the yeast.
- The enzymes are active during the fermentation period. The sugars are broken down to release heat which causes the temperature of the dough to rise. This rise can be controlled by the speed of fermentation and the storage temperature.

Knock Back

The fermented dough is punched down to knock off the air bubbles that had developed during the intermediate proving. This is so done to redistribute the yeast and the other ingredients evenly all through the dough. After knocking back the dough is allowed to rest for a while as the gluten tends to stretch and it will be difficult to mould the bread. This stage is called intermediate proving. It is important not to over knead the dough in the machines as the gluten will loose its resilience. The knock back is also done to equalize the temperature in the dough.

Dividing and Scaling

This is used to portion dough into pieces of the required weight. As discussed in the above paragraph, it is important to rest the dough before dividing and shaping is done. The scaling of the bread that needs to be baked in a mould will depend upon the size of the mould. Though there is no particular formula to calculate the weight, normally a loaf is calculated by pounds, so one pound loaf of bread will be baked in a one pound mould (the moulds are sold by volume they can hold). The scaling of rolls will depend upon the final usage of the product. Table 19.4 shows the weights of certain breads. This should be used only as reference and the weights could change with regards to the usage.

Shaping/Panning

Divided pieces of dough are shaped in the form of loaves or rolls. It should be done in a sparingly floured surface, handling the dough gently and placing it for final proving.

Table 19.4 Sizes of breads

Bread	Size
Breads baked in loaf	Made in moulds of specific weight like 1 lb/2 lbs, etc.
Bread rolls	25–30 g per roll
Burger buns	90–100 g
French baguette	300–350 g per loaf
Bread loaves not made in moulds	450–650 g loaves

After a few minutes of resting, the dough reaches its optimum ripening. Thus the dough is scaled and then shaped. As the dough is deflated during knock back, it has to be carefully manipulated again as it becomes more resilient. Machine dividing can destroy the structure of the dough. The dough is rested slightly before shaping, to allow for shaping without pressure. This final moulding is essential as the shape of the product and the crumb structure is affected. This step is also known as panning, which means to shape the bread and put in a pan.

Final Proving

As the dough is being shaped it is temporarily 'degassed' and the gluten tightens. If the dough is mature and the moulding done correctly, the skin surface will be smooth. The objective of the final proof is to allow the loaf to expand completely before baking. The production of the gas and the breakdown of the sugars must be vigorous and the gluten should be in such a condition, that it is strong enough to hold the gases and expand.

The condition under which the final proof is carried out is important. If there is a lack of humidity, the dough surface will be dry and there will be a lack of bloom on the crust of the bread. Skinning is the result of draughts of air, and will show as grey patches. Excessive humidity will result in a tough leathery crust, a wrinkled surface, and holes under the top crust of the loaf.

The final proving is usually done in equipment known as proving cabinet or proving chamber. Proving chambers have a temperature of 30°C and are maintained at humidity levels of 90 per cent which is the ideal condition for the yeast to work and ferment the dough. In case one does not have a proving cabinet, it is advisable to place the bread in a warm place sprinkled with water or covered with plastic to avoid the formation of scales on the dough which will then cause a fault in the bread.

Scoring

It is the process of giving marks on top of the dough with a sharp blade or a knife. It helps the bread to expand during baking without cracking. This step is not mandatory and chefs can choose to do scoring to give a rustic look to the breads. However, certain breads, such as classical French baguette, have scoring marks on them. Some chefs score the breads after shaping and some choose to do it just before baking. The look of the bread is different in both the cases.

Baking

The bread is ready to be baked once it has proved to optimum. Under proving of the dough will yield in cracked loaf and over proving will make the bread collapse in the final baking process. The bread is said to have proved well, if it springs back when depressed slightly. During baking, dough goes through the following three stages.

First Stage

The oven spring occurs and the gas bubbles in the dough expand and it rises rapidly. The yeast activity increases rapidly in the oven and the activity of the yeast stops as it kills the yeasts at 60°C. The gas in the dough expands and so do the steam and the alcohol vapour pressure. This causes a sudden burst in the volume of the bread and is called the 'oven spring'. Some of the starch is gelatinized to make it more susceptible to the enzymic activity.

Second Stage

The dough solidifies because of the coagulation of proteins and transforms into bread. Here the gases escape out of the dough leaving a dispersion of holes, which are responsible for the sponginess of the bread.

Third Stage

The dough gets its colour and crust. Enzymes are active till about 80 to 90°C producing sugars even beyond the yeast activity. This helps in the colouring of the crust. The enzyme activity helps in the crumb, crust colour, and bloom of the bread. As the baking proceeds, weight is lost by the evaporation of the moisture from the crust. As the moisture is driven off, the crust takes on a higher temperature, reaching the temperature of the oven. The sugars caramelize and the breakdown of the soluble protein blends to form the attractive colour of the crust. The sugars caramelize at 140°C.

The texture of the bread can be altered by regulating the heat at this stage. Crusty bread would require lowering the temperature after the bread is baked to 80 per cent. This would help the top crust to get dry thereby giving a crisp crust to the loaf. The crust of the bread can also be altered by giving various types of glazes, which is mostly done in case of bread rolls. Table 19.5 shows various glazes and toppings for bread.

Table 19.5   Various glazes and toppings for breads

Glaze/Topping	Use
Egg wash glaze	This gives a darker colour, shine to the roll, and also adds a nutritional value. It is advisable to use the egg yolk only; but the whole egg can be used. Beat the egg with a little amount of water and strain it to get smooth flowing egg liquid so that it can be applied with a soft brush or with a piece of cloth.
Salt water glaze	This gives a rustic whitish appearance to the bread. Prior to baking the saline water is brushed on top of the bread. Care should be taken as this could make the bread salty.
Starch glazes	The breads are sometimes glazed with corn starch slurry. This also provides shine to the bread.
Honey glazes	Honey is boiled prior to applying as a glaze, for it is a viscous substance and boiling will allow it to set into a glaze that will stick to the surface of the bread. This glaze is applied after the bread is taken out of the oven. Ensure that the bread is still hot, when this is applied. It is mostly done for sweet breads such as gingerbreads, *zopf*, etc.
Seeds as toppings	Various seeds such as cumin, fennel, poppy, sesame, *nigella*, etc. are sprinkled on top of the bread to improve the look of the bread and add nutritive value and assortment to the rolls on the buffet or in a bread basket. Care should be taken to sprinkle the seeds only after the wash or glaze has been applied to ensure the sticking of the seed. The amount of seed to be used will depend upon the intensity of the flavour of the spice or the seed.

Contd

Table 19.5 (Contd)

Glaze/Topping	Use
Nuts as toppings	Various nuts can be sprinkled on breads prior to baking. Following the same principles as in the case of seeds, one must also ensure that the nuts are chopped evenly to be able to be used as topping.
Herbs as toppings	Various chopped herbs can be used as toppings. It is advisable to use dry herbs as fresh herbs will anyways loose their colour when baked in oven. One can create different types of crusts by combining herbs, seeds and nuts to create unique toppings.
Vegetables as toppings	This is used in specific breads such as Italian *focaccias*. The bread is sprinkled with assortment of grilled or sautéed vegetables such as onions, bell peppers, olives, etc. which are spread along with olive oil and rock salt on top before baking. This should not be confused with flavoured breads, as in case of flavoured breads the ingredients are mixed along with the dough and not used as topping.
Flour as topping	Many types of bread are dusted with large amounts of flour prior to baking. It is important to first glaze the bread with plain water as this will allow the flour to stick to the bread. In case bread has to be scored, it is always done after the dusting of flour has been done.
Cereals as toppings	Many cereals, such as oats, bran, germ, bulgur, etc. also are used as toppings after applying the water wash. In these cases the top surface of the bread is rolled onto the topping so that the entire surface gets coated. This is done right after the shaping and the bread is allowed to prove with the topping, which disperses evenly when the bread is proved and is ready for baking.

Cooling the Baked Bread

When the bread has been taken out from the oven, it is essential that it should be demoulded and cooled reasonably quickly, as insufficiently cooled bread when sliced will be subject to mould formation and spoilage. The bread must be cooled on a wire rack because if the bread is placed on a flat surface, the heat from the base will condense and the humidity will let the moulds grow into the bread. Also, proper cooling allows for evaporation from the surface of the loaf which would otherwise condense on the crust, known as 'sweating'. This will show as moist patches on the crust.

Here we also came to know a little history regarding some breads and the various uses that these breads can be put to. The following points summarize production of bread.

1. Keep everything warm during fermentation. Yeast requires warmth so that fermentation may take place.

2. Mix to a soft dough. Some flours are more absorbent than others and require more liquid. The dough must be elastic. If it is too dry it becomes hard and not elastic in texture.

3. Knead the dough thoroughly to distribute the yeast throughout the dough.

CHEF'S TIP

Some breads involve application of steam to make them crisp. A layer of moisture on the bread when subjected to heat will turn into steam and evaporate. When it evaporates it will take large amounts of moisture away from the surface of the bread thereby resulting in a crusty roll.

The standard test for yeast bread is to take the loaf from the baking tin and tap/knock the base. If it sounds hollow it indicates that the bread is done. The colour should be an even brown from all sides.

4. Place the dough in a warm position to rise until twice its original size. In the presence of moisture, sugar and warmth, fermentation takes place; the gas produced expands causing the bread to rise.

5. Knead lightly the second time to expel the gas which escapes as the dough collapses and to reduce the volume of this dough and yield a dough of a closer texture.

6. Allow to rise or prove a second time to continue the action of fermentation, which has been retarded by the kneading.

7. Do not over prove. The bulbs of gas may expand until they break through the dough, which will collapse and cause a heavy loaf that is sour in flavour.

8. Place in a very hot oven (235°C). This will kill the yeast and stop the bread from continuing to rise. It will enable this starch to gelatinize and proteins to coagulate giving a stable structure to the loaf/rolls.

9. Reduce the heat after the roll or load is set so that the loaf cooks through to the centre without burning.

10. Cooked bread should sound hollow when tapped at the bottom.

BASIC FAULTS IN BREAD MAKING

Before discussing the various faults in breads it is vital to look at the various points on which the bread can be judged. They may not necessarily be in the order of importance as each one has a vital role to play in guest satisfaction. These points are as follows:

Volume

The volume here refers to large bread that is light for its weight. But there are many types of bread that are very heavy for their weight, and they are traditionally made in this way, because the recipes call for them.

Bloom of Crust

This refers to the colour and texture of the crust of the bread. However, we can alter the texture of the crust by giving various types of glazes (refer to Table 19.5) to the bread prior to baking.

General Shape

This refers to the hand rolled bread which is shaped round, oval or is given any other shape as required by the bread.

Colour of the Crumb

The colour of the crumb will depend upon lots of factors such as types of flour used in dough, oven temperatures, and the ingredients in dough. Bread high in sugar will have a darker colour as compared to the one with less or no sugar at all.

Evenness of Texture

The texture here refers to the equal distribution of the holes in the bread loaf.

Sheen of the Crumb

The sheen refers to the shine in the dough and this is achieved by kneading the dough to optimum.

Moistness

The amount of moisture in the bread will determine its texture and keeping quality. Some breads turn out dry when over baked or when there is less water used while making the dough.

Flavour

Many ingredients such as salt, sugar, and yeast affect the flavour of the bread. Table 19.6 shows the faults in breads.

Table 19.6 Common faults in bread making

Fault	Reasons
Flaked crust also known as flying tops	If fermented dough is left uncovered in an atmosphere which is not saturated with moisture (80–85 per cent), water evaporates from the surface of the dough leaving the skin dry. This skin, once formed, is difficult to eliminate and when a skinny dough is knocked back, scaled, and moulded; the dry skin, breaks off and some which remain on the exterior will get folded into the dough and show as whitish coloured patches which are hard and knotty. When moulded dough pieces become skinned it will give an unsatisfactory bloom of the crust. Also there will be a number of bursts or 'flying tops'.
Lack of volume	Bread not fermented enough has a lack of volume. This fault can be said to be a direct effect of the insufficient ripening of the gluten. It has already been discussed in detail how fermentation affects the gluten structure and the final flavour of the bread. Over fermentation may also be a reason for lack of volume in bread. Longer fermentation time increases the acid production giving a very sour taste. This activity will weaken the gluten for lack of volume and large holes. It will also give a bad structure to the baked bread which will begin to crumble easily. Other reasons for lack of volume are: • breads not proved for required length of time; • due to improper mixing of the dough, the gluten does not develop, which is directly responsible for the volume of the bread; • too much salt in dough; • less yeast in the dough; • too high oven temperatures.
Uneven texture, showing large irregular holes	• When the dough is not fermented long enough the gluten will not reach its maximum extensibility. As the gluten is not fully extended, the loaf will be smaller in volume. Also, some of the smaller gluten strands will break down under the expansion pressure of the gas, creating irregular large sized holes in the baked product. • Use of over fermented dough • Under proved bread may show a crack on the base, thereby giving an irregular shape to the bread.
Lack of shine on the crust	• Under fermented bread. The sheen of the crumb depends upon the structure of the gluten formation, as kneading increases number of fine glossy cell surfaces to reflect the light. Greater the web like structure of the gluten greater will be the reflection of the light. • Use of over fermented dough.
Lack of flavour and aroma	• Use of under fermented bread. • Use of over fermented dough, which also gives a slightly acidic flavour to the bread.
Stales rapidly	• Bread not fermented for required time. • Not enough salt used in dough. • Over proved bread.

Contd

Table 19.6 (Contd)

Fault	Reasons
Crumbly bread	• Use of over fermented dough. • Use of over proofed dough. • Not enough fat used in dough.
Lack of colour on crust	• Use of over fermented dough. • Insufficient sugar in the dough.
Raw inside	• Under baking of the bread. • Baking done in high temperature, whereby the crust has got a colour but is doughy in the centre.
Rope	Rope is one of the main diseases that affect the bread. The spores of *bacillus mesentericus vulgatus*, the microorganisms, are responsible for the development of the rope. It is usually present in the flour itself. This is not apparent until the bread is some hours old. This develops in the form of patchiness and the crumb becomes sticky. At the same time a peculiar odour similar to that of pineapple develops. This will occur only when the spore is given suitable conditions for it to develop, increase, grow, and so produce an attack of the disease. These conditions include warmth, moistness, and a deficiency of acid in the medium. Spores cannot develop in an acid medium. Also as the spores require warm weather rather than cold, it becomes even more important to cool the bread quickly and completely. This can be prevented by using sour dough in the making of the bread, as sour dough will have sufficient acid content to prevent the formation of rope. This is also called the 'mature parent dough' method.

EQUIPMENT USED IN BREAD MAKING

Many large and small types of equipment are used in production of bread. It starts from sieving the flour to mixing in dough machines and proving and baking. Table 19.7 shows the most common equipment used in production of bread.

Table 19.7 Equipment used in bread making

Equipment	Usage	Photograph
Sieve	Drum sieve is mostly used to sieve the flour and the size of the mesh through which it will be sieved will depend upon the type of flour being used. The wholemeal flour will be sieved through a coarse mesh as one does not want to sieve the bran and other nutritious things away from the flour. Industrial flour sifters are also used in many hotels which produce breads on large scale.	
Weighing scale	Preferably a digital weighing scale is better, as the accuracy of the ingredients is very important.	
Baking trays	Often known as sheet pans, these can be of iron, or teflon coated for non stick.	

Contd

Table 19.7 (Contd)

Equipment	Usage	Photograph
Bread moulds	These are containers of various shapes and sizes. These are often sold by the volume they are intended for. So one can easily procure moulds of 1 lb, 2 lb, etc.	
Proving cabinets	Electric, gas, and pressure steam models of proving cabinets are available. In a proving cabinet, water is heated with an element. It maintains the temperature of 25°C and humidity of 90 per cent.	
Retarder prover	A retarder prover is an equipment that controls the rate of proving for the bread to be made. The baker can shape the bread in the evening, while the baking can be done at a later stage, i.e. as and when required. A retarder prover can be automatically adjusted so that the breads are ready to be baked at programmed timings. This equipment can freeze the shaped bread to stop the functions of yeast and then brings up the temperature whereby the dough comes back to a temperature whereby the dough is cold but not frozen. The machine then brings up the temperature to the required degree and humidity so that the bread can be proved.	
Dough mixers	Various kinds of dough mixers are used to knead the dough. Spiral dough mixers are used in which the dough hook and the bowl both move in opposite directions, so that the dough is automatically scraped while making the dough. They are also tuned to two speeds—slow and high—as most of the bread recipes call for mixing doughs at particular speeds for the optimum development of gluten.	
Dough divider	Equipment used for dividing the dough into equal sizes and portions. It is also used for shaping them into round balls. It is mostly used in large establishments.	
Ovens	Baking oven, with steam attachment for better baking.	
Dough scorers	A piece of equipment having a sharp surgical blade in the end, to score the breads at an angle before baking.	
Dough scrapers	Available in plastic or steel, they are used to scrape the doughs and also to cut the dough for scaling.	

Contd

Table 19.7 (Contd)

Equipment	Usage	Photograph
Wooden table top	Traditionally maple wood is used, as it is non porous and very hygienic. The wooden table top allows the bread dough to be at a required temperature as wood is a bad conductor of heat and hence does not take or give heat to the dough and also the dough rarely sticks to wooden surfaces. However, because of its non availability and expense, people also use granite or marble table tops. Metal is not preferred as it can react with the dough and discolour it.	
Bench brush	A large hard bristle brush to clean the table top and to brush away excess flour.	
Spray bottle	Used for spraying water onto the breads, if the ovens are not equipped with steam injections.	

BREADS OF THE WORLD

During researches on different civilizations, the ways of life of the people who comprised the civilizations have been evidently found. We have come to know about their food habits, their rituals, and their ways of earning livelihood.

Food has always been a very important part of these researches, as it was the sole most important objective every single individual wanted to strive for and this remains a universal truth, that they have and will always continue to do so. However, the invention of food started with the discovery of grains, specifically cereals, which opened a different world of food products in front of them. Before that, people led a nomadic life where they survived on natural food available in the form of fruits as well as hunted animals to satisfy their hunger.

Cereals have always been an integral part of human diet, specifically, bread formed an integral part of it in different parts of the world. Tables 19.8, 19.9, 19.10, 19.11, 19.12, 19.13, and 19.14 show some common breads from different parts of the world and the way they should be served.

Table 19.8 Breads of France

Name	Description	Serving Techniques	Photograph
Baguette, also known as French bread	This was invented around 1930, and slowly but surely gained popularity. It has a sharp contrast of a crisp crust with a wonderfully soft interior. It is highly influenced by the soft	This bread has a variety of uses in the current culinary world. • It can be used as loaf bread. • It can be sliced to make a *crostini*, and can also be used for making a garlic bread as well as *bruschettas*.	

Contd

Table 19.8   (Contd)

Name	Description	Serving Techniques	Photograph
	French flour, the long kneading and rising which is given to the bread by the skilful baker. Baguettes are always around two feet in length and will always have six scorings on the top.	• It can be used as a base bread for open sandwiches and canapés.	
Brioche	There are two concepts regarding brioche—one is that it was made by using brie, so it is called brioche, second is that it has been derived from the word 'brier' which means to pound. Traditionally made as a *brioche a tete*, or a Parisian brioche in which two balls of dough, a smaller on top of a larger one is placed and baked.	• It is used as a breakfast pastry, but also can be baked in the form of a loaf and sliced to make toasts or French toasts. • It can also be made as in a savoury one, generally with the inclusion of cheese.	
Croissant	It was first made in Hungary by a group of bakers who were working in the night and thus became aware of a potential attack on their country by the outsiders and raised the alarm. They were asked by the king to prepare something to make the occasion memorable, hence they came up with the idea of making a croissant which resembles the Hungarian flag. It has a wonderful contrast of a sharp, flaky texture and a soft buttery interior which just melts in the mouth. The dough is laminated with butter and rolled several times to bring perfection to the product.	It is widely acceptable as a breakfast pastry. It can also be used sliced lengthwise and made into a sandwich. Leftover croissants can be stuffed with almond cream and dipped in sugar syrup and baked with almond cream on top. It forms a great dessert. Many leftover breakfast rolls are cut into cubes and made into a bread and butter pudding by using cream, milk, eggs, and sugar.	
Epi	It translates as 'wheat ear' and is traditional harvest bread. Its main feature is its unusual shape. Primarily it is rolled like a baguette and then insertions are given with scissors at regular intervals and the dough is pointed to alternate directions from the insertions.	It is used as a very popular option in the lunch or dinner bread basket to have a good contrast of the shape and the texture. It can also be used to the same effect in the case of a buffet.	
Cereale	A torpedo-shaped loaf which is dusted with flour. This is made with eight different cereals which include wheat, corn, rye, millet, oats, malted wheat, sunflower and sesame seeds.	It is mostly used as a loaf bread in a buffet. It can also be sliced and made into toasts, or can also be used to make sandwiches, canapés and *crostini*.	

Table 19.9 Breads of Italy

Name	Description	Serving Techniques	Photograph
Ciabatta	Ciabatta means 'slipper', which is justified by the shape of the bread. The dough of the bread is very soft and becomes difficult to handle and hence lots of flour is used in the dough to give it a shape. This bread is traditionally topped with lots of flour before baking.	• Mini ciabatta can be served in the bread basket. • It can be used for making sandwiches. • It can be sliced to particular thickness and used for open faced sandwiches and *bruschettas*.	
Focaccia	This bread has a lot of olive oil, thus it requires to be kneaded for a longer time, which enhances the taste as well as the texture of the bread. The texture is contrasting with a hard and sharp crumb and a soft interior. There are different types of flavouring, such as tomatoes, olives, cheese, nuts, etc. which can be added to the bread. Prior to baking, spread lots of olive oil on the surface and dig your fingers into the dough to create lots of indentations. This helps the oil to seep in and give more flavour to the bread.	• Frequently used as a component of the bread basket, as well as used to give a contrast and variety of shapes in any bread display. • Also used for making *crostini*, *bruschettas*, and sandwiches.	
Biova	It is made with lard which gives a very different flavour. It has a cylindrical shape like a loaf but the outer crust is generally serrated to give an appearance of horns. The bread has a contrast of a hard crust and a soft interior which adds on to the excellent flavor.	This is a regional bread which is widely consumed in a particular part of the country.	
Panettone	Panettone is usually made in tall cylindrical earthenware pots and is often sold in them. This bread is enriched with eggs, milk, butter, and raisins.	It is very festive bread and is made during Christmas.	

Contd

Table 19.9 (Contd)

Name	Description	Serving Techniques	Photograph
Grissini	These are thicker versions of a breadstick, which are made by incorporating olive oil in the dough. These sticks are shaped and then baked in a moderately hot oven and dried till they get hard. It results in a very crunchy breadstick.	• It is mostly served along with soups. • It can also be rolled in poppy seeds or sesame seeds or coarsely grounded sea salt and then baked. • It can be rolled with *parma ham* and served as *antipasti*.	
Pagnotta	The loaves can be of different shapes and sizes. But mostly pagnotta means a standard wheat loaf, which may be made using wholemeal flour. It is a very country style bread.	Generally used as family bread which is made in bulk, cut and shared.	
Pugliese	This bread comes from the region Puglia, which is very famous for excellent olive oil and wheat. It is white, with a floury surface and a dense interior texture with soft crumb.	• It is used as a bread loaf. • It can be sliced and used individually.	

Table 19.10 Breads of Germany

Name	Description	Serving Techniques	Photograph
Kastenbrots	It means 'box bread' which is derived from the way the bread is enclosed in a tin and steamed. The bread is made by rye flour or wholemeal flour and is leavened by a natural sourdough. White flour is also added in the dough, as a malty taste, dark colour, and a chewy texture.	• It is best complemented with a German beer. • It can be served thinly sliced, butter or a soft cheese can be used as a spread. • It can also be served along with smoked salmon, herring, and sausage.	
Pumpernickel	It is normally made with rye flour but small amounts of wheat flour are seldom added to the dough to lighten the texture. The bread has a dark colour, dense texture, and a sour and earthy flavour.	It is generally consumed with smoked sausages, marinated fish, and cheese.	

Contd

Table 19.10 (Contd)

Name	Description	Serving Techniques	Photograph
Landbrot	It is made in different shapes and sizes, but all are generally made with rye flour and buttermilk, which gives a sweet and sour taste to the bread. The crust is sometimes coated with flour before baking and sometimes not. The bread is chewy and not dense in texture.	Because of the chewy texture and the sour after taste, the bread is particularly good when served along with soups or stews.	
Pretzel	It is always made from wheat flour, milk, and yeast. After proving they are dipped in a caustic soda liquid also known as lye, then sprinkled with sea salt and then baked till the crust acquires a dark golden colour. The crust is salty and the interior is contrastingly sweet.	This bread is traditionally made during October Beer Fest in Germany and is hanged on pretty wooden stands often known as pretzel stands. As the caustic soda in the bread creates a drying effect in the throat, a person would drink more beer.	
Stollen	It is a very rich bread which has raisins, sultanas, mixed candied peels of fruits, eggs, butter, and milk. Nuts are also frequently added. The bread is heavy with a dense texture and a moist interior and a brownish crust with an oblong shape and a tapered end.	This bread is particularly associated with Christmas, and is consumed all over during that festival. During Christmas, the bread is often stuffed with marzipan and baked. It is called *dresden stollen*. The *stollens* are dipped in clarified butter after baking and dusted with icing sugar.	
Kugelhopf	The bread is believed to have originated in Alsace in France, Austria, and Germany. The name is derived from the type of mould which is used for baking, called *kugel* which means 'ball'.	The bread is consumed when it is little old and stale and is paired well with wines from Alsace.	
Zopf	This is a plaited loaf which can trace its roots to the Jewish community, which started the tradition of plaiting the loaves. These plaited loaves are almost always enriched with yeast, made with white flour and plaited using three or more dough. In most of the cases the appearance of the bread is excellent, with a glossy finish achieved by egg wash and sometimes brushed with hot honey.	It is used as a good component in any bread display, where it adds on to the appearance. It is also used to a certain extent in lunch and dinner bread baskets.	

Table 19.11 Breads of the UK

Name	Description	Serving Techniques	Photograph
Cob	It means 'head', as the appearance resembles one. It is made with whole meal flour and baked as a plain loaf. It can also be made with white flour or granary.	• It is basic loaf for family consumption. • It is used as a loaf bread in the buffet.	
Bloomer	It has a crusty exterior, complemented with soft interior. Characteristic 5 or 6 scores on the crust. It can be made from refined, rye or multi grain flour.	It is mostly used as a loaf, but also can be sliced and eaten.	
Danish	It has a firm crust and fine interior. The loaf is cylindrical in shape and is dense in volume.	It is used as a regular loaf.	
Hovis	When people strived for more nutritious food, the idea of hovis evolved. It is actually the flour made with the inclusion of the wheat germ which is separated from the bread during milling. Now, it is a proprietary product, which has the name 'hovis' imprinted on the side of the loaf.	It is used as a regular loaf.	
Hot cross bun	The bread has a traditional cross mark, which is said to be evolved from the belief that it would ward off evil. This bread is flavoured with raisins, currents, nutmeg, and cinnamon and enriched with yeast and eggs.	It is a popular celebration bread, especially during Easter.	
English muffin	These are traditional flat discs of breads which are enjoyed in the winters. It is a very light bread and develops a thin skin like crust when cooked. They are not baked in oven, instead they are cooked on hot plates.	It can be slit, toasted, buttered and mostly preferred to be served along with eggs in a classical dish called egg Benedict.	
Pikelet	It originated in counties such as Leicestershire, Yorkshire, Derbyshire, and Lancashire. It is cooked on the grill like English muffins. The surfaces of the bread are dotted with minute holes.	It is used as base for sandwiches or simply eaten with butter.	

Contd

Table 19.11 (Contd)

Name	Description	Serving Techniques	Photograph
Stotie	It is a rustic bread which comes from the northern part of England. It used to be the last one to be baked at the end of the day and was believed to be ready only when it was dropped on the floor and seen to bounce back.	It is used as a loaf.	

Table 19.12 Breads from the Middle East

Name	Description	Serving Techniques	Photograph
Lavash	It is a flat bread which can be very large. It can be oval, round, or rectangular in shape. It can be made with or without the addition of yeast, but the unleavened variety is more popular. Generally it is baked in a low heat till it becomes very crisp and brittle. In the Middle East the oven in which it is baked is called *furunji*. It is also baked in a dome covered oven called *saroj*.	This bread gives a good variation of texture when used in bread baskets. It can also be flavoured with ingredients such as poppy seeds, black onion seeds, chilli flakes, and oregano. It can be brushed with olive oil and served warm as a snack along with dips.	
Khoubiz	It is very similar to the pizza base, the only difference is that it is made with wholemeal flour. But it can also be made with refined flour.	Commonly consumed in the Middle East as part of daily diet.	
Pita	Pita is very similar to Indian *phulka*; the only difference is that unlike phulka, pita is made in oven. Pita is baked on hot stones and is ready to eat as soon as it puffs up.	Pita is served as a snack with assorted dips. In the Middle East, pita is served with falafel fritters stuffed inside the pita pocket with a salad called *fatoush*.	

Table 19.13   Breads from the USA

Name	Description	Serving Techniques	Photograph
Burger bun	It is made with regular leavened dough, but it is normally kept bland in taste, as it will contain the fillings which will be flavourful. Normally they are sprinkled with poppy or sesame seeds.	One of the most consumed breads in the world in the form of different burgers. Hot dog buns are made in the same way, the only difference being that they are rod shaped.	
Banana bread	The bread is more like a cake. Ripe bananas are used for making it along with baking powder for the required rise. The interior has a very dense texture. It is generally baked in tins.	It is a very popular tea bread and can also be served in breakfast.	

Table 19.14   Jewish breads

Name	Description	Serving Techniques	Photograph
Bagel	The bread is made with both refined flour, wholemeal, and multigrain flour. It can include vegetable oil, margarine, or butter along with eggs and yeast. The bread is poached or steamed for a minute or two and then it is given an egg wash and baked, which helps in the formation of a glossy crust and dense interior. This bread resembles a doughnut with a hole in the centre.	Bagel is served commonly in breakfast, but it can also be served as sandwiches. Smoked salmon in mini bagels are very famous canapés.	
Challah	It is made with white flour and is leavened. It also has eggs, sultanas or raisins, which makes it rich. It has a deep brown crust and a sweet white crumb. It can be round or plaited with 3, 6, or 12 strands of dough.	This is a festive bread which is consumed on the Sabbath in Jewish families.	

CONCLUSION

This chapter dealt with basic principles of bread making. Making of bread is a combination of art and science. Care has to be taken while producing breads, as badly made bread can spoil the experience of great lavish meals. We spoke in detail regarding the various ingredients that are used in bread making and how each ingredient lends a particular texture and feel to the bread. We understood the effect of heat, mixing techniques, and other factors that affect the gluten, which in turn helps in the volume of the bread. We read about different types of flour in Chapter 18 and also read about gluten-free breads. In this chapter we discussed how water is the key element in the consistency of the dough. More liquid turns the dough into a batter, which is also essential in making certain breads. We learned about dried and fresh yeasts and understood the importance of yeast foods such as sugar. We could check the freshness and life of yeasts by doing certain tests. We also came to know about various temperatures suitable for the yeasts and saw that high temperatures like 50°C can kill the yeast completely. While we discussed ingredients, we also touched upon the comparison of the products having less and excessive quantity of a particular ingredient and how it affects the texture of the final product. We also read about bread improvers, commonly used in bread making in hotels around the world.

After discussing the ingredients and the way they react in different combinations and mixing techniques, we also talked about the principles behind bread making. We discussed the making of breads in ten crucial steps starting from collecting ingredients and weighing them to cooling of the bread. We learned about the various ways in which the dough is mixed. The dough can be mixed with hands, but machines provide better results, especially in cases where one would like the gluten to fully develop in dough. We studied about straight dough, salt delayed and ferment sponge methods of mixing dough, and how each method differs from the other. We studied about proving of breads and understood why it is done in that manner. We discussed the various shapes in which breads can be made. We discussed various kinds of glazes and toppings that are used to create a variety in bread baskets.

We also discussed the faults in breads and understood how an ingredient or even the process of bread making can result in poor quality bread. We also discussed the range of small and big equipment used in bread making.

We studied about the various breads from around the world and listed few which are the most common ones made in hotels in India. We also saw the basic production of these famous breads.

KEY TERMS

Antipasti Italian for starters.

Baba A dessert made from flour, eggs, milk, and butter, where a very soft dough is made and baked in special moulds. The bread is then soaked in rum-flavoured sugar syrup and decorated with fresh fruits. It is also called *savarin* in France.

Baguette Rod-shaped bread from France, often called French bread.

Biga Starter used in Italian breads.

Bloom The shine on top of the bread.

Bread rolls Small buns of different shapes eaten with meals.

Brie Soft French cheese.

Brioche Bread from France often eaten in breakfast.

Bruschettas Italian preparation, where a toasted, grilled, or pan-fried thick slice of bread is topped with various toppings. It is served as a snack or starter.

Compressed yeast Fresh yeast available in the form of a cake.

Croissant Breakfast pastry from France.

Crostini A thin slice of bread, topped with various toppings and can be toasted, pan-fried or grilled. It is served with salads or soups.

Crumb The netted structure of the bread, concealed in a crust.

Demoulding Removing the bread from the mould in which it was baked.

Dresden stollen *Stollen* with a stuffing of marzipan.

Dry yeast Yeast available in dry form. It looks like tiny round globules of poppy seeds.

Egg Benedict Classical preparation of poached eggs on English muffin.

Exothermic reaction A reaction whereby the heat is released.

Falafel   Dish from the Middle East. Fritters made from chickpeas.

Fatoush   Lebanese salad served along with pita and falafel.

Ferment   Mixture of yeasts with sugar and little flour to prepare a batter, which is used in dough making. It is also known as starter.

Flying tops   Cracked patches on the bread due to drying out of the skin of the dough.

Gassing power   The trapped air in the bread is often known as gas and the strength of the flour dough that holds the air inside is called gassing power of dough flour.

Improver   A chemical or natural ingredient used to improve the quality of the bread. It is available commercially in the form of powder.

Jalebi   A dessert from India, where a fermented batter is piped in circles in hot fat and then soaked in sugar syrup.

Khameer   Indian term for fermented dough.

Knock back   The dough is punched down after the proving. This is also known as degassing the dough.

Lactose   Natural sugar present in milk.

Levain   Starter used in French breads.

Lye   Mix of caustic soda and water.

Maltose   Natural sugar present in the flour.

Marzipan   A dough made from almond flour and sugar which is widely used in confectionery.

No time dough   Bread dough made with high amounts of yeast. This yields bread with good volume, but the fermented flavours are missing.

Pancake   A disc-shaped breakfast preparation, where a sweet batter is poured onto hot plates.

Panning   Moulding the dough and putting in the bread mould or pan.

Parma ham   An air dried ham from Italy.

Phulka   Indian flat bread that is served puffed up.

Proving chamber   An equipment used for proving of bread.

Proving   The dough left to ferment so that it is double in size.

Rope   A disease caused by the bacteria present in the flour. It is commonly found in bread production.

Scoring   Giving a slit with a sharp blade on the bread for decorative purpose.

Sour dough   Dough left overnight or for many days to ferment naturally. It is often known as mature parent dough.

Sponge   Same as ferment, but the only difference is that it contains more flour and is more dough like.

Straight dough   Method of mixing dough where all the ingredients are combined and kneaded to a dough.

Waffles   Crisp pancake made in special equipment called waffle machine. It is had for breakfast and is very popular in America.

Windscreen test   Dough stretched to check if the gluten is developed.

CONCEPT REVIEW QUESTIONS

1. Bread making is a combination of art and science. Justify the statement.
2. Apart from the recipe, what are the other factors that influence the making of good bread?
3. Define a sour dough starter. What are its uses?
4. What is rope and how does sour dough starter help to prevent the same?
5. Why should the oven be used to its optimum capacity with regard to humidity?
6. What points should be kept in mind with regard to baking of bread in an oven?
7. What is the role of water in bread making?
8. What care should be taken while making bread with hard water?
9. Why is ice used in certain bread recipes?
10. How many types of yeast are there and how does one test the freshness of yeast?
11. What range of temperature kills the yeast?
12. Analyse the difference between less salt and excessive salt in the bread.
13. Compare the results of less sugar and excessive sugar in the bread dough.
14. Write at least five impacts of sugar on the bread dough.
15. What is the effect of fats and oils on the bread dough?
16. What are bread improvers and what is their role?
17. List the 10 steps of bread making in the particular order.
18. List the jobs that you would take care of in the first principle of bread making.
19. What are the various ways of mixing bread?

20. How does salt delayed process help in giving better flavour to the bread?
21. What is no time dough method and what are its advantages and disadvantages?
22. What is a windscreen test and why is it done?
23. Why is knock backing done after the first proving of the bread?
24. Define the process of proving. Why is it essential?
25. List the sizes of bread rolls, burger buns, French baguette, and bread loaves.
26. What is panning?
27. Write down the salient features of retarder prover.
28. What are the three stages in baking?
29. List at least five glazes and the effects of the same.
30. List the reasons of flaked crust in a baked bread.
31. What are the reasons if the baked bread lacks volume?
32. A bread has uneven texture and large holes in it. What could be the fault?
33. When would bread stale rapidly?
34. Why is a wooden table top preferred to metal ones?
35. List at least three French breads.
36. Write the name and uses of at least five Italian breads.
37. Which is the dark rye bread from Germany?
38. What is the significance of pretzels during the beer fest in Germany?
39. What is unique about the bagels?
40. Name at least three breads from the Middle East.

PROJECT WORK

1. In groups, make dough using at least five different kinds of flour. Check the windscreen of each dough and wash the dough under running water until it washes no further and the chewy structure is obtained. This is the gluten. Compare the percentage of the gluten in each flour and record your observation. Now make bread using the same flours and compare the results in volume of each bread. Record your observation and now using the same flours make the recipe again, this time put the gluten obtained from the first test into the recipe of bread. Make sure to add the gluten obtained from the respective flours into the same type of flours. Bake the breads and compare with the ones baked the first time. Compare your results and see how gluten percentage in dough affects the volume of the bread.
2. In groups, try out the recipes as provided in the ORC for the same bread and mix the dough by different mixing methods such as straight dough, sponge, ferment, salt delayed, and quick dough. Compare the results.
3. Divide yourselves into seven groups. Six of the groups shall make at least one fault in the dough and one of the seven groups shall make the perfect bread. Compare the results and record the observations.
4. Visit various hotels and restaurants and do a study of various kinds of breads and breakfast rolls served in those outlets.
5. Research about various other breads made around the world and make the breads for tasting and get evaluated by experts.

CHAPTER 20

BASIC SPONGES AND CAKES

Learning Objectives

After reading this chapter, you should be able to:
- understand the different principles of sponge and cake making
- know various steps involved in production of sponge
- prepare various kinds of sponges and cakes from the recipes provided in the online resources
- understand the various techniques used in preparing pastry goods and the importance of each with regard to the texture of the product
- know the role played by different ingredients in sponge making
- know about various types of international cakes and their usage and serving techniques
- appreciate the usage of equipment used in making sponges and cakes

INTRODUCTION

In the previous chapter we read in detail about the techniques for bread making. In this chapter we shall discuss the techniques for baking sponges and cakes. Just as we cannot form sentences if we do not know the words and we cannot make words without knowing the alphabets, we cannot cook food without the knowledge of ingredients to be used.

It is very important to learn about the various pastry techniques before we go on to create basic sponges and cakes. The previous chapter on breads dealt with various pastry techniques such as kneading, proving, knock back, fermenting, baking, etc. These are very typical to bread making; the pastry techniques have a much larger area of usage.

Sponge is a light and airy cake which contains three basic ingredients namely eggs, sugar, and flour and is leavened solely by aeration which occurs by beating or whisking the eggs. Aeration is a term which means the incorporation of air in a batter which helps to lighten and increase it in volume.

A basic sponge cake is made by whisking the eggs and sugar until thick and fluffy. Sifted flour is then carefully folded into the eggs to make a sponge which is poured into a greased and lined mould and baked. A basic sponge does not contain any fat and therefore it is important to handle the batter with utmost care. This type of sponge is generally termed as a fatless sponge and is mainly used to make layered cakes and gateaux. Sponge cakes that are suitable for vegan, lacto-ovo intolerant, and low cholesterol diets can also be made. Most often this is done by using plant-based milk instead of dairy such as rice or soya and vegetable oil instead of eggs, although many alternatives to eggs are used such as flax seeds, bananas, and commercial eggless cake powder. Sponges are high on both fat and sugar. Sponges are very versatile and can be used for a number of purposes from the confectioner's point of view. They can be presented in many forms, such as sheets for decorative purposes to be used in making fancy or so-called designer cakes and gateaux. Cakes are the richest and the sweetest of all bakery products. Baking is a skill which requires a lot of precision in terms of measurement of ingredients and the quality of the ingredients being used. Sponge cakes are so called because of their texture which resembles a sponge. Well-distributed holes are a result of air trapped in eggs and finally when the cake bakes, the gluten in the flour helps the cake to retain its shape and the air escapes leaving those holes behind. Many sponges can be served on their own but many are used as basis for classical cakes and other desserts. In this chapter, first we shall deal with all the techniques used in pastry, some of which are also used in sponge making.

PASTRY TECHNIQUES AND PRINCIPLES

We read in the previous chapter that the first step in making pastry is to collect the entire *mise en place* and get the work counter organized so that the chefs do not have to run around too much. As a pastry work deals with eggs, the chance of contamination is very high and so it is important to have a well-sanitized area for pastry work. But before making sponges it is imperative to understand the various techniques involved in making pastry and various other products. Most of these techniques are basically followed for aeration of the product to make it light and soft. The eating qualities and appearance of baked goods depend very much on the lightness of the products. To achieve aeration, we use leavening agents. They may be physical, chemical, or biological and maybe used individually or in combination to incorporate steam, gas, or air cells into the mixtures.

Physical

Air and steam are incorporated into mixtures in a specific way, involving physical or mechanical action. Air is introduced into the mixture by a variety of methods, such as sifting of dry ingredients, creaming of fats and sugar, and whisking of eggs and sugar.

Steam is a leavening agent found in all baked products. For example, in puff paste, the laminated fat incorporated into the paste in layers melts during baking, producing some steam which lifts the layers.

Chemical

Baking powder, baking soda, ammonia, etc. are the chemical leavening agents. When moistened and heated they react to produce carbon dioxide. To test whether baking powder is still active, stir some powder in warm water. If it gives off bubbles freely, it is active.

Biological

Yeast is a biological leavening agent. Refer to Chapter 19 for more information on yeast.

In Chapter 17 we discussed various types of raising agents—both biological and chemical. Let us now discuss various pastry techniques which are used to provide physical aeration to the pastry products.

SIFTING

Sifting is associated with flours. Flour is put into the drum sieve and shaken so that the flour sifts from the small meshes. This is done to incorporate air into the flour and also to get rid of any physical impurities present in it. Sifting also helps to mix certain ingredients in the powder form. For example, mixing of raising agents with flour or sifting cocoa powder and flour for chocolate-flavoured sponge.

CREAMING

Creaming is a method of mixing foods with high fat content in order to incorporate air and make the mixture lighter. Mixing may be either mechanical or manual. There are three attachments of dough mixers—balloon whisk, flat paddle, and dough hook. The creaming is always done using a flat paddle (refer to Table 20.2). Creaming refers to beating the butter and sugar together to incorporate air and to make it light and fluffy. The type of sugar used will depend upon the usage of the final product. For making butter cream which will be eventually used for icing cakes, one would use icing sugar and the icing used for sponge will have castor sugar. Certain factors which need to be kept in mind while creaming are as follows:

- Care should be taken while choosing the equipment for creaming. One should always use stainless steel as using aluminium deteriorates the colour of the butter and sugar making it grey.
- Take care not to over-cream as the final product might not be able to hold the volume resulting in the collapse of the cake.
- Fats used must be soft and not oily.
- Use a bowl large enough for rapid movement of the paddle as creaming is always done at high speed.

WHISKING

Whisking is a method which is very much like creaming. It uses fast movement to incorporate maximum air into liquid ingredients, achieving foam. Like creaming, whisking can also be done mechanically or manually in a suitable bowl. Once the mixture has started to foam, whisking needs to be continued until the desired stage has been reached. Usually this stage is referred to as 'ribbon stage' as the emulsion when lifted and dropped falls like ribbons. Most of the basic sponges are made by this method. When whisking egg whites, make sure there is no egg yolk present, since yolks contain fat, and the presence of fat hinders the formation of meringue. The equipment one uses must also be fat-free. Egg whites at room temperature will whisk to a foam quicker and produce a greater volume than those straight from the refrigerator. If an acid such as lemon juice is added, it will help stabilize the foam.

Egg white foam starts to break down when it is whisked past the medium-firm peak stage. If it is whisked too much, and then used in cooking, for example, to make poached meringue quenelles the whole structure will collapse because the bubbles have expanded too much.

For best results when whisking dairy cream, use old cream from the refrigerator, and a cold bowl.

CHEF'S TIP

If you put dairy cream into a piping bag, do not whip it too much, or the extra manipulation and squeezing in the bag can turn the cream to butter.

RUBBING IN

The rubbing in method is generally used for making short or sweet pastry. The product containing fat and flour is said to be short when it snaps off or crumbles when pressed. First, cut the fat into small pieces. Then, using your fingertips rub the pieces of fat into the flour, all the time lifting the ingredients and allowing them to fall back into the bowl. The fat will reduce to small particles of the size of bread-crumbs, each with its own coating of flour.

The purpose of rubbing in is to make a lighter pastry. During baking, the moisture from the fat becomes steam, which makes the pastry expand. For best results, all the ingredients should be cold, with liquid ingredients added in all at once to the flour and fat mixture. Do not over mix, as this will toughen the pastry; mix just enough to bind all the ingredients. Cover and rest the pastry in the refrigerator before you use it. The other method to make a short pastry is to cut the butter into 1 cm dices and mix with flour. Now roll both the things with a rolling pin until the butter forms flakes. Now combine the dough very lightly and add the required amount of cold water to make the dough into a light dough. Over kneading will result in stretchy dough, which will hamper the shortening effect of the flour. Rubbing in is done for the following reasons.

- Fat and flour are rubbed together.
- The aim is to reduce the fat to breadcrumb-size particles.
- The fat particles melt during baking, giving off steam which makes the pastry expand and rise.

FOLDING IN

This is a method of combining other ingredients into the aerated mixture so that there is little reduction in lightness or volume. This is achieved by turning the mixture over gently, using a large spoon or the hand while adding the other ingredients gradually. The mixture must be lifted and folded over gently. Make sure you reach to the bottom of the bowl. Take care not to over mix the mixture. This is also known as cutting and folding method. When folding in mixtures with different consistencies (for example, adding whisked egg white to a cake batter), soften the heavier mixture first by adding a portion of the softer mixture, then fold in the rest of the soft mixture. Dry ingredients should always be sieved and added gradually, ensuring they are dispersed evenly throughout the mixture. This technique is very crucial for making a basic fatless sponge. Certain factors to be kept in mind while folding in are as follows:

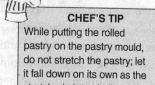

CHEF'S TIP
While putting the rolled pastry on the pastry mould, do not stretch the pastry; let it fall down on its own as the stretched gluten in the pastry will try to come back, thereby creating shrinkage in the pastry.

- Dry ingredients are added to a creamed or whisked mixture.
- The ingredients should be added and turned gently with open hands and fingers so as to disturb the air bubbles as little as possible.
- One can always fold while another person can help him/her to add the sifted dry ingredients in a continuous flow.
- Proficient chefs can also mix in the dough mixer itself, but should keep in mind that the machine is on the lowest speed possible and the batter does not over mix.

DOCKING

This means making small holes in pastry goods to allow the steam to escape during the baking process. Docking prevents the pastry from lifting and going out of shape. One can use a docker (refer to Video 21.1) or a fork for docking.

BLIND BAKING

This is the process of baking empty pastry shells which would eventually be used for any fillings to make pies or tarts. Pastry moulds are lined with a short crust pastry or sweet paste and a greaseproof paper is placed on top of the pastry and filled with dried beans or rice, etc. This is to prevent the pastry from either rising, developing bubbles, or shrinking during baking. The pastry may be baked half done or completely done according to the recipe requirements. If you want to bake the pastry completely through, remove the paper and beans when the pastry edges are set and lightly browned so that the base can also get coloured.

PINNING OR ROLLING

Rolling dough or pastry to the thickness and size required is a very important process. You can use a rolling pin or a pastry brake—a roller-type machine turned either by hand or electric motor. It is commonly known as dough sheeter. Best results are achieved when pastry is rolled out on a smooth, cool surface. Avoid over handling the pastry by preparing the pastry into the shape you want before rolling and rolling out only the amount of pastry you want for immediate use. To roll out dough or pastry, start from the centre. First roll upwards, then downwards, through the centre to the bottom. Rotate the dough a quarter turn and roll again. Repeat the process, rolling and turning, until you reach the required thickness. You should occasionally lightly dust with flour or cornflour to prevent the dough or pastry from sticking. What you use for dusting will depend on the type of dough or pastry you are rolling, and what it will be used for. During rolling, take care to maintain an even thickness. Maintain an even pressure on the rolling pin. Adjust your pressure to suit the type of dough or pastry being rolled.

Never stretch the pastry, as this will cause it to shrink and lose its shape during cooking.

PIPING

This is the process of forcing and piping various mixtures through a piping bag. The bag may be fitted with a piping nozzle to achieve certain decorative effects. Before you fill a bag, fit a nozzle if you want one. If the mixture you will be using has runny consistency, twist the bottom of the bag just above the nozzle, and push it firmly into the nozzle. This will prevent the mixture from running out while you are filling the bag. Also while filling the bag, fold the top of the bag inside out to form a cuff, which will prevent the mixture from spilling onto the outside of the bag. With one hand inside the cuff, fill the bag using an appropriate spoon or scraper. Avoid trapping air in the mixture. Do not overfill the bag, or you will be in trouble when you start squeezing to do the piping. Twist the top of the bag to close it, making sure it is tightly filled with the mixture.

CHEF'S TIP

Remember that if you are piping with cream, it is best not to use large quantities, because the manipulation and forcing through a nozzle can turn it to butter.

For writing with a piping bag, make a small piping bag with a butter paper (refer to Video 20.1) and make sure the piping mixture is free of any lumps.

Hold the bag so that it lies in the palm of your hand, holding the nozzle closed with your index finger and thumb. Apply an even pressure with your remaining fingers, to force mixture from the bag. Use your other hand only as a guide and allow the mixture to drop onto the surface you are decorating.

LAMINATING

Laminating is a word used to describe the incorporation of fat between the layers of dough and this is the base for preparing croissants and puff pastry. The chilled fat block is placed between the rolled dough and is encased within the dough. It is now rolled out to a long rectangular shape and folded like a handkerchief. It is chilled and rolled out again. We shall discuss these methods in detail, when we discuss laminated pastries in Chapter 22.

Thus, laminating is always to do with fat or oil.

ICING

Icing is the term used for cakes and pastries only. The basic sponge is sliced horizontally and then layered with flavoured sugar syrup and desired filling. The cake is then covered with the same filling or whipped cream and decorated and garnished with fruits, etc. This whole process is known as icing and various types of fillings which can be used for icing the cakes are discussed in the next chapter. Cakes are iced on a cake turn table also known as 'Lazy Suzanne' and a flat palette knife is used for spreading the icing on the cake (refer to Video 20.2 on blackforest cake to understand the technique).

INGREDIENTS USED IN SPONGE MAKING

'Cake' in trade terms refers to a cake made from flour, sugar, fat, and eggs. It may also contain milk, baking powder, fruit, nuts, etc. Cake is usually heavier than sponge. However, 'cake' has a broader interpretation which includes 'gateau' (French) and 'torte' (German). These are made of layers of sponge, Genoese, meringue, cream, and pastry; the name given to the cake usually refers to the filling and the main flavour used, for example, lemon cream gateau.

The foundation of a good cake begins with the base. Every effort and care needs to be taken in preparing the base, as there is no advantage in decorating poor quality bases in an attempt to make them look better. The purpose of decorating a cake is to make it more appealing to the eye and to the palate. The decoration of a cake is wholly satisfying because it enables a chef to express himself/herself in a creative manner. In this chapter we will consider only sponges and genoise bases, which can be used to prepare and decorate a variety of basic cakes and 'petit four glaces'. Petit four glaces, is a generic title which covers all small bite size pastries and cakes that are 'iced'. Petit four glaces are served with coffee after a meal, particularly for special functions, buffets, etc. You will have the option of making petit four glaces with a genoise base or a rich marzipan sponge base.

The main ingredients you require to prepare cakes are eggs, flour, fat, baking powder, and emulsifiers.

Eggs

When using eggs in cake preparation, you should warm the eggs either by placing the eggs in hot water or by warming them along with the weighed sugar with a gentle heat over a bain-marie. The reason for doing this is to produce strong whisked foam which has the stability to withstand the additional mixing of other ingredients. If the foam loses its incorporated air, the result will be a heavier cake.

Warming the eggs will also prevent the curdling of mixtures when fat, sugar, and eggs are creamed together. Eggs can be separated and the whites whisked separately to increase the lightness of the cake.

Sugar

When preparing a sponge batter, use castor sugar because it readily dissolves in the batter.

Flour

All cakes of a light nature need a weaker soft flour (one with low gluten) to obtain a more crumbly result. If this type of flour is not available, all-purpose flour can be used with the addition of some cornflour to make it softer. Usually, 20 per cent of the cornflour or cocoa powder is replaced with the amount of flour.

Baking Powder

This is used to aerate the cake. Make sure that it is weighed correctly and sieved several times with the flour to ensure that the cake is not over or under aerated and the distribution is even throughout. Cake mixtures should be cooked immediately or the gases emitted from the baking powder will start to develop and break out of the batter.

Fat

The use of butter is recommended. For creaming, butter should be soft, not oily and the amount of fat that is added to a sponge batter will determine its texture. The more the fat, the heavier will be the sponge.

Emulsifier

Commercially prepared stabilizers are used in sponge batters to help keep the batter from breaking down, thus forming a perfect, light emulsion. It is available in powder forms or even gel forms. These types of cake batters have a different recipe as it involves putting everything together into a mixing bowl along with warm water and whisking the entire thing to a stable emulsion, which can be held for a long duration of time. Refer to recipe of vanilla or chocolate sheet under the pastry section (basic sponges) in the accompanying ORC.

Flavouring Ingredients

Many other types of ingredients can be added to the sponge mixes, depending upon the usage that the sponge will be put to. For example, if the sponge is being made for chocolate cake then it is advisable to substitute 20 per cent of the flour with cocoa powder to give a dark rich chocolate-flavoured sponge. For a coffee flavoured cake, a paste of coffee with water can be used; for honey and almond cake one could use flaked almonds and honey and so on.

PRINCIPLES BEHIND MAKING OF SPONGE

The aim of mixing cake batters is to combine all the ingredients into a smooth uniform, stable emulsion, i.e., water in fat. It may seem very easy but the process requires a thorough understanding of the principles involved in making a sponge. For example, sometimes an experienced baker becomes too impatient to get on with his/her tasks and may just increase the speed on the mixer while creaming fat and sugar, thinking that high speed will do the job faster only to later realize that due to the high speed no air cells were formed, which resulted in poor texture of the product. So it is very important to be extremely careful with combining the ingredients which is the first crucial principle of sponge making.

Combining of Ingredients

Careful attention has to be given to the mixing process. The sponge mixture has to form a uniform emulsion, so that the water is held in suspension surrounded by fat and other ingredients in the batter. A batter can curdle if the mixture changes to fat in water, with small particles of fat surrounded by water. Curdling can occur due to the following factors.

- The quantity of butter should be measured accurately in the given recipe, so as the formula has a balance of both fat and water. Whole eggs, if ever used, will help the batter hold the liquids in the mixture.
- Ingredients should not be too cold; a temperature of 21°C will enable an emulsion to form best.
- Mixing of ingredients in the first step too quickly will not be able to incorporate a good quantity of air into the batter.
- Adding of liquids too quickly may also cause the batter to curdle hence, they have to be added in steps and a little at a time.
- Preparing of the moulds prior to baking sponges is of utmost importance and is an art in itself. Many chefs lightly grease the cake tins with oil and fill up the tin with cake flour and pour out excess by tapping it slightly. This ensures a thin film of flour on the cake tin and prevents the batter from sticking to the mould. The other method is to line the cake tin with grease roof parchment paper.

Formation of Air Cells

Formation of air cells in a batter is of great importance since they give the sponge its texture and also act as a leavening agent. The air trapped in the batter expands when subjected to heat and this acts as a natural leavener giving the sponge a good raise even if no chemical agent is used.

Correct temperature of ingredients and a suitable mixing are vital for the formation of good air cells in the batter. In the case of foam cakes the egg and sugar mixture should be slightly warmed to approximately 38°C. Whipping should be done at high speed first, then at moderate speed to retain the formation of air cells.

CHEF'S TIP
When liquids such as eggs are added to creamed fats, the mixture may curdle. This is caused by the temperature of the liquids, or the liquid being added too quickly. If this happens, warm the outside of the bowl a little by placing it in hot water. Or you can add a small quantity of your measured flour that should smooth the mixture and return it to an even texture.

Texture

Another important principle in sponge making is the texture of the sponge. The development of gluten in the batter is responsible for the texture of the end product. A very little amount of gluten is required in cake making; hence weak flour will be a better choice. In some sponge recipes, corn starch replaces some of the flour requirement, thereby reducing the gluten content even more. On the other hand certain rich fruit cakes require more gluten to hold the structure and the fruits in the cake.

Since the amount of mixing affects the gluten, the flour in the recipe is always added towards the end of the mixing process after all the ingredients have been added, thus ensuring that there is very little development of gluten. If the batter is mixed for too long after the addition of flour, then the cake is likely to be tough.

Formula and Balance

Ingredients and quantities can be changed only to a certain extent in a given recipe. A formula in which the ingredients fall within these limits is said to be in balance.

For the purpose of balancing ingredients can be classified into the following four functions.

Tougheners   They provide structure, for example, flour, eggs (white and yolks).

Tenderizers   They provide softness or shortening of protein fibres, for example, sugar, butter, and chemical leavener.

Driers   These are the ingredients that absorb moisture, for example, flours and starches, cocoa powder, and milk solids (powder), etc. A sponge may require formula balancing if even after following all the steps, it has not come out correctly.

Moisteners   They provide moisture to the batter, for example, water, milk, liquid sugar, eggs, etc.

The formula would be balanced if tougheners equal tenderizers and driers equal moisteners. In other words a balance has to be maintained between the given ingredients. Egg yolk contains fat which is a tenderizer and at the same time contains protein which is a toughener. A common practice in balancing a formula is to decide the flour and sugar ratio, then balance the rest of the ingredients against this combination as follows:

- If liquid is increased, reduce the eggs and the shortening.
- If eggs are increased, increase the shortening.
- If extra milk powder is added as enrichment, add an equal weight of water.
- If large quantities of moist ingredients such as apple sauce, mashed bananas are added, then the batter may require an increase in the quantity of flour and eggs.

BAKING AND COOLING OF SPONGES

In the previous chapter we discussed the importance of the salient features of baking while baking breads and the same should be followed for baking cakes. In addition, the following points would be useful.

- Preheat the oven. The sponge needs to be given an instant shock of heat as this will help create the oven spring. Cool ovens will result in dry and crumbly sponges.
- Make sure that the oven shelves are even. The cake batter is very soft and if the shelves are uneven, the batter will tend to flow with the slant, thereby resulting in a thick and thin cake. While the thick will cook, the thin might burn or become crisp.
- Do not let pans, tin trays, etc. touch each other. There should be an even circulation of air, as it creates humidity, which helps to bake the products with uniform colour.
- Bake at the correct temperature. Baking at low temperatures will give dry and pale cakes, and baking at high temperature will colour the cake too fast resulting in burning it.
- Do not open the oven door and disturb the sponge, until it has finished rising and is partially browned. Opening of the door of an oven might result in collapse of the sponge, as when the oven is opened, the steam formed in the oven tends to come out of the oven with a force thereby creating a vacuum in it, which results in the collapse of volume.

Test for Doneness

Doneness can be tested by the following.

- The sponge will be springy, the centre of the cake on the top will spring back lightly.
- A cake tester or a wooden skewer/toothpick when inserted into the centre of the cake should come out clean.

Cooling and Removing from the Pan

CHEF'S TIP
If it is difficult to remove the sponge sheet from the paper, then brush water on the back of the paper and let it stand for a couple of minutes. The sheet of paper can be easily peeled off now.

The following points should be remembered for cooling and removing the sponge cakes from the pan.

- Cool the sponge cakes for 15 minutes in the pans and then turn out when slightly warm. If removed from the moulds when just baked they will be too hot and break.

- Place the sponge onto cooling racks for proper circulation of air. If they are not cooled on the cooling racks (refer to Table 20.2), the moisture will accumulate in the base resulting in a soggy cake.

BASIC SPONGES

Different types of sponges are made by using various principles as listed above. Each sponge is used for a different purpose. A plain vanilla sponge can be used for making fruit based cakes, such as pineapple, mango, kiwi, etc., whereas chocolate sponge can be used for making chocolate cakes. Sponges can also be used as bases for mousse cakes and pastries, or simply crumbled and mixed with fruits such as apples to make apple pie. The sponge in this case helps to absorb the juices coming out of the apple and results in a crisp apple pie. Table 20.1 discusses various types of basic sponge and its uses.

Table 20.1   Basic sponges and their uses

Name	Description	Use	Method
Genoise	Genoise is named after its place of origin, Genoa in Italy. It belongs to the family of light and airy sponge cakes. While the technique for making this batter is similar to that of a basic sponge cake it does differ in that it contains melted unsalted butter. The adding of melted butter produces tenderer and flavour-ful sponge cake. The chocolate genoise is made by adding cocoa powder. Genoise is the basic sponge cake and can be made by omitting butter, in that case it will be known as fatless sponge.	Genoise can have a variety of fillings, such as gateau mocha—with coffee butter cream from France, wiener orange torte—filled with curacao butter cream from Austria, dobos torte—chocolate butter cream and finished with caramel sugar from Hungary.	Whole eggs and sugar are whisked together to a ribbon stage. This is done over a double boiler to get a greater volume. When the temperature reaches 32°C, remove from bain-marie and continue whisking until temperature reaches 24°C. Flour is carefully folded in melted butter; it is folded in by cut and fold method. The sponge is baked at once in order to retain as much air in the sponge as possible.

Contd

Table 20.1 (Contd)

Name	Description	Use	Method
Chiffon cake	Chiffon cake is a very light and airy sponge that has characteristics of both butter cake and a sponge. Its texture of a sponge makes it much desirable to a sweet tooth. Though unlike butter cakes, chiffon cakes use oil and whipped egg whites are used to provide the aeration, along with baking powder that gives it a sponge based texture. It is fairly quick to make this cake as a batter is made combining oil, raising agent, and flour and whipped egg whites are folded in.	The high oil content in this cake does not allow the cake to set firm, as oil remains liquid at room temperatures therefore the cake is more moist than the butter cakes. This makes them ideal choice for cakes and gateaux which need chilling or freezing. Chiffon cakes are also healthier choice as they do not have saturated fats and rely solely on oil. The disadvantage here is that the cake lacks the real buttery flavour that people often look, when it comes to cakes. This is the reason why these cakes are teamed up with lots of fresh fruits and served with accompanying sauces such as chocolate or fruit-based sauces.	Sift all the dry ingredients together. Whisk the egg yolks with sugar until light and creamy. Add oil and fold in the flour. Make a meringue with egg whites and remaining sugar and carefully fold in the flour and oil mixture. Traditionally chiffon cakes are baked in tube shaped round mould.
Angel food cake	Angel food cake is airy and because of its lightness and its pure white colour it is said to be the 'food of the angels'. This cake has no egg yolks, fat, or artificial leavener therefore it relies totally on stiffly beaten egg whites for leavening. Its sole ingredients are egg whites, cream of tartar, sugar, flour, salt, and flavouring such as fruit extracts and essences. Angel food cake has the highest sugar content of all the sponge cakes and this added sugar is needed to support and stabilize the whipped egg whites. As the egg whites give the cake its volume and structure, care must be taken when adding them to the dry ingredients so that they do not collapse.	Angel food cake is served whole with fruit-based sauces or sugar glaze pored over on the top. Angel food should preferably be sliced with a serrated cake knife as a normal knife would compress the cake as it is very soft.	Egg whites are stiffly beaten with sugar. Cream of tartar is sifted with flour and carefully folded in the mixture. This is traditionally baked in tube shaped ring mould.

Contd

Table 20.1 (Contd)

Name	Description	Use	Method
Victoria sponge	Victoria sponge is named after Queen Victoria, who popularized this cake in her afternoon teas. Unlike basic sponge, this sponge is made by creaming method and is usually sandwiched with jam and whipped cream. The top is not iced and can be dusted with icing sugar.	Used in afternoon tea as a dry cake. The cake is sliced into two halves and sandwiched with jam and whipped cream. It is not iced but can be topped with sifted icing sugar.	Cream butter and sugar until light and fluffy. Add eggs one by one until a stable emulsion is formed. Sift the flour with raising agent and carefully fold in the mixture to form a smooth batter. Bake in a greased mould line with paper. Bake at 220°C for initial 10 minutes and then bake at 180°C for 30 minutes.
Devil's food cake	This is a rich dark chocolate cake and is made by creaming method which is very similar to Victorian sponge. The only difference is that instead of melted chocolate, this sponge relies upon cocoa powder, which makes it more profound and rich in chocolate flavour. The cake also uses hot boiling water as the main liquid to bind the flour and cocoa powder as it contains less eggs.	This cake is cut into two, layered with rum flavoured sugar syrup and with dark chocolate truffle (refer to chapter on filling).	Cream butter and sugar. Add eggs and sifted flour and cocoa powder. Add hot water and mix. Bake in a round cake tin at around 180°C for 40 minutes.
Swiss roll sponge	This is a very soft sponge and it is baked in thin sheets and at high temperatures as low or medium heat will bake it into a biscuit. One could add flavouring depending upon the usage of this sponge. This sponge is also called *roulade*.	It is used for making Swiss rolls. Flavoured cream is spread on the sponge sheet and rolled into a tube.	Separate the eggs and whisk the egg yolks with half of the sugar in the recipe. Whisk the egg whites with the remaining sugar to form stiff peaks. Fold in flour into the egg yolk mix and carefully fold in egg whites. You can also add melted butter in the last stage to add richness to this sponge.

Contd

Table 20.1 (Contd)

Name	Description	Use	Method
Madeira sponge	This is slightly different from Victorian sponge. Here the butter and sugar are creamed with egg yolks and the egg whites are stiffly beaten and folded in. The resulting cake has a good volume and a very spongy texture.	It can be served plain as tea cake or some candied and dry fruits can be added to the same.	Cream butter and some sugar to form a light and fluffy mixture. Add egg yolks to the mixture and meanwhile mix the egg whites and remaining sugar into a stiff meringue. Fold in the flour into the butter and sugar mix and carefully fold in the meringue.
Jaconde sponge	This is a decorative sponge and is mostly used for lining the sides of the cakes. This is quite a modern invention and is made in two stages. The stage one is to make a deco paste. This is made by combining egg whites, equal parts. It can be coloured to give designer effects. Spread on a silicon baking mat as it sticks to other surfaces. Freeze the deco paste and then spread the special sponge mix onto the design in a layer and bake at high temperatures. The resulting sponge gets the design from deco paste printed on the sponge.	Various kinds of templates and stencils are available to imprint the designs onto the sponge, which are used as a side collar for designer effect.	Make deco paste (see recipe in the ORC) and spread on a silpat and freeze. Whisk egg yolks with sugar and separately whisk the egg whites with remaining sugar to form a meringue. Add the almond powder and flour to the yolk mix and fold in the meringue. Lastly fold in the melted butter, spread on the deco paste and bake at 230°C for 5 minutes.
Butter cake sponge	Butter cake sponge as known in the USA and commonly known as an English pound cake around the world is called so because it contains a pound (450 g) of butter, flour, sugar, and eggs.	This cake is often eaten during afternoon tea. It is never iced. At the most it can be topped with sifted icing sugar. Candied dry fruits can also be added to this cake. It will then be called fruit cake.	Add cream, butter, sugar, and eggs one by one. When the mixture is fluffy, fold in flour and bake at 180°C for 30–40 minutes.

20.3

POINTS TO BE KEPT IN MIND WHILE MAKING SPONGES AND CAKES

The following points should be kept in mind while making sponges and cakes.

1. Weigh and measure ingredients correctly.
2. Sieve flour to aerate and remove impurities.
3. If using baking or cocoa powder, sieve it several times with the flour to ensure even distribution.
4. Tins, frames, hoops, and baking trays should be properly cleaned and prepared. The paper that is used as a liner should be free from creases.
5. Dried fruits should be washed and well drained. This is done not only for hygienic reasons but to increase the moisture content in the dried fruits, giving the cake a moist quality. Another way of achieving this is to macerate the dried fruits in spirits or liquors.
6. Remember to get the entire necessary equipment ready (for example, moulds, tins, oven pre-heated) before starting to prepare the cake. Whisked mixtures will collapse if left too long before baking.
7. Cakes that are large or heavy (such as fruit cakes) require longer cooking times at lower temperatures. Smaller or low density cakes require shorter cooking times at higher temperature.
8. To prevent cakes from over-colouring on the top during the baking process, place them under a greaseproof paper and reduce the top heat.
9. To check for doneness in small cakes and sponges, press lightly on the surface. The impression made should spring back immediately. For heavy fruit cakes, insert a clean skewer, on withdrawal, it should not have any moist mixture clinging to it.
10. Allow cakes to stand in the moulds they were baked in for a few minutes prior to removal.
11. Cakes are turned upside down on cooling racks and allowed to cool. Castor or icing sugar or cornflour is sprinkled on the greaseproof or parchment paper to prevent the cake from sticking. Sponge cakes may be cooled in the mould, turned upside down. This will give the cake a flat top and also prevent drying out.
12. Do not remove the paper that was used to line the cake until you are ready to use it. This will prevent the cake from drying out.
13. Cover cakes properly for storage either in the refrigerator, freezer, or a dry and cool place, depending on the cake and your personal needs.
14. In the case of frozen decorated cakes, it is advisable to cut and portion the cakes while they are still frozen.

EQUIPMENT USED IN SPONGE AND CAKE MAKING

Many large and small types of equipment are used in production of sponge cakes. Many types of equipment such as sieves, weighing scales, ovens, dough mixers, sheets, and moulds are used for baking sponges (refer to Table 19.7 in Chapter 19). We shall discuss some of the other small equipment that are typical to making sponges in Table 20.2.

Table 20.2   Equipment used in sponge making

Equipment	Usage	Photograph
Cake mould	It is used for baking cakes and is available in various shapes and sizes. Traditionally cake moulds are round, but nowadays various shapes such as triangles, ovals, and even pyramids are available.	
Savarin mould	A mould used for making *baba au rhum*. The hole in the centre of the cake is filled with fresh fruits after baking.	
Tube-shaped round mould	It is used for baking angel food cakes and chiffon cakes. Moulds are seldom greased for a better finish and hence it will be advisable to use non-stick moulds.	
Serrated knife	It is used for slicing sponges into layers, which can then be layered with assorted fillings to create cakes and gateaux.	
Palette knife	A flat knife used for icing the sponges.	
Piping bag and nozzle	It is made from material similar to that of shower curtain. Piping bag is used to pipe designs on top of cakes to decorate them.	
Turn table	Also called Lazy Suzanne or cake turn table, it is used for icing and finishing the cakes. It rotates on an axis thereby allowing the chef to evenly spread the cream and other fillings.	
Flat paddle	An attachment of a dough mixer, used for creaming butter and sugar to make sponges by creaming method.	
Balloon whisk	A balloon shape wire whisk used for whipping the ingredients.	
Wire rack	A piece of equipment used for cooling the baked goods.	
Silpat	Non-stick silicone baking sheets used for baking things that tend to stick on other surfaces.	

CONCLUSION

Sponges are the base of all cakes. A gateau is French for cake and in Germany it is referred to as torte. To make any good quality cake, the first step is to make a good sponge and the next step is to layer and ice the cake with good quality filling. We discussed sponges in this chapter and we shall discuss in detail various kinds of pastes and fillings in the next chapter. Prior to baking sponges, it is very essential to know about various techniques of mixing, as these mixing methods decide the texture of a product. It will be very surprising for you to know that sometimes from the same recipe and ingredients, one could make sponges or cookies and the entire thing will be decided by the method of mixing ingredients involved in making the product. In this chapter we discussed various methods of mixing such as beating, whisking, rubbing in, etc. and also various methods of baking such as blind baking. Unlike bread which relies on yeast or any other raising agent for aeration, the sponge solely relies on trapping of air by physical aeration and this could be achieved mostly by whipping or whisking. In most of the sponge recipes, the flour will be lightly mixed in or folded in and never mixed vigorously as done for bread. In case of sponges we do not want the gluten to develop and that is the reason that weak flours or cake flours are used for production of sponges. Since the method of sponge making relies mostly on physical aeration, eggs are often beaten separately and combined together as egg whites have the capacity of incorporating more air in them as compared to egg yolks. Certain cakes also use the chemical and biological methods of aeration by using raising agents and yeasts and it is done in sponges that are made without egg.

In this chapter we also read about the ingredients used in sponge making and understood the role of each ingredient. We saw how eggs form an integral part of sponge making and how oil or fats affect the texture of the sponge.

We understood the ingredients and how to control the curdling or the over mixing of the ingredients. We also read about balancing the formula in a recipe to create sponges with the desired texture.

We discussed various kinds of sponges and their uses in pastry and confectionery. We understood the different methods involved in sponge making and how they affect the textures of the same. Some precautions are to be kept in mind while making sponges. In this chapter we also dealt with very basic points that can make a good or a poor quality sponge. Certain small equipment used in making and icing sponge were also discussed and now the next chapter shall deal in various kinds of pastes and fillings.

KEY TERMS

Batter   Mixture of flour and liquids. The consistency would depend upon the amount of liquid in the batter.

Beating   Creaming butter and sugar in a mixer using a paddle attachment.

Blind baking   Usually refers to biscuit shells baked without any filling.

Butter cream   Creamed unsalted butter and icing sugar in the ratio of 2:1 used for layering and icing cakes.

Cocoa powder   A residual powder left behind in the production of chocolate.

Docking   Making small holes in pastry goods to discourage the puffing up.

Folding   Mixing two ingredients together without loosing volume.

Gateaux   French for cakes. Usually a reference to whole uncut cakes.

Icing   Layering and decorating a cake.

Lazy Suzanne   Another name for cake turn table.

Leavened   The incorporation of air into a dough or a batter.

Petit fours   Bite size decorative sponge or pastries, served with coffee.

Pinning   Rolling the dough with a rolling pin.

Piping   Forcing out the mixture from a bag.

Pound cake   Another name for butter cake.

Ribbon stage   Whipping of eggs and sugar to a stage where they become thick and pale.

Rubbing in   Mixing in flour and butter with finger tips to resemble breadcrumb texture.

Short   A baked good that snaps when broken; also known as *khasta* in Hindi.

Silpats   Silicon baking mats.

Sponge   A light and airy product, made with eggs, flour and sugar.

Torte   German for cakes. Sometimes referred to individual slices of cake.

Turn table   An equipment used for layering and icing cakes.

Whipping   Incorporating air into liquids, by using wire whisk.

CONCEPT REVIEW QUESTIONS

1. Define a sponge. How is it different from a cake and bread?
2. What are the various ways of aerating a product?
3. Define sifting and list two important reasons why sifting is done.
4. What do you understand by creaming and what considerations are to be kept in mind while creaming?
5. Define whipping and list down the factors to be kept in mind while whipping cream.
6. What happens if the egg whites are overwhipped?
7. Define rubbing in. How does it create shortness in the product?
8. What factors should be kept in mind while rubbing in?
9. Define cutting and folding method. What care should be taken while folding in?
10. Why does docking prevent rising of the dough?
11. What is blind baking and what is its importance?
12. What care should be taken while pinning the dough?
13. Define icing. What equipment are used for the same?
14. List the role of ingredients used in sponge making.
15. What precautions will you take while combining the ingredients for sponge making?
16. Why does a sponge batter curdle and how can you fix it?
17. What do you understand about balancing the formula in a sponge or batter?
18. What care should be taken while baking the sponges?
19. Why is cooling of cakes so crucial?
20. List at least five basic sponges and its uses.
21. What sponge would you use for the following cakes: blackforest, pineapple, dark chocolate mousse cake, and dark and white chocolate mousse cake?
22. How is chiffon cake different from Angel food cake?
23. What is the difference between Victorian sponge and *madeira* sponge?
24. Write down the procedure of making *jaconde* sponge.
25. What is a silpat and what are its uses?
26. List at least five points to be kept in mind while preparing sponges.

PROJECT WORK

1. In groups, make at least five different sponge recipes from the recipes provided in the ORC and compare the sponges and write your observation in the chart below.

Name of Sponge	Texture	Softness	Taste	Colour

2. In groups, try out the recipes as provided in the ORC for the basic genoise sponge and alter the mixing techniques. One group shall make the sponge with chiffon method while two other groups will make the genoise by *madeira* and Victorian methods. Compare the results of the cake and see the difference in the product.
3. Visit the various pastry shops in the city and write down the various types of cakes and pastry products available. Now do a list of sponges which were used for making those cakes and record your observations.
4. Prepare eggless sponge and compare the results with the genoise sponge.

CHAPTER 21

PASTES, CREAMS, FILLINGS, AND SAUCES

Learning Objectives

After reading this chapter, you should be able to:

- know the different types of pastes and creams that can be used as fillings for the cakes and pastries to create desserts
- know various steps involved in production of pastes and the usage and storage of each paste
- prepare various kinds of creams and understand the various desserts made from them
- understand the various techniques used in preparing sauces and the faults of sauces
- know the role played by different ingredients in making pastes and creams
- know about various types of international desserts

INTRODUCTION

This is one of the most important chapters when it comes to pastry and confectionery. In the previous chapter we learnt about various sponges. In order to turn the sponges into cakes and pastries we need to know various kinds of creams that can be used to fill them. Cream in this context has a much broader meaning than just whipped bakery cream. We shall discuss various types of creams such as pastry cream, creme Ciboust, crème Chantilly, etc.

Pastes form the base of many pastries, while many are used as fillings prior to baking. For example, a classical cake called gateau pithivier is made with puff pastry (refer to Chapter 23) and filled with frangipani paste or almond paste. Even the pastes in bakery and pastry do not refer to things that are grounded into pastes with addition of liquids; instead many basic pastes such as choux paste, are mostly used for making various types of pastries and sweet paste is used for making cookies and tarts. Probably pastry came from the word paste, where the paste of water and flour was used to encase meat while cooking. As the paste baked, it got flavoured with the juices from the meat and was eaten along with the meat. Even today, certain meats are wrapped with doughs or puff pastry and then roasted in ovens. In bakery and confectionery there are many such pastes, which actually resemble dough (sweet paste) but are grouped into this broad group of pastes.

We shall also discuss various types of sauces used in pastry kitchens. Many pastry cooks believe that a good sauce is the main feature of good cooking because of the skill needed to prepare it and the variety it gives to food. Sauces are basic but versatile commodities in the pastry kitchen and the professional pastry chef knows the importance of making good sauces. There is no doubt that sauce making requires skill. Over the years culinary techniques have been developed to produce consistently good quality products. Various types of hot and cold sauces are used in pastry kitchens and the role of these sauces is similar to those discussed in Chapter 10. A sauce provides texture, nutrition, and contrasting flavours to desserts and cakes. Many sauces are also used to add colour on the plated dessert.

PASTES

In this section we shall understand various types of pastes and their uses in the pastry kitchen. It is important for chefs to understand the role played by the ingredients in each paste, as it will help them to critique their product and enable them to provide a different texture by altering the recipe. As discussed above, many pastes used in confectionery appear to be like dough, but are classified into pastes. The reason for this is that when they are freshly made they appear like a paste, but when refrigerated, the butter sets with flour to give the texture of dough. It must be understood that dough refers to kneading of flour with water to develop the gluten. However, while making certain pastes, such as short crust paste, care is taken so that gluten does not develop in flour otherwise it will lose its 'short' properties. Here short refers to the flakiness of the pastry product. In Hindi it is called *khasta*.

SHORT CRUST PASTE

Short paste or short crust paste is crisp and brittle. It is not elastic and resilient like dough and this shortness in a cookie or many pastry products is much desired to alter the textures and mouth feel of a product. Short crust paste is usually used for making savoury products such as turnovers and pies. Table 21.1 shows the role of ingredients in making short crust paste.

Table 21.1   Ingredients used in short crust paste

Flour	Fat	Liquid
Soft flour is used to avoid elasticity in the product. The product should be resilient and brittle.	The fat used is usually butter; margarines also give a good product, but since margarines are trans fats we should avoid the usage of these.	Cold water should be used as warm water tends to melt the fat, thereby giving a stretchy texture to the pastry.
Flour is sifted well to provide aeration to the final product.	Fat is cut into smaller pieces and rubbed in with flour.	Cold liquid, preferably ice water, is sprinkled on the top of rubbed flour and fat is allowed to be absorbed by the paste.

Steps in Making Short Crust Paste

The steps in making short crust paste are given below. Usually the first two steps are used for making short crust paste; but if pies or flans are to be made then the latter steps are used.

Step 1

Mix the butter and flour. This could be done by using the following methods.

Rubbing in Method The cut pieces of butter are rubbed into the flour to produce coarse bread-crumb size particles. Cold water is sprinkled over and the dough is lightly mixed to form short crust paste. Chill in the refrigerator before using.

Pinning Method Cut the butter into small cubes and roll with the flour until the butter flakes. Collect in a bowl and sprinkle cold water. Collect the paste together to form a short crust paste.

Step 2

Chill the paste in the refrigerator. It will be easy to roll later, if the paste is put in a plastic bag and flattened out with the tip of fingers before refrigerating.

Step 3

Pin the dough with rolling pin. If lining a pie dish, roll into a circle, roll it back on to the rolling pin, and unroll it in the pie dish. Usually 240 g of dough lines a pie dish of 8 inches diameter. There is no need to grease the pie mould as short crust paste contains lots of butter. While making tarts or tartlets in bulk, place the tart mould close to each other to form a large rectangle. Sheet out the dough and follow the same procedure of rolling it over rolling pin and unrolling it on to the tart moulds. Then press the short crust over the tarts with the palms and then finish each one individually. The trimming left over should not be kneaded together; instead just collect it and roll it again if needed.

Step 4

The pastry is docked to allow the steam to escape from it while baking.

Step 5

Blind bake the shell. Line with greaseproof paper and fill with beans. The paper does not let the beans stick to the pastry. When half done remove the beans and bake the pastry again so that it turns golden brown. At this stage, you can add minced meats and vegetables along with cream and eggs to make a pie called quiche.

Uses of Short Crust Paste

The uses of short crust paste are as follows:

- Short crust can be used for making tarts, pies, and flans. Table 21.2 shows the difference among tarts, pies, and flans.
- Short crust can also be rolled and cut into shapes and used as savoury biscuits.
- It can be crumbled and used as a topping on pies to give a rustic crust.
- Certain meats are encased in short crust prior to baking.

Table 21.2   Difference among tarts, pies, and flans

Tart	Pie	Flan
Tart is a mould which is 3–4 inches in diameter with raised edges. The smaller version of 1 inch diameter is called tartlet.	This is a dish made in a flan mould. Pies are usually sweet or savoury fillings baked in a flan.	Flan is a mould that is 6–8 inches in diameter with raised edges.

Contd

Table 21.2 (Contd)

Tart	Pie	Flan
Tarts are baked like shells and then cooked fillings and creams are added to them. They are open and never closed.	A pie is usually covered on top with another piece of crust.	Sometimes large tarts which are open are also referred to as flans; we cannot call them pies as pies are always covered.
Tarts are blind baked and cooked fillings are filled.	The pie is blind baked till half done and the filling is put inside and then baked again.	Flans are blind baked and cooked fillings are filled inside the shell.
Tarts are usually sweet.	Pies can be sweet or savoury.	Flans are usually sweet.
Examples—fresh fruit tart, lemon curd tart, etc.	Examples—apple pie, Australian leek pie, etc.	Examples—fruit flan, custard flan, etc.

SWEET PASTE

Sweet paste is sweet in taste. It is short and brittle like the short crust paste; but its usage is restricted to sweet products. Table 21.3 shows the role of ingredients in making sweet paste.

Table 21.3   Ingredients used in sweet paste

Flour	Fat	Liquid	Sweetener
Soft flour is used to avoid elasticity in the product. The product needs to be resilient and brittle.	The fat used is usually butter. Margarines also give a good product, but as margarines are trans fats the use of these should be avoided.	Eggs are used as liquid in pastes.	Castor sugar or icing sugar is recommended to be used as sweetener. They are more readily soluble, which has a softening effect on the gluten in the flour, in turn influencing the shortening properties of the sweet paste.
Flour is sifted well to provide aeration to the final product.	Fat is cut into smaller pieces and either creamed with sugar or rubbed in with flour like short crust paste.	Cold eggs are used as liquid.	Sift the icing sugar to avoid any lumps in the sweet paste. Do not use grain sugar as it will leave brown specks after baking.
If chocolate flavoured sweet paste is desired, then substitute 20 percent with cocoa powder. Other flavourings such as lemon zest can also be added to the flour.		Sometimes milk is added if almond flour is added to the sweet paste.	

Steps in Making Sweet Paste

The steps in making sweet paste are given below.

Usually the first two are used for making short paste; but if pies or flans are to be made then the latter steps are used.

Step 1

Mix the butter and flour. This could be done by using the following methods.

Creaming Method   The butter and sugar should be creamed well. The idea is to make it lighter by incorporation of air. This is the most commonly followed method to make sweet paste. Then add eggs one by one until all the eggs are added. Use a flat paddle to beat the mix. Now remove from the mixer and carefully fold in the sifted flour to obtain sweet paste. Do not overmix as the sweet paste will lose its shortening effect.

Rubbing in Method   The cut pieces of butter are rubbed into the flour to produce coarse bread-crumb size particles. Beaten eggs are lightly mixed to form sweet paste. Chill in the refrigerator before using. This method is not very commonly followed.

Step 2

Chill the paste in the refrigerator. If the paste is put in a plastic bag and flattened out with the tip of fingers before refrigerating, it will be easy to roll later.

The other three steps are similar to that of short crust paste.

Uses of Sweet Paste

The uses of sweet paste are as follows:

- Sweet paste is used for making tarts, pies, and flans. To understand the difference among the three, refer to Table 21.2.
- It is rolled, cut into various shapes, and baked as cookies and biscuits.
- It is used as base for certain cakes and pastries.
- Thin cut out sheets of sweet paste can be used as decorations.

CHOUX PASTE

Choux means cabbage in French. Here baking results in a shape that resembles cabbage and probably that is the reason why it has been given this name. Choux paste has a consistency between a dough and a batter and is used in both savoury items and desserts. This paste may or may not contain sugar, depending upon the usage of the final product. Choux is a versatile, partially precooked paste that can be baked for use in pastries and gateaux, fried for use in potato dishes and fritters, or boiled in gnocchi dishes. Pastry products made from choux paste include èclairs, Paris Brest, gateau St Honore, profiteroles, and many others. There are many recipes with varying formulae, each giving a product of a different consistency depending on its purpose. When the choux paste is baked, a steam is formed inside which pushes the paste out giving it a hollow texture. This is then baked at low temperature to dry out the pastry to keep it firm and crisp. Otherwise the product will collapse and will be chewy. The tunnel thus created in the pastry is filled with different types of flavoured fillings and decorated. Choux paste is not only baked, but also deep-fried. Table 21.4 shows the role of ingredients in making choux paste.

Table 21.4   Ingredients used in choux paste

Flour	Fat	Liquid
Medium to strong flour should be used, as the gluten is required to provide good elasticity and volume to the paste.	The fat used is butter as it gives a better flavour to the product.	Water and eggs are the primary liquids used in choux paste. Water is boiled with butter and flour is cooked until it leaves the sides of the pan. It is removed from fire and eggs are incorporated one by one until a paste is obtained.
In India we would use the normal flour.		The quantity of eggs will depend upon the size of the eggs, degree of cooking of flour, and the amount of flour and fat used.

Steps Used in Making Choux Paste

The steps used in making choux paste are given below.

Step 1

Place the fat and water in a pan and heat until the fat is melted and the water boils. The fat and water should boil simultaneously. The fat should be cut into small pieces to help it melt quickly, preventing the loss of water through evaporation. If water loss occurs, the pastry would contain too much fat, making it heavy.

Step 2

Add the sieved flour all at once to the fat and water emulsion and stir continuously with a wooden spoon. This will prevent lumps from forming in the paste. The flour should be added only when the fat and water have come to a boil. This paste, called *panada*, is cooked until it leaves the sides of the pan without sticking.

Step 3

Add the eggs to the *panada* when the mixture has cooled to approximately 60°C. This is done to prevent the eggs from completely cooking in the paste, which would result in a heavy paste. Add the eggs one at a time, working the paste to a smooth consistency before the next egg is added. The final consistency of the paste for pastries should be that it falls off the back of the spoon. Take care when adding the last of the eggs as different flours have different adsorption characteristics. The paste should have a good, smooth sheen. It should be soft and should be able to retain its shape when piped.

Step 4

This paste does not require a resting period. It can be piped immediately and then baked. Choux paste can be piped onto lean baking trays that are lightly greased. They can also be lightly dusted with flour after they have been greased.

Step 5

Bake at a high temperature initially (200 to 220°C, depending on the size). The baking of choux pastry requires a lot of care. While baking, the oven door should not be opened too frequently because the loss of heat may cause the pastry to collapse. Also, if the pastry is not baked thoroughly until it is properly dried, it may collapse. The development of colour is not sufficient indication that the item is cooked properly. If the 'shell' is not firm and crisp, reduce heat and bake further to dry out. The choux paste, after baking, should be light for its weight and when sliced open it should be hollow from inside.

Uses of Choux Paste

Choux paste is a versatile paste and can be used in savoury or desserts. There are many uses of choux paste and some of these are shown in Table 21.5.

Table 21.5   Preparations of choux paste

Item	Description	Photograph
Savoury items	Used for making savoury items such as fritters, cheese balls, gnocchi, etc.	
Chocolate eclairs	Piped in tube shape usually 4 inches long and after baking it is filled with flavoured cream or custard and glazed with melted chocolate or fondant.	
Profiteroles	Round-shaped balls of choux paste baked and filled with flavoured creams and glazed with chocolate, caramelized sugar, sifted icing sugar, fondant, etc.	
Croquembouche	Profiteroles are filled with custard, flavoured with 'grand mariner' and glazed with caramelized sugar. It is built in a height and is used as a traditional wedding cake in France.	
Profiterole suchard	Profiterole shells are filled with vanilla ice cream and glazed with chocolate sauce.	
Gateau St Honore	It is a classical gateau from France. It is made by piping a ring of choux paste on a thin disc of puff pastry and baked. The ring of choux paste is then sliced from top and filled with a creme Ciboust. The ring is decorated with filled profiteroles glazed with caramelized sugar and the centre of the gateau is piped with alternate swirls of creme Ciboust and pastry cream.	
Paris Brest	A ring-shaped choux paste, baked and piped with whipped cream and decorated with fresh fruits and berries. It is decorated with sifted icing sugar.	

Contd

Table 21.5 (Contd)

Item	Description	Photograph
Swans	Choux paste is piped in the shape of a teardrop and baked to make swans. The top is cut and then split in half lengthwise to make the wings of swans. The neck is piped in a thin curved shape and baked separately. Swans can be filled with crème Chantilly and assembled to resemble swans.	
Cream slices	Thin sheets of choux pastry are layered with cream and fruits and garnished with sifted icing sugar.	

MARZIPAN

> **CHEF'S TIP**
> *Persipan* is a mock marzipan made from various kernels of stone fruits such as apricots and peaches. It is less expensive than marzipan.

Marzipan is a paste composed of ground almonds and sugar mixed in varying proportions. Better quality marzipan has more almonds and less artificial flavouring. It is commercially made as it has better taste and texture if made that way. The homemade marzipans are rarely of the consistency and texture as that of the commercially available ones.

Marzipan is mainly used as covering for wedding cakes and other rich cakes. It is widely used to make flowers, fruits, and figures as the dough like texture allows the chef to mould it into various forms and figures. Marzipan is also used as filling, for example, Dresden Stollen (refer to Table 19.10), and in the production of some high quality cake mixtures.

Table 21.6 shows the role of ingredients in making marzipan.

Table 21.6   Ingredients used in marzipan

Nuts	Sugar	Water	Liquid Glucose
Coarsely ground almonds without the skin are used.	Good quality refined sugar is to be used. It is advisable to use high quality castor sugar.	Water is used as a binding agent for almonds and liquid glucose.	It contains dextrin gum that retards the crystallization of sugar.
It is important to check the taste of almonds, as bitter almonds would spoil the taste.	Though the sugar will be boiled, the impurities in the grain sugar will impact the colour of the marzipan. So it is advisable to avoid the use of grain sugar.	Water is boiled along with liquid glucose and heated up to 121°C.	Liquid glucose keeps the product pliable, and allows you to work with it for a longer time without crystallizing the sugar.

Contd

Table 21.6 (Contd)

Nuts	Sugar	Water	Liquid Glucose
Mock almonds, such as kernels of stone fruits, are used to make a paste called *persipan*.			

Techniques Used in Making Marzipan

Marzipans are better if procured from commercial shops and there are many good reasons to do so—quality of the product, consistency, etc. In case of non availability one could make marzipans by following the steps given below.

> **CHEF'S TIP**
> Check whether working tables, tools, containers, and storage facilities are spotless prior to commencing work with marzipan. People suffering from excess sweating of their hands will need to wear gloves. It is unhygienic and there is a danger of fermentation.

Step 1

Blanch almonds and skin them, if you wish to get a marzipan which is creamy white in colour. Dry the almonds well and coarsely grind them, ensuring that they do not become oily. You can avoid almonds from becoming oily by not grinding them in a heavy duty mixer grinder.

Step 2

Combine sugar, water, and liquid glucose and boil to 121°C. Make sure that the sugar does not colour and this can be done by ensuring that the sides of the pan are constantly brushed down with a wet brush.

Step 3

Take the liquid off the fire and add coarsely ground almonds to the mixture. Spread the mixture onto a cleaned marble surface and let it cool down.

Step 4

Grind the mixture into a paste. This paste will form into a pliable dough when it is cooled down.

Uses of Marzipan

The uses of marzipan are as follows:

- It is used for making flowers, decorative figures, and moulds.
- It is used for covering wedding cakes and rich cakes, to make a smooth base for spreading 'royal icing,' which is a thick paste made by beating egg whites with icing sugar and little lemon juice.
- It can be used for modelling purposes and in that case one part of marzipan is mixed with one and a half part of icing sugar. The paste thus made will remain more firm and hard when creating modelling structures.
- Marzipan can be used to make high quality cakes. When adding marzipan to a light density mixture, such as egg yolk, it is necessary to first break down the marzipan to avoid lumping. If the preparation uses sugar, blend the sugar with the marzipan first. Sugar is an abrasive which will gradually grind down the marzipan and allow it to disperse when the egg yolks or other liquids are gradually added.

ALMOND PASTE

Almond paste is commonly known as 'frangipani paste' and composed of ground almonds or marzipan mixed with butter, flour, and eggs. This is used in fillings of many classical cakes such as gateau pithivier. This paste unlike marzipan is always baked before eating. It can also be used for making pies and tarts. As it is enriched with marzipan, it produces pies and tarts of rich quality. Table 21.7 shows the role of ingredients in making frangipani paste.

Table 21.7 Ingredients used in Frangipani paste

Marzipan	Sugar	Butter	Egg Yolks	Flour
Marzipan gives body and texture to the paste.	Castor sugar is better as it forms an abrasive and helps in creaming the marzipan. Icing sugar can also be used for a finer paste.	Butter helps to cream the almond paste and adds flavour and texture to the paste.	Egg yolks are used for adding creamy texture and flavour.	Soft flour is carefully folded in and helps to bind the paste together when baked.
It is crumbled into smaller pieces and mixed with eggs and sugar.	Sugar helps in providing the browning effect to the paste when baked or gratinated.	Only butter has to be used for quality.	Only egg yolks are to be used. If the paste is too thick then egg whites can be added for thinning down the mix.	Do not overmix the flour in the paste as the paste will become chewy.

Steps in Making Frangipani

The steps in making frangipani are given below.

Step 1

Crumble the marzipan and beat it with flat paddle with sugar and butter until a creamy mixture is formed.

Step 2

Add yolks one by one until a creamy mixture is obtained.

Step 3

Add flour and fold with hands and store in refrigerator. This paste can be frozen for later use.

Uses of Frangipani

The uses of frangipani are as follows:

- It is used for filling in cakes such as gateau pithivier, almond croissant, etc.
- It is used for filling in breakfast pastries such as Danish pastry and almond croissant.
- It is also used as filling for pies and tarts. In a flan or tart, add frangipani to the tart base and arrange sliced fruit such as pear on top and bake at 180°C until golden brown. The almond paste rises up encasing the fruit.

TUILE PASTE

Tuile is derived from the French word 'tile' and this name is probably given to it, because it is used for making thin flat cookies for garnishes and decorations. A design is first drawn onto a cardboard (2–4 mm thick) and then cut out. For regular work, it is best to use plastic or heavy aluminium stencils which are available in the market for this purpose. Lay the stencil on lightly greased and floured baking trays and spread the mixture in the centre of the cut out. When spread, lift the stencil carefully. Bake evenly at approximately 190°C until golden brown. When the paste is just baked, it will be soft and can be moulded into various shapes such as curls, twists, etc. to give a dimension to the garnish. There are many types of *tuile* paste used for garnishes in pastry and each has a different texture and mouth feel, though the purpose of each is same—garnishing and decoration. *Tuile* paste can also be used to make cups and cones for serving desserts or ice cream in them.

Table 21.8 shows the different types of *tuile* and their uses.

Table 21.8 Different types of *tuile* and their uses

Name	Description	Uses
Basic *tuile*	It is made by creaming butter and icing sugar. Eggs are added and folded in flour to form a paste.	It is used for making cones, swirls, and various garnishes.
Almond *tuile*	It is made as per the recipe of basic *tuile* but 20 per cent of the weight of flour is replaced with almond powder or marzipan.	It is used for garnishes and decorations. It can also be served with ice creams.
July pan	It is made by mixing equal amounts of milk, icing sugar, and flour which form a paste.	It is used as garnish for eggless desserts. It is used for making springs as it takes longer time to set to brittle texture as compared to other *tuile*.
Brandy snaps	This *tuile* is very different from the above mentioned *tuile*, both in terms of making the mix and the appearance. This bakes to very thin dark brown sheets with large holes. The butter, sugar, and honey are cooked until they come to a boil and flour is added to make a paste. It is then cooled and baked on silpats. It is usually spread out on a sheet.	It is used for decorations and served as crisps with ice creams.

PUFF PASTRY

Puff pastry consists of laminated structure built up with alternate layers of dough and fat. This is achieved by rolling out the dough and giving it sufficient turns until there are many layers of dough and fat. When this pastry is baked, the expanding air and water vapour 'puff' separate layers apart from each other, resulting in a delightful, crisp, light, flaky pastry called puff pastry. Though it is a form of dough which is laminated, it is still classified into pastes as it is not baked directly. We shall discuss about puff pastry in detail in the next chapter on laminated pastries.

CREAMS

Traditionally creams are butterfat separated from fresh milk. This can be done manually by allowing the milk to churn in centrifugal machines, which help the cream to rise up. In Chapter 18 we discussed the various types of creams (Table 18.10) and their uses. But in this chapter when we refer to creams, it is not just dairy cream. In pastry kitchen, smooth thick viscous liquids are classified into a broad heading and referred to as creams. For example, butter cream, custard cream, etc. Let us discuss some creams commonly used in the pastry kitchen.

PASTRY CREAM

Pastry cream or custard cream or creme patisserie as called in French is one of the most common creams used in cakes and pastry products. This cream can also be baked and hence is used in both hot and cold desserts. Pastry cream is basically a mixture prepared from vanilla flavoured milk, egg yolk, sugar, and starch. It is boiled until a thick mixture is obtained and butter is added on top to avoid the skin formation. This cream has the same value as the white sauce used in Western cooking. Some chefs believe that the mix cooked with milk, sugar, egg yolks, and starch is called custard cream and only when whipped cream is added to the custard cream it is known as crème patisserie or pastry cream. Pastry cream, however, has a variation as well. When a part of whipped meringue is added to the pastry cream it is known as creme Ciboust or St Honore cream, as it is used as a filling for gateau St Honore. Here when we refer to pastry cream we will also refer to custard cream which can be enriched with a part of whipped cream folded in to make the mixture more delicate thereby making it more useful in fillings for cakes and pastries. Adding of the cream to the custard cream however renders it useless for baking purposes.

The pastry cream is used as a filling for choux buns, fruit flans, èclairs, etc. without the whipped cream being mixed. It can also form the base of many hot desserts such as 'cobblers' and pies. It is also used in filling breakfast Danish pastries.

Table 21.9 shows the role of ingredients used in making pastry cream.

Table 21.9   Ingredients used in pastry cream

Milk	Sugar	Egg Yolks	Starch
Usually whole fresh milk is used for better flavour.	Granular sugar is used as it will be dissolved when cooked with milk.	Egg yolks are beaten with sugar. This is done to break the yolks, so that they cook uniformly.	Commonly proprietary custard powder is used, but one can use a mixture of corn flour and custard powder in ratio of 1:1. Some recipes just use plain flour but in that case one has to cook for long time to get rid of the raw flavour of flour.

Contd

Table 21.9 (Contd)

Milk	Sugar	Egg Yolks	Starch
After the pastry cream is made it can be cooled and mixed with a part of whipped cream.		Whole eggs could be used, but it will give a much lighter product.	

Steps in Making Pastry Cream

The steps in making pastry cream are given below.

Step 1

Boil milk and half the sugar with scrapings from vanilla bean. This is done to test the quality of the milk. If the milk is old or sour it will curdle and second hot milk will form the emulsion of eggs fairly quickly and will also help the starch to swell faster.

Step 2

Beat egg yolks and remaining sugar together along with the starches to form a smooth creamy mixture. If making in large quantities, one can also whip this mix to form a smooth emulsion.

Step 3

Temper the egg and flour emulsion. Tempering means adding little amount of hot liquid into the egg mix whilst mixing continuously. This helps to bring both the mixes at almost same temperatures and hence when added, the egg and starch mixture will disperse equally and will not form lumps.

Step 4

Boil the pastry cream until it starts to bubble.

Step 5

Remove from fire and add butter and mix well.

Step 6

Use immediately or wrap with plastic and keep refrigerated until further use.

Uses of Pastry Cream

The uses of pastry cream are as follows:

- Pastry cream is used for fillings in tarts, pies, and flans.
- It is used for filling between sponge cakes to create gateaux and pastries. This is a neutral cream flavoured with vanilla. It can be flavoured with any desirable flavour.
- It is used as a base for hot desserts, as it can be baked also.
- It is also used for filling choux pastry items to create desserts (refer to Table 21.5).

CRÈME CHANTILLY

Crème Chantilly is a basic bakery cream that is whipped with sugar and vanilla flavours to be used in various desserts, cakes, and pastries. These days imitation creams are available (refer to Table 18.10) and there is no need to add sugar to the same. This cream finds its uses in almost all the cakes and pastries in one way or the other. Crème Chantilly is one of the most important ingredients used in a mousse and soufflè. Mousse and soufflès are cold desserts of French origin and have a smooth texture, which is imparted by whipped cream. Table 21.10 shows the differences between a mousse and a soufflè.

Table 21.10   Differences between mousse and soufflè

Mousse	Soufflè
Mousse comes from the French word *moussir,* which means to froth.	Soufflè comes from the French word *souflir,* which means to fly.
The texture of the mousse is provided by the whipped crème Chantilly and forms the most important ingredient in the mousse. It may or may not contain eggs.	Soufflè is a light and airy textured dessert and the texture is provided by the whipped yolks and flavourings carefully folded in with meringue. It may or may not contain crème Chantilly; but eggs are the main ingredients.
Mousse can be made savoury and the principle is same—mince the meat and add with whipped cream.	Soufflè traditionally was made savoury and always had a starch base, usually béchamel sauce, mixed with egg yolks and whipped egg whites folded in.
Mousse is always served cold.	Traditional souffle was baked and hence served straight from the oven, but chilled soufflè which is a very recent addition to confectionery is served cold and is sweet.
Mousse can be set and presented in any mould.	Soufflè is traditionally made in soufflè mould. Chilled soufflè is prepared in such a way that it seems as if the mixture is coming out of the mould, as happens naturally in case of baked soufflè.

Table 21.11 shows the role of ingredients in making whipped cream.

Table 21.11   Ingredients used in whipped cream

Cream	Sugar	Flavouring
Bakery cream with at least 40 percent fat content is best suited for whipping.	Castor sugar is used as it will dissolve well.	Vanilla pods are scraped and used to flavour the cream.
Cream should be cold and even the bowl and other equipment used in whipping cream should be cold so that the cream whips faster and to a good volume.	Usually 200 g of sugar is used to sweeten 1 litre cream.	Other flavourings such as citrus fruit zests and fruit purees are added to flavour. It is always advisable to use natural flavourings as against synthetic ones for better quality.

Steps in Making Crème Chantilly

The steps in making crème Chantilly are given below.

CHEF'S TIP
1. If the cream gets over whipped, do not throw it away, and instead whip it further to get butter, which can be used for various other purposes. 2. Mix a part of crème Chantilly with a part of imitation cream to get a good textured and flavoured dessert.

Step 1
Collect all the ingredients and make sure the cream is chilled and the bowl and whip are at a cold temperature.

Step 2
Add sugar and scrapings from the vanilla pod and whip the cream at a medium speed.

Step 3
Keep the consistency of the cream slightly under whipped, as it will be mixed or folded again into desserts or for baking cakes. If it is whipped to the right degree, it might split to form butter and water while mixing.

If you are using imitation cream, make sure the cream does not have any ice crystals, as this cream always comes frozen. Whip normally to the right degree and not under whip, as it will give a synthetic taste if not whipped to the right consistency.

Step 4
The imitation cream can be stored refrigerated for upto two days, but freshly whipped crème Chantilly must be used immediately.

Uses of Crème Chantilly

The uses of crème Chantilly are as follows:

- It is used for filling between sponge cakes to create gateaux and pastries.
- It is used for making various dessert such as mousse and soufflès.
- It is used for serving with coffee and milkshake.
- It is used as a condiment with fresh fruits such as strawberries and mangoes.
- It is used as a condiment with scones in afternoon tea.

CAPRICE CREAM

Caprice cream is a variation of crème Chantilly. The only difference is that it is whipped without sugar and when the cream is whipped, one fifth of whipped cream is replaced by broken pieces of meringue. The cream is left refrigerated until further use. It is mostly used in the filling of cakes and pastries.

BUTTER CREAM

Equal quantities of unsalted butter and icing sugar are creamed to produce a smooth aerated creamy mixture called butter cream. It should be stored in a sealed container in a cool place. One should refrigerate it because of the perishable nature of butter; but it needs to be creamed back to a creamy consistency before using. Butter creams are used as fillings and toppings for sponge to make cakes and pastries. This is one of the oldest creams used in cake fillings and can be flavoured with any kind of flavourings. One can also add colours to the butter cream to make fancy shapes for children's birthday cakes. The ingredients used in butter cream are very simple—only butter and sugar; but the quality of the ingredients is crucial, as this filling is never cooked and will be used as it is.

Table 21.12 shows the role of ingredients in making butter cream.

Table 21.12   Ingredients used in butter cream

Butter	Sugar	Flavouring
It is important to use only unsalted butter. Fresh white butter that is home churned sometimes has an odd flavour and too much water in the mix. So care must be taken to beat the butter alone to remove any water present.	Only good quality icing sugar is to be used. It is important to sift the sugar so that there are no small granules of sugar in the mix. As this cream can also be used for writing on the cakes, it becomes a nuisance when the small sugar crystals get in the way and block the flow of icing.	Butter cream can be flavoured with melted chocolate, vanilla pods, zest of citrus fruits, or even synthetic flavours, as combining it with fresh fruit puree will not form an emulsion and the cream will split. Liquors and spirits, such as rum, whisky, etc., can be added to the butter cream. Before adding powdered colour, mix the colour in few drops of water and then mix it with the butter cream.

Steps in Making Butter Cream

The steps in making butter cream are given below.

Step 1

Collect all the ingredients. Make sure that the butter is at room temperature, as it creams well if it is not very cold. Cut the butter into smaller pieces and beat it on slow speed with a flat paddle for 5 minutes.

Step 2

Sift the icing sugar and mix it into the butter. Make sure that the speed gear of the machine is turned to one as the icing sugar might fly out, if the mixer is run on the medium or high speed. Let the icing sugar incorporate and then run the machine on high until the cream is light and fluffy.

Step 3

Remove from the mixer and scrape out into a clean plastic bowl and store in a cool place.

There is another variation of butter cream known as 'Italian butter cream'. It is made by making Italian meringue. Italian meringue is made by whipping egg whites with a little amount of sugar and the rest of the sugar is cooked to 118°C and slowly added to the egg whites to form Italian meringue. This type of meringue is more stable than the French or the Swiss meringues. The French meringue contains plain egg whites whipped with sugar to form a thick fluffy mixture whereas Swiss meringue is a cooked meringue. The recipe of the Swiss meringue is similar to that of French meringue, the only difference is that Swiss meringue is whipped over a double boiler and hence, the meringue is more stable than the French one. A part of Italian meringue is folded in butter cream to make Italian butter cream, which is lighter as compared to the French butter cream.

Uses of Butter Cream

The uses of butter cream are as follows:

- It is used for filling between sponge cakes to create gateaux and pastries. This is a neutral cream flavoured with vanilla. It can be flavoured with any desirable flavour.
- It is used in piping bags to decorate the cake and pipe wordings on the cakes.

LEMON CREAM

Lemon cream is commonly known as lemon curd and is a smooth, creamy paste that has a balanced taste of sweet and sour. It is often known as curd because of the slow poaching of eggs in butter and lemon juice with sugar. Lemon curd was invented for usage with scones instead of jam in England, but slowly people started to use it as topping and filling for cakes, muffins, and tarts as well. When piped into a tart, the mixture is very sharp, so to cut down on the sharpness, whipped meringue is topped onto the tart and gratinated. This is very commonly eaten as lemon meringue pie in the USA and the UK.

Table 21.13 shows the role of ingredients in making lemon cream.

Table 21.13   Ingredients used in making lemon cream

Butter	Lemon Juice	Egg Yolks	Sugar
Unsalted butter is used. Butter helps the curd to set into a creamy texture when it becomes cold. If oil is used instead of butter, the mixture will be runny.	Lemon juice is used traditionally to make lemon cream. Various other fruit juices might become bitter after cooking.	Egg yolks help to form a creamy texture. The proteins present in eggs coagulate with the application of heat and it gels up together with butter and juice to form a cream.	Granular sugar will be a good choice as it will get dissolved in the lemon juice.

Steps in Making Lemon Cream

There are two methods of making lemon curd. One of the methods is to combine the butter, sugar, and reduced lemon juice on a low heat until the sugar has dissolved. Then the mix is added into a machine with wire whisk and an egg yolk is added one at a time until all the eggs are incorporated and the mixture has become thick. The mixture is removed from the bowl and put it in a plastic container and chilled until set. Metal containers might react with the acid present in the lemon curd and hence it is preferable to store it in plastic containers.

The steps of the other method commonly followed in making lemon cream are given below. It is one of the easiest and hassle-free methods.

> **CHEF'S TIP**
> When filling tart shell with any cream, brush the inside of the tart shell with melted chocolate. This will form a film and prevent the tart shell from getting soggy from the liquid in the cream.

Step 1

Combine butter, egg yolks, sugar, and lemon juice and mix well to break the egg yolks with sugar.

Step 2

Cover the bowl with aluminium foil and place over a bain-marie. Remove at intervals of 15 minutes and mix again to form a homogeneous mix. Place it back on the double boiler and cook for another 1 minute.

Step 3

Remove from the fire and allow cooling. It will settle into a thick cream which can be piped into tart shells.

Step 4

Use immediately or wrap with plastic and keep refrigerated until further use.

Uses of Lemon Cream

The uses of lemon cream are as follows:

- It is used for fillings in tarts, pies, and flans.
- It is used for filling between sponge cakes to create gateaux and pastries. This is a neutral cream flavoured with vanilla; it can be flavoured with any desirable flavour.
- It is used as a condiment with scones and dry cakes in place of jam.

GANACHE

Ganache and truffle are made with chocolate and cream. Chocolate is an indispensable product in sweet preparation. There are two types of chocolates available to chefs—coverture and compound chocolate. Chocolate is used in a variety of ways—for dipping sweets, pastries, marzipan shapes, moulding of eggs and figures, flavouring creams, icing cakes, etc.

Coverture chocolate is a high-class product—smooth and refined. It is manufactured from cocoa mass, sugar, and cocoa butter. To produce a milk coverture, full-cream milk solids are added to cocoa butter and sugar. The cocoa butter gives the impression of cooling the mouth as it melts because the melting point is just below body temperature.

Since coverture contains cocoa butter it needs to be tempered prior to using. The process is necessary because cocoa butter has two types of fat crystals with two different melting points. If the coverture is not tempered, the cocoa butter would set slowly when melted, separate and then rise to the surface, giving the finished product a grey, patchy appearance. Tempering is a process where the coverture chocolate is melted to specific temperatures and then cooled to enable the cocoa butter fat crystals to bind together. Because this process requires considerable skill, time and effort, most establishments do not bother with it, preferring to use compound chocolate that only requires melting to be ready for use.

To melt chocolate, chop it into small pieces and place in a double jacketed container or bain-marie with water not exceeding 50°C. Do not let the chocolate become contaminated by moisture as it will have a harmful effect on the chocolate by changing its consistency and gloss. Excess heat will make the chocolate separate, lose its gloss and granular texture.

The truffle and ganache have the same ingredients but they are slightly different from each other in preparation and usage. Though many chefs use these words—truffle and ganache—interchangeably, there are differences between the two (refer to Table 21.14).

Table 21.14 Differences between truffle and ganache

Ganache	Truffle
To make ganache, cream is brought to a boil and chopped chocolate is added in the ratio of 1 part cream and ½ part chocolate. The mixture is cooled down and then whipped to creamy consistency.	To make truffle, cream is brought to a boil and chopped chocolate is added in the ratio of 1 part cream and 1½ part chocolate. Truffle sets into a dark creamy paste.
Ganache is used for fillings of cakes and pastries.	Truffle is warmed slightly till it becomes flowy and is used to glaze the top of chocolate cakes. It can be used as a filling to produce a dark chocolate truffle cake which is very rich.

The chocolate needs to be mixed together with dairy cream that has been brought to the boil and allowed to set to firm paste. This product is called truffle. A cheaper variety can be made by using a mixture of cream and milk, or by using milk alone. The consistency of truffle may be adjusted by decreasing or increasing the proportion of chocolate. Both ganache and truffle can also be used for piping purposes. Truffle can be mixed with a part of whipped cream to make ganache.

Table 21.15 shows the role of ingredients in making ganache.

Table 21.15   Ingredients used in ganache

Cream	Chocolate
Bakery cream with high fat content is used as it helps to whip to a thick mousse like texture which is desirable for filling in the cakes.	Dark, milk, or white chocolate can be mixed, depending upon the end use of the chocolate.
Milk can also be used, but it will not whip properly and give a very thin and flowy ganache.	Use couverture chocolates only, as compound chocolates do not have quality taste.

Steps in Making Ganache

The steps in making ganache are given below.

Step 1
Boil bakery cream and take it off the fire.

Step 2
Add chopped chocolate and mix with wooden spoon till all the chocolate has melted to a smooth paste.

Step 3
Cool the ganache and whip it into a creamy mousse texture.

Step 4
Use immediately or wrap with plastic and keep refrigerated until further use.

Uses of Ganache

The uses of ganache are as follows:

- It is used for filling sponge cakes to create gateaux and pastries. It can be flavoured with any desirable flavour such as whisky and rum.
- It is used for filling choux pastry items to create desserts.

SAUCES

We read about sauces in Chapter 10. The sauces discussed in this chapter are only sweet sauces or sauces that can be served with desserts or cakes and pastries. The purpose of serving sauces with desserts is to add colour to the desserts and provide the necessary moisture to dry cakes or pastries.

It also enhances the nutritional value of the particular dessert and more often gives a contrasting taste to a particular dish.

While we discuss various sauces, we will come across many terminologies, such as crème Anglaise, etc. Though literally it translates to English cream, it is not classified into creams listed above, but it is classified under sauces. When the crème Anglaise is mixed with whipped cream and set in the moulds with the help of gelatine, the dessert is then known as crème Bavarois or Bavarian cream. The name of the dessert will be coined on the basis of the fruit or ingredient used, for example, mango Bavarois, chocolate Bavarois, etc.

These sauces, made by using milk or cream and thickened with eggs, come under the category of custards. These should not be confused with custard cream or pastry cream as discussed above. Table 21.16 shows different types of custards and their uses.

Table 21.16   Types of custards

Name	Description	Use
Basic custard	This is the uncooked mixture of milk or cream with eggs and sugar. Usually 1 litre of cream or milk, or both depending upon the usage, are combined with 200 g sugar and 10 whole eggs.	There are various ways to use this custard: • It can be poached in moulds lined with caramelized sugar to produce the famous dessert called crème caramel. • Basic custard can be poured onto diced, leftover breakfast rolls and baked to produce a baked dessert called diplomat pudding or bread and butter pudding.
Crème Anglaise ●● 21.2	Milk or cream is boiled with sugar. Egg yolks are beaten with a whisk and then tempered with hot milk or cream. The egg yolk mix is then mixed back to the liquid, stirring all the time with a wooden spoon, until the custard is thick and coats the back of the spoon. The ratios are same as that of basic custard and only difference is that instead of whole eggs, only yolks are used.	This can be used for the following. • Making a base for many desserts such as Bavarios. When the Bavarios is filled in a mould lined with slices of Swiss roll it is called 'Charlotte royal' and when it is lined with sponge fingers, it is called 'Charlotte russe'. • Served as a sauce. Crème Anglaise can be flavoured with many flavours such as citrus zests, caramelized sugar, and spirits such as brandy, rum, and whisky.
Sabayon	Egg yolks are mixed with sugar and few tablespoons of liquid such as milk, cream, or liquor and are whipped over bain-marie until they form ribbons.	• Used as a base for deserts such as mousse and souffles. Served as a dessert all by itself, for example *zabaglione* from Italy. • Used also for topping upon fruits and berries and gratinated.

Sauces used in pastry do not have any classifications like Western sauces, but the sauces are broadly classified as shown in Table 21.17.

Table 21.17 Classification of pastry sauces

Classification	Description	Use
Puree based	Various types of fruit purees, such as 'coulis', are fruit based. The fruit is stewed with some amount of sugar and cooked till soft. The sauce is either left coarse or passed through a sieve resulting in a coulis.	Berry coulis is served with desserts and ice creams. Coulis folded in with whipped cream results in a dessert called 'fool'. It is used as a topping or filling for cakes and pastries.
Custard based	Refer to Table 21.16.	Refer to Table 21.16.
Chocolate based	Truffle can be thinned down with milk, cream, or sugar syrup to obtain chocolate sauce. The colour of the chocolate sauce can be regulated by adding the milk and cream in required proportions.	Served with ice creams, desserts, cakes, and pastries.
Cream based	Reduced cream with flavourings is widely used in confectionery. Caramel sauce, butterscotch sauce, etc. are examples of cream-based sauces.	Served with ice creams, desserts, cakes, and pastries.
Miscellaneous	These sauces are basically reductions of liquids such as wines and fruit juices.	Served with ice creams, desserts, cakes, and pastries. Mixed with whipped cream to make fillings for cakes and other desserts.

ADDING FLAVOUR TO THE PASTRY SAUCES

Flavour is the total sensory impression formed when food is eaten. It is a combination of the sensations of taste, smell, and texture. Many of the sweet sauces are based on specific flavouring agents, for example, vanilla, fruit pastes, nuts, spirits, etc. Skilful blending of flavouring agents with the basic sauce is the key to successful sauce making. The result is an almost endless range of sauces that have individual and characteristic flavours. If you have a natural alternative, it is best to avoid artificial flavours, which can be overpowering.

Condiments, herbs, spices, and flavourings are used to modify, blend, or strengthen natural flavours. The use of these materials may make the difference between a highly palatable food and a drab, tasteless one. Some flavourings used in pastry sauces are discussed in Table 21.18.

Table 21.18 Flavourings used in the pastry sauces

Flavouring	Description
Salt	Salt is the most widely used ingredient for the seasoning of food; but its application in the preparation of sweets is limited. It is used to bring out the natural flavours in the food.
Acids	Acids in the form of vinegar and lemon juice are commonly used to flavour food.

Contd

Table 21.18   (Contd)

Flavouring	Description
Extracts	These are derived from the natural flavouring material. The flavour is extracted by macerating the natural source, for example, vanilla in ethyl alcohol. These extracts are the best flavouring materials, but also the most expensive.
Herbs and spices	The art of skilfully adding the right amount of spice or herb to a food is basic to successful pastry cookery. Frequently used herbs and spices are cinnamon, nutmeg, cloves, and mint.
Essential oils	They contain the principal flavour of all fruits, nuts, and flowers. Lemon and orange oils are most useful in patisserie work, as they will withstand high temperatures without deterioration.
Essences	Many flavours can now be made artificially. Used with discretion, they are very useful when the natural substitute is not available.
Blended flavours	These are compounded from both natural and artificial sources. Such essences have the true bouquet of the natural flavour, reinforced with the strength of the artificial essence.
Fruit pastes and concentrates	These are products that impart the true flavour of the fruit.
Spirits and liqueurs	These are expensive and therefore should be used with discretion. Since they are volatile substances that can evaporate when heated, their use should be confined to creams, icings, sauces, etc. They should rarely be used as ingredients in goods that are to be baked. Some of these are available as concentrates and are ideal for cooking purposes.

COMMON FAULTS IN SAUCE MAKING

Many a time the dishes do not turn out to be as good as desired. The reasons could be many. In case of bakery and confectionery, where each process is weighed and scaled, the temperature is to be taken into account apart from various other factors that influence the outcome of the product. The quality of flour varies in different countries because of the soil and processing techniques. So chefs have to adjust their recipes according to the ingredients available. Some errors in bakery and confectionery could be technique and process related and chefs can fix them, once they know the source of the problem. The most common faults that could occur during making sauces for pastry are as follows:

Lack of Flavour (poor base)   This happens when the flavouring used was insufficient, or the quality of flavouring was old and lacked the desired flavour.

Too Strong a Flavour   This is the result of using too much flavouring.

Starchy Flavour   The insufficient cooking of sauces gives it a starchy flavour.

Too Thin   The sauce is too thin when insufficient thickening is added, or the mix is not cooked to the desired consistency.

Sauce Ferments   This happens when the sauce is too old or it is not stored properly.

Incorrect Colour   The use of too much artificial colour causes this.

CONCLUSION

In this chapter we discussed various kinds of pastes and creams that can be used for fillings and decorating the cakes. After the completion of this section, you could make cakes of international standards. As baking of cakes calls for a lot of skill and patience, it can be perfected by practice.

We discussed various kinds of pastes used in confectionery and read about the versatile uses of the pastes. We saw how some pastes, such as choux paste and puff pastry, form a base for many desserts, while other types of pastes, such as frangipani, are used in the fillings of gateau pithivier prior to baking. Some pastes are used for making garnishes and decorations (*tuile* paste).

We discussed short crust paste and sweet paste and understood the uses of the same. We also saw how techniques can affect the final product. If the ingredients in the sweet paste or short paste are kneaded, then they will form into dough; but just by treating the ingredients differently, we get the product that can be used for making cookies and pies. In each category we also discussed the role played by the ingredients in making of these products and how we can alter them to change the textures. While discussing creams, pastes, and sauces, we also discussed various other pastry-related things such as differences among tarts, pies, and flans; differences between mousses and souffles. We focused on the making and usage of choux pastry and various desserts made from the same.

We also discussed various types of creams apart from whipped cream and understood how the creams can be flavoured to enhance the textures in a cake. The crunchy texture of caprice cream will be different from creme Chantilly and the usage of each in a cake will provide a different mouthfeel or texture to the guest. We also discussed in detail butter cream that is the most commonly used filling in cakes around the world. Lemon cream and ganache were also discussed. The difference between ganache and truffle completed our understanding of the uses of each. While dealing with ganache and truffle we also touched upon another versatile commodity used in pastry kitchens— chocolate. We discussed various types of chocolates and their uses and also briefly discussed tempering of chocolate and the purpose of the same.

We discussed various types of sauces and their classification and once we are able to produce cakes with fillings and toppings, we can then present them on plate with accompanying sauces and garnishes to create plated desserts for à la carte and buffets. We also discussed some common faults in sauce making and how to rectify them.

In the next chapter we will discuss laminated pastries and this will complete our basic and elementary knowledge about bakery and confectionery.

KEY TERMS

Bavarois A chilled dessert made by folding crème Anglaise into whipped cream and setting with gelatine in a mould.

Brandy snaps A type of crisp cookie used for decoration.

Caprice cream Variation of crème Chantilly, where sugar is replaced by broken pieces of meringue.

Charlotte royal Bavarois lined with sliced Swiss roll.

Charlotte russe Bavarois lined with sponge fingers.

Chocolate cobbler An English dessert where a sponge sheet is topped with pastry cream and topped with cooked fruit and baked with crumbled sweet paste on top.

Compound chocolate Chocolate stabilized with stabilizers and cocoa fats.

Coulis Stewed fruits with sugar and pureed to a sauce.

Coverture Cooking chocolate available in blocks or pellets.

Crème Anglaise Cooked custard of milk, sugar, and egg yolks.

Crème caramel A dessert made by poaching the basic custard in a mould lined with caramelized sugar.

Crème Ciboust Pastry cream mixed with a part of meringue, also known as St Honore cream.

Crème patisserie French for pastry cream.

Danish pastry A type of breakfast pastry.

Diplomat pudding Dessert made by using toasted buttered bread or leftover breakfast rolls, soaked with basic custard and baked.

Fool Dessert made by folding in fruit purees or coulis with whipped cream.

Frangipani   Paste made from almonds, sugar, eggs, butter, and flour.

French meringue   Egg white whipped with castor sugar.

Ganache   One part cream and one and a half parts of chocolate cooked together and whipped to a cream when cold. It is used for fillings.

Gateau pithivier   A disc-shaped cake made by sandwiching two discs of puff pastry with almond paste and baked in the oven.

Gratinate   To colour the top of the cake under the radiated heat source.

Italian Meringue   Egg whites whipped with sugar syrup cooked at 118°C.

July pan   A type of decorating paste.

Lemon zest   The outer skin of lemon, without the white pith.

Muffins   Breakfast pastry made by the pound cake method and baked in tart shells.

Panada   Starch thickened with liquid to resemble a thick paste or dough.

Pastry cream   Milk cooked with eggs, sugar, and starch to produce a creamy product.

Persipan   Marzipan made by using kernels of stone fruits instead of almonds.

Quiche   Savoury pie baked in a short crust paste.

Royal icing   Egg whites beaten with icing sugar and cream of tartar till it becomes like a thick paste.

Swiss meringue   Egg white and sugar whipped over a bain-marie.

Temper   Used in different contexts. Tempering of liquids means adding a hot liquid to a cold one, to bring them to the same temperature. In case of chocolates, it refers to heating and cooling the chocolate to allow it to be used for garnishes.

Truffle   One part of cream cooked with one and a half part of chocolate to a dark creamy texture.

Turnovers   Small snacks or desserts made by rolling the puff pastry into thin shapes like square or circle and filled with a filling and turned over to encase the filling and baked.

Vanilla pod   A small pod that contains vanilla flavour.

Zabaglione   Italian dessert made by cooking egg yolks and sugar with Marsala wine.

CONCEPT REVIEW QUESTIONS

1. Define a paste in the context of confectionery. How are these different from pastes used in Western cooking?
2. What do you understand by cream used in pastry kitchen?
3. What is the difference between a short crust paste and a sweet paste?
4. How would you differentiate among a tart, flan, and a pie?
5. What is the role of flour, fat, and liquid in short paste?
6. Short crust paste and sweet paste need to be short. What do you understand by this and how would you ensure that the final product is short?
7. List the steps used in making short crust paste.
8. What are the factors that cause shrinkage of short crust paste and sweet paste?
9. Define the role of ingredients in making sweet paste.
10. List five products made by using choux paste.
11. Give names of at least three classical desserts made from choux paste.
12. What is the principle behind rising of the choux paste and the hollow structure?
13. List the steps involved in making choux pastry.
14. What is the difference between a marzipan and a *persipan*?
15. How is marzipan different from almond paste?
16. Give at least three uses of marzipan.
17. What are the steps involved in production of marzipan?
18. Name at least two products made by using frangipani.
19. What is a *tuile* paste? What is it used for?
20. List at least five types of creams used in pastry kitchen.
21. What is the difference between a pastry cream and crème Ciboust?
22. How would you differentiate between crème Chantilly and caprice cream?
23. List at least three differences between a mousse and a soufflé.
24. What care should be taken while whipping cream?
25. How do you make the butter cream and how should it be stored?
26. What kinds of flavourings can be added to butter cream?

27. What are three different types of meringues and what is an Italian butter cream?
28. What is lemon curd? How is it made?
29. What is a ganache and a truffle? What are its uses?
30. Write a short note on chocolate.
31. What do you understand by the word 'tempering' of chocolates?
32. Write a short note on sauces used in pastry kitchens. How are they different from Western sauces?
33. List three types of custards and their uses.
34. What is Bavarian cream? List down at least two classical products made from it.
35. Define *zabaglione*.
36. How would you classify the Western sauces?
37. How is coulis different from fruit puree?
38. List at least three broad categories of flavours used in pastry sauces.
39. What are the common faults in sauce making?
40. What are the differences between essences and extracts?

PROJECT WORK

1. In groups, make at least five different paste recipes from the recipes provided in the ORC and compare the pastes and write your observation in the chart below.

Name of Paste	Texture	Consistency	Baking Effect	Colour

2. In groups, try out the recipes as provided in the ORC for the basic creams and make the cakes and pastries with the same. Compare the results of the cakes and see the difference in the products.
3. Visit the various pastry shops in the city and write down the various types of cakes and pastry products available. Now make a list of fillings that were used for creating those products.
4. Make pastry cream and add various flavours to the same. Compare the results and add different textured products to the cream such as roasted chopped nuts, chopped chocolate, broken pieces of meringue and taste the products. Make a creative cake with this cream and compare with each other.

CHAPTER 22

LAMINATED PASTRIES

Learning Objectives

After reading this chapter, you should be able to:

- understand the different types of laminated pastries
- know the various methods of making laminated pastries
- know various steps involved in production of laminated pastries and usage and storage of each pastry
- prepare various kinds of products made from the same ingredients
- understand the various techniques used in preparing Danish pastry and croissant
- know the role played by different ingredients in making laminated pastries

INTRODUCTION

The word 'lamination' means covered with thin film, but when we refer to lamination in cooking, it means a dough layered fat in such a manner that, layers of dough are separated by fat. This is achieved by encasing the dough with butter and rolling and folding it several times to get the desired effect. Laminated pastries can be made with plain dough to produce puff paste or fermented yeast dough can be laminated to produce croissant and Danish pastry that are the most common breakfast rolls eaten around the world. Sometimes dough is rolled out very thin and then laminated with fat, for example, strudel from Austria. Very thin sheets of 'phyllo pastry' are laminated with fat and used for various purposes.

Whatever may be the name and use, there are two things that are present in any laminated pastry and these are—dough and fat. The dough could be of various types such as leavened or plain. Making laminated pastries requires lots of skills and technical knowledge. That is the reason why we are dealing with this group in the last section of pastry work. It is very important to be aware of the commodities and pastry techniques to be able to make laminated pastries.

PUFF PASTRY

Puff pastry consists of laminated structure built up of alternate layers of dough and fat. This is achieved by rolling out the paste and giving it sufficient turns until there are hundreds to thousands of layers of dough and fat. When this pastry is baked, the expanding air and water vapour 'puff' the separate layers apart from each other, resulting in a delightful, crisp, light, flaky pastry. When the thinly sheeted piece of puff pastry is baked in the oven, the puff goes through the following stages.

- The fat present in the layers melts and creates gaps between the sheets of dough.
- The dough sheets start to harden and maintain their shape as the gluten present in them coagulates and holds the shape.
- As the heating continues, the liquid in the dough turns to steam, which pushes the sheets of the dough apart and pushes them upwards.
- The end result is the pastry which is almost 10 times thicker than the original size that was put in the oven.

Puff pastry is also known as *mille feuille* in French and literally it translates to thousand layers. The number of layers in the pastry is not fixed—the layers could be between 700 and 1500 depending upon the rolling of the puff pastry. It is also known as leaf pastry, as when it bakes each layer of dough resembles a crisp leaf. Puff pastry is rolled in such a way that the layers of dough and fat form naturally.

Table 22.1 shows the ingredients used in puff pastry.

Table 22.1 Ingredients used in puff pastry

Flour	Fat	Liquid	Acid	Flavouring
Hard flour or bread flour is preferred for puff pastry. The gluten must be well formed and elastic so that it can hold the layers of fat when rolled.	The fat should be elastic and should have a high melting point. The quantity of fat should be equal to the dough for best results. It is best to use butter; but care has to be taken as it has low melting point.	Cold water should be used as warm water tends to make a warm pastry. Also when warm water is used, the fat encased will melt, thereby making an oily puff pastry.	Usually lemon juice is added. This is done for two purposes—to have a bleaching effect on the dough and to strengthen the gluten strands.	Salt is the only flavour added to the puff pastry. Salt gives a colour and taste to the end product. It also helps in the keeping quality of the product, as it prevents staling.
Eighty per cent of the flour is made into dough with water, acid, and salt.	Remaining 20 per cent of the flour is creamed with butter and the butter is set into a rectangular shape to make 'butter block'. This helps the butter to stretch while rolling, thereby ensuring even distribution.	Cold liquid, preferably ice water, is used to make the dough. The dough needs to be kneaded until the gluten is formed. Do the wind screen test (refer to Chapter 19).		

Types of Puff Pastry

There are four commonly known types of puff pastry—'half,' 'three-quarter,' 'full,' and 'inverted puff'. The terms describe the amount of fat to the weight of flour. 'Full' denotes equal weights of fat and flour, 'three-quarter' means three-quarter of the weight of the fat to the flour, etc.

Inverted puff is a special way of making puff pastry, where butter is not encased in the dough; but on the contrary, dough is encased into butter block. This might seem very difficult, but with a little practice, one can achieve the desired result. Inverted puff gives much better results. Croissant and Danish pastry are made from laminated dough that is prepared in much the same way as puff pastry, but uses yeast in the dough base. The dough is laminated by rolling or pinning. It is encased with butter and then folded in various ways depending upon the method used.

METHODS OF MAKING PUFF PASTRY

Commercial chefs commonly use one of three methods to incorporate fat into the dough. These are known as the French (Continental), English, and Scotch methods. The difference among these methods is the way that the fat is encased into the dough during the stages of rolling and folding. Before we go on to describe the methods of lamination, let us talk about the methods of rolling and folding the pastry to give it the laminated layers. These methods are as follows:

Three Fold or Single   Roll out the paste into a rectangle and fold it equally into three parts.

Book Fold or Double Turn   Roll out the paste into a rectangle, fold the ends in first, and then fold together like a book.

Combination Method   This method involves using the combination of single and double folds.

It is important not to give the pastry too many turns because too many layers will break down the layers of fat and dough, making it more like short crust pastry. However, too few layers will result in coarse layers of pastry that will perhaps also give uneven lift and allow the fat to run out during baking.

Let us now discuss all the three methods of laminating fat into the dough to create a laminated pastry. These will be clearer once we discuss each of them step by step.

French Method

[22.1]   Let us understand the making of puff pastry by French method from the following steps.

Step 1

The main agent for success in the preparation of puff pastry is the dough. It is essential to have the fat and dough of equal consistency, if even laminations are to be attained. If the fat is harder than the dough, the fat will ooze through the dough during rolling. Preparing the dough can be done either mechanically or manually. Use low speed when preparing the dough mechanically to allow slow development of the gluten. When preparing manually, knead the flour, cold water, salt, and lemon juice together on a work bench until the dough is smooth and elastic. A potion of the weighed fat (10 per cent) can be added when making the dough to help in the rolling out of the pastry. Take care when adding the water, as different flours have different absorption characteristics.

Step 2

Roll the dough into a ball and let it rest before incorporating the fat. The dough can be chilled in the refrigerator.

Step 3

Make the butter block. Cream the butter and add 20 per cent of the flour into the creamed butter and shape it to form a rectangular piece of the same thickness as that of the first rolling of the dough. Refrigerate until set, but not too hard. Flour is added to the butter so that it becomes elastic and rolls along with the dough. The consistency of the butter has to be the same as that of the dough, to allow the fat to slide and form a 'film' during rolling and folding. If it is too hard, the fat will break down the laminated structure. On the other hand, if it is too soft the fat will squeeze out and shorten the pastry.

Step 4

Take the ball of rested dough and cut a cross into the top of the dough by cutting the dough half way through with a knife as shown in Fig. 22.1.

Step 5

Pull the cut corners of the dough and roll to form it into the rough shape of a star as shown in Fig. 22.2.

Fig. 22.1 Ball of dough with a cross

Fig. 22.2 Dough rolled into the shape of a star

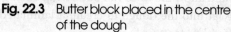

Fig. 22.3 Butter block placed in the centre of the dough

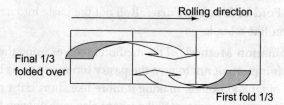

Rolling direction

Final 1/3 folded over

First fold 1/3

Fig. 22.4 Folding of puff paste

Step 6

Now roll each corner of the star to form a shape of '+'. The thickness of the corners should be one fourth of the thickness of the centre (refer to Fig. 22.2).

Step 7

Now take the butter block and give shape to it. Trim it to the same shape as the centre of the dough and place the fat on the centre part of the dough (Fig. 22.3).

Step 8

Fold all the four flaps to the centre, so that the entire fat is encased in the dough. Now you will understand as to why the flaps were rolled to one-fourth of the thickness of the centre part of dough. The four flaps when folded over ensure that the butter is encased exactly in the middle of the dough. Wrap the pastry with a damp cloth or plastic film and rest the pastry in the fridge for 20 minutes. The pastry is wrapped this way to avoid the formation of scales on top of the pastry, which will result in 'flying tops' (refer to Table 19.6.).

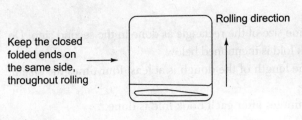

Keep the closed folded ends on the same side, throughout rolling →

Rolling direction

Fig. 22.5 Completed single turn

| 2/3 dough is covered with fat | 1/3 dough is clear |

Fig. 22.6 Two-third of dough covered with fat

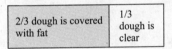

A   B

Fig. 22.7 First fold of the dough

Step 9
Roll the dough into a rectangle. The length of the rectangle should be three times the size of the width and the thickness should be around 7 mm. Now fold the dough to form a block again as shown in Fig. 22.4. This is said to have completed 'a single turn'.

After folding, the puff paste will look as shown in Fig. 22.5. Chill for 30 minutes after this step is done.

Step 10
Roll out the dough and repeat the ninth step five more times to achieve six single turns. The dough is rolled by using minimal amount of dry flour, and this should be brushed off when turning folding the dough to give turns. It should be ensured that the closed folded ends are on the same side when pinning or rolling the dough. It does not matter whether the closed end faces the right or the left; but while rolling it must be at one side during all the six turns.

English Method
The difference in the French and the English methods is essentially in the rolling of the dough. The making of the dough is same as the first step of the French method. Let us understand the making of puff pastry by English method from the following steps.

Step 1
Make the dough as in the French method; but instead of making a ball, sheet out the dough and rest it.

Step 2
Roll out the dough to a rectangle which is almost three times of the width and 7 mm thick.

Step 3
Roll the butter block so that it is the size of two third of the length of the rectangle. See Fig. 22.6.

Step 4
Fold the clear part onto the dough. The shape will now resemble Fig. 22.7.

Step 5
Now fold the part marked 'A' in Fig 22.7 onto the part marked 'B' and the shape obtained will be same as that of Fig. 22.5. Same care needs to be taken about the folded end and the rolling direction as mentioned in Fig. 22.5.

Step 6

Pin out the pastry again to 7 mm thickness and same size of the rectangle as done in the second step. Do five more single turns or four book turns. The book fold is mentioned below.

Roll the dough to form a rectangle, where the length of the dough is at least four times the size of the width. Fold the dough as shown in Fig. 22.8.

The dough should be rested for at least 30 minutes after each book fold is done.

Scotch Method

This method is very different from the French and English methods and is also commonly known as 'rough puff pastry' method. There are certain products that need the short and laminated effect, but do not require the structure of a real puff pastry. In such cases, the scrapings left over from the puff pastry are used, or if possible, rough puff pastry is made as it is a fairly quicker method and does not require as much care as puff pastry.

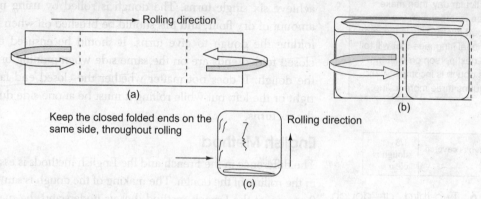

Fig. 22.8 Book folds for puff pastry

Let us understand the making of rough puff pastry by the Scotch method from the following steps.

Step 1

Sift the flour, cut the hard butter into 1 inch cubes and dissolve lemon juice, water, and salt together.

Step 2

Mix flour and cubes of butter together so that the butter is well coated with flour. Now add the liquids to the flour and mix together to form dough. Care should be taken that the butter remains intact during the entire process and so it would be advisable to make the dough manually.

Step 3

Rest the dough in a fridge for 10 minutes and then give six single turns or four double turns as mentioned in the last two steps of the French method.

INVERTED PUFF PASTRY

As discussed earlier in the chapter, this is a reverse style of making puff pastry where the dough is encased in a butter block. The ingredients used in this method are same as that of French or English method; but the quantities of some ingredients and methods of mixing might differ (refer to recipes provided in the ORC). Let us discuss the steps used in the production of inverted puff dough.

Step 1

Make dough with chilled water, lemon juice, salt, and some amount of fat and chill in the refrigerator.

Step 2

Make a butter block. Here the amount of flour added to butter is at least 35 per cent and not 10 per cent as in the case of the French and English methods. The amount of flour is high, because the fat needs to be rolled. Roll out the butter block to a rectangle and this can be easily done by placing the block between two sheets of plastic and pinning it.

Step 3

Dough

Butter block

Fig. 22.9   Rolling of inverted puff

Dust the work table with flour and place the rolled out dough on top of the butter block. Please ensure that the size of the butter and the dough are same. Press the dough down onto the butter block by knocking it down with a rolling pin and roll out to a rectangle where the size is three times the size of the width (Fig. 22.9).

Step 4

Now proceed for five single turns as done in Steps 9 and 10 in French method. This method does require lots of flour initially while rolling. Care should be taken to brush off all the excess flour with the pastry brush before giving the folds.

USES OF PUFF PASTRY

Puff pastry is a versatile pastry and can be used in savoury or desserts. There are many uses of puff pastry such as in snacks, main courses, and even in making pastries and cakes. Some of these are discussed in Table 22.2.

Table 22.2   Preparations of puff pastry products

Item	Description	Puff Pastry	Photograph
Vol-au-vent 22.2	Literally it translates to 'gone with the wind' in French, as it is very light in weight. A disc of 2 inch diameter is cut 7 mm high, and the centre is cut separately with a smaller disc but only to a depth of 5mm. The baked hollow shell looks like a well. The centre part is taken off and the cavity is filled with savoury fillings such as creamy vegetables, etc. The mouth-sized vol-au-vent is called bouchee.	Both puff pastry, made with English or French method can be used.	

Contd

Table 22.2   (Contd)

Item	Description	Puff Pastry	Photograph
Palmiers	It is often used as a breakfast pastry and is known as French hearts because of the heart shape. It is also called by other names such as pig's ears or papillon. The puff pastry is rolled with castor sugar into a rectangle and it is given folds inwards from both the ends to meet in the centre. It is finally folded just like a book fold and sliced 1cm thick. The open ends are twisted to resemble a 'T' shape and it is baked to form a heart shape.	Rough puff pastry/Scotch pastry or scrapings of puff pastry can be used. For selling in pastry shops one could also use real puff pastry.	
Napolean cake/pastry	The puff trimmings are rolled out into thin sheets and docked well before baking. The baked sheets are then filled or layered with pastry cream. This cake is traditionally decorated with melted fondant and feather icing. This is also referred to as cream slice.	Rough puff pastry/Scotch pastry or scrapings of puff pastry can be used.	
Eccles	These are sweet snacks eaten often during afternoon tea. Roll the puff to 4-inch circle and 6 mm thick. Place the filling of raisins, brown sugar, mixed spice and little softened butter. Bring all the sides of circle together and form a ball. Now roll the ball into a 3-inch circle, wash with egg white, and dip in castor sugar and give two to three small slits on top and bake at 180°C till golden brown.	Rough puff pastry/Scotch pastry or scrapings of puff pastry can be used.	
Turnovers	The puff is rolled into discs or squares and filled with savoury fillings and turned over or folded over. These are then glazed with egg yolk and baked. Some shapes of these turnovers represent the following—a semicircle represents chicken, a triangle represents vegetable, and a square represents lamb.	Real puff pastry is used.	

Contd

Table 22.2 (Contd)

Item	Description	Puff Pastry	Photograph
Gateau St Honore	Refer to Table 21.5 in the previous chapter. The puff pastry is rolled into a disc and docked before baking. This is used as the base for gateau St Honore.	Rough puff pastry/Scotch pastry or scrapings of puff pastry.	
Beef Wellington	A classical roasted dish often used on carving table, beef Wellington is a tenderloin fillet wrapped with mushroom duxelle and covered with puff pastry. It is glazed with egg yolks and baked. Some snacks such as sausage puffs, etc. are made by rolling marinated sausages in the thinly rolled out puff and baked in oven.	Rough puff pastry/Scotch pastry or scrapings of puff pastry can be used.	
Branbury cake	Same as eccles as mentioned above. The only difference is that it is rolled into boat shapes or oval shapes and baked in the same way.	Rough puff pastry/Scotch pastry or scrapings of puff pastry can be used.	
Jalousies	These are thin broad strips of puff pastry, filled with cooked fruit and covered with another strip of puff and baked. They are glazed with melted apricot jam and served as dessert.	Real puff pastry is used.	
Garnishes	Sometimes the puff pastries are also cut into different shapes, usually half moon and are used as an accompaniment with grilled fish in French cooking. They are also known as 'fleurons'.	Puff pastry is used.	
Cheese straws	These are savoury cheese twists. Puff paste is rolled out thin and sprinkled with grated cheese and paprika. It is then covered with another sheet of puff and small strips are cut and twisted into long twists and baked.	Rough puff pastry/Scotch pastry or scrapings of puff pastry can be used.	

PREPARATION OF PUFF PASTRY

Making puff pastry requires lot of skill and knowledge about the commodities used in the same. The handling techniques and the science behind each must be kept in mind. Some of the points that need to be kept in mind are as follows:

- Keep all ingredients cool.
- Sieve the flour to aerate it and remove impurities.
- When using the English and French methods, the dough should be smooth, elastic, and rested prior to adding the fat. Dough and fat should always be of equal consistency.
- The smaller the quantity of pastry being made, the less it needs to be pinned out in thickness.
- Rest the pastry after every two turns for a minimum of 20 minutes in a cool place. Cover the dough to prevent the formation of a crust. You can keep track of the number of turns you have made by making an indentation in the pastry with your finger each time you turn it.
- When rolling out the pastry, use minimum amount of flour for dusting.
- Cut and pin out only as much puff pastry as you need at the time. When trimming, use a sharp knife or cutter to prevent pressing the layers together at the edge.
- Scrap pastry that remains after items have been cut out can be added to the pastry that has never been pinned out, and reused later.
- When egg washing the pastry, be careful not to drip onto the side of the product or it will not rise evenly.
- Rest prepared products before baking, to avoid shrinkage and distortion.
- Bake at a high temperature—200 to 220°C.
- The trays or the baking sheets do not need any greasing when baking puff products.
- While roiling the dough, lift the dough and let it relax. Otherwise the products will shrink in the oven. Rest the cut out shapes like vol-au-vents, etc. before baking, as they will shrink and become oval in the oven.
- Always brush the egg yolk on the vol-au-vents and then cut in the centre. If you brush the egg after cutting then it will prevent them from rising in the oven.

DANISH PASTRY AND CROISSANT

Danish pastry and croissant are the two most important breakfast rolls served in the breakfast of any high end hotel. The quality of the croissant is very important. It is said that the impression of the hotel is created or broken by the quality of the croissants served in the breakfast. Danish and croissant doughs are used interchangeably in hotels, but traditionally, the Danish dough is richer than the croissant dough. Danish dough contains eggs and more butter than the croissant dough and that is the reason why sometimes croissant dough is referred to as lean dough. The recipes of both the doughs are given in the ORC.

The style of laminating the Danish or croissant dough is same as that of French or English methods of making puff pastry; the only difference is that the Danish and croissant dough are leavened with yeast. There is nothing called rough Danish or rough croissant dough. The Danish and croissant are flaky to eat, but their texture is soft and not crisp as in case of puff pastry. The other difference is that both the Danish and croissant are proved like any other bread before baking in the oven.

Table 22.3 shows the ingredients used in Danish pastry and croissant.

Table 22.3   Ingredients used in Danish pastry and croissant

Flour	Fat	Liquid	Leavening	Flavouring
Medium flour is preferred for Danish pastry. The gluten must be well formed and elastic but not overmixed as the pastry will toughen.	The fat used is only butter as it gives a better taste in the product. Unlike puff pastry, the quantity of fat is half in case of Danish dough and croissant dough.	Cold milk or water or a combination of both should be used as warm water tends to activate the action of the yeast (refer to Chapter 19). Eggs are added to the Danish dough for a richer effect.	Yeast is the leavening agent used for Danish and croissant, and the process of making the doughs could be same as any other bread dough. Care has to be taken to let the gluten develop and not to overmix.	Salt and sugar are the only flavours added to the Danish and croissant doughs. Salt gives a colour and taste to the end product. It also helps in keeping the quality of the product, as it prevents staling.
Eighty per cent of the flour is made into a dough with milk, eggs, sugar, yeast, and salt.	Remaining 20 per cent of the flour is creamed with butter and the butter is set into a rectangular shape to make butter block. This helps the butter to stretch while rolling, thereby ensuring even distribution.	Cold liquid, preferably ice water, is used to make the dough. The dough needs to be kneaded until the gluten is formed. Windscreen test is not required as we will not overmix the dough to get that texture.		

Techniques Used in Making Danish Pastry

The techniques of laminating the Danish pastry and the croissant are similar to that of puff pastry with some differences. Let us discuss the steps in making Danish pastry.

Step 1

Combine the flour, salt, sugar, milk, and yeast together in a mixing bowl and knead it to smooth dough. Take care not to overmix the dough, so that the end product is soft yet flaky.

Step 2

Cover the dough and keep it for intermediate proving (refer to Chapter 19).

Step 3

Do the knock back (refer to Chapter 19).

Step 4

Rest the dough and meanwhile prepare the butter block.

Step 5

Roll the dough and laminate the butter in English style as in English method. French method can also be used, but English method is more preferred. Follow the same steps as shown in puff pastry; but in case of croissant and Danish pastry, only three single folds are given unlike six in puff pastry.

Step 6

Cover the rolled pastry with plastic and freeze in the freezer until use. This is so done so as to avoid the proving of the dough in refrigerator. Before rolling for final use, one can let the Danish pastry rest at room temperature for some time, which will bring it back to rolling consistency.

The shaping of croissant is very different to that of Danish pastry. A croissant can be plain or stuffed. A Danish pastry is garnished with fruits or various other fillings such as pastry cream, frangipani cream, etc. and glazed with boiling apricot jam to give a shiny glaze. It can be drizzled with melted fondant to enhance the look. A croissant on the other hand is glazed with eggs and baked. It is served plain or dusted with icing sugar if it is filled with sweet filling such as chocolate or almond paste. Table 22.4 discusses different Danish pastries.

Table 22.4   Various types of Danish pastry

Type	Shape and Filling	Photograph
Custard Danish	Roll out the pastry to 7 mm thickness and cut out 4-inch squares. Fold two opposite corners to meet in the centre and press them with your fingertips. Pipe custard cream in the centre and bake after proving. Glaze with melted hot apricot jam and drizzle with melted fondant. A fondant is available ready made. To make it, cook sugar with little amount of water and cream of tartar to 118°C. Spread on a clean marble table and work it with a spatula to make shape '8' on the table. When it cools down, knead it to smooth in room temperature white dough. This can be stored until further use at room temperature.	
Cinnamon roll	Roll out the pastry to a rectangle 6 mm thick. Spread it with custard cream and sprinkle cinnamon powder and some raisins. Roll it like a Swiss roll and chill. Slice it 1 cm thick and prove prior to baking. Glaze with hot apricot jam.	
Bear's paw	Roll out the Danish to 6 mm thick and 4-inch square pieces. Place almond paste into the centre and fold over one side to form a small rectangle. You can use egg wash to stick the sides. Now slit the end at small equal intervals, where the joint is and spread into a crescent for it to resemble a bear's paw. Prove, bake, glaze with jam and sprinkle roasted almond flakes.	

Contd

Table 22.4   (Contd)

Type	Shape and Filling	Photograph
Pin wheels	Roll the Danish to 6 mm thick and 4-inch square pieces. Make a slit from each corner till centre of the pastry, taking care that the slits do not meet each other. Pipe custard cream in the centre and place a slice of stone fruit such as peach, prune, apricot, or fig and fold each ear of the slit corner to meet in the centre.	
Flower Danish	Roll the Danish to 6 mm thick and 4-inch square pieces. Fold the corners to meet each other in the centre. Pipe custard cream in the centre and place a slice of stone fruit such as peach, prune, apricot, or fig.	

Techniqes Used in Making Croissant

The method of making and rolling the croissant dough is same as that of the Danish dough; but the final shaping of this crab-shaped pastry is different from the Danish pastry. Let us discuss these in the following steps.

Steps 1–6

The first six steps in making croissant are exactly similar to those of Danish pastry.

Step 7

Roll the pastry to about 7 mm thick, and around 6 to 7 inch wide strips.

Step 8

Cut the strips into triangles. Ensure that the base of a triangle is 4 inches.

Step 9

Make a small slit in the base of the triangle and roll the triangle like a cigar, from the base till the tip.

Step 10

Place onto baking sheets and prove until double in size. Glaze with egg wash and bake until golden brown and crisp.

Croissants are usually baked plain, but sometimes they could be stuffed with grated chocolate to make chocolate croissant. They can be glazed with melted chocolate after baking or dusted with icing sugar.

STRUDEL

This pastry comes from Austria. The cross section of the strudel shows the layers of crisp flaky dough to resemble circles and that is why it is called strudel meaning 'whirlwind'. There is a German variation of strudel that uses puff pastry rather than strudel dough. Strudel is made by pulling and stretching the dough to a thin sheet and laminating it with fat. This pastry is synonymous with the famous Austrian dessert, apple strudel. Obtaining a crisp pastry depends on the correct preparation and working of the dough. It is essential to get the correct consistency to allow the dough to be 'pulled out' or stretched out satisfactorily. The basic ingredients used to prepare strudel are shown in Table 22.5.

Table 22.5   Ingredients used in strudel

Flour	Fat	Liquid	Flavouring
Hard flour or bread flour is preferred for making strudel dough. The gluten must be well formed and elastic so that the strudel dough can be stretched to form thin sheets.	Ten per cent of the fat, such as butter, etc. is used in the dough. When the strudel dough is made, it is soaked in oil for a couple of hours. The oil prevents the formation of scales on top of the dough and also softens the gluten so that it can be stretched to a thin sheet.	Water is used to make the dough. Egg yolks are also used along with water. Eggs enrich the pastry and help to hold the dough while stretching. The yolks help to give a creamy yellow colour to the dough.	Salt is the only flavour added to strudel dough. Salt gives a colour and taste to the end product. It also helps in keeping the quality of the product, as it prevents staling. A chef must be careful in using salt, especially if he/she wants to use this for desserts.
Do not over knead the dough.	Melted butter is used for lamination of the dough.	Milk can also be used to enrich the dough, but usually water is used.	Many sweet or savoury fillings are used. The filling gives its name to the strudel, such as apple strudel, vegetable strudel and so on.

Method of Preparation

Strudel dough can be prepared mechanically or manually, following the guidelines used in preparing the puff pastry dough. Knead the dough thoroughly to develop the gluten and make it easier to pull out. It should be slightly soft, smooth, pliable, and elastic. The strudel dough is stretched with hands till it is so thin that one can read a newspaper under it. This dough is very fragile and thus care must be taken to avoid any cracks and cuts, while stretching. Let us understand the making of the strudel dough from the following steps.

Step 1

Collect all the ingredients and make sure that they are cold, as it will help to get better strudel dough. Use clean equipment to make the dough, as smallest of particle in the dough can cause it to break while stretching. Sieve all the dry ingredients to remove any particles.

Step 2

Knead the dough. Once the dough is nice and smooth, form it into a round ball, and rub some oil around it and put in a bowl. Cover it with remaining oil until the strudel dough sinks in the oil. Rest it in a cool place for at least two hours. Do not reknead the dough after it has been rested as the extra working will develop the gluten further and make the dough too elastic for pulling out.

Step 3

Cover the bench with a tablecloth dusted evenly with flour. Lay the dough on the cloth, pull and stretch it out to form a rectangle. You may use a rolling pin to assist in this process. The dough must be pulled out until it is thin enough for you to read through it. This will make the pastry light and crisp to eat (refer to Fig. 22.10).

Step 4

The stretched dough will dry out very quickly, and so must be used immediately. Rub it with butter and fill according to the requirements of the recipe. The filling is prepared by cutting the fruit into dices and combining it with sugar, nuts, spices, and crumbled plain sponge. The pieces of sponge help to absorb the moisture from the fruits during baking process, thereby keeping the end product flaky and crisp.

Step 5

Place the filling on one edge of the pulled dough leaving a space of 3 to 4 inches. The thickness of the strudel desired will decide the amount of filling to be used in strudel. Now fold over the left 3 to 4 inches of the sheet on top of the filling to lock the filling in the strudel. The dough should be sheeted on top of a cloth, as it will be difficult to roll the strudel with hands, since the sheeted dough becomes very fragile. So lift the ends of the cloth and roll the strudel like a Swiss roll pastry. The brushed melted fat thereby laminates the pastry along the way (refer to Fig. 22.10).

Fig. 22.10 Making of strudel

Step 6

Lift the strudel and carefully place it on the baking sheet. Brush with melted butter and bake in the oven till crisp and golden brown (refer to Fig. 22.10).

Step 7

Slice the strudel with a serrated knife and present with desired accompaniments. Apple strudel is served with vanilla sauce and ice cream.

PHYLLO PASTRY

Also known as filo, this pastry is mainly prepared commercially and traditionally and is used in Greek and Turkish sweets, savoury pies, and pastries. This pastry is plain, with paper thin leaves made from flour and water. These leaves are usually kept frozen and used as and when required. In case one has to make homemade phyllo, pulled strudel could be used for the same.

The phyllo sheets must be laminated before they are made into a product. These are separated with melted butter or oil, and sometimes sprinkled with breadcrumbs or nut meal to give varying characteristics to the finished product. Filo can be baked and also deep fried. Manufacturer's recommendations need to be followed when using this pastry. This pastry can be used in variety of ways.

Table 22.6 discusses the various products made from this pastry.

Table 22.6   Products made from phyllo pastry

Product	Description	Photograph
Spanakopita/Borek	It is a Greek dish using filo sheet and has a filling of spinach and garlic combined with pine nuts and feta cheese. Cut a 12-inch long and 3-inch wide strip of filo. Brush with melted butter and place filling at one corner. Fold the corner to form a triangle in a way that the filling is encased completely. Now keep turning the pastry to form triangles until it meets the other end of the sheet. It can be deep fried or baked.	
Bourekakia	Cut a filo sheet into 6 inch square. Brush with melted butter and place the filling on one end. Roll into a pipe taking care of folding the sides to encase the filling. It is usually deep-fried.	
Lebanese baklava	In a pan, layer a sheet of filo and brush with melted butter, place another sheet on top and repeat this process until at least six sheets are used. Sprinkle with crushed nuts and sugar and repeat the process until baklava is 2 inches thick. Brush the top with melted butter. Traditionally it is cut into diamond shapes, prior to baking, as the crisp pastry will be difficult to cut after baking. Bake till golden brown and while it is hot pour in rose-flavoured sugar syrup. The cuts help the syrup to percolate down. Serve the pastry warm or cold.	
Turkish baklava	Cut a rectangle of 10-inch length and 6-inch width. Brush melted butter and place the filling of nuts and sugar and roll like a cigar. Now coil up the cigar and brush with melted butter. Bake and pour hot rose flavoured sugar syrup over it.	
Phyllo cups	These are used for garnishes and sometimes used as main course, where the shell can be filled with creamy vegetable or meat filling. Cut a square of the desired size of the cup and layer at least 4 squares with melted butter. Put this on top of inverted metal mould so that is falls down like a sheet. Bake in oven till golden brown. When cool, take it off the mould and it will resemble a mould.	

COMMON FAULTS IN LAMINATED PASTRIES

As we saw in the above paragraphs, laminated pastry production is a combination of skill and science. It requires lots of practice to be able to make good puff pastry, Danish pastry, and croissant. We shall discuss some of the faults of laminated goods and when we are doing that we will specifically be talking about puff pastry and croissant/Danish pastry. Table 22.7 shows the faults in making laminated pastries.

Table 22.7 Faults in making laminated pastries

Faults	Reasons
Uneven lift	• Incorrect rolling technique • Fat too hard while rolling, so uneven distribution of fat • Uneven distribution of heat in the oven
Poor lift	• Too many folds given to the laminated dough • Pastry rolled thinner than 7 mm during rolling • Less fat, or too much fat used
Excessive shrinkage of the pastry	• Pastry not rested for sufficient time • After rolling, the stretched pastry is not lifted up and allowed to come back to normal
Baked product is very oily	• Oven temperature is cold • Sufficient rolling not given to the puff • Too much fat used • Improper rolling, where the fat has oozed out in the rolling process • Pastry was not rested between rolling
Dough tears while rolling	• Weak flour used instead of strong flour • Dough is over kneaded and the gluten has become elastic

CONCLUSION

In this chapter we discussed in detail the various laminated goods made in the pastry kitchen. We saw that pastries are used both in sweet as well as savoury preparations. The most commonly used laminated pastries in the kitchens are puff pastry, Danish pastry, and croissant. The other pastries such as strudel and phyllo are specific to only certain cuisines; but puff and croissant are used internationally.

The plain dough or the leavened dough is stuffed with fat, rolled and folded many times in a particular way to get thin alternate layers of fat and dough. This process is called lamination.

We talked about puff pastry and saw the role that ingredients play in making the same. The effect of acid on the protein of the dough and the softening effect of the butter on gluten make the dough more elastic so that it can be stretched along with the fat. Puff pastry is used in many ways. It is used for making breakfast pastries, desserts, cakes, and snacks. We also discussed production of many products that can be made from puff pastry. The videos of the products and the recipes are in the ORC. We also discussed various methods, such as French, English, and Scotch methods, of making puff. We discussed about another unique way of doing puff, where the dough is encased in butter and then it is rolled and folded to produce inverted puff pastry. We also saw the single and book folds and how they influence the product.

We discussed the steps in the production of puff pastry. We also discussed the salient features of making puff pastries, where we became aware of the finer aspects that need to be taken care of while making laminated doughs.

Danish and croissant doughs were discussed and the difference between both the doughs was highlighted as well. We discussed the production of Danish and croissant and the shapes and fillings of Danish served in hotels.

We also discussed strudels from Austria, since it is a laminated pastry which is very different from the conventional puff and croissant. We talked about the role that ingredients play in the production of strudel. We also discussed the step by step production of strudel.

Last but not the least we touched upon the Greek pastry called phyllo or filo. We discussed few products made from phyllo and its importance in the pastry kitchens.

We also discussed the various faults in making laminated doughs and what care should be taken to avoid such faults in laminated doughs.

KEY TERMS

Baklava   Dessert from the Middle East, made from phyllo sheets and nuts.

Book fold   The dough is folded into a rectangle and the outer ends are folded in to meet in centre and then folded over like a book.

Boreks   Snacks made from phyllo—stuffed and shaped into triangles. It can be deep-fried or baked.

Bouchee   Mouth-sized vol-au-vent.

Butter block   Creamed butter and flour prepared for rolling between the dough.

Continental method   Same as French method.

Croissant   Crescent-shaped pastry served in breakfast.

Danish   A type of laminated pastry served in breakfast.

English method   Style of rolling puff dough, where the fat is placed on two thirds of the dough and then folded for rolling.

Feather icing   Piping straight lines of chocolate on to fondant and running toothpick vertically to give the designs that resemble feathers.

Feta cheese   Cheese from Greece.

Filo   Refer to phyllo.

Flaky   Layers of crisp dough which are short in texture.

Fleurons   Crescent-shaped puff pastries used for garnishing in French cuisine.

Fondant   Sugar boiled to 118°C and cooled and kneaded to form dough.

French hearts   Another name for palmiers. Other names are papillon and pig's ears.

Full puff   The amount of the fat rolled in dough is equal to the dough.

Half Puff   The amount of fat to the dough is half.

Inverted puff   Type of puff pastry, where the dough is encased in butter and folded.

Laminated   Rolling and folding of dough and fat to form alternate layers of dough and fat.

Mille feuille   French for puff pastry; literally means thousand layers.

Paprika   Hungarian red pepper powder.

Phyllo   Commercially available thin sheets, made with water and flour, often used in Greek cooking.

Scotch puff paste   Also known as rough puff paste. Cold butter is kneaded with the flour and rolled and folded like a puff pastry.

Single turn   Puff dough is rolled to a rectangle and folded into three parts.

Strudel   Thinly sheeted dough laminated with butter, filled with a filling and rolled like a Swiss roll.

Three quarter puff   The ratio of the fat to the dough is 3:4.

CONCEPT REVIEW QUESTIONS

1. Define laminated pastries.
2. Name the most common laminated pastries used in the kitchen.
3. What is the principle behind laminated pastries? How is the lamination achieved?
4. What are the stages that a puff goes through when it is baked in oven?
5. What is the role of the hard flour in puff pastry?
6. What is the role of acid in puff pastry?
7. What is a butter block and why is it made?

8. Why is ice water used to make puff dough?
9. What are different methods of making puff pastry?
10. What is the difference between regular puff and inverted puff pastry?
11. What do you understand by single turn of puff pastry?
12. Describe the book fold.
13. List the steps involved in making puff pastry.
14. What care should be taken while rolling the pastry?
15. List five products made from puff pastry and its uses.
16. What is the difference between eccles and branbury?
17. What is the difference between Danish dough and croissant dough?
18. What is the role of ingredients in Danish dough?
19. List down the steps involved in making Danish pastry and croissant.
20. What is the difference between puff dough and Danish dough, when it comes to folding the pastry?

21. List at least three shapes of Danish.
22. List at least three types of croissant.
23. How are Danish pastries traditionally finished?
24. Define a strudel. How is Austrian strudel different from German strudel?
25. What is the role of ingredients in strudel dough?
26. Why is the strudel dough dipped in oil for a couple of hours?
27. List the steps in making of apple strudel.
28. How is the lamination of strudel different from that of puff pastry?
29. What is phyllo pastry?
30. List at least three products that one can make from phyllo pastry.
31. If there is an uneven lift in the puff pastry, what could be the possible reasons?
32. The puff pastry shrinks too much while baking. What could be the fault?
33. How should we avoid the tearing of the dough while rolling?
34. What will happen if the laminated pastries are baked in low heat?

PROJECT WORK

1. In groups, make 10 different types of puff pastries by using butter for lamination. Use a range from 100 g of butter per kg dough to 1 kg butter per kg dough. Bake the product and compare the results and write your observation in the chart below.

Type of Puff	Texture	Consistency	Baking Effect	Colour

2. In groups, try out the recipes as provided in the ORC for the laminated pastries and make the products as listed in tables. Compare the results with each other and critique each other's products.
3. Visit the various pastry shops and restaurants in the city and write down the various types of laminated products served there.
4. Research about various dishes, apart from the ones mentioned in this chapter, which are made using phyllo sheet. Cook these recipes under the supervision of your professors and compare the results.

CHAPTER 23

INTRODUCTION TO INDIAN COOKING

Learning Objectives

After reading this chapter, you should be able to:
- understand the history and evolution of Indian cuisine
- be aware of the philosophy behind Indian cooking and the thoughts on Indian food
- know the influence of conquerors and travellers on Indian cooking and how it has had an impact on the Indian cuisine of today
- understand the role of ayurveda in our cooking and theory behind the same
- be aware of religious influences on Indian cooking
- know the various cooking equipment used in Indian cuisine
- comprehend and use the different techniques for cooking
- understand the concept of slow cooking and organic food

INTRODUCTION

Around 5000 years ago, when the Aryans first came to India, it was known as *Aryavrat*. They settled around the Indus valley, which did not receive much rainfall as required for the cultivation of rice, so they started harvesting wheat. The carvings and the paintings of that time depict the stone mill which was used for grinding of wheat. From simple wheat to lavish buffets laid out today, Indian food has made a complex journey. Indian food has gone through simple times, rough times and today with globalization, Indian food is still struggling to find its niche in the world. The reason for this is very simple—to an uninitiated, Indian food is just like a normal curry eaten with rice or bread. Indian food has an impression of being spicy and hot; but Indians know that this is far from true. Contemporary Indian food is constantly being refined and chefs are devising new and modern ways to prepare and present the same. India has a glorious tradition of healthy food. People in India have always respected food and this can be seen even today. In Indian tradition, food is placed next to God and that is the reason why food forms a main part of ceremonies and religious celebrations. Indian food is born from the concept of ayurveda. Ayurveda comprises two word—*ayus*, meaning life and *vedas*, study or knowledge; hence ayurveda means knowledge of life, which begins from the basic ingredient 'food'. Indians had a very healthy lifestyle—the most common profession was farming and hence fresh produce would be cultivated for consumption. It was the advent of Mughals and trades that brought in spices from the world. Vasco da Gama is believed to have brought chillies to India from Chile in South America. We really do not know how India was like thousands of years ago and there are many reasons for it. The first one is that with the coming of Mughals, the chefs working in the royal kitchens were scared to hand over the knowledge to even their next of kin, as they were very afraid of losing their jobs and the status of being close to kings and their men. As most of the cooks were not educated, written scripts and recipes were not available to research and dwell upon. Some old books do have references to Indian food, but do not talk about the recipes and methods of cooking the same. Even if mentioned in the texts, the common man does not have any knowledge of the script and the language they have been written in.

With colonization, Indian food lost its glory and all kinds of changes came in. People started to use recipes that suited them and use of more aromatic spices and flavours started to prevail and hence Indian food started to lose its originality. Along with many other arts and crafts, food too went into a decline. Cooking skill was passed down from generations to generations, from mother to daughter and from chefs to their juniors. The famous *guru-shishya parampara*[1] also prevailed in the kitchens. The recipes were never recorded and they were only memorized. The drawback here was that every generation reduced something or added something to the recipes and put their own stamp on the same. Thus many delicacies became imbalanced and they also lost their medicinal value. Most of the herbs and spices used in Indian food had lots of medicinal value and that is true for most of the Asian food. Our ancestors were very careful in choosing food. Some foods are 'complete proteins', which means that they contain all the important amino acids needed by our body, for example, eggs, meat, and fish. Yet some food needs to be combined with something else to make it into a complete protein, so that our body is not devoid of any nutrition. That is the reason for combinations such as *rajma chawal, kadhi chawal,* paneer parathas, etc. The food was so detailed that it was listed down for people from childbirth till they became adults. One would have heard about women in pregnancies and post birth being given special dietary requirements to heal up internally. These concepts are unheard of in the Western world, where after childbirth some people would celebrate with chilled wine, whereas in some parts of the world women are usually not allowed to partake cold water for the first forty days of childbirth.

Nowdays, we eat anything and everything and are not bothered about combinations of food and that is the reason why food-related diseases are constantly on the rise. Indian food had a very strong philosophy and we have completely destroyed it because of our own selfish needs and also due to ignorance. Today the Western world is trying to research and follow what our ancestors did, but we are happily devising combinations of Indian and Western dishes.

In this chapter we shall discuss the techniques of Indian cooking, the philosophy of Indian cuisine, the medicinal value of spices, flavouring agents, and basic gravies.

PHILOSOPHY OF INDIAN FOOD

Indian food is as varied as its culture and people. In every region, food changes its flavours and techniques of cooking. As Indian food is influenced by its religions, customs, and traditions, it is not easy

---

[1] The old tradition where the guru imparted knowledge to only few selected students.

to group it like the food of the Western world. There cannot be mother gravies like mother sauces and its derivatives. Each curry has it its own gravy. It is a country where the climatic conditions are also so varied, that the availability of produce too determines the way the food is cooked. For example, in Rajasthan, which is an area mostly covered by deserts, vegetables do not grow and hence in the cuisine, one would see usage of dried berries and yoghurt-based gravies. Similarly in south India, where the land is apt for cultivation of rice, the vegetable dishes and curries have more liquid than compared to the vegetables in the north.

The strongest influence on Indian food is from ayurveda. Here we are not talking about kebabs and biryani, as this food was brought in by Mughal emperors who came into India from Persia and Iran. Ayurveda deals with areas concerning the healthy and long life of human beings. The origins of ayurveda date back to 1000 BC. Ayurveda does not only deal with natural medicines, it covers the whole aspect of life. It discusses the purpose of being born. It preaches that we live to eat and not eat to live. Ayurveda talks about the aspects of mental as well as physical health and it suggests that the way to salvation is through healthy living, which includes eating good and living life on the basis of an ethical code of conduct. The very famous phrase 'you are what you eat' probably is the origin of the teachings of ayurveda. Life, in ayurveda, has been described as the combination of mind, body, and soul. A healthy body will practice and preach healthy thoughts, which in turn will reform a soul and thus will help reach salvation.

Researches show that people who are depressed, eat more and many diseases emerge from the fact that people keep the emotions bottled up inside them. This leads to coronary problems and hypertensions. Ayurveda talks about foods that regulate the health and even emotions of human beings. For example, the usage of the oil and ghee—oil is believed to heat the body whereas ghee is supposed to provide a cooling effect. Another speciality of ghee is that it assimilates all the nutritional properties of the food that it is mixed in, without losing any of its own. The usage of oil in winters and ghee in summers is still widely followed in many Indian homes. Ayurveda revolves around the concept that our body is made up of three elements—fire, air, and water, commonly referred to in Sanskrit as *vatta, pitta*, and *kappa*. Ayurveda believes that rise of ailments and diseases is due to the imbalance in these elements. Even today, when doctors diagnose us with any ailment, we are advised to refrain from some kinds of food.

Ayurveda classifies Indian food into the following six tastes.

Sweet

This type of taste gives strength to tissue elements and harmonizes the mind. Probably that is the reason why people always share sweets when they are celebrating. Ayurveda does not classify only sweet products into sweet taste; but also foods such as rice, ghee, and fruits form this category.

Salty

Salt stimulates digestion; it clears the obstructions in the nervous system of the body, thereby cleansing the body by the function of sweating. Excessive salt in the food tends to give rise to wrinkles and greying of hair.

Pungent

The pungent taste helps in digestion. Foods such as onions, garlic, pepper, etc. form this category. These types of food also help improve the metabolism of our body.

Bitter

These kinds of food help purify the blood and are easy to digest. Bitter gourd and some spices have this taste.

Astringent

These kinds of food help treat ulcers in the body and also help in healing wounds. Green vegetables, apples, potatoes, etc. are grouped into this category.

Sour

The sour taste aids in digestion and it is believed that it also helps the heart to function well.

Ayurveda believes that any deviation from these six tastes in food will bring about the imbalance in the elements of body and each element is responsible for certain diseases. This is where the concept of *thali* evolved. It became imperative to serve foods to balance all the elements in our body. A healthy Indian meal would consist of rice, salad, lentils, vegetables, proteins in forms of paneer or meat, yoghurt, pickle, and sweet.

Ayurveda also classified food on the basis of its characteristics. These are as follows:

Rasa

Food is categorized based on its tastes as listed above. *Rasa* is a Sanskrit word which means the taste.

Veerya

This is the food that provides potency to the body. Meat provides energy and vigour and hence, was given to warriors and kings. Brahmins and godly people were given food that did not provide heat to the body, thereby letting them meditate and stay in touch with God.

Prabhav

This is the food that has some special action on the body. This is known as *tehseer* in Urdu and it implies the hot and cold effects of food on body.

Two types of foods could be very similar in their *rasa* and *veerya*, but they still might differ in their *prabhav* on the body.

The human states of mind are also classified into three types. These states of mind of a person make them different to each other's response to stimuli. That is the reason why some people are so quiet and some so aggressive. Ayurveda classifies them as the following.

Satyavik *Satya* means pure and hence, a *satyavik* person will be a highly intellectual person, with a curious mind. He/she will always strive for more knowledge and will try to live life on fair means and his/her hard work.

Rajasik This person will be the one who is basically a doer. He/she will use almost any means to succeed.

Tamasik *Tamasik* on the other hand will have no desire to learn or expand his/her knowledge. He/she would also lack the intellectual capacity.

Ayurveda is the base that forms the philosophy of Indian cuisine. Ayurveda even talks about how the food tastes different when cooked with love and when cooked in angry mood. Probably that is why home cooked food made with respect and love is always the best and tasty. In many Indian homes, in spite of having cooks at home, some women prefer to cook food for the family with their own hands.

Meats and spicy food have a force of violence in it; hence, it is believed that people will be more aggressive if they live on these diets. Ayurveda teaches humans to follow the purity of life, to respect food and the times it should be eaten. It teaches us to eat nutritious food and practice yoga and meditation to purify our mind, body, and soul.

We maintain some of the sanctity of the ayurvedic concepts of food till today. Indian food has a very simple philosophy—'cook what is in the season, with love and respect and share the food with people and enjoy life to its fullest'. Even at most of the hotels, there is a section in the menu called *ghar ka khana*, which means home style food. The moment we talk about home made food; we think of food that is cooked with less oil, less spices, and is fresh and seasonal.

Indian food has been influenced by many factors—one of them is religion, another is the regional influence on the food. One would frequently hear about terms such as cuisines of royal *gharanas*, cuisines of Mewars, cuisines of Iyengars, or Kashmiri Pundit cuisine, and so on.

INFLUENCE OF THE INVADERS AND TRAVELLERS ON INDIAN CUISINE

As we saw above, cuisine of India is as varied as its people and culture. Each ethnic group has its own blends of spices and special cooking equipment and spices to produce the most delectable gourmet fare. India is a land that was invaded by many conquerors and rulers. Indian heritage, culture, and arts were the major attraction to the rulers. India was invaded by Arabs, Central Asians, Mughals, and also Greeks. Some travellers from China also left a huge impact on the Indian cooking styles and flavours. Let us discuss the influence of these invaders.

Influence of Greeks

Greece is very famous for its fresh herbs such as oregano, mint, coriander, and essential oils such as oils from olive, etc. With the conquest of India by Alexander the Great in 350 BC, the Greeks brought their profound herbs and spices. Greek cuisine itself is a blend of cuisines of the Roman empire and Turkey. During the time of Alexander, trade flourished between India, Rome, and other Mediterranean countries and this brought the usage of saffron and many herbs and spices to Indian cooking. Greek cuisine till date is very famous for its fruits, cheese, nuts, grains, and oils. Greece contributed to India vegetables such as egg plant and zucchini. A pan-fried *kalhari* found in Jammu, India, is similar in taste to the Greek cheese called *kasseri*.

Spices such as fenugreek and fennel were the influence of Greeks and these two are the most commonly used spices in Indian cooking. Greek influence gave elegance to Indian food, along with lots of flavours and nutrition.

Influence of Mughals

The Muslims from western Asia brought with them the rich heritage of *Mughlai* cuisine. This was during the twelfth century, when the Mughals invaded India and covered a large area of the country. Mughals enjoyed wining and dining and their large and sumptuous buffet fares influenced Indian traditions to a large extent. On special occasions, such as marriage, it was common to see a range of minimum 50 dishes. A variety of food on the table described one's royalty and richness.

The Mughals brought with them the usage of aromatic spices such as cardamom, mace, nutmeg, and variety of nuts such as almonds and pistachios. The cooking styles too changed with the advent of spit fire roasts and this give birth to kebabs and tandoori items. The spices were quite exquisite and can be referred to as exotic spices. The usage of milk and cream along with nuts and dried fruits gave rise to biryanis and pilafs that are still an integral part of Indian culture. Meats were the preferred choice of Mughals unlike Aryans, who preached vegetarianism. Meats started to be included in the main meals of Indian society. Each Mughal ruler added his own preferred food items to the Indian cuisine. It is believed that when Babur invaded India, he brought with him the tradition of grilled meats and dried fruit. His son Humayun continued the traditions and introduced a range of pilafs and biryanis. Certain Mughal rulers married Rajput princesses and this added a new dimension to the *Mughlai* cuisine.

Influence of Mongolian and Chinese

The parts of India closer to the borders of China and Mongolia, such as the states of Mizoram, Arunachal Pradesh, Nagaland, and Manipur have more influence of Mongolian and Chinese cooking. The Mongolians introduced cooking styles of hot pots and stews and also introduced new ingredients into cooking.

Traditional Mongolian cooking uses lots of meat and dairy products along with rice and this can be seen in eastern India, where rice is the most preferred staple food. The style of hot pots, where the mince meat is formed into dumplings and prepared like a soup, is very commonly seen in these states. Mongolians also introduced simplicity into Indian cooking. It can be seen that in these states, the food is very simple and not very elaborate. The cooking methods employed are steaming and frying and this is again the influence of Mongolia. Mustard oil is the influence of Mongolia and so is the usage of sugar. Eastern India is very famous for its sweets, especially Bengal.

Chinese also gave certain methods of cooking to Indian cooking. Stir-fries used in Indian cooking are the result of Chinese influence. The *kadhai*, which is the most common utensil found in every Indian home, resembles the Chinese wok.

Influence of Portuguese

When Vasco da Gama came to India, Indian food went through a sea change. The advent of spices and the most common ingredient chilly was brought in by Vasco da Gama in the year 1498. Today one can hardly find any food on the table without this main ingredient. The influence of the Portuguese food can be seen in the Goan cuisine, where the names of dishes too are very Portuguese. The dishes of Goa are tangy, spicy, and immensely flavoured. The introduction to seafood such as prawns and meats such as pork and beef, has been the influence of the Portuguese. Portuguese brought in fruits, such as cashew nuts, tomatoes, pumpkins, pineapples, guavas, and passion fruits, from the tropical Caribbean islands.

Influence of British

During the reign of the British in India, many elaborate cooking techniques were introduced to the Indian kitchens. Flexibility and diversity were introduced in Indian cooking and people started to use many of the European cooking styles in the menu. The concept of grilling on the cast iron pans and roasting has been given to us by British people. The English brought whisky and tea to India. The concepts of curries with rice are a concept that started during the British time. Many preparations such as *murgh makhni* are said to be invented by the British. Today people in London claim that *murgh butter masala* is a national dish of the English people. Many of the dishes made during the British time are classified as *Raj cuisine*, which mostly comprises of Anglo-Indian cuisine. There was no concept of soups in Indian cuisine, but all the various types of *shorbas* are the influence of the British. Soups such as mulligatawny are also concepts given by the British people.

REGIONAL AND RELIGIOUS INFLUENCES ON INDIAN CUISINE

Religious practices vary widely in India. In the olden times when the caste system was prevalent, many sects classified food accordingly and also posted prohibition and restrictions on certain castes, for example, Brahmins could not eat meats, etc. India has always been a secular state with people practicing different religions and faiths. As we read earlier, ayurveda believes that food provides purity of mind, body, and spirit. Thus many religions have demarcated food accordingly so that the people eat healthy food and lead a simple and pure life. Natural variations are also common; further individual adherence to a religious diet is often based on personal degree of orthodoxy. Today the growing awareness towards animal rights and health consciousness has given birth to vegetarianism (vegetarian food). It is easy to digest and builds strong immune system. It is only by keeping the preferences of the target market in mind, can the product be satisfactorily delivered.

EQUIPMENT USED IN INDIAN COOKING

Indian cooking demands a wide range of small and large equipment. Various regional cuisines require unique cooking utensils and aids, which are often not used in other regions of the country. Hence, the list of Indian cooking equipment can be very long. Each equipment has a purpose to serve. For example, iron vessels help to impart iron into the food, which is a natural mineral required for the growth of the body. Ancient people used to eat in gold and silver plates for the same reasons. Table 23.1 shows some important and common equipment used in Indian kitchen.

Table 23.1 Commonly used equipment in Indian kitchen

Name	Description	Photograph
Chakla and belan	*Chakla* is a small marble or wooden platform and *belan* is a rolling pin, usually made of wood. They are used for rolling dough to make various Indian breads such as chapattis and *puris*.	
Chimta	In simple term, *chimta* means tongs used for holding food during the process of frying, griddle cooking, etc.	
Chalni	Round, deep utensil with lots of holes meant to drain liquid and retain the residue for further processing.	
Kadoo khas	Equipment that has sharp grooves of different sizes meant for the purpose of grating.	
Masala dani	Commonly known as a spice box, it contains the commonly used dry spices—both whole and powdered.	

Contd

Table 23.1 (Contd)

Name	Description	Photograph
Pauni	A perforated spoon used for frying food commodities.	
Kadchi	It is a ladle used for stirring and mixing.	
Sil batta	These are two pieces of stone—one of flat form called *sil* and the other is *batta*, which is much smaller and looks like a large rolling pin without handles.	
Hamam dasta	Mortar and pestle used to pound dry masala, usually made of iron.	
Tandoor	Clay oven chamber which is lit with live charcoal. It is used for baking various Indian breads, kebabs, and other tandoori items.	
Tandoori sariyas/ jodi	Long iron skewers sharpened at one end and curved at the other in the shape of a hook. These are used to skewer kebabs before they are cooked in the *tandoor*. These are also used to remove rotis from the *tandoor*.	
Gaddi	It is used for sticking breads in *tandoor*. It is made by stuffing hay into cloth.	

Contd

Table 23.1 (Contd)

Name	Description	Photograph
Mathni	Special wooden equipment, which is used to churn yogurt to make *lassi* or butter milk. It is more prevalent in Punjab and other parts of north India.	
Bhagona or patili	A *bhagona* or *patili* is generally made of brass. It comes with a lid. It is used when great deal of sauté is required, or even for boiling and simmering. It is also required for making gravies and cooking in bulk. It is available in various sizes.	
Deg or degchi	It is a pear-shaped pot with a lid made of brass, copper, or aluminium. The shape of this utensil is ideally suited for the *dum* method. It is used for cooking *pulao*, biryani, *nehari*, etc.	
Kadhai	It is a deep, concave utensil made of brass, iron, or aluminium and is used for deep-frying.	
Lagan	It is a round and shallow copper utensil with a slightly concave bottom. It is used for cooking whole or big cuts of meat or poultry, especially when heat is applied from both top and bottom.	
Lohe ka tandoor	It is an iron tandoor, as distinct from the clay tandoor and is more commonly used in Lucknow and Delhi. It is a kind of dome-shaped iron oven used for making most of the breads such as the *sheermal*, *taftan*, *bakarkhani*, etc.	
Mahi tawa	It is the Awadh version of the griddle, shaped like a big round, flat-bottomed tray with raised edges. It is used for cooking kebabs and also used for dishes where heat is applied from both ends, when covered.	
Seeni	It is a big *thali* or round tray usually used as a lid for the *lagan* or *mahi tawa* when heat is to be applied from the top. Live charcoal is placed on it and heat is transmitted through it to the food. Thus the indirect heat has the desired effect of browning and cooking the ingredients. All the copper and brass utensils are almost always used after *kalai* or tin plating the insides.	

Contd

Table 23.1 (Contd)

Name	Description	Photograph
Tawa	It is a flat-base equipment usually made of cast iron, used for making Indian breads such as *rotis, parathas,* etc. It is available in various sizes, depending upon the usage.	
Palta or khoncha	Flat metal spoon used for stir-frying or sautéing the ingredients.	
Jhara	It is basically a strainer used for straining excessive oil or ghee from *boondi* or any other sweets such as *jalebi.*	
Jalebi tawa	It is a flat cast iron utensil with raised edges. It is specifically used to make *jalebis.*	
Roomali roti tawa	Made of cast iron, this equipment resembles a *kadhai* without handles and is used inverted on the fire. It is used for making *roomali roti,* which is rolled thin like a handkerchief and baked on the inverted side.	
Grinders	Various types of grinders, such as wet grinder, are used for making pastes. This equipment is mostly used in south India, whereas in other parts of India, *sil batta* is most commonly used for making pastes.	
Sigri	It is used for grilling kebabs. It is an open fire grill, where coal is the only medium of fuel used.	

TECHNIQUES EMPLOYED IN INDIAN COOKING

Indian cooking is vast and varied and each state has a special technique of cooking which could be very different from each other. *Tandoor* is widely used in north India and one can hardly find its usage in east or south India, with an exception of Hyderabad, which is famous for its *Mughlai* food. The methods of cooking also change with the type of equipment used. Steaming is very common in east, west, and south; but in north Indian cooking, steaming is unheard of.

The methods of cooking are same around the world. The only difference is that according to the cuisine, the methods could be a combination of few or might be known by different names depending upon the country. Indian methods of cooking are simple and combinations of many methods used together.

Table 23.2 shows the different techniques used in Indian cooking. These will be referred to very often in the coming chapters.

Table 23.2 Techniques used in Indian cooking

Techniques of Cooking	Description	Use
Dum	*Dum* means cooking an item in its own steam. This is carried out in an enclosed container, where the steam is not allowed to escape out. The heat is applied from top and bottom. This is done by putting the utensil on top of smouldering charcoals and some burning charcoal is also put on top of the lid to allow the heat from all the sides. The meat or vegetable is partially cooked and then the lid is sealed with the help of dough and placed to cook on *dum. Dum* cooking is usually done for dishes that have aromatic and flavourful spices and this method will help to retain all the aromas in the dish.	It is usually used for cooking biryani. Some vegetarian curries such as *dum aloo* from Kashmir are also cooked by this method. The dishes cooked by this technique might have its name associated with the technique such as *gosht dum masala, dum ki batak, dum ka karela,* and so on.
Bhunao	This is the combination of sauté, stir-frying, and stewing. This is a process of cooking where the heat is regulated from medium to high, adding liquid such as water, yoghurt, or as required by the dish. This continuous stirring or *bhunana* is done to prevent the dish from sticking to the bottom of the utensil. *Bhunao* is not a complete process of cooking, but is a part of the entire process that is used for preparing a dish. Sometimes when the dish is made using only this technique then it is known by the name of the technique used such as *bhunna gosht.*	It is used for preparing most of the Indian dishes, where the ingredients are sautéed, prior to finishing them into curries and dry preparations. *Bhunna murgh, bhuneey dum key bateyr,* etc. are dishes that are made using this technique.
Talna	This is term used for frying in Indian cuisine. Usually done in *kadhai* as the shape of the *kadhai* facilitates the frying of large quantities in less amount of oil.	It is used for frying of snacks or even main course and desserts.
Baghar	It is also known as *tadka* in north, *chonkna* in central India, *baghar* in west. Tempering is done using hot oil, as the hot oil has properties of extracting flavours and colours of spices. This process can be done in the beginning of the dish or in the end. Sometimes a dish is also tempered in the beginning and finished with tempering in the end. For example, *kadhi* from north or *salan* from Hyderabad.	The major use of tempering is to flavour a dish with aromatic spices. Sometimes tempering is also done to get rid of the strong smell of mustard oil or even ghee. In Lucknow, people remove the smell of ghee by tempering it with aromatic spices such as green cardamom and mace and the procedure is known as *ghee durust dena.*

Contd

Table 23.2   (Contd)

Techniques of Cooking	Description	Use
Dhungar	*Dhungar* or smoking means to impart the smoky flavour to the dish. Unlike Western smoking, which involves burning of wood such as oak, chicory, etc. Indian *dhungar* uses oil or ghee, poured over smouldering charcoals. Sometimes spices such as cloves are used to impart a flavour, which mainly goes well with meat dishes. The item to be smoked is put in a *handi* and small piece of coal is placed in a bowl and placed in the *handi* among the food to be smoked. Pour ghee and spices over it and cover the *handi* with a lid and let the smoke impregnate the food.	It has many uses. The Rajasthani specialities, such as *maans key sooley, laal maans*, etc. are smoked with cloves. Kebab mixes such as *galouti* kebab from Lucknow is also smoked with cloves.
Sekna	*Sekna* means baking or roasting. One could use this method for Indian breads such as chappati, *phulka paratha*, etc. over the *tawa*, or even breads, meats, and vegetables in *tandoor*. This term is often confused with *bhunao;* but it can be referred to as *bhunao* only when in context with *tandoor*. Certain parts of India use cow dung as a medium of fuel for *sekna*. For example, *Rajasthani baati* is baked over smouldering dry cow dung cakes called *kanda*.	It is used for Indian breads made on *tawa*. It is used for kebabs, breads made on *sigri* or *tandoor*.
Bhapa	*Bhapa* refers to cooking with steam. It is used mostly in all parts of India, with the exception of north. Steaming is carried out in a simple pot or sometimes things are steamed in bamboo tubes. Things can be steamed in banana leaves also.	It finds its uses in the breakfast to main meals and even desserts. Idli from south, *macher paturi* from Bengal, and *modak*, a dessert from Maharashtra, are few examples of steamed dishes.

CONCEPTS OF SLOW FOOD AND ORGANIC FOOD

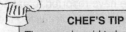

CHEF'S TIP
The curry is said to be ready when the oil or fat comes on top of the surface.

This concept is very old and is believed to have originated in Italy. It is important to discuss this here, as most of the concepts of Indian cooking revolve around this concept. As the name suggests, slow cooking does not mean that food is cooked on slow heat or low heat; it is a concept that has much broader application. Slow food talks about various aspects of food, such as respecting the food, treating the ingredients the way they are supposed to be treated to retain maximum nutrition and goodness in food. Slow cooking is interrelated to organic cooking. In the olden times, all the vegetables and grains were grown without any usage of chemicals and pesticides. Natural plants, such as strong garlic, etc., were planted alongside other vegetables and grains to repel the insects and animals which fed upon those commodities. This resulted in food that we refer to as organic in today's world. This term did not exist many years ago, as most of the farming was organic. It is only with the advancement of science that people started using chemicals and pesticides to prevent the crops from getting spoiled. Also some chemicals boost the size and growth of commodities, which yields more money in the market. People have started making hybrid fruits and vegetables that have enormous shapes and sizes. It is not uncommon to see square-shaped tomatoes, grown especially in the West.

The concept of the slow food has come into limelight in today's world, while we are partaking fast food. There are many government bodies around the world which are running slow food and organic movements to educate people to eat healthy and nutritious food. The philosophy of slow food is to eat good food that is healthy and fair. It should be prepared in a clean and hygienic manner and there should not be abuse of food such as wastage, improper processing of raw commodities, and cooking things only for required time to retain all the nutrition. Slow food movement lays emphasis upon cooking food in a way that it does not harm the environment, animal welfare, or our health. It also ensures that the agriculture farmers get a fair remuneration for their produce. Slow food movement is also trying to save healthy grains and other commodities that are diminishing from the world due to the convenience food now available as a result of agribusiness.

One should properly cook organic food which is healthy and nutritious.

CONCLUSION

This is the first chapter on the introduction to Indian cooking. It is important to know the basis and foundation of any cuisine when we need to master the same. Indian cuisine is vast and varied. It has evolved from the time of the ancient Aryans and was constantly used and abused by certain people. Ayurveda laid the basis of Indian food many years ago. It not only talked about the type of foods but also dealt with eating the right kinds of food and living healthy. Over the times, with the colonization of India the food also went through a reform. Various invaders and travellers influenced Indian cooking in a big way. Slowly the concepts as mentioned in ayurveda faded away and influence from other countries set in. The flavour and the taste of food changed drastically. Indian food is regarded hot and spicy; but strangely most of the spices used in Indian cooking today were brought in by travellers and invaders from various countries such as Iran, Persia, Greece, the Middle East, and South-east Asia. Vasco da Gama brought in chillies from Chile in South America in the year 1498.

There are no texts that depict the recipes of food cooked in India before the invaders came in and influenced the rich heritage and culture of our Indian gastronomy. Since the cooks were uneducated and the recipes were handed down verbally, there are no written recipes available to us. Someone who does not know about the varied cultural heritage of India, does not know about Indian gastronomic rapport.

Like for an Indian there is a dish associated with gravies and we understand that each preparation has a name; but for a foreigner they are all curries and nothing more than that. In India every state has its own cuisine and surprisingly, it is sometimes very different from other parts in India, both in terms of tastes, textures, and the way they are cooked or even served. The cuisine of Kashmir is quite different from the cuisine of Jammu. Even being parts of the same state, the cuisines have no resemblance with each other. Certain spices used in Kashmiri cuisine are unheard of in cooking of Jammu food and there are many such examples when we compare all the states of India. In this chapter we discussed how religions too influence the eating habits and cuisines of India. Equipment used in Indian cooking has also come a long way. In the olden times, people used gold and silver cutlery, but now people have resorted to eating in various kinds of metal available. We discussed the common equipment used in making Indian food and saw how they are unique in themselves.

We also touched upon various cooking techniques used in Indian cooking. The classical methods of cooking, such as boiling, poaching, stewing, etc., are same around the world, but the techniques employed could vary according to the country or cuisine. We saw the usage of these techniques in Indian cooking, which will help us to understand the terms when we will use them in the coming chapters.

The concept of slow cooking and organic food has been touched upon as well. This concept was first started in Italy; but the roots of Indian cooking or any cuisine lie in the concept of slow food. In the coming chapters we will discuss the various spices, their medicinal use, etc. along with basic Indian food.

KEY TERMS

Aryavrat   Ancient name of India.

Ayurveda   Ancient Indian system and document regarding knowledge of life and traditional medicine.

Baati   Dumplings made from wholewheat and ghee from Rajasthan.

Bakarkhani   Flaky bread from Lucknow and Hyderabad.

Bateyr   Indian for quails.

Chapatti   Flat Indian bread made from wholewheat on a *tawa*.

Chicken tikka   Morsels of chicken marinated in yoghurt and spices and broiled in *tandoor*.

Complete protein   Food that contains all the 22 amino acids necessary for the human body.

Cuisine of Iyengars   Cuisine of royal families from Tamil Nadu.

Cuisine of Mewars   Cuisine of royal families from Rajasthan.

Cuisine of royal gharana   Cuisine of royal families.

Galouti   Speciality of Lucknow, lamb kebabs with a consistency of a paste.

Ghar ka khana   Home cooked food.

Ghee   Clarified butter used in Indian cooking.

Ghee durust dena   Tempering ghee with aromatic spices to get rid of the foul smell.

Jalebi   Indian dessert made from *besan*.

Jhatka   Style of slaughter of meat, where the animal is slaughtered in one stroke and not sacrificed.

Kadhi   Curry made from yoghurt and gram flour.

Kalai   Tin plating done for brass utensils.

Kalhari   Cheese from Jammu that is chewy and stringy like mozzarella.

Kanda   Dry cow dung cakes used as fuel in Indian cooking.

Kappa   Water element in the body as per ayurveda.

Karahi   Cast iron concave vessel used in Indian cooking.

Pundit cusine   Food made by Kashmiri Brahmins, which is devoid of onion and garlic.

Kasseri   Type of cheese from Greece.

Lassi   Drink made with curd churned with water, which can be sweet or salted.

Macher paturi   Speciality from Bengal, fish steamed in banana leaf.

Modak   Steamed rice flour based dessert from Maharashtra.

Mulligatawny   Literally means pepper water. It is a soup made with *besan* and chicken stock, flavoured with curry leaves.

Murgh makhni   Chicken tikka cooked in tomato-based gravy flavoured with butter and fenugreek leaves.

Nehari   Stew of meat from Hyderabad.

Paneer paratha   Wholewheat Indian bread stuffed with cottage cheese.

Phulka   Indian wholewheat bread griddled on *tawa* and puffed up on an open fire.

Pitta   Air element in the body as per ayurveda.

Prabhav   Classification of ayurveda, where the food is categorized on the basis of its effect on the body.

Puri   Deep-fried wholewheat bread.

Raj cuisine   Anglo Indian cuisine of India.

Rajma   Hindi for red kidney beans.

Rajasik   Terminology used in ayurveda to classify state of mind of a person who is a doer.

Rasa   Classification of ayurveda where the food is categorized on the basis of its taste.

Roomali roti   Thin Indian bread, folded like a handkerchief.

Satyavik   Terminology used in ayurveda to classify state of mind of a person who is pure in his/her thoughts.

Sheermal   Leavened bread from Hyderabad.

Shorba   Indian for soup.

Taftan   Leavened bread from Kashmir.

Tamasik   Terminology used in ayurveda to classify state of mind of a person who lacks intellectual capacity.

Thali   A metal plate in which array of dishes are served as a complete meal.

Vatta   Fire element in the body as per ayurveda.

Veerya   Classification of ayurveda where the food is categorized on the basis of its potency.

CONCEPT REVIEW QUESTIONS

1. How old is the Indian civilization?
2. How do we know that the oldest cereal eaten was wheat?
3. How has Indian food evolved over the years?
4. What is ayurveda and what are its thoughts on food?
5. Why do you think there are no written recipes of ancient Indian cuisine?
6. What were the drawbacks of the *guru-shishya parampara* in relation to Indian cooking?
7. What do you understand by complete proteins?
8. What is the philosophy of the Indian cuisine?
9. What are the factors that influenced Indian cooking?
10. What is the concept of ayurveda?
11. How does the ayurveda classify the food on the basis of *rasa*?
12. What are the foods classified into *veerya* and *prabhav*?
13. According to ayurveda, how is the mental state of the human mind classified?
14. What is the concept behind *ghar ka khana*?
15. What was the influence of Portugese people on Indian food?
16. Write a short note on influence of Greeks on Indian food.
17. How have Mughals influenced the cuisines of India?
18. Which parts of India were influenced by Mongolia and China and what has been the impact?
19. Did British Raj have any impact on Indian cuisine?
20. What is Anglo Indian food?
21. What equipment is used to roll the rotis?
22. What is *chalni* and what are its uses?
23. What is *pauni*?
24. What is the usage of *sil batta*?
25. What is *hamam* dasta and what is it used for?
26. What is the difference between *jodi, sariya*, and *gaddi*?
27. In context to usage, how is iron *tandoor* different from clay *tandoor*?
28. Why is the *degchi* suited for *dum* cooking?
29. What is *lagan* and how is it different from *mahi tawa*?
30. Write the uses of *sigri*.
31. Write a short note on *dum* cooking.
32. What is *bhunana* method and why is it used?
33. What are the various terminologies used for tempering in Indian cooking?
34. What is *dhungar*? Name at least three Indian dishes associated with it.
35. What is slow food? How is it healthy?

PROJECT WORK

1. Visit various restaurants and hotels and write down the various types of food served in the *ghar ka khana* section. Critique the selection and write down the salient features of the dish that makes it apt for this section.
2. Divide yourself into groups of four to five and visit the various locations in your city that are famous for food joints. Visit the areas that are influenced by a typical religious sect and taste and sample the food. Write down your observation and try to classify the tastes and textures and make a common flavour profile of each sect and see how the region and the religion influence the food.
3. Visit some organic farms in your city and observe how various fruits and vegetable commodities are grown. Why do you think these foods are more expensive? Justify your statements.
4. In groups, research about the medicinal values of Indian herbs and spices and share the information with each other. You might have to visit various ayurveda shops, etc.

CHAPTER 24

CONDIMENTS, HERBS, AND SPICES USED IN INDIAN CUISINE

Learning Objectives

After reading this chapter, you should be able to:
- understand the various condiments and spices used in Indian cuisine
- appreciate the medicinal value of the spices
- know the philosophy behind classification of Indian spices
- understand the usage of spices in Indian cuisine
- know how to take care of storage of spices and their shelf life

INTRODUCTION

Spices are strongly flavoured fresh or dried roots, bark, shoots, or even seeds of plants. These are aromatic substances that have distinct fragrance, taste, or flavour and hence, they are used to season food. Apart from seasoning food, spices are used extensively all around the world to flavour wines; to produce perfumes, cosmetics, and even medicines. The major spice producing countries are India, China, Indonesia, Malaysia, Thailand, South Vietnam, France, Greece, Morocco, Lebanon, the UK, etc. However, most of the spices come from the eastern parts of the world as the tropical climates are apt for growing these. The first spice that was introduced to Europe was pepper from India and then the major trade between the countries started due to the need of spices. This was also the reason that Portuguese navigator Vasco da Gama sailed to India. At the same time Christopher Columbus briefed the investors about availability of many new spices on discovering the 'New World', which led to travellers coming to India in search of different spices. Many spices have been used extensively in medicine since ancient times. Spices were used in the ancient times to preserve the food or to replace the lost flavour by long periods of cooking, as most of the food was in the form of stews that required moist method of cooking for longer duration of time. The usage of spices is not restricted to Indian cuisine only. It is used in cuisines from all over the world but in different ways. The main purpose still remains the same and that is to enhance the flavour of the food.

SPICES USED IN INDIAN COOKING

India is known as the home of spices. In India when we refer to food being spicy, we mostly refer to it being hot but the interpretation of spicy food does not mean hot food. Food being spicy and food being hot are two different things. Spices come from various parts of plants—seeds, barks, flowers, etc. Table 24.1 shows spices and condiments from various parts of plants.

Table 24.1   Spices and condiments from various parts of plants

Part of Plant	Examples of Spices
Bracket or the aril covering the seed	Mace
Barks	Cassia, cinnamon
Berries	Allspice, pimento, black pepper, juniper
Buds	Cloves
Bulbs	Garlic, leek
Floral parts	Saffron
Fruits	Cardamom, chillies
Kernels	Nutmeg
Leaves	Bay leaf, curry leaves, marjoram, coriander, etc.
Underground stems or rhizomes	Ginger, turmeric
Latex or resins	Asafoetida
Seeds	Aniseed, caraway, cumin, coriander, mustard, etc.

There are many ways in which spices can be classified. One of the ways is to classify them on the basis of parts they come from, for example, bark spices, leafy spices, etc; but in kitchens, chefs usually classify spices into fresh spice or dried spice. The dried spices are further classified into whole spices or ground spices. There could be another classification of spices whereby they are classified on the role they play in the food. The classification could thus be as follows:

Hot spices   Chilli, pepper, etc.

Mild spices   Paprika

Colouring spices   Annatto, saffron, turmeric, etc.

Aromatic spices   Allspice, cardamom, cloves, etc.

The role that Indian spices and condiments play in Indian cuisine is discussed below (*see also Plate 16*).

Cubeb pepper

Hindi: *Kebab cheeni* **English:** Java pepper **Other:** Tailed pepper
As the name suggests, this spice has the flavour of many spices and thus the name. Allspice or cubeb pepper is not very commonly used around India, but is very common in certain Mughlai dishes. Cubeb pepper, also known as tailed pepper is so called due to a protruding stak from the berry. This spice is native to Java and Sumatra Islands and is widely used in Indonesian cooking. In India, we use this pepper to flavour kebabs and biryanis and hence the Hindi name *kebab cheeni*.

Usage in Indian Cooking

In Indian cooking *kebab cheeni* is used in flavouring lamb based kebabs. The sweet fragrance of this spice also finds its uses in certain pickles and chutneys.

Medicinal Features

Apart from cooking this spice has many medicinal properties. The allspice oil is believed to cause an irritation on the skin, allowing the blood vessels to expand, thereby causing flow of blood to make the skin feel warm. The tannins found in the allspice acts as a mild anaesthetic and is thus applied on sore joints of arthritis and is used for muscular pains for immediate relief.

Storage Techniques

Cubeb pepper will keep good for two years if stored in airtight jars. Keep the jar in a cool dark place for long lasting flavours.

Ajowan

Hindi: *Ajwain* **Other:** Carom seed, bishop's weed
This spice was probably carried to India from the Mediterranean belt and is widely used in Indian cooking for flavouring curries, pickles, and deepfried products. A close up of this spice reminds one of a bishop's hat and that is why it is also called bishop's weed. It is greyish green in colour and resembles the seeds of thyme or parsley. It is always sold whole. The largest growers of ajowan are Rajasthan and Gujarat in India. It is also grown in Iran, Afghanistan, and the Middle East.

Usage in Indian Cooking

This is a famous Indian spice particularly used in parathas, an Indian bread made on a griddle. It is also used in savoury snacks such as samosas, pakoras, and breads. It goes very well with beans and pulses, as it has an affinity for starchy food. It has powerful thyme-like aroma, which can be simply found by crushing the seeds in the hands. It is one of the main ingredients of pickling spice and is used with fish too. It can be seen that this spice is associated with foods that are deep-fried. Cooking reduces its bitter and strong aroma. It is usually used in tempering.

Medicinal Features

Ajwain contains an essential oil known as thymol. Simply chewing of the seeds causes numbness on the tongue and freshens up the breath. For decades it has been used for gastric troubles as it aids in digestion. It also is believed to have aphrodisiac properties.

Storage Techniques

Ajowan will keep good indefinitely if stored in airtight jars. Keep the jar in a cool, dark place for long lasting flavours.

Aniseed

Hindi: *Saunf* **Other:** Anise

Aniseed has a similar flavour to that of fennel seed and the spices can be used interchangeably. Grown in the Mediterranean countries, this spice is green coloured, oval, and has a sweet fragrance. It is used in pickles and is available in both whole and powdered form. It is extensively used in flavouring breads and cookies in Europe. It resembles the grain of rice while still in husk, and the only difference is the green colour of aniseed. It is also used in curries and spice mixes. It is one of the spices that is extensively used in Indian desserts.

Usage in Indian Cooking

Aniseed is used extensively in curries and gravies from Kashmir. In other parts of India it is used as an ingredient for pickles and often eaten as a mouth freshener post meals along with crystal sugar candies. It also forms one of the ingredients for a tempering spice from Bengal known as *paanch phoran* (*saunf, jeera, sarson, kalonji, methi,* and sometimes *radhuni*). Roasting or broiling of aniseed increases its flavour and makes it easier to grind into a powder.

Medicinal Features

Saunf is eaten after meals in India. It is used as a mouth freshener and also aids in digestion of food. It is a mild expectorant and thus aids in soothing the throat. Few seeds taken with water often help to relieve hiccups.

Storage Techniques

This spice loses its flavour very quickly, so it is advisable to store it whole and ground to a powder as required. They keep well for six months if kept covered in an airtight jar in a cool and dark place.

Asafoetida

Hindi: *Hing* **Other:** Devil's dung

The name of this condiment comes from two words—aza meaning gum or resin and *foetidus* meaning stink in Latin. It is extracted from the root or rhizome of certain species of ferula tree and three of those species are found in India. It is believed to have been found in the Mediterranean belt. Then it moved to central Asia and Afghanistan, which is the largest producer of *hing* today. *Hing* came to India from Afghanistan. It is used to flavour foods that do not have any onion and garlic. This condiment stinks and that is the reason it is not very popular around the world. It stinks because of the high content of sulphur present in the resin and the flavour goes away while cooking. It is available in solid as well as powder form. The powdered version is refined and available commercially.

Usage in Indian Cooking

There are two types of *hing* available and both have different procedures of usage. The whitish coloured one is called *kabuli hing* or *safed hing* and the other variety is *lal hing*. The *kabuli hing* is dissolved in water and then added to oil or gravies and the *lal hing* is oil soluble so it is made into a paste with oil and then used. It is extensively used in vegetarian cooking which do not use onion and garlic in the food.

It pairs up well with lentils and curd-based gravies and is extensively used in Kashmiri, Rajasthani, and in west and south Indian cuisine.

Medicinal Features

Asafoetida helps relieve colic pain, and is used as an antiflatulent. It is also advised for patients suffering from asthma, bronchitis, and whooping cough. It is also believed that the strong flavours of this condiment keep germs and evil at bay and hence it is very common to see small children wearing it around their neck tied in a black cloth.

Storage Techniques

It is important to store asafoetida in an airtight container as its strong flavour would affect the other food ingredients in the kitchen. It will last for indefinite period, if stored away from sunlight in a cool place.

Bay Leaf

Hindi: *Tej patta* **Other:** Laurel leaf

Ancient Greek emperors and heroes used to wear wreaths of this leaf on their heads as it symbolized courage and laurels they had achieved. This is the reason why this leaf is also called laurel leaf. There are many varieties of this spice which can range from small musty olive green colour to large beige coloured dry leaves. It is an aromatic leaf of an evergreen shrub that belongs to the Mediterranean belt such as Rome and Greece. The leaves are bright green and when dried turn from olive green to beige colour depending upon the species. It is always included in bouquet garni to flavour soups and stocks. It is available as whole dried leaf or also in crushed form.

Usage in Indian Cooking

This is a famous Indian spice and is not cultivated in India for commercial purposes. The *tej patta* used in Indian cooking is the leaf of the tree, from which the cassia bark is obtained. In Indian cooking, it is used for flavouring stocks, rice pilafs, and meat dishes. It is tempered in oil and removed prior to service as it is inedible and woody. In India it is always available in dried form.

Medicinal Features

Bay leaf has lauric acid, which acts as a repellent for moths and is used for its insecticidal properties. Bay leaf is also widely used in India as a medicine for migraines, bacterial and fungal infections, and gastric ulcers. Bay leaves are also used to reduce blood sugar levels.

Storage Techniques

Bay leaves will keep good for at least two years if stored in an airtight jar. Keep the jar in a cool, dark place for long lasting flavours.

Green Cardamom

Hindi: *Choti elaichi*

It is obtained from an aromatic plant from the Malabar region of south western India. The pods contain seeds which are dried and used as a spice. As the name suggests, it is green in colour and it is so named to distinguish between the green and the black cardamom. It is an essential spice which is also the main ingredient of the whole as well dried ground garam masala. It has a pleasing and slightly warm lemon-like flavour. India contributes 80 per cent of the total cardamom grown the world over.

Usage in Indian Cooking

It is used extensively in curries, desserts, and most commonly in flavouring tea. It is also used as a mouth freshener and sometimes bleached to look white in colour or sometimes coloured silver to be served after meals. Cardamom can be used whole or ground to a powder that can be used to flavour food. It is mostly used in *Mughlai* dishes such as kebabs, biryanis, and curries. It also finds its application in milk-based sweets.

Medicinal Features

Cardamom is used as a mouth freshener and also in making perfumes. It is believed to be aphrodisiac in nature and the ancient Indians regarded it as a cure for obesity.

Storage Techniques

Cardamom has volatile oils that lose flavour very quickly. Storing in airtight containers is advisable for making it last up to six months. Store in a cool, dark place and keep the pods whole. Crush and powder them as and when required.

Black Cardamom

Hindi: *Moti elaichi/Badi elaichi*

It is dark brown in colour and so it is called black cardamom. It has an entirely different flavour from the green cardamom and its usage is more in meat dishes than sweets. The only similarity between green and black cardamom is that they are from the same family. India is the largest producer of this spice in the world and the main areas of production are Sikkim, Darjeeling, and other parts of West Bengal. This cardamom enhances the taste of other ingredients when used with them. It has a wrinkled skin.

Usage in Indian Cooking

This spice is usually crushed and tempered in oil. It is also an important ingredient in garam masala. Unlike green cardamom, it has a smoky and camphor-like flavour which pairs up very well with meats and rice pilafs.

Medicinal Features

Black cardamom is an ancient cure for colic pain, fatigue, and stress and also helps to cure headaches. Black cardamom oil is also used in aromatherapy to stimulate energy.

Storage Techniques

Black cardamom has volatile oils that lose flavour very quickly. Storing in airtight containers is advisable for making it last upto six months. Store in a cool, dark place and keep the pods whole. Crush and powder them as and when required.

Cinnamon

Hindi: *Dalchini*

Cinnamon is known as *dalchini* and the name comes from the Persian word *dar-el-chini* which means wood of China. This spice is of Chinese origin and is abundantly grown in Sri Lanka. Real cinnamon or the Ceylon cinnamon, is a muddy orange to dark coloured bark of wood that is rolled up like a pipe. In India we do not always get the real cinnamon. The cinnamon used in Indian cooking is the cassia bark. The cassia bark is readily available and is less expensive as compared to real cinnamon.

It is darker in colour and thicker than the true cinnamon. The cassia bark is also known as bastard cinnamon. It is available in whole or in powdered form.

Usage in Indian Cooking

In Indian cooking, it is another ingredient in garam masala. It is also used in rice dishes and in curries. It is usually tempered in oil. In Bengal, the cinnamon is made into a paste before being used. It is also used in certain desserts.

Medicinal Features

Cinnamon is said to cure blood sugar and cholesterol. It is mixed with honey and used as a medicine to cure arthritis. Cinnamon is also used to cure diarrhoea and cough-related ailments. It is believed to be a reliever of nausea and vomiting and aids in digestion.

Storage Techniques

Cinnamon too loses it flavour quickly and hence should be stored in airtight containers away from sunlight. Powdered cinnamon will lose flavour more quickly so grind only when required. Broiling the cinnamon intensifies the flavour.

Clove

Hindi: *Lavang, laung*

The name comes from the Latin word *clavus* which means a nail. The shape of this spice resembles a nail and hence the name. Clove has been used as a medicine in the South-East Asia for many centuries now. It is believed that people in China used it as a mouth freshener prior to having a meeting with the emperor. It is a native to the Moluccas in South-East Asia, but it is now widely grown in Caribbean islands, Indonesia, Malaysia, Madagascar, and Sri Lanka. Cloves are picked when the flower buds are small and as they are dried in the sun the colour changes to a darker brown. Clove is of great culinary significance. It is included in Indian garam masala and Chinese 'five spice powder'. It is also used in the onion cloute for flavouring the béchamel sauce. Cloves are arranged in a diamond pattern for flavouring and appearance on the baked ham and they go excellent with apples and meats.

Usage in Indian Cooking

This spice is usually crushed and tempered in oil. It is also an important ingredient in garam masala. It is used for smoking meats for curries and kebabs. It is also used in curries and biryanis. It is used whole or can be used in powder form. It is difficult to grind clove into a powder with *hamam dasta*, as it has high oil content. Small spice grinders, such as coffee grinders, are the best equipment to grind cloves.

Medicinal Features

Old folk stories say that sucking on two pieces of cloves, helps curb the desire for alcohol. Clove oil has been used for curing tooth ailments since ages. Chinese use cloves to cure indigestion, hernia, athlete's foot, and other fungal infections. In ayurveda, clove is used for curing respiratory and digestive ailments.

Storage Techniques

Clove has volatile oils that lose flavour very quickly. Storing in airtight containers is advisable for making it last upto six months. Store in a cool, dark place and keep the pods whole. Crush and powder cloves as and when required.

Coriander Seed

Hindi: *Dhaniya*

The name comes from the Greek word *kopis*, which means bugs in Greek. The shape of coriander seed resembles a bug and hence the name. Coriander is used as a spice whereas the fresh leaves of the plant are used as herb in many countries including India, where it is used for garnishing. The seeds look like tiny, pale, creamy-brown peppercorns. When they are dried and fried they have heady, slightly burnt orange aroma, which is very appealing. They are also used in the cuisines of the Middle East. In Germany they are used in game marinades and also for flavouring and seasoning cabbage. Whole or coarsely ground coriander are a classical ingredient in the à la grecque dishes. The fresh leaves of this plant, called cilantro or *hara dhaniya*, are used as herb in Indian cooking.

Usage in Indian Cooking

It is used in Indian cuisine as one of the main ingredients, finding its place in almost any curry prepared in India. This spice is usually crushed and tempered in oil when used whole. The powdered form is made into a paste with water and added to the oil. This is one of the spices that provides thickening to the gravies. Coriander seeds can be used in pickles, snacks, yoghurt gravies, and meat preparations. As this spice is very mild, it is used in fairly large quantities than other spices.

Medicinal Features

Coriander is generally beneficial to the nervous system. It is used in digestive ailments and also against piles and headaches. The paste of the coriander seeds is believed to cure ulcers in the mouth.

Storage Techniques

Ground coriander loses its flavour quickly, so it is advisable to buy it whole and store in an airtight jar away from sunlight. This way it can keep indefinitely. It can be powdered after broiling and stored for two months in an airtight jar.

Cumin

Hindi: *Jeera/Zeera*

Cumin is a spindle-shaped seed that is used around the world for various purposes. It is used for flavouring certain cheese and is also used in pickling meats. It is a spice that is widely used in north India in tempering and in curries. Cumin is used either whole or ground. It has a piquant and slightly bitter taste combined with warmth that is evident in its aroma. It is used in the Middle Eastern dishes in *dolmas* and rice preparations. It is a classic flavouring for breads in Europe and also used in the preparation of cold meats and cheese, such as *munster* cheese. It has two varieties—brown and black. The black variety is also called *shahi jeera* in India and is more powerful than brown cumin. It is available whole and in powdered form. It is of the same shape as caraway seeds, but the flavours of the two spices are entirely different. Sometimes people refer to black cumin or *shahi jeera* as caraway.

Usage in Indian Cooking

It is of great use in the Indian cuisine and is used for tempering curries and dals and for making rice preparations such as pilafs. This spice is usually broiled and crushed to extract the flavours and is used as a sprinkling on *dahi bhalla* and other Indian chaat items. It is also tempered in oil or ghee and mixed with boiled lentils. It is also an important ingredient in garam masala. It is combined with other ingredients to make *jal jeera*, which is an aperitif drunk during summer months. Jeera is a mild to strong spice so it is used with half the quantity of coriander powder in the curries. It also gives the dark colour and thickening to the curry. Jeera can be broiled and made into a powder. *Shahi jeera* is always used whole and seldom powdered.

Medicinal Features

It has been used as a medicine in the East for many decades. Cumin is believed to increase the milk in lactating mothers and reduces nausea during pregnancy. It is often eaten in the early morning as it is believed to cure morning sickness. Cumin stimulates the appetite and aids in digestion too.

Storage Techniques

Broil and powder the cumin and store in airtight containers away from sunlight in a cool place. The whole unroasted grains can be stored in an airtight container away from sunlight almost indefinitely.

Chilli

Hindi: *Mirch* **Other:** Cayenne pepper

Chillies were brought to India in the year 1498 from Chile and ever since it has become an integral part of Indian cuisine. There are many types of chillies found in India and some of them are associated with a particular dish, thus giving the uniqueness to the dish. From the mild Kashmir chillies to the spiciest chilli in the world—*dhani chilli* from Mizoram, India is famous for its chillies. The chemical present in the chillies is called capsaicin and the intensity of the chillies is mentioned in Scoville units. The Scoville units can range from zero as in case of bell pepper to almost half a million as in case of habanero from Mexico. It varies from species to species. The higher the capsaicin content in chillies, the higher the pungency. There are sweet chillies to fiery hot chillies. They are available in several different forms all around the world. The birth place of chilli is Mexico, where one can find the maximum varieties. Chillies are available in all the supermarkets around the world in dried form as whole or powdered. They are largely used to provide pungency to the dishes. They are used in Indian cuisine in powdered as well as fresh form to provide colour and pungency. Some other varieties of chillies are *ancho, nyora, serrano, cascabel,* bird's eye, and *passado*—all having a fiery hot taste. *Serrano* chillies are used to make the classic guacamole in Mexico.

Usage in Indian Cooking

Whole chillies are also used for tempering several south Indian dishes. There are many types of chillies found in India. They range from Kashmiri to Goan, from Guntur chillies in Andhra, etc. Some chillies are speciality of certain dishes such as mathania chillies are used for lal maas from Rajasthan and malwani chillies are used for some Maharashtrian food. The red chilli powder is added in the oil along with onions and garlic to sauté until the colour deepens. The red chillies are also boiled and made into a paste for various curries and kebabs. Fresh green chillies are eaten as salad and also used in dals, biryani, kebabs, curries and many other dishes. Another variety of chilli used in cooking is deghi mirch, which has a deep red colour but not much of pungency. It is therefore used in dishes that require a deep red colour. Another chilli from south India is a small chilli that is pickled and sun dried. It is often known as curd chilli and it is fried and served with meals.

Medicinal Features

Apart from the hot flavour, green chillies have the highest amount of vitamin C in them. Red chillies are full of beta carotene. Capsaicin present in chillies helps to reduce cholesterol in the body.

Storage Techniques

Red chillies are available in dried or powdered form. Only green chillies are available fresh. The red chillies and powder can be kept indefinitely if stored in an airtight jar away from sunlight.

Long Pepper

Hindi: *Pipli* **Other:** *Pipul*

This spice is native to Indonesia and has travelled to the world through India. It went through to Europe via Vietnam and India. It is not commonly used like black pepper, but finds its uses in certain kebabs and spice mixes. It is hotter than black pepper and has various medicinal uses. This spice is like a cluster of small baby peppercorns on a small 1.5 inches long stick that resembles mulberries. It is available whole and can be crushed as per requirement. It is quite spicy and hot when milled.

Usage in Indian Cooking

This spice is usually used in the Mughlai cuisine. It forms part of the *potli masala* that is used in preparation of biryani in Hyderabad. Pipli is also powdered and used in various kebabs. It is not a very commonly used spice in every day kitchen.

Medicinal Features

Long pepper has been regarded as diuretic, expectorant, stimulant, and antiseptic. Ancient Indonesians used it as a cure for chronic bronchial patients. It also has a stimulating effect on the membranes of urinary tract. It is considered to be an aphrodisiac, as it is used to cure many sexual disorders such as gonorrheae.

Storage Techniques

Pipli has volatile oils that loose its flavour very quickly. Storing in airtight containers is advisable for making it last upto 6 months. Store in a cool and dark place and keep the pods whole, crush and powder them as and when required.

Dill

Hindi: *Sowa* **Other:** Garden dill

Dill comes from the word *dilla*, which means 'to lull' in Latin. Native to Europe, this spice is the favourite of the Russians. Both the leaves and seeds are used in both Indian and Western cooking. Dill is used around the world as a herb and its flavour goes well with fish. It is also used in pickling. The dill leaves are paired with fish whereas the seeds are mostly used in breads, cookies, and in pickling meats. It is quite commonly used in Kashmir. The dill seed is light brown in colour and it has a flat and an oval shape. It exudes a sweet aroma and is extensively used in Kashmiri cooking. It is available whole, but can be coarsely crushed and used. It is used in both curries and breads.

Usage in Indian Cooking

This herb is usually crushed and tempered in oil. It is grown in cold climates and hence, very commonly used in Kashmiri cuisine. It has a sweet flavour like caraway seeds.

Medicinal Features

Dill seed contains essential oils known as carvone. It is believed to stimulate the body and help cure stomach aches. Dill water is given to children for digestion and eating dill can stop hiccups.

Storage Techniques

Dill has volatile oils that lose flavour very quickly. Storing in an airtight container is advisable for making it last upto six months. Store in a cool, dark place and keep the pods whole.

Fenugreek

Hindi: *Methi daana* **Other:** Bird's foot

As the name suggests, it is a spice that was given to Indians by the Greeks. It was brought to India by the Greeks, during the reign of Alexander the Great. Fenugreek seeds resemble small yellow stones with angular structure. They are of different shapes—some are square, some rectangle, and some have the shape of distorted rectangle. Fenugreek seeds are highly aromatic and are used in a wide range of curry powders. They are also used dried and whole for tempering. They have a very good digestion property, which is the reason it is used in Indian cuisine to add it to the curries and dals to aid digestion. The seeds can also be used to make sprouts and can be used in sandwich fillings. They are usually available dried.

Usage in Indian Cooking

This spice is usually tempered in oil. It is a bitter spice and the bitterness usually goes away to a large extent after broiling. The leaves of fenugreek plant are used as vegetable in Indian cooking and the dried leaves are used as a condiment to flavour dishes. The best dried leaves come for *Qasoor* in Pakistan and are called *kasoori methi. Kasoori methi* is slightly toasted and crushed to powder before adding into curries. In Rajasthan the dried fenugreek seeds are soaked in water overnight and prepared like a vegetable dish. This spice is used in the commercial curry powders.

Medicinal Features

Fenugreek is a great digestive and some people recommend drinking water infused with a teaspoon of fenugreek on an empty stomach in the morning. Fenugreek seeds help in reducing sugar levels in the blood. It also helps in reducing blood pressure. It relieves nasal congestion and helps in fighting the infection.

Storage Techniques

Storing in airtight containers is advisable for making it last up to two years. Store in a cool, dark place and keep the pods whole. Crush and powder them as and when required.

Wild Mangosteen

Hindi: *Kokum*

Commonly known as *kokum* or *cocum*, it is an integral part of Konkani cuisine and other regions around the Western Ghats of India. *Kokum* comes from slender evergreen trees found on the tropical rain forests of the Western Ghats. The fruits of these trees are purplish in colour, sweet, fleshy, and are concealed inside in an acid pulp. *Kokum* is available as a dried rind, resembling the skin of thick black plum. The ripened rind is commonly used in cooking, whereas the dried and salted *kokum* is used as a condiment in curries. It is essentially used to add a souring flavour to the dishes, especially fish curries.

Usage in Indian Cooking

This condiment is used to make red coloured sherbets called *sol kadhi* and is a great thirst quencher in the hot and humid summer months. The *kokum* skin is soaked in hot water and then added to the curries. It imparts a sour and tangy flavour to the food and is much advised for fish curries. It is mostly used in Konkani cuisine of the coastal regions of Goa, Maharastra, Karnataka, and Kerala. The deeper the colour of the dried *kokum*, better is the flavour and quality.

Medicinal Features

Kokum is used for allergies on the skin. When applied to burnt skin, it relieves pain.

Storage Techniques

Kokum is usually available dried from the market. Always select darker coloured one and store in an airtight jar away from sunlight. It will keep good for over a year if stored in this way.

Liquorice

Hindi: *Mulethi* **Other:** Black sugar, sweet root

Liquorice is the rhizome of a root plant, which resembles a piece of wood when dried. It is hard like wood and contains many fibres. It has a brown outer skin and beige to yellow coloured interior. Liquorice is also candied in Europe and served as a mouth freshener. It is cultivated in England, where it usually finds its uses in confectionary items. It is usually used to flavour tea. In India, this spice is particularly used in north and central India and has medicinal value attached to it.

Usage in Indian Cooking

This spice does not usually find its place in curries, savoury foods, or desserts in India; but is often used in making milk tea or masala chai.

Medicinal Features

Liquorice is the best known spice to cure sore throat and its related infections. In India tea made with *mulethi* is considered very good for cough. It is also believed to treat stomach ulcers and reduces the acid in the stomach.

Storage Techniques

If dried liquorice is stored in airtight containers, it will last almost indefinitely. Store in a cool and dark place and keep the spice whole. Crush and powder it as and when required.

Mace and Nutmeg

Hindi: *Javitri and Jaiphal*

Both the spices come from the same tree but are different from each other in terms of flavour and aroma. Mace is an aril that surrounds the berry called nutmeg in a green coloured fruit that resembles an almond. Mace has a sweet flavour whereas nutmeg has a flavour like camphor. Mace is like a webbed structure. It is pressed, dried, and used as it is or reduced to powder. It has a slightly stronger flavour than nutmeg. Both are included in garam masala used in Indian cuisines. In Europe, it is a popular seasoning for sausage meats and force meats, and is used in pork dishes. It can also replace nutmeg in potato puree and soufflés but precaution has to be taken to add in minimum quantity. When the mace dries up it loses its deep red colour but the flavour intensifies. So this spice is always sold in a dried form.

Usage in Indian Cooking

It is used a great deal in *Mughlai* dishes such as kebabs, biryanis, and curries. Since it has a nutty aroma, it pairs up very well with meats. Mace and green cardamom pair up very well and they are usually combined to make a smooth powder, which is used in many dishes. Nutmeg has the same culinary usage as mace. Mace is stronger than nutmeg and hence should be used in less quantity.

Medicinal Features

Both nutmeg and mace aid in digestion. They also help in removing the feeling of nausea. Nutmeg contains an enzyme called 'myristicin' and large dosage of nutmeg can result in death. However, generous usage in culinary preparations is harmless.

Storage Techniques

Mace and nutmeg should be stored separately and will keep for one year if stored in airtight jars. Keep the jar in a cool, dark place for long lasting flavours.

Mustard

Hindi: *Sarson* **Other:** *Rai*

The word mustard comes from the Latin word mustum—the name for the grape juice used to mix the ground mustard seeds to a paste, known as *mustum ardens*, which translates to 'the burning paste'. Mustard is quite a versatile spice. Mustard seeds are used for tempering and are also used to produce mustard oil, which is healthy as it contains monounsaturated fatty acids. There are three different types of mustard seeds—white (*alba*), brown (*juncea*), and black (*nigra*). The seeds have a very mild smell or almost no smell. The real taste comes when they are crushed and mixed with a liquid. Mixing powder mustard with a liquid activates an enzyme present in the seeds and it reacts with other constituents to develop the essential oil, which gives mustard its characteristic taste. Black seeds have the sharpest taste followed by white and then the brown. Black seeds have several culinary uses, apart from being used as condiments. It is also one of the main ingredients in salad dressings. In India its oil is used for cooking and in marinades. It is also used in lot of south Indian dishes as a tempering. It goes excellent with steaks giving them that extra tang as some people find the taste of steaks to be a little bland.

Usage in Indian Cooking

Mustard on its own is quite bland. To bring about the pungency in the mustard, it has to be ground into a powder and mixed with water. It is then rested at least for 10 minutes to let the enzymes do their work and make it pungent and flavourful. Indian cuisine uses mustard seeds for tempering, especially in east and south India. The leaves are used for making a famous dish called *sarson ka saag* in Punjab. In the east it is used as a paste. It is advisable to grind soaked mustard on a *sil batta*, or else it will get bitter. It is used in curries, pickles, and tempering.

Medicinal Features

The ancient Greeks believed that this condiment was created by God for the healing of mankind. Mustard is used as a remedy for snake bites, scorpion bites, toothaches, and stiff neck.

Storage Techniques

Mustard seeds should be stored in an airtight jar away from sunlight. This way they can be kept indefinitely. The fresh mustard paste should be stored in a refrigerator and should be used in a day or two.

Nigella

Hindi: *Kalonji* **Other:** *Onion seeds*

The term nigella is derived from the Latin word *nigellus*, which means black. Nigella seeds are small and black in colour and are shaped like a triangle. Some people get confused between this spice and black sesame seed. Some people refer to it as onion seeds, but it has no connection with onions. When these seeds are rubbed together, they give a flavour close to oregano.

Usage in Indian Cooking

Nigella is used in Indian breads and curries. *Kalonji* is used whole and is mostly used in tempering. It forms one of the main tempering ingredients of *paanch phoran*. It is used in curries such as kormas and in pickles. It is also used as a topping for naan. It is also used in Indian breads. It should not be consumed in very large quantities, as one of the chemicals found in it is not good for health as it can cause paralysis.

Medicinal Features

Nigella is used to improve bowel movement and has been used in curing indigestion since decades. It is believed to aid in lactation of feeding mothers.

Storage Techniques

Nigella will keep good indefinitely if stored in airtight jars. Keep the jar in a cool, dark place for long lasting flavours.

Pepper

Hindi: *Kali mirch*

Pepper or peppercorn has been cultivated in India for centuries now, and it is from here that the spice spread all across the world. There were times when this spice was traded like money. This spice is called the 'king of spices' and is widely grown in south India. There are three types of peppercorns—black, white, and green. The black pepper is a dried unripe berry. It is round and wrinkled and its appearance is similar to that of papaya seed, which is often used as an adulterant into this spice. The white pepper comes from the same plant, but is allowed to ripen more on the plant. The green pep-percorn used in Western cuisines is pickled and not dried. The pink peppercorn, which is very popular in the Western world, is however not a real pepper.

Usage in Indian Cooking

In Western food, black pepper is directly milled on to the food. In Indian cooking it is used whole to flavour curries and biryanis and is also powdered and used in masalas. It is abundantly used in the spice mixes in south India. Pepper is one of the ingredients in garam masala.

Medicinal Features

Pepper has been used as a medicine since ages. It is believed that pepper stimulates the taste buds and helps in secreting the gastric juices. It helps in curing nausea and also increases sweating in the body, thereby allowing cooling down during summers.

Storage Techniques

Pepper is best purchased whole and it will almost keep indefinitely if stored in an airtight jar away from sunlight in a cool and dark place. Fresh pepper should be powdered as and when required, as the powdered form tends to lose its flavour and aroma due to the volatile oils present in it.

Poppy Seeds

Hindi: *Khus khus* **Other:** Maw seed, opium poppy

Also known as *khus khus* in Hindi, it is used in *Mughlai* dishes. It imparts more of thickening properties rather than the flavour itself. Poppy seeds are available in two colours—black and white. The flavour of both is similar. Black poppy seeds are mostly used in Western cuisine and white is used in Indian cooking. The seeds resemble tiny round hard grains. It is believed to induce sleep. The Oriental poppy seeds are said to have narcotic properties, which led to the cultivation of this condiment in the Western world. This condiment is in demand, as people believe that it contains opium drug. The opium drug is found in the latex of the poppy plant and rarely reaches the mature seeds.

Usage in Indian Cooking

In India, this condiment is used in the east and in *Mughlai* dishes. It is usually used for the purpose of providing thickening to the curries and gravies. Its nutty flavour makes it ideal choice for kormas and other rich gravies. Poppy seed is also used as a crust for certain kebabs. Poppy seeds are very hard to grind, so it is advisable to soak them and grind with *sil batta*. Electric mixer and grinder can be used for grinding large quantities.

Medicinal Features

Western poppy seed or the black variety is used as a remedy for cough, whereas the Oriental or the Eastern poppy seeds are believed to be narcotic. An infusion of seeds is believed to comfort the ear and relieve toothache.

Storage Techniques

Poppy seeds are available in dried form and they will almost keep indefinitely if stored in an airtight jar away from sunlight in a cool and dark place. Do not store in a humid place as it tends to get infected by fungus and moulds.

Saffron

Hindi: *Kesar* **Other:** *Zafran*

The name saffron comes from the Arabic word *zafran* which means yellow. Saffron is grown around the world, but the most famous varieties are Spanish saffron and Kashmir saffron from India. The Pampore region in Kashmir produces the finest saffron in the world after Spain. It is an expensive condiment and it is so because it takes 75000 flowers to produce half a kg of saffron. Each stigma is hand-picked from a purple flower. There are three to four stigmas in a single flower. The spice is used for a subtle flavour and providing a pleasing yellow colour to the food. The saffron colour is considered sacred and is the religious colour of the Hindus in India. It is used to flavour and provide a pleasing yellow colour to the food.

Usage in Indian Cooking

Indian desserts and *Mughlai* dishes such as biryanis, kebabs, and certain kormas and other gravies such as *rizflla*, etc. use saffron for its strong colouring properties. Most of the Kashmiri dishes utilize saffron as it is believed to provide warmth and it is apt for the winter climate. Saffron is infused in hot water or milk. It can be slightly broiled and crushed to a fine powder with mortar and pestle. This way the colour will be darker. Since it is a strong condiment, a pinch is more than enough. Being an expensive condiment, it is often just sprinkled over the desserts for being flaunted.

Medicinal Features

Saffron is believed to be an aphrodisiac. Large dosages can however be fatal. It helps in warming the body and hence helps in curing cold and cough.

Storage Techniques

Saffron is best purchased in strands and not powdered, as the powdered variety is adulterated most of the times. It will keep for almost a year if stored in an airtight jar away from sunlight in a cool, dark place.

Star Anise

Hindi: *Badalphool, chakri phool* **Other:** *Badian*
This star-shaped spice is native to China and is the primary ingredient in the Chinese five spice powder. It also grows in Vietnam and Arunachal Pradesh in India. It gets its name from its star-shaped pods that contain small brown to orange seeds. Eight to ten boat-shaped pods join in the centre to form a star. Star anise has a sweet and nutty aroma and hence pairs up well with fruits and meats. It is used as a spice and can be used whole or in powdered form.

Usage in Indian Cooking

The whole star anise can be used to flavour curries, rice, and lentils and the powder form can be used in kebabs. Broil the star anise before pounding it. Star anise also forms the basis of certain masalas. It is mostly used in western coast and south India. It can also be added to sweets as it has a pleasing sweet aroma.

Medicinal Features

Star anise is used to cure colic pain and is a good digestive. It is commonly found as flavouring in medicines and used for curing cough.

Storage Techniques

Star anise is best purchased whole and it will almost keep indefinitely if stored in an airtight jar away from sunlight in a cool, dark place. It should be powdered as and when required, as the powder tends to lose its flavour and aroma due to the volatile oils present in it.

Tamarind

Hindi: *Imli*
The name tamarind comes from the Arabic word *tamar-e-Hind* or fruit of India. During trade, this fruit must have gone to the Middle East where it was given this name. It is a curved brown coloured bean pod from the tamarind tree. A sticky pulp is concealed in the pod with a shiny black seed that is square in shape. It is available in various forms such as pastes, dried pulp, etc which is sold in slabs. It can be either with seeds or seedless. Tamarind helps to clean the copper utensils in the Indian kitchen.

Usage in Indian Cooking

Tamarind is used as a thickening agent and also souring agent in Indian cuisine. It is mostly used in south Indian dishes. The slab is steeped into water and then it is passed through a sieve leaving behind a fibrous husk. The puree thus obtained can be used in curries and gravies. It is also used for making chutneys and pickles and used in chaats.

Medicinal Features

Tamarind is believed to aid in digestion and is a good cure for bronchial infections. It is recommended for sore throat. Being high in acid content, tamarind is cooling and hence a good cure for fever.

Storage Techniques

Tamarind is best purchased in the slab form, without the seeds and it will keep for one year if stored in an airtight jar away from sunlight in a cool, dark place.

Zedoary

Hindi: *Aam haldi* **Other:** *Kentjur, wild turmeric*

Zedoary root is a rhizome that resembles ginger and garlic. It has a tanned skin that conceals either a white flesh or a yellow flesh. Zedoary is found extensively in India and Indonesia. It is not very commonly used in all parts of India; it is quite extensively used in east India. It is also grown in Maharashtra and Gujarat in India that have moist and humid climate. It is a close relative to the rhizome of turmeric and is available in two colours. The white one is more like Thai 'lesser galangal' or mango ginger and is also known as *aam ada*. It is available in dried form and during the months of October and November it is available fresh. It is extensively used in medicines in ayurveda.

Usage in Indian Cooking

In east India, it is used in the form of paste to make curries. It is commonly used in pickles and chutneys.

Medicinal Features

Zedoary is used for purification of blood. It is also antiseptic like turmeric and is applied locally to cuts and wounds.

Storage Techniques

Zedoary root is best purchased fresh but when it is available dried, it can be stored for a year, if stored in an airtight jar away from sunlight in a cool, dark place. It should be broiled and crushed to powder as and when required.

Turmeric

Hindi: *Haldi* **Other:** *Curcuma, yellow ginger*

Turmeric is a rhizome that is used as a colouring agent in Indian cooking. It is antiseptic in nature and is used widely in spiced dishes, to counteract the ill effect of spices on the stomach. This spice is hugely popular in India. It can be said that it is a less expensive alternative to saffron. It has a peppery aroma and adds a warm, musky flavour to food. It is used in its dried form and commonly available in the form of powder. It resembles ginger but it is rare to see it fresh as most of the turmeric produced in India is sold dried. In Bengal and other parts of eastern India, however it is available fresh and is used in the form of paste. Turmeric is also used in other South–East Asian countries such as Indonesia and Malaysia. It is also used in rituals and religious ceremonies.

Usage in Indian Cooking

It is largely used in curries to provide colour and to lend a flavour. It is also used in various pickles and said to have great medicinal uses also. While cooking, turmeric can be put directly on the hot oil, but care has to be taken as it tends to burn quickly. It will be a good idea to make a paste of turmeric and water and add it to hot oil after the onions have been browned. Turmeric if added in the curry directly gives a raw flavour.

Medicinal Features

Turmeric is an antiseptic and that is the reason it is widely used in Indian cooking. It nullifies the effect of other spices and ingredients in the stomach. It aids in digestion and is often dissolved in milk and given to a person who has had an injury. Turmeric paste is also applied on the skin to give a golden hue to the skin and this is seen during Indian weddings.

Storage Techniques

Turmeric is available in the powdered form and it will almost keep indefinitely if stored in an airtight jar away from sunlight in a cool, dark place. Fresh turmeric should be made into a paste and used immediately.

Celery Seeds

Hindi: *Radhuni*

Celery was considered to be a holy plant in Greece. In the West it is mostly used in flavouring soups and stocks. It is also eaten raw as crudites and salads. Celery seeds are used as flavouring condiment in Indian cooking.

Usage in Indian Cooking

Radhuni is grown in Punjab and Uttar Pradesh, where it is used as a flavouring spice for pickles and chutneys. In Bengal, it is an important condiment of the *paanch phoran* spice. It is tempered with oil to bring out the flavours. It is also used in pickles.

Medicinal Features

Celery seeds have been recommended for weight loss since decades. It is believed, that it is difficult to digest the celery seeds and thus it acts as a negative calorie, as human body burns more calories in trying to digest it.

Storage Techniques

Radhuni will almost keep indefinitely if stored in an airtight jar away from sunlight in a cool and dark place. It is used whole and not powdered as the spice will get bitter if powdered.

Curry Leaf

Hindi: *Kari patta*

It is the leaf of the curry tree and is used as seasoning in curries. This leaf spice is one of the ingredients of Madras curry powder and is used in all parts of India. It usually is added for flavour and aroma and it is used mostly in south India and Bengal. It is available fresh and sometimes it is also used in dry form.

Usage in Indian Cooking

This spice is usually used as a tempering in most of the south Indian curries. It is also ground into a paste and used as the base for curries or served as chutney to accompany the snacks. In north India it is used in the tempering of *kadhi*. It is also used to make pickles.

Medicinal Features

The leaves and stems of the curry tree are used as a tonic. Both the leaves and stems are applied on skin eruptions or bites to cure them. Fresh juice of the curry leaves combined with lemon juice and sugar is prescribed for digestive disorders. It is believed that eating 10 curry leaves every morning for three months cures hereditary diabetes.

Storage Techniques

Curry leaf can be bought fresh from the market and used straight away. It should be stored in a perforated basket covered with a damp cloth. In most of the Indian homes, the curry tree is grown at home and the leaves are plucked as and when required.

Marjoram

Hindi: *Marwa*

Marjoram is a herb that is used in Western cooking. Though it is not commonly used in India, it is quite popular with the people living in the mountain regions of the north such as Himachal Pradesh. Its flavour is quite similar to that of oregano but slightly sharp and bitter.

Usage in Indian Cooking

Marwa is used in preparation of lamb dishes and is also commonly ground into chutney and eaten as an accompaniment. The dried leaves and the flower tops are used as spices in Indian cooking.

Medicinal Features

This herb is believed to cure asthma, cough, and bronchial related problems. It is also used to strengthen the stomach and the intestinal walls and prevents the formation of ulcers.

Storage Techniques

Fresh marjoram is not readily available in Indian markets and hence the dried one is preferred. It will keep for a year if stored in an airtight jar away from sunlight in a cool and dark place.

Oregano

Hindi: *Mirzanjosh, sathra* **Other:** *Mountain mint*

Oregano is a creeper with woody stems. The flowers are pale purplish in colour and are available between June and August. It is a very special herb known to Italians for years. It travelled to India probably through the Greeks, who also use it abundantly in their cuisine. Originally cultivated in the Mediterranean region, it is available on the mountain regions of India from Kashmir to Sikkim.

Usage in Indian Cooking

In India the dried berry of this herb is used as a spice. It is used in chutneys and also in meat preparations.

Medicinal Features

This herb is believed to cure lung disorders. It also helps in perspiration, thereby bringing the temperature of the body down. It helps to reduce fevers and is believed to be a cure for cold and coughs. According to Indian folklore, it helps to relieve menstrual discomfort of women.

Storage Techniques

Mirzanjosh is stored fresh in the refrigerator in a perforated basket covered with a damp cloth.

Pomegranate Seeds

Hindi: *Anardana*

Pomegranate is widely consumed in Iran and Afghanistan and it probably came to India through this belt. It is used as a souring agent and is available in a dried form. Its usage is limited to northern parts of India, especially Punjab, where it is extensively used for adding a sour flavour to various dishes. It is pounded into powder and used in stuffing or tossed with potatoes.

Usage in Indian Cooking

Anardana is only used in dried form and it is purplish black in colour. The darker the colour, the better is the flavour of the *anardana*. It is usually pounded to a coarse powder and used in stuffing in bitter gourd, etc. or is used for adding to curries to give a piquant taste. It is difficult to grind the seeds in a mixer grinder so it is advisable to crush it with mortar and pestle.

Medicinal Features

Pomegranates do not contain any sodium, cholesterol, or fat and are an excellent source of vitamin C. It is believed that eating pomegranate can combat cancer-related diseases, high blood pressure and can prevent the hardening of arteries.

Storage Techniques

Anardana will almost keep indefinitely if stored in an airtight jar away from sunlight in a cool and dark place.

Capers

Hindi: *Marathi moggu* **Other:** *Kiari, kabra, himachali laung*

Capers are mostly used in Western cooking. These are immature flower buds which are pickled in vinegar or salted to preserve them. In Indian cooking they are allowed to dry and are used in south India as a spice. The flavour of this spice may be related to that of mustard and black pepper.

Usage in Indian Cooking

This spice is mostly used in south India and is unheard of in other states of India. This spice has a piquant and peppery taste and is one of the important ingredients of *chettinad* masala from the south.

Medicinal Features

Marathi moggu is a very rare spice found in India. It is used only in south India. It is believed to cure arthritis.

Storage Techniques

Marathi moggu is best purchased dried and it will almost keep indefinitely if stored in an airtight jar away from sunlight in a cool and dark place. It should be crushed as and when required, as the powdered form tends to lose its flavour and aroma due to the volatile oils present in it.

Stone Flower

Hindi: *Pathar ka phool* **Other:** *Kalpasi,* lichen flower

As the name suggests, this spice grows on stones and trees. In this flower, an alga is surrounded by a fungus and they live in a symbolic association. The fungus protects the alga and alga on the other hand prepares food by photosynthesis. It is available around the world and can grow even in extreme conditions. It is available in dried form.

Usage in Indian Cooking

Stone flower is widely used in Hyderabad and other south Indian states. It is used as the major flavouring spice of the famous *pathar ka kebab* from Hyderabad. It is also one of the ingredients in potli masala or Indian bouquet garni which is used for flavouring biryanis. It is also an ingredient for the *chettinad masala.* It can be used in powdered form.

Medicinal Features

Recent research says that stone flower has a poly saccharide compound that suppresses the HIV antigen expression in cells.[1]

Storage Techniques

They are dried and will almost keep indefinitely if stored in an airtight jar away from sunlight in a cool, dark place.

Cobra Saffron

Hindi: *Nag kesar* **Other:** Indian rose chestnut, *mesua ferrea*

This herb is native to India. The root of this herb is used as a remedy against snake bites, and probably that is why it is called cobra saffron. It is also used to cure cobra stings. This entire plant is used for medicinal purposes. Due to lack of awareness, many young people are unaware of the uses of this plant.

Usage in Indian Cooking

All the parts of this plant are used as condiments in Indian cuisine. The flowers of this plant are dried and powdered and are used as the main ingredient in *goda masala* from Maharashtra.

Medicinal Features

All the parts of this plant are used for medicinal purposes with each part having its own medicinal properties. The oil extracted from the flowers is used for curing bacterial and fungal infections. The root is used as an antidote for snake bites. The fresh flowers are useful for preventing dehydration, cough, and indigestion. The stamen of this flower is used for preventing piles.

Storage Techniques

It is available from the ayurveda stores in a dried powder form and will almost keep indefinitely if stored in an airtight jar away from sunlight in a cool, dark place.

---

[1] *Source: The Hindu,* 22 September, 2001

Basil Seeds

Hindi: *Subza* **Other:** *Tootmalanga*

The seeds of basil plant are used as condiment only in India. It is strange that even in Italy, where this herb is popular; the seeds do not have any use in its cuisine. In India, these seeds are found in Maharashtra.

Usage in Indian Cooking

When soaked in water, basil seeds swell up to a jelly-like structure with a black dot in the centre resembling small eyes. These are served with Indian frozen desserts called kulfi as a garnish. These seeds are very commonly seen in Maharashtra and are not widely used in other states.

Medicinal Features

It is believed that basil seeds ward off common flu. They can be combined with some peppercorns and sugar to clear the breathing problems caused by blocked nose.

Storage Techniques

The seeds can be stored indefinitely if kept in an airtight jar away from sunlight in a cool and dark place. Once the seeds are swollen, they should not be stored for more than one day.

Dried Ginger

Hindi: *Saunth* **Other:** *Sunthi*

This condiment is the dried rhizome of ginger plant. There is a reference of saunth in ancient ayurveda for various purposes. The whole ginger is left to dry in the shade and is sold either whole or in a dried form.

Usage in Indian Cooking

This spice is crushed and boiled with liquids to let the flavours infuse or is used in powder form to flavour curries, breads, desserts, and aperitifs. It is also used in making medicinal teas. Saunth powder is the principal ingredient in *saunth* chutney that is served with chaats.

Medicinal Features

Ginger helps to relieve the discomfort of nausea and has been used in digestive disorders since centuries. It also helps in lowering the cholesterol levels in the blood and helps to relieve the pain in joints.

Storage Techniques

It is available whole or powdered. The whole saunth can be stored indefinitely if kept in an airtight jar away from sunlight in a cool, dark place. The powdered saunth should be kept for not more than six months as it will lose its flavour.

Black Salt

Hindi: *Kala namak* **Other:** *Sanchal*

This condiment is a type of volcanic salt and is procured from the volcanic regions of India and Pakistan. Black salt smells of sulphur and is available in rock form or powdered form. The rock crystals appear to be black and hence the name. This salt is named so because of its black coloured appearance, which when powdered turns to a pink or purple colour.

Usage in Indian Cooking

It is used to flavour many Indian savoury snacks, such as chaats and curd-based products such as raitas. It is also a principal flavouring agent in *jal jeera*, which is an aperitif served as a thirst quencher in hot summer months.

Medicinal Features

It is a salt that is regarded to be of high medicinal value, due to the presence of mineral compounds in the same. It lowers blood pressure as it does not contain sodium. *Kala namak* also aids in digestion and in improving eyesight.

Storage Techniques

It is available as a crystal or in powdered form. It can be stored in whole form indefinitely if stored in an airtight jar away from sun in a cool, dark place.

Root of Betel

Hindi: *Paan ki jadh*

The betel plant is a creeper and the leaves are cultivated for chewing. The root is used in the dry form and is used as a flavouring spice for meat dishes. The root of betel plant is a very common condiment used in *Mughlai* dishes of Hyderabad. This plant is indigenous to India, but is widely used in other South-East Asian countries such as Malaysia and Indonesia. This condiment is typically used in Hyderabadi cuisine.

Usage in Indian Cooking

The dried root of this plant which resembles bundles of thin straws of dried wood are used in *potli masala* to flavour biryanis and meat dishes. The usage of this spice is seen mostly in Hyderabadi cuisine.

Medicinal Features

This is considered to act as a digestive and has stimulant properties. It is also used as a cure for bronchitis and helps to cool the body.

Storage Techniques

It is available in the dried form and can be stored for six months if stored in an airtight jar away from sunlight in a cool and dark place.

VARIOUS WAYS OF USING THE SPICES

There are many ways in which spices are used in Indian cuisine. Many times the entire dish is based on the predominant flavours of the spices used. Let us see the various ways in which the spices are used in India.

Dry Roasting

It is also known as broiling of spices. Dry roasting is done in the case of seed spices such as cumin, coriander, fennel, mustard, ajowan, poppy seeds, etc. It is done for two main reasons—to bring out the flavours of the seeds and to expel the moisture, which helps to grind them into powder.

Frying in Oil

Whole spices are usually fried in oil or ghee. There are volatile oils present in spices that bring about the real flavours of the spices when they come in contact with warm oil or fats. That is the reason why spices are tempered in oil.

Grinding

Spices contain moisture and so it becomes difficult to crush them to powder. It is advisable to broil the spices on low heat to expel the moisture without the loss of any flavour. Certain large spices, such as cinnamon and black cardamom, are ground to avoid biting them while eating.

Grating

Spices such as nutmeg, horseradish, and ginger are grated before use.

Bruising/crushing

Spices such as cardamom, *anardana*, and zedoary root are crushed to get better aroma.

Shredding and Chopping

Fresh spices are mostly chopped before being used in cooking. Curry leaves, fresh coriander leaves, etc. are usually shredded or torn with hands.

Infusing

Few spices are infused in warm liquid before use, for example, saffron, tamarind, etc.

STORAGE AND USAGE TIPS FOR SPICES

Some storage and usage tips for spices are as follows:

- Store the spices in airtight jars in a cool, dark place away from sunlight. Humidity, light, and heat cause the volatile oils to evaporate, thereby making the spices lose their flavour more quickly.
- Whole spices may last up to many years and proper storage will enhance their medicinal value.
- Ground spices should not be stored for more than six months as they tend to lose their flavour more quickly than whole spices.
- Try to grind whole spices to powder close to the usage time. Toasting the spices over low heat will make them easier to grind.
- For a darker coloured spice powder, roast the spices for longer duration of time, taking care not to burn them.
- To keep the spices for a longer duration of time, store them in the freezer in tightly sealed containers. Care should be taken as certain spices get affected when kept for long in the fridge.
- Use a light hand while seasoning or spicing a dish.
- While boiling the spices for infusion, tie them up in a muslin bag to form *potli masala* for easy removal after the flavours have infused.
- Keep the food simple. Do not complicate the dish by using many strong spices at the same time. The general rule is to use light coloured spices for light meats such as green cardamom, mace, etc. and darker spices such as cloves, black cardamom, pepper, etc. for more robust dark coloured meats.

- Aromatic spices, such as garam masala, etc., are sprinkled on the dish after it has been cooked. Spices such as black pepper can be added by the guest at the table.
- Cinnamon, nutmeg, clove, and allspice have a special affinity towards sweet dishes.

COMMON ADULTERTANTS USED IN SPICES

With the advent of modernization, many greedy people want to make extra money by unfair means, hence it is not uncommon to find a range of inexpensive ingredients in the market. Many of such ingredients are cheap, because either they are adulterated by cheaper substitutes or are duplicate products of an established brand. Spices are one of the most consumed items in India and it is also true that many spices of high quality are expensive due to their production and farming. It is also difficult for a common man to identify the adulterants in the spices and the only way to check that is to either have them tested in laboratories or procure spices from a reputed source, who can give you an authentication for the quality of their products that they are supplying. In many cases, whilst the adulterants can be hazardous to consume, some inexpensive ingredients are added as an adulterant to a particular spice to increase its quantity. For example, low quality semolina mixed with poppy seeds will also be considered as an adulterant. Whilst it is almost impossible to see an adulterant in powdered spices, when buying whole spices, a few checks can be conducted randomly in one's kitchen that do not require expensive laboratory equipment and chemicals. Let us discuss some of the common adulterants used in spices and how to identify them in your kitchen (Refer to Table 24.2).

Table 24.2 Common adulterants used in spices

Spice	Adulterant	How to Identify
Saffron	Many a times an imitation saffron is created by soaking maize corn hairs in colours flavoured with artificial saffron.	Soak the saffron in the water. A real saffron will not change its colour. If after a few hours the saffron strands become white, it means that the saffron is adulterated.
Black pepper	As good black pepper is an expensive spice, people often use dried papaya seeds to adulterate the pepper as they look alike and are difficult to identify in first glance.	Soak the pepper in water. A real peppercorn will sink to the bottom, whereas the seeds of papaya will float on the surface. You may also crush a few seeds of peppercorn. The papaya seeds will be hollow and will not have a peppery taste.
Red chilli powder	Many colourants such as Sudan red and even powder of brick are used for adulterating red chili powder. Sometimes, cheap and falvourless dried chilies are also ground to a powder and then coloured with a dye to resemble a bright coloured red chilli powder.	Fill a glass tumbler with water and sprinkle a small quantity of red chilli powder on top. As and when the particles settle to the bottom, if they leave a streak of red color, it signifies that the red chilli powder is adulterated.

Contd

Table 24.2 (Contd)

Spice	Adulterant	How to Identify
Turmeric	Many artificial colours such as lead chromate and melanin yellow dye and even products such as corn flour and other cheap starches are used for adulterating turmeric.	To test the presence of a chemical, you will have to get it tested in laboratory.
Cumin	Many kinds of wild seeds and even coloured and flavoured sawdust is used as an adulterant in cumin seeds.	Fill a glass tumbler with water and sprinkle a small quantity of cumin seeds on it. The sawdust will remain on surface and cumin seeds will settle down.
Mustard seeds	The most commonly found adulterant in mustard seeds is argemone seeds. They both look very similar to each other but one can differentiate if one looks at them very closely.	The mustard seeds have a smooth skin, whereas the argemone seeds have a rough and matty finish. The mustard seeds upon splitting will reveal a yellow interior, whereas the argemone seeds will have a black interior.

CONCLUSION

This chapter dealt with various condiments, spices, and herbs used in Indian kitchen. The knowledge of Indian spices and condiments is very essential and it is the basic step towards preparing curries and other dishes. In this chapter we saw how spices are so varied and distinct to each other. We saw the classification of the spices and understood how spices come from the various parts of plants. Some spices are the bark of the tree such as cinnamon and cassia, some are flowers such as saffron, some come from the latex or the resinous parts of the plant, etc. We also discussed the ways to classify the spices based on the roles that they play in cooking such as hot spices, colouring spices and so on, or the parts of the plant they come from.

It would come as a surprise to many students that the herbs that are used in Western cuisine are actually grown in India and are also used in its cuisine in one way or the other. We also discussed the regions that they grow in and this was done to explain that the usage of these spices will be predominant in the state

that they come from. We also read about the various ways in which the spices are used in a particular dish. For example, saffron can be boiled in milk or water or broiled and crushed to powder form to extract the colour and flavour. It is interesting to see how certain spices are broiled to bring about the flavours in them and how some of them are tempered in oil to get the flavours out.

We also discussed the medicinal values of the spices. It should also be understood that for each spice there are many species of the same herb. For example, cinnamon and cassia are from the same family with almost the same flavour; but their medicinal properties could be different from one another. We also read about the storage of these condiments and spices and came to know that some spices last for an indefinite period, while some keep well only for six months. Care should be taken while handling spices as overcooking of these can lead to loss of flavours and in some cases it might have an adverse effect on the dish.

Aam aada Type of zedoary root that resembles ginger.

Ajwain Hindi for Ajowan.

Anardana Dried pomegranate seeds that are purplish to black in colour.

Capsaicin Chemical present in chilli, which makes it hot.

Carom seeds Another name for ajowan, also known as bishop's weed.

Cassia bark Bark of cassia tree; it is from the cinnamon family.

Chaat Tangy and spiced savoury meals served as snacks.

Chettinad masala Blend of spices used in south India. There are secretly guarded recipes of every household.

Chinese five spice Powder of five spices used as an aromatic spice in Chinese cooking.

Chotti elaichi Hindi for green cardamom.

Dahi bhalla Fried dumplings, soaked in yoghurt and served as an accompaniment with Indian meals.

Dalchini Hindi for cinnamon.

Deghi mirch Type of red chilli that is not hot but has a red colour.

Devil's dung Another name for hing or asafoetida.

Dhani Chilli Type of chilli from Mizoram in India and is considered to be the spiciest chilli in the world.

Dhaniya Hindi for coriander.

Garam masala Flavouring masala made up of seven to eight spices, which is used for aroma.

Goda masala Blend of masala from Maharashtra.

Jal jeera Indian aperitif made with broiled cumin.

Juniper Type of berry used as a spice in Europe.

Kabuli hing Type of hing from Kabul in Afghanistan. It is also known as safed hing and is white in colour.

Kadhi Dish made by boiling yoghurt with spices and tempering with more spices. *Kadhi* is made in Punjab, Rajasthan, and Gujarat in India.

Kasoori methi Dried leaves of fenugreek that come from Qasoor in Pakistan. They are used as flavouring condiment.

Kebab cheeni Hindi for cubeb pepper.

Kebabs Small morsels of meats or vegetables marinated and cooked in *tandoor* or griddle-fried to be served as appetisers or snacks in the Indian meal.

Korma Type of Indian gravy made with yoghurt.

Kulfi Frozen milk-based Indian dessert.

Lal hing A kind of hing that is red is colour. It is oil soluble.

Laung Hindi for cloves, also known as *lavang*.

Laurel leaf Another name for bay leaf or *tej patta*.

Lesser galangal Type of ginger used in Thai cooking, also known as *krachai* or mango ginger.

Mathania chilli Type of chilli found in Rajasthan and is used to flavour *lal maas*.

Maw seed Another name for poppy seed.

Munster cheese Cumin-flavoured cheese from England.

Panch phoran Mixture of five whole spices used in tempering of certain Bengali dishes. Traditional recipes use *radhuni* as the sixth spice.

Pakoras Batter-fried vegetable or meat. The batter is made of besan.

Parathas Indian flat bread made from wholewheat and griddle-fried.

Pipli Hindi for long pepper. Also known as java pepper.

Potli masala Spices tied into a sachet or a bag for flavouring dishes.

Rizala Mughlai dish based on yoghurt and flavoured with saffron.

Sarson ka saag Dish made in Punjab, India from fresh mustard leaves.

Saunf Hindi for aniseed.

Scoville Unit to measure the hotness of a chilli.

Shahi jeera Black cumin, often referred to as royal cumin.

Sol kadhi Pink coloured aperitif from Maharashtra, India, made from kokum and coconut.

Sowa Hindi for dill.

Subza Hindi for basil seeds, also known as tootma-langa. These are served as garnish with kulfi.

CONCEPT REVIEW QUESTIONS

1. Define spice. How are spices different from each other?
2. On what basis are the spices classified?
3. Name at least five parts of plants that are used as spices. Give examples.
4. Which are the major spice producing countries in the world?
5. What is bishop's weed and how is it used in Indian cooking?
6. Define *saunf*. How is it used in Indian food?

7. List the ingredients of *paanch phoran* spice.
8. What is the spice that comes from certain types of ferula plant?
9. How many types of hing are there? What is the uniqueness of hing?
10. How is laurel leaf used in Indian cooking?
11. What is the difference between *choti elaichi* and *moti elaichi*?
12. What spice is known as wood of China and which are its major producing countries?
13. Describe the usage of cloves in Indian cooking.
14. Differentiate between a *shahi jeera* and *jeera*.
15. What are the various ways in which *jeera* is used in Indian cooking?
16. Describe the role of chillies in Indian cooking and name at least five Indian chillies.
17. How is the hotness of chilli measured?
18. Describe cubeb pepper. How is it used in Indian cooking?
19. In which Indian cuisine is dill seed used the most?
20. What are the various ways in which fenugreek is used in Indian cuisine?
21. What is *kasoori methi*?
22. How is *kokum* used in Indian dishes?
23. What is so peculiar about *javitri* and *jaiphal*?
24. How many types of mustard are there? What are the uses of mustard?
25. Why should you not use large quantities of *kalonji* in food?
26. Why is pepper known as the 'king of spices'? How many types of pepper are there?
27. What role does poppy seed play in Indian cooking?
28. Why is saffron so expensive? What are the various ways of using saffron?
29. Write down the Hindi names of star anise.
30. What is the difference between *aam haldi* and *aam aada*?
31. What is *Marathi moggu*? What are its uses?
32. Describe stone flower. What are its uses?
33. What is *naag kesar*? What is so peculiar about the same?
34. Write a short note on basil seeds. How is it used in Indian cooking?
35. What are the various ways in which the spices are used in Indian cooking?
36. What care should be taken while storing spices?
37. What should be kept in mind while using spices?

PROJECT WORK

1. In a group, visit the spice market in your area and write down the observations regarding the same.
2. In two groups, select at least 10 different types of spices. Now one group should broil the spices and other group should fry them. Now compare the results and record your observations.
3. In groups, visit the ayurvedic stores and enquire about various herbs used in medicines. Research on these and see how they can be incorporated into cooking.
4. Broil five of the following spices for the mentioned minutes and record your observation with regard to colour and flavour in the chart below.

Spices	1 minute	2 minutes	3 minutes	4 minutes	5 minutes
Cloves					
Cumin					
Coriander					
Fennel					
Fenugreek					

CHAPTER 25

MASALAS AND PASTES

Learning Objectives

After reading this chapter, you should be able to:

- understand the usage of various kinds of spices and the traditional blends of spices to create masalas
- know various kinds of masalas, their usage, and storage
- understand various kinds of dry masalas and wet masalas and their usage in Indian cuisine
- get acquainted with various types of pastes used in Indian cooking, their usage, and storage

INTRODUCTION

In the previous chapter we read about various kinds of condiments and spices that are used in the Indian kitchen. Apart from the condiments mentioned in the previous chapter, there are many more which are used in cooking in India because of the diversity of Indian cuisine. We shall discuss most of these in separate chapters. But the basic ones discussed in the previous chapter will form a strong base towards cooking Indian food.

Indian cuisine is one of the most unique cuisines in the world. The uniqueness of the Indian cuisine is due to the successful combination of spices. As mentioned in the previous chapter, spices are used not just as flavouring agents but also for their medicinal values. The spices used in home style curries are limited in number, yet using the same combination of spices in varying proportions one can achieve numerous curries with very distinct characters. The main factors in using spices are the following.

- The choice of spices to be used in the dish is important. Usually a dish contains not more than five to six different spices, out of which one or two are predominant.
- The sequence in which each spice is used in a recipe and the length of time allowed for each spice to release its full aroma and flavour are different.

- There are many ways in which the spices are used in different cuisines. Some spices are used whole, some are fried, and some are powdered and blended.

This chapter will deal more with the blending of spices, regional specialities such as *paanch phoran, chettinad masala, goda masala*, etc. We shall also discuss various kinds of pastes made in Indian kitchen. These could range from nut pastes to vegetable pastes that form a strong foundation of Indian food.

BLENDING OF SPICES AND CONCEPT OF MASALAS

Spices are dried seeds, fruit or flower parts, bark, roots, or leaves of plants. They are usually from the tropical regions. Spices form important ingredients of various dishes, especially in the Indian cuisine. The effect of each spice on the recipe is very complex since each spice has its very own distinctive nature, flavour, and aroma. Since each spice has its own peculiar taste and flavour, great care has to be taken while blending the spices. Blending can be done through various ways such as broiling, frying in oil, grating, bruising, and grinding the spice to a powder or crushed form. Blending can also be done by combining few whole spices such as the *achari masala* or the Indian pickling spice. Spices are extensively used for cooking in all parts of India starting from north to south and east to west. Each region has its own spices that are influenced by the climatic condition of that particular region. The blending of spices needs great understanding of the flavours as they are central to the dishes.

There are different categories of spices such as:

Sweet spices   Cinnamon, mace, etc.

Hot spices   Black pepper, red chilli, etc.

Regional spices   Spices which are used in particular regions, such as saffron and *ratanjot* from Kashmir in north India; mustard oil, mustard seeds, and poppy seeds which are used in east India; and peppercorn and star anise which are used in the south.

A regional spice is mostly used in its respective region because of its relation to climatic condition, as spices have some therapeutic effect on the body. In cold regions, such as Kashmir, people use lots of *desi ghee*, saffron, and warm spices such as aniseed, which help in keeping the body warm. Similarly, in Rajasthan where the temperature is generally high, people eat lots of chillies, which actually help in keeping the body cool by excessive perspiration. A masala can be used in many contexts in Indian cuisine. It can be used to describe a single spice or a blend of spices. It is also used in reference to the base for a curry that has been sautéed in oil with spices.

CONCEPT OF DRY AND WET MASALAS

As discussed above, a masala in Indian cuisine is used in varied forms, depending upon the role it will play in a particular dish to get the required flavour and aroma. In this section we shall only deal with masala in context to spices. Whether the masala will be used in dry form or wet paste form is based on the culinary application such as cooking technique and the equipment used in that region. In a curry where the base of the thickened gravy is spices, it is better to use the masala in the paste form. Masalas are used in certain ways in dishes to extract their flavour. The usage of masalas depend on its cooking time. Dry masala is more or less fried in hot oil or tempered. If a masala requires more cooking or frying in oil to extract the flavours and colours, it is advisable to use it in the paste form as dry powder might burn in oil giving an unpleasant taste to the dish. Masalas are generally categorized into the following two types.

Dry Masala

Dry masalas are those which are in their dry form and no additional liquid component is added to them. These masalas may be whole or broiled and powdered. They might also include those ingredients which are specifically dried. Some ingredients might also be specifically dried in order to be blended with other dry spices. For example, mint is dried and powdered to be used as an ingredient in kebab masala. Let us discuss some of these dry masalas.

Aamchoori Masala

Ingredients: *Aamchoor*, aniseed, cumin, ginger powder, coriander, red chilli, salt, ajowan

This is the regional masala of Punjab. It adds a tangy flavour to the dish as it uses *amchoor*, which is the powder of raw mangoes. We shall discuss more about *aamchoor* in chapter 26. It is a souring agent and is used to add piquancy to the masala. *Aamchoori masala* is mostly used in stuffing okra, baby brinjal, or bitter gourd to create dishes. The names of these dishes are suffixed or prefixed with the name of the masalas, for example, *aamchoori bhindi*, etc.

Potli Masala

Ingredients: Cubeb pepper, clove, stone pepper, allspice, cinnamon, mace, green cardamom, root of betel

This can be classified as Indian *sachet d'epices*. In Hindi it literally means a pouch of spices. It is a bouquet of spices tied up in a muslin bag and left in a curry or liquid to let the flavours infuse. There can be many variations of this masala. This masala is added for easy removal of the spices after they have infused with the liquid. It has various uses in Indian cooking. It is used to flavour curries and also to flavour the water or stock for biryani.

Khada Masala

Ingredients: Cinnamon, cloves, green cardamom, black cardamom, cumin

Khada means whole spices. There can be various types of whole spices that are tempered into hot oil, to allow the flavours to infuse with the oil. The purpose of this could also be to remove the unwanted flavours from the fat such as from ghee. In Lucknow it is a ritual to add some *khada masala* to *desi ghee* to remove the unwanted flavour—this procedure is known as *ghee durust dena*. If the same spices are tied up in a bag they will be called *potli masala*. The main purpose of the *khada masala* is to flavour the oil. The spices contain volatile oils that help release their flavour better in warm oil. The spices can be slightly crushed before adding to hot oil or fat.

Garam Masala

Ingredients: Cumin, black cardamom, black peppercorns, coriander seeds, cloves, cinnamon, mace, bay leaf, green cardamom

Garam masala is a blend of aromatic spices, which is used to give flavour to the dish. This masala is usually sprinkled on top of the dish before being served, to retain all the aromas. There could be a slight variation in the ingredients used in different parts of the country. The individual spices are broiled separately, as they all have different degrees of cooking. They are broiled over low heat until a pleasing aroma comes out. In hotels the spices are arranged in a tray and kept under the 'hot lamp' or 'pick up counter'. The heating is done to expel the moisture from the spices and to bring out the flavour. Awadhi garam masala may include some more aromatic spices such as rose petals and star anise in addition to the ones mentioned above.

Chaat Masala

Ingredients: Cumin, peppercorns, black salt, dry mint leaves, *kasoori methi*, green cardamom, cloves, cinnamon, asafoetida, tartaric acid, mace, dried mango powder

As the name suggests this masala is used with the Indian street food called chaat. Chaats are mostly associated with sour, piquant, and spicy titbits that are eaten as snacks. The piquancy in chaats is created by adding chaat masala. Chaat masala is rarely made in hotels or homes, as it is readily available in the market as a condiment. Chaat masala can be added to various other spices to create more masala blends such as kebab masala. The masala is cooked by broiling the spices individually and grinding them to a smooth fine powder. Some masalas are sifted through muslin cloth to get the smooth powdery texture.

Chana Masala

Ingredients: Coriander, cumin, red chillies, ginger powder, dried mango powder, cinnamon, black pepper, black cardamom, garlic, cubeb pepper, nutmeg, clove, mace, dried pomegranate seeds

This is a regional masala from Punjab. Many people make it at home, but mostly it is available in the market as a condiment. This masala is used as a flavouring agent in chickpea curry from Punjab. This masala gives the required amount of spice and piquancy to the dish. *Chana masala* is also used to flavour the dark *pindi chole*—a famous dish from Rawalpindi in Pakistan.

Pudina Masala

Ingredients: Dried mint, chilli powder, dried pomegranate seeds, chaat masala, garam masala, black salt

This masala is a blend of dried mint powder with other spices. It is predominantly flavoured with dried mint powder. It is used for making Indian flat bread called *pudina paratha*. This masala is also used with kebab masala for kebabs or as sprinkling for various chaats.

Kebab Masala

Ingredients: Chaat masala, chilli powder, black salt, garam masala, *kasoori methi*

There can be varieties of masalas made for kebabs. Usually kebabs are sprinkled with chaat masala, but one can create unique kebab masalas. Toast the *kasoori methi* and crush to a powder. Mix all the ingredients and keep it in a container. Though this can be made in bulk in the hotel, it is advisable to make it once a week. You can be creative and combine spices with chaat masala as a base and create different types of kebab masalas.

Amritsari Machli Masala

Ingredients: Cumin, coriander, mint powder, ajowan, asafoetida, *aamchoor*, ginger powder, chilli powder

This is again one of the regional masalas from Punjab and is typically used in the preparation of *Amritsari machli*. The spices are individually broiled and crushed to a powder. The fish is marinated in this spice blend and then batter-fried. This spice is also added to the batter that is made of *besan*.

Paanch Phoran Masala

Ingredients: Anise, mustard, cumin, fenugreek, nigella, *radhuni*

This is a regional spice from Bengal and is used in tempering of vegetables, lentils, and fish. As the name suggests, this is a blend of five spices. *Radhuni* is optional, but most of the classical recipes include it as a part of *paanch phoran*. This spice adds a peculiar taste to the Bengali dishes and is a mixture of whole unroasted spices. It is always tempered in hot oil.

Achari Masala

Ingredients: Anise, mustard, cumin, fenugreek, nigella seeds
It is also known as Indian pickling spice. This masala blend is used in flavouring pickles and hence the name. This is a combination of the above mentioned spices and is used in pickles. It is also tempered in hot mustard oil to create *achari gravy*. These spices are used whole and rarely crushed, as they will give a bitter flavour.

Bhatti Da Masala

Ingredients: Ajowan, green cardamom, black cardamom, clove, bay leaves, cumin, coriander, mace, nutmeg, black pepper, ginger powder, *kasoori methi*
This is another regional spice from Punjab. This masala is made by broiling the spices individually and grinding them into a fine powder. In this masala, the spices are broiled until they attain a darker colour. This masala is used in kebabs grilled over open charcoal grill or *sigri*. *Sigri* is also known as *bhatti* in Punjab and thus the name of the spice blend.

Dum Ka Masala

Ingredients: Anise, ginger powder, green cardamom, black cardamom.
It is a blend of aromatic spices and is suitable for *dum* cooking, where the dish is cooked covered with heat applied from both top and below. The slow heat in *dum* cooking helps to extract the flavours out of this masala. This masala is usually used for *Mughlai* foods cooked by the method of *dum* cooking. There could again be several variations of *dum masala* depending upon the regions they are used in.

Gunpowder Masala

Ingredients: Dried curry leaves, red chillies, *chana dal, urad dal*, sesame seeds, black pepper
The name suggests that this masala is very hot. This is a regional masala from Tamil Nadu in south India. The ingredients are broiled separately and ground to a fine powder. The powder is used to flavour boiled rice with ghee, which is eaten as a meal. It is also served with idlis in the breakfast. This masala is served as a condiment rather than used as a spice in cooking.

Bafat Masala

Ingredients: Red chilli, coriander seeds, cumin, mustard, pepper, turmeric, clove, cinnamon
This masala is used both in vegetarian and meat dishes. It is a very popular regional masala of Mangalore region of south India.Traditionally the spices used in this masala are dried out in the sun for almost a week and then ground into powder. This can keep up to several weeks and can be used as a base for vegetable, fish, and meat preparations. It gives a pleasing aroma and colour to the dish.

Goda Masala

Ingredients: Coriander seeds, sesame seeds, dry grated coconut, dried red chillies, cumin, cinnamon, asafoetida, star anise, bay leaf, pepper corn, cobra saffron
Goda masala is a regional masala from Maharashtra. It is used in many Maharashtrian preparations and also used in Konkani cuisine. Heat a very little amount of oil and lightly fry the spices in oil one by one. Keep draining the spices on a paper napkin to soak excess oil. Now grind them into powder and

keep the powder in an airtight jar in a cool dry place. This spice blend can be used to flavour meats and vegetables. In case of non availability of cobra saffron, turmeric can be used.

Kolhapuri Masala

Ingredients: Red chilli, coriander, sesame, cumin, mace, cinnamon, dry ginger, green cardamom, black cardamom, mustard, clove, bay leaf, dry coconut, garlic, poppy seeds, peppercorns, asafoetida, turmeric, fenugreek, star anise, nutmeg, oil

This is one of the regional masalas from Maharashtra. It is a reddish coloured masala and is very hot due to large amount of red chillies in it. Heat a very little amount of oil and lightly fry the spices in oil one by one. Keep draining the spices on a paper napkin to soak excess oil. Now grind the spices into a powder and keep the powder in an airtight jar in a cool dry place. This spice blend is normally used to flavour meats.

Rasam Masala

Ingredients: Red chilli, coriander, cumin, peppercorns, turmeric

Rasam masala is a regional masala from south India and is used for flavouring a hot aperitif called *rasam.* This aperitif can be described as spiced lentil water, tempered and flavoured with *rasam masala* and other ingredients such as tamarind, lemon, pineapple, pepper, etc. Broil the spices individually until a pleasing aroma comes out. Grind into a fine powder and store the powder in an airtight container. There could be many variations of the *rasam masala* depending upon the *rasam* and the region it is from.

Wet Masala

Wet masalas are those masalas which are actually made by soaking the spices in liquid and grinding them into a paste. They might also use fresh ingredients which tend to yield wet masalas. For example, usage of fresh turmeric, ginger, and garlic with other spices will yield wet masalas or masala pastes.

Malabar Masala

Ingredients: Coriander, anise, fenugreek, nutmeg, star anise, clove, cinnamon

This is the regional masala from south India. It is used as a base for fish curries. The spices are soaked in water and ground into a paste. Malabar masala is fried in oil to get the flavours infused into the curry. This spice also lends a thickening to the curry.

Sambhar Masala

Ingredients: Coriander, cumin, *chana dal, urad dal,* asafoetida, dry red chilli, peppercorn, grated coconut, fenugreek seeds, curry leaves, turmeric

This is a regional masala from south India. It is used to flavour the lentil curry called *sambhar.* Sambhar is served as an accompaniment with south Indian meals such as idli, dosa, and vada. It is also served as a dal with rice. *Sambhar* can be taken any time of the day. The dry spices are broiled separately. *Chana dal* and *urad dal* are sautéed in a minimum quantity of oil and the entire ingredients are ground along with fresh curry leaves to a fine paste. This paste can be cooked in oil in the beginning or can be added to the boiled lentils towards the end as a tempering.

Chettinad Masala

Ingredients: Red chilli, black cardamom, cinnamon, fenugreek, anise, cumin, coriander seeds, star anise, poppy seeds, pepper corns, green cardamom, cloves, stone flower, nutmeg, capers, curry leaves, mace, grated coconut

The merchant community known as *Chettiars* live in the dry region of Tamil Nadu. They used to travel to South–East Asia and brought back many ingredients and spices that can be seen in their cuisine called *chettinad* cuisine. *Chettinad masala* is one of the famous wet masalas. Except the grated coconut, all the ingredients of the masala are broiled one by one. Lightly fry the grated coconut in coconut oil until it turns brown. Grind the masala into a paste and this can keep well for months if stored in cool, dry place. It usually goes well with chicken and the name of the masala will suffix the ingredient used, for example chicken c*hettinad.*

Goan Masala

Ingredients: Red chilli, garlic, coriander, cinnamon, clove, green cardamom, toddy vinegar, sugar
As the name suggests, this masala is the regional masala of Goa and is predominantly used in Goan cuisine. This masala pairs up well with fish and seafood. All the spices can be placed in a blender and made into a paste with vinegar. This masala does not require broiling of spices as it will be cooked in oil along with onions and tomatoes to create a piquant and hot Goan curry.

Xacutti Masala

Ingredients: Sliced onions, grated coconut, coriander seeds, cumin, clove, black cardamom, green cardamom, cinnamon
This is a regional masala from Goa and pairs up well with chicken. The spices are broiled seperately and then made into a paste with sliced onion. The roasting of the masala is done to give a stronger flavour to this masala. This masala is cooked in oil to form the base of curries that go well with chicken.

Rechado Masala

Ingredients: Red chilli, garlic, peppercorn, coriander, cumin, turmeric, star anise, sugar
This is a regional masala from Goa and is used in fish preparations. This masala goes well with a fish preparation called pomfret *rechado* masala, where this masala is stuffed into the belly of the fish, which is then pan-fried. All the ingredients are made into a paste with some vinegar. This masala is stir-fried with chopped onions and fresh herbs such as chopped coriander.

Balchao Masala

Ingredients: Red chilli, cinnamon, green cardamom, peppercorn, cumin, fresh ginger, garlic
This is a regional masala from Goa and is used in preparation of seafood, such as prawns and shrimps. This masala is used to marinate the seafood which is then cooked in oil until the dish is cooked. The spices are ground into a paste along with ginger and garlic.

Soola Masala

Ingredients: Coriander seeds, clove, green cardamom, fennel, garlic, mustard oil, coriander roots, peppercorn
This is the regional masala from Rajasthan and is used in kebabs called *soola*. The name *soola* refers to kebabs made on thick iron skewers on a *sigri*. The mustard oil is smoked and brought to a lower temperature. The spices are then fried in the oil, except fennel, which is added to the oil in the last. The mix is then cooled and blended into a fine paste. This paste is used for marinating meats, which are then grilled on open fire.

Tandoori Masala

Ingredients: Hung yoghurt, red chilli, turmeric, garam masala, ginger-garlic paste, malt vinegar, salt
This is a regional masala from Punjab, but this is used around India to marinate the tandoori kebabs such as tandoori chicken, tikkas, tandoori fish, etc. The red chillies are made into a paste and all the

ingredients are mixed together to form a masala. The meats or vegetables are marinated with this masala for at least six to eight hours and then skewered on to the *seekh* and cooked in *tandoor*.

Salan Masala

Ingredients: Sesame seeds, cashew nut, peanuts, desiccated coconut, peppercorn, red chilli, turmeric

This is the regional masala from Hyderabad used in making gravy called *salan*. The spices are broiled individually and ground together into a fine paste. This spice blend is cooked in hot mustard oil along with ginger garlic paste and other spices. *Salan* is usually a vegetable preparation that is used as an accompaniment with biryani.

Ver Masala

Ingredients: Garlic, shallots, red chilli powder, anise seeds, black cumin, black cardamom seeds, green cardamom seeds, cinnamon, cloves, fenugreek seeds, black peppercorns, mace, star anise, nutmeg

This is the regional masala from Kashmir and the method of preparation of this spice blend is very unique. It is made into a paste and then sewn on a string and hung to dry. This spice is then crushed and used in Kashmiri curries and dishes. It is also known as Kashmiri masala *tikki* as it is sold in form of dry cakes.

PASTES USED IN INDIAN COOKING

A paste, when referred in Indian cuisine, means the ground mix of either one ingredient or combination of ingredients for use in gravies and curries. There is a very thin line of difference between a wet masala and a paste. The pastes usually do not have combination of spices, for example, spinach paste, ginger and garlic paste, red chilli paste, etc. Many varieties of pastes are made in Indian cuisine and mostly it is for the purpose of adding a thickness to the gravies. Some of the most common pastes used in Indian cooking are discussed in Table 25.1.

Table 25.1 Various kinds of pastes used in Indian cooking

Name	Description	Use and Storage
Ginger paste	This paste is made from fresh ginger. Ginger is first scrapped to remove skin and then washed so as to remove dirt. Then it is cut into small pieces so that it can be ground easily into a paste in a mixer.	This paste can be used in many ways. It can be used as a base for curries and even for marination of kebabs. It is normally used in conjunction with garlic paste and both together make ginger–garlic paste—60 per cent garlic and 40 per cent ginger is good for this paste. Sometimes, only ginger juice is used to flavour the curries. It can be stored for a week under refrigerated conditions.

Contd

Table 25.1   (Contd)

Name	Description	Use and Storage
Garlic paste	This paste is made from peeled garlic in a grinder along with some water so as to get smooth texture. Some *Mughlai* dishes use fried garlic paste. This can be made by frying the garlic until golden brown and then grinding it into a paste.	This is widely used in curries and gravies along with the ginger paste. It can be used alone when the dish is to be flavoured only with garlic. It is also used along with ginger in marinating kebabs. It can be stored for a week in a refrigerator.
Cashew paste	To make this paste, cashew nuts are soaked in hot water or left overnight to soften. They are then ground in a mixer to get a smooth paste. Some recipes call for brown cashew paste and in this case the cashews are deep fried until golden brown and then ground to a paste.	It is used as an ingredient in kebabs to enrich them. It is also used in gravies to provide thickening and base. This paste can be used for creating desserts such as *kaju ki burfi*. This paste is quite perishable and should be made fresh and stored in a refrigerator. The fried paste has a better shelf life than the raw one.
Coconut paste	The coconut is cracked open and the fleshy part is removed from the shell. Then the skin is scrapped off and the coconut is grated. This can also be done by using coconut scraper. It is then put in the blender to make paste. Coconut water can be added to get a smooth paste.	This is used as a base for gravies and curries and can be used for making chutney. It is perishable and should be made fresh and stored in a refrigerator. It can be made once in two days.
Poppy seeds paste	Poppy seeds are first soaked in warm water for 30 minutes or in room temperature water overnight. It is then ground into a paste. It is easier to grind on *sil batta* as it becomes difficult to do smaller quantity in electric blender because of the small size of the seed. It can be done in a grinder if the quantity is more.	This is used as a base for gravies and curries and is used as a thickening agent. It also provides flavour to curries, for example, *aloo posto* from Bengal. This is also highly perishable and should be stored in a refrigerator and made fresh once in two days.
Tamarind pulp	This is not a paste in real terminology as it is not ground. But it is a paste which we get by soaking tamarind in water for over an hour and then taking out the pulp. The pulp thus obtained is strained through sieve to remove unwanted particles.	This is used as a souring agent in curries and also provides thickening. It is widely used to make chutneys and pickles too. It can be stored for a month in a refrigerator.

Contd

Table 25.1 (Contd)

Name	Description	Use and Storage
Red chilli paste	To make this paste, whole red chillies are cleaned of any superficial dirt. The stalks are removed and the chillies are deseeded. The chillies are soaked in warm water or boiled with a little vinegar, which helps in the deepening of the red colour. When the chillies become soft, they are ground into a fine paste. The paste can be cooked in oil until a deep red colour is obtained and the oil starts to float on top. Frying of chilli paste helps to remove the raw taste and tones down the hotness of the chilli and increases the shelf life.	Red chilli paste is used for marinating and flavouring kebabs to give them a deep red colour. It is also added to curries for the same reason. Since the job of red chilli paste is mainly to provide colour to the dish, it is advisable to use milder chillies which are deep red in colour such as Kashmiri chillies. This paste can be stored for over a month if kept in the refrigerator with oil floating on top. The chilli oil can be used to garnish a curry.
Char magaz paste	This is a combination of four seeds and hence, the name *char magaz*. Squash seeds, melon seeds, cucumber seeds, and pumpkin seeds are soaked in water till they soften and then ground into a fine paste.	This paste is used as a thickening agent in white gravies such as kormas and *shahi gravy*. This is also highly perishable and should be stored in a refrigerator and made fresh once in two days.
Almond paste	This paste is made from skinless almonds. The skin is removed by blanching the almonds or soaking them in water overnight. Soaking makes the nuts softer and easier to grind.	This paste is used for thickening some of the exotic gravies and also used as a base to make *badaam halwa*—an Indian dessert, where the paste is cooked with ghee and sugar. This is also highly perishable and should be stored in a refrigerator and made fresh.
Chironjee seed paste	*Chironjee* is a seed of a nut found in a fruit that abundantly grows in Chattisgarh region of India. Also known as cudapa almond in English, this seed is extensively used as a thickening agent.	This paste is used as a thickening agent in white gravies such as kormas and *shahi* gravy. It is also used as an ingredient in kebabs. It is highly perishable and should be stored refrigerated and made fresh once in two days.
Spinach paste	Boiled and refreshed spinach is ground to a paste and is kept as a basic *mise en place* in the Indian kitchen.	This paste is commonly used with other gravies such as onion tomato masala to make *hariyali* gravy (refer to Chapter 27) or the famous *saag gosht* –lamb cooked with spinach.
Masala paste	Various types of wet masalas discussed above fall into this category of paste.	It is used in marination, curries, grills, etc.

Contd

Table 25.1 (Contd)

Name	Description	Use and Storage
Boiled onion paste	Onions are boiled with *khada masala* and a small amount of vinegar till soft. They are then ground into a paste.	The boiled onion paste is used as a base for light coloured gravies such as *Mughlai*, yellow, and white gravies, etc. Boiled onion paste can keep well up to a week if stored in a refrigerator.
Fried onion paste	Thinly sliced onions are deep-fried until brown and ground into a paste. If a liquid needs to be added to make a paste, yoghurt has to be used.	This is used as a base for curries and gravies. It is also used as an ingredient in kebab mix.

CONCLUSION

In the previous chapter many common Indian spices were discussed in detail. This laid the foundation for understanding the blend of masalas discussed in this chapter. These masalas are used in various parts of India for cooking. Some of them are specific to a region and are known as regional masalas. The other masalas mentioned in this chapter are used all over India as a base for Indian cuisine. Apart from the ones discussed in this chapter, there could be many more regional masalas used in Indian cuisine, because of the cultural and gastronomic diversity of our country. The combination of spices involves a lot of understanding of the flavours that can complement each other. The selection and knowledge of each spice is the first step to create blended masalas. Each spice reacts differently to heat and each spice has its unique flavour and aroma. The sequence of using spices in a recipe is also very important. The spices that are used for colouring are tempered in oil, whereas

the aromatic masalas, such as garam masala, are sprinkled over the top to finish a dish. We also saw how some spices are used whole in combination or broiled and blended to form aromatic powder. Some spices are soaked in liquid and ground to a paste. The usage of each type of masala, whether in a paste or dry form, has its own uniqueness. Some are used for flavouring and some are used for thickening or giving the base to the gravies and curries.

Apart from the dry and wet masalas, this chapter also discussed various other kinds of nut and vegetable pastes, such as cashew nut paste, *char magaz* paste, spinach paste, and ginger–garlic paste, that are used as a base in many Indian foods. In the next chapter we shall discuss various other kinds of spices and condiments that provide certain characteristics to the gravies such as colouring agents, souring agents, spicing agents, etc. This will make our foundation to get started to cook different varieties of Indian dishes.

KEY TERMS

Aamchoor Dry mango, either whole or in powder form.

Achari masala Indian pickling spice made by blending whole spice.

Aloo posto Curried potato preparation from Bengal flavoured with poppy seed paste.

Amritsari machli Batter-fried fish from Punjab, eaten as a kebab or main meal.

Badaam halwa Dessert made by cooking almond paste in ghee and sugar.

Bhatti Open fire charcoal grill, also called *sigri*.

Black salt A type of salt rich in minerals, used in cold savoury dishes.

Char magaz Blend of four seeds ground into a paste.

Chaat masala Blend of sour spices to add piquancy to the chaat dishes.

Dosa Griddle-fried fermented rice and lentil pancakes from south India.

Ghee durust dena Adding tempering of aromatic spices to *desi ghee* to get rid of unwanted flavours from the ghee.

Hariyali gravy Green coloured gravy made with spinach paste.

Idli Fermented and steamed rice and lentil cakes from south India.

Kaju ki burfi Dessert made from cashew nut paste and dried milk solids.

Khada masala Blend of whole spices used for tempering.

Pindi chole A dark coloured preparation of chickpeas with spices.

Pudina paratha Wholewheat flaky bread flavoured with dried mint masala and baked in *tandoor*.

Ratanjot Bark of a tree used to colour food in Kashmiri cuisine.

Saag gosht Lamb and spinach dish from Punjab.

Saunth Dried ginger, either whole or in powder form.

Sigri Refer to *bhatti*.

Tikkas Small bite size savoury snacks usually served as kebabs.

Toddy vinegar Fermented vinegar from Goa made from cashew nut liquor.

Vada Deep-fried dumplings of fermented lentil paste served as snack.

CONCEPT REVIEW QUESTIONS

1. Differentiate between a spice and spice blend.
2. What do you understand by the term 'masala' and how many types of masalas are there?
3. What factors will you keep in mind before selecting a spice blend?
4. What do you understand by the term regional spices?
5. What is the difference in terms of usage of dry and wet masalas?
6. Name at least five regional masalas from north, south, and west India.
7. List the ingredients of *chettinad masala*.
8. What is a *potli masala* and how is it used?
9. Define a *khada masala*. What are its uses?
10. What is a garam masala and what are its uses?
11. List the ingredients of chaat masala and its usage.
12. What is *pudina masala?*
13. What is a kebab masala and what is its use?
14. Write a short note on *achari masala*.
15. What is gunpowder and where is it used?
16. Write a short note on *goda masala*.
17. What is the difference between wet masala and paste?
18. List the ingredients of *sambhar masala*.
19. List at least four famous Goan masalas.
20. What is the difference between a *soola masala* and a tandoori masala?
21. List at least three nut pastes used in Indian cuisine.
22. What is *char magaz and* what are its ingredients?
23. What is the ratio of ginger and garlic paste used in curries?
24. How would you prepare coconut paste?
25. What is the correct way of preparing red chilli paste and what factors should be taken into account while making it?
26. What is the difference between boiled onion paste and fried onion paste? What is the difference in their usage?
27. What is *bafat* and which region does it belong to?
28. Write down the uses of almond paste.
29. How would you make tamarind pulp?
30. What is *salan masala* and how is it used?

PROJECT WORK

1. The recipes of the spice blends and masalas are provided in the ORC. Divide yourselves into two groups. One group makes the standard recipe and the other group makes the deviations. For example, if one masala is to be made by broiling spices, then you do it by frying method and vice versa. Now compare the results with the standard recipes and record your observations.
2. Make the same dish using two different forms of dry, wet, or blended masala to create two individual dishes. Taste and critique the dish and see if your choice of masala was apt. Remember to use light coloured masalas with light meat and dark ones with dark meat.
3. In groups, visit the various hotels and restaurants and sample the food. Analyse and critique the kinds of masalas used.

CHAPTER **26**

UNDERSTANDING COMMODITIES AND THEIR USAGE IN INDIAN KITCHEN

Learning Objectives

After reading this chapter, you should be able to:
- understand the role and important features of souring agents used in Indian cooking
- know about various colouring agents used in Indian cooking
- list various thickening agents used in Indian cooking, their usage, and limitations
- critique the usage of spicing agents, flavouring agents, and aromatic agents in Indian cooking

INTRODUCTION

In Chapter 24 we discussed about the individual spices, their usage, medicinal value, and storage. In the previous chapter we discussed about the blending of spices either in whole, dry, powder, or paste form to be used in a dish. Apart from the spices we also discussed the various kinds of paste used in Indian cooking that form the base of Indian food. This chapter will deal with other condiments that are used for specific role in Indian food. The spicing agents, flavouring and aromatic spices were discussed in detail in the previous two chapters but in this chapter they are shown in tables for better understanding. It should be noted that many spices have multiple roles to play in a dish. They can act as colouring agents, flavouring agents, and aromatic agents too.

Before we get into understanding the role of ingredients let us see what a basic dish or food is all about. There are various steps involved in cooking any dish around the world and these steps are not mere *mise en place* and cooking methods, but much beyond that. It is important to understand the flavour profile of food products of a country or a region. The food of a country may be influenced by its geographical location, climatic conditions conducive to produce available, eating habits, social and religious parameters of the country, and so on. Let us now look at the Indian curry for example. The first and the foremost step will be to understand the fuel that we use. Slow heat from simmering wood charcoals is the best medium to cook Indian food, as the flavour of the spices is infused into the dish, maintaining the nutritious value too. The second step is to understand the equipment that we will cook the dish in. We read in Chapter 23 about the Indian equipment and the role played by them in Indian cuisine. The dish cooked in the iron *kadhai* will be darker in colour as compared to the dish cooked in *lagan* and so on. This means if we need darker coloured *bhunna gosht*, we will cook it in *kadhai* and if white korma gravy is required, we will cook in tinned brass utensil. The next thing would be to choose the flavour, texture, colour, taste, and finally the look of the dish. All these are achieved by use of condiments and spices in order to get the desired results. The spices and condiments are categorized into various categories such as spices that help to colour the food, flavour the food, thicken the food, etc. In this chapter we will discuss about the various condiments that are used for the particular purpose such as adding tang and sourness, colouring, tenderizing, thickening, spicing, and flavouring the food. We shall discuss the role played by them in Indian food and how we can combine and monitor the different tastes, textures, and the look of the dish. Let us discuss each of these condiments that we will refer to as agents.

SOURING AGENTS USED IN INDIAN COOKING

Ayurveda specifies that the mind and body are satisfied when all the taste buds of the tongue are satisfied. The tongue can taste various tastes. Apart from sour, sweet, bitter, hot, and salty, there is another taste—the sixth taste—that is undefined. It is called the kinaesthetic taste and in modern terms it is referred to as *unami*. This is a Japanese word for 'sixth taste' and the chefs and food lovers are researching on the *unami* flavours. Many of the condiments and spices that will be discussed here have probably been dealt in the previous chapters. Here we shall discuss in detail only those commodities that we did not discuss in the previous chapters. Let us discuss various types of souring agents in detail.

Vinegar

Vinegar is the acid obtained by fermenting alcohol. The usage of vinegar has been probably handed over by the Portuguese travellers, which explains the amount of vinegar used in Goan and south Indian food. North India also utilizes vinegar for marinating kebabs. Vinegar is also used in Mangalore, Kerala, as well as in Anglo Indian and Parsee cooking. Table 26.1 discusses the various vinegars used in Indian cooking.

Table 26.1 Types of vinegar

Name	Description	Use
Synthetic vinegar	This is a chemical known as acetic acid. It is quite strong in its pure form and is diluted to around 4–8 per cent for usage in food or 18–20 per cent for pickles.	It is used to make various types of spice pastes in Goan cuisine. In the north it is used for making pickles.
Toddy vinegar	It is basically palm vinegar that is made from the sap of coconut palm. The sap thus yielded is allowed to ferment naturally and this process takes about 4–6 months. The resulting liquid is then distilled to yield toddy vinegar. It has a faint yeasty and a musty taste.	It is popularly used in Goan cuisine to make spice pastes and to add sour tastes to the food.
Malt vinegar	This is quite popular in the north, especially Punjab. This vinegar is a by-product of beer. Malted barley is used to brew and then left to ferment or age in wooden barrels, which deepens the colour and improves the flavour.	It is used for marinating tandoori kebabs. It is also used for pickling onions and lime.

Tomato

Tomato is widely used in Indian cuisine. It is popularly used for its colour, flavour, and a touch of sourness. There are two types of tomatoes used extensively in Indian cooking (Table 26.2).

Table 26.2   Types of tomato

Name	Description	Use
Salad tomato	These are the red tomatoes that are available in the market and they could be any variety such as roma, plum, etc. However in Indian cuisine, tomatoes are not classified as in Western cuisine.	This tomato can be used in various forms. It can be cut up or pureed and used for adding a piquant taste and colour to the dish. The usage of tomato is more popular in north India.
Sour tomato	This tomato is commonly known as *desi tamatar*. This is the type of tomato which is not as bright coloured as salad tomato but is sour and used abundantly in Indian cooking.	The main purpose of the use of this tomato in cooking is to give a sour taste to the curries and dishes.

CHEF'S TIP
Whisk small amount of gram flour to the curd as this will avoid the curdling in gravy. Plain curd can be whisked and flavoured to make raita, which is served as an accompaniment.

Curd

Curd is very popular in India and has various uses. It is used as souring agent and also has many medicinal values attached to it. It is believed that drinking yoghurt with biryani does not allow the fats in biryani to assimilate in the body. Apart from being used as a souring agent in curries, it is also used as a thickening agent for gravies. Both plain curd and hung curd are used in Indian cooking (Table 26.3).

Table 26.3   Curd as a souring agent

Name	Description	Use
Plain curd	The curd culture is added to tepid milk and allowed to rest for 6–8 hours to set into curds.	Curd is used in the *Mughlai* gravies and many north Indian gravies and curries, especially from Kashmir. The curd used in the gravy has to be fresh.
Hung curd	The plain curd is hung in a cheese cloth to drain away the whey and hung curd or curd cheese is obtained, which can be used in various ways.	Hung curd is used for marinating kebabs. It can be tempered to make curd curry also known as *chonka dahi*. Hung yoghurt can be used for making creamy raitas.

CHEF'S TIP
The curd gravies are to be made only in tinned brass utensils as they tend to curdle in steel or aluminium utensils.

Tamarind

Tamarind is a pod, which is deseeded and dried. It is soaked and the pulp is squeezed and used as a souring ingredient in Indian cuisine. It is mainly used in south Indian cuisine (refer to Chapter 24).

Lime/Lemon

The juice of a citrus fruit, such as lemon and lime, is added to the dishes at the end to make them tangy. It is also used as a tenderizing agent. Lime is round in shape, whereas lemon is large and oval-shaped (Table 26.4). Lemon is also known as *gandharaj* (king of perfume) in Bengal, due to its pleasing aroma.

Table 26.4   Lime/lemon as a souring agent

Name	Description	Use
Lime	In Hindi it is referred to as *nimbu* and has many uses. It is round in shape.	Used as a souring agent, the lime juice is used to add sour and piquancy to the curries and dishes. It should be added only in the last and never cooked as it would get bitter and also prolong the cooking time. Lime is used in making pickles and the lime juice is also used for making an aperitif called *shikanji* or *nimbu pani*. Lemon is also served as an accompaniment with fish. In hotels, half a lime is tied up in a muslin cloth to make lime *podlum* which makes it easy for the guest to squeeze, without having to bother about the seeds.
Lemon	It is larger than lime and oval in shape.	It is popularly used in Bengali cuisine and is mostly used as a souring agent and also as an aromatic agent. It is also used as an accompaniment to salads.

Mango

Mangoes are grown in almost all the parts of India and each mango has a special name and regional specialities attached to it. Mango is eaten like a fruit and used in dessert preparation just like it is done in the West, but many other forms of mangoes are used for souring purposes of gravies and curries and other applications. Table 26.5 shows the role of mango as a souring agent.

Table 26.5   Mango as a souring agent

Name	Description	Use
Kairi or raw mango	*Kairi* is an unripe green mango that is very sour and is used as a souring agent in many Indian dishes.	*Kairi* is used for making pickles and chutneys. It is cut and added to curries and lentils to give a sour taste. It is also roasted in *tandoor* and the pulp is used for making a refreshing summer drink called *aam ka panna*.
Aamchoor	This is the dried mango cheek of raw mango. *Aam* means mango in Hindi and *choor* means powder. *Aamchoor* is available both as whole dried cheeks and as powder.	*Aamchoor* is used as an ingredient for various blended masalas (refer to Chapter 25).
Aam papad	This is available as a condiment from the grocery stores. It is commercially made from dried ripe mangoes.	It is used as an ingredient to make chutneys and pickles and is also added to the dishes as a souring agent.

Kokum

This deep purple fruit with a large seed is grown in the western coast. This is mainly used in coastal regions of the west. It is also used in gravies, curries etc. in south India. *Kokum* is mainly used in fish preparations and aperitifs. *Kokum* also gives a pleasing pink colour to the dish and can also be used as a colouring agent (refer to Chapter 24).

Pomegranate Seed

Dried pomegranate seeds are available whole or in crushed form to be used in Indian cooking (refer to Chapter 24).

Gamboge

Gamboge is also known as fish tamarind or Malabar tamarind. It is a tropical fruit that grows in Kerala. It is also used commonly in Malaysia and Sri Lanka. Table 26.6 shows the usage of gamboge as a souring agent.

Table 26.6   Gamboge as a souring agent

Name	Description	Use
Gamboge	This is also known as fish tamarind as it is used as a souring agent in fish curries. It is a fruit containing a berry that has a tough rind. It is dried and used in the cuisine of Kerala.	It is used as a souring agent in fish curries from Kerala.

Kachri

Kachri is a small fruit that is mostly available in Rajasthan. Table 26.7 shows *kachri* as a souring agent.

Table 26.7   Kachri as a souring agent

Name	Description	Use
Kachri	This is a small oval-shaped fruit that is bitter when raw, but gets a sour taste as it ripens. It is found in Rajasthan. When fresh, it resembles a watermelon.	*Kachri* is used as whole to make chutneys or is added to potato curries. When powdered, it is used to add sour taste to meats and vegetables. Fresh *kachri* is also prepared as a vegetable. The dried *kachri* is used as a souring agent in Indian dishes. *Kachri* is also used as a tenderizing agent in Indian cooking.

CHEF'S TIP

Add the souring agents in the later stage of cooking, as they delay the cooking process and meats, etc. take longer time to cook.

Organic Salts

Certain chemicals or organic salts are used as souring agents in Indian cuisine. Chemicals such as tartaric acid and citric acid are very commonly used, though not so much for the purpose of souring, but various other uses as discussed in Table 26.8.

Table 26.8 Chemicals as souring agent

Name	Description	Use
Tartaric acid	It is also known as *tantri* in Hindi. It is a white coloured salt that is used as a preservative and also to add sourness to certain products. It is a natural organic acid that is extracted from fruits such as grape, banana, and tamarind. It is also known as cream of tartar.	It is used in chaats and other savoury snacks. Tartaric acid is added to pickles for sourness and also as a preservative. It is also used in production of Indian sweet called 'milk cake'.
Citric acid	As the name suggests, these are organic salts extracted from citrus fruits, especially lemon and lime.	It is used in pickles as a preservative.

COLOURING AGENTS USED IN INDIAN COOKING

When we talk about colouring agents in Indian cooking, we do not refer to artificial colours added to food. The spices and herbs that are used in Indian cooking impart a natural colour to the curries and make them appetizing. This also helps in the planning of menus for buffets and *thalis*, where many dishes of different colours can be arranged and served to guests. The use of artificial colours in Indian food is not a norm; but some local caterers seem to do it to save costs. Various kinds of spices, herbs, and condiments are not added just for the sake of adding colour but also to add flavour and aroma. These spices and condiments have natural colouring properties. Table 26.9 shows the colours obtained from various spices and condiments.

Table 26.9 Use of spices and condiments as colouring agents

Spices and Condiments	Use
Saffron	The use of saffron in the white gravy gives it a pale apricot hue.
Turmeric	The use of turmeric in white gravy gives it a bright yellow hue.
Red chilli powder	Red chilli powder gives an orange tinge to gravies.
Red chilli paste	The colour given by the red chilli paste will depend upon the variety of chilli used. Kashmiri red chilli paste will give a reddish brown tinge to the gravy.
Green paste—coriander, spinach, curry leaves, etc.	It gives a bright green colour if added in good quantity and cooked for short time; longer time will make this turn into an olive green colour.
Tomato	It will give a pink colour if combined with any white paste such as curd, nut paste, or boiled onion paste and deep red if used on its own.
Boiled onion paste	It gives a light pinkish hue to gravies.
Fried onion	It gives a deep dark brown colour to the gravies. If added along with white nut pastes and yoghurt, it gives a beige colour.
Coriander powder, cumin powder	If cooked for short time they give a light brown colour, but if fried for longer duration of time these tend to give a dark brown colour to the gravy.

Contd

Table 26.9 (Contd)

Spices and Condiments	Use
Garam masala powder	If fried in oil, garam masala gives a deep dark brown colour to the gravy.
Brown and dark coloured blended spices such as *chettinad, chana masala*, etc.	They tend to give dark brown colour if fried in oil for some time.

Apart from spices and condiments, there are various other colouring agents used in Indian food, for example, tomato, *kokum*. Some other colouring agents are discussed as follows.

Spice Powders

Many spice powders and blends, as discussed in the previous chapter, give respective colours to the curries. The reddish coloured spices such as *Kolhapuri masala*, etc. will give reddish brown colour, whereas dark coloured spice blends such as garam masala, *bhatti masala, chettinad masala*, etc. will give a deep dark brown to black colour to the curries and dishes (refer to Chapter 25).

Onion

Onions are also used to provide colour to the dish. The degree to which the onion is browned in oil or fat also has a deep impact upon the colour of the curry. Pastes such as boiled onion pastes and fried onion paste have different colouring properties (refer to Chapter 25).

Turmeric

This is a rhizome which is yellow in colour with similar looks as that of ginger. It has a very pungent smell and is mostly available in powder form. But in some parts of Bengal it is used whole and ground into a paste. It not only acts as a colouring agent, but also has antiseptic medicinal properties (refer to Chapter 24).

Red Chilli

Apart from making the dish hot, it also has a unique colouring property. The colour provided by chillies depends on the form they are being used in. Using them in paste form provides a brighter colour, while the powder form gives a reddish brown to orange colour. There are different types of chillies, such as Kashmiri red chillies, *deghi mirch*, which give better colour as compared to other chillies (refer to Chapter 24).

Saffron

It is the most important colouring and aromatic agent in Indian as well as Western cuisine and one of the most expensive spices. It originates from Kashmir in north India. For better results, we need to broil the saffron first and then make it into fine powder—this gives better colour and the quantity used is also very less (refer to Chapter 24).

Green Pastes

Various kinds of green leafy vegetables and herbs are often used to give a body and colour to the curries and dishes. Table 26.10 shows few of these.

Table 26.10 Herbs and vegetables as colouring agents

Ingredient	Description	Use
Spinach paste	Boiled and refreshed spinach is ground to a paste (refer to Table 25.1).	Sometimes spinach paste is combined with dill leaves to flavour certain popular dishes of Uttar Pradesh such as *subzi nimona*.
Coriander paste	Fresh coriander can be ground to a paste of its own or sometimes in conjunction with curry leaves to form a base for curries. It is quite popular in south India.	It is used as base for curries such as *kothmir kodi*—chicken cooked with coriander and curry paste. It can also be used for marinating kebabs to give a green colour to kebabs such as *hariyali fish tikka*, etc.
Fenugreek leaves	Fresh fenugreek leaves are cooked or boiled and then made into a paste. These should be pounded, as grinding them in electric mixers make them bitter sometimes.	They are used as a base for certain curries such as *methi tsaman* from Kashmir, where fresh cottage cheese is cooked in paste of fenugreek. They also team up well with lamb dishes.
Mustard leaves	These are used in conjunction with amaranth leaves called *bathua* and this is done to cut down on the bitter and sharp flavour of the mustard leaves. These are boiled and ground into a paste, preferably in a mortar and pestle.	They are used for preparing the famous *sarson ka saag* in Punjab.
Ratanjot	It is a root of an alkanet tree and is used to give a deep red colour to the lamb dishes from Kashmir. It is believed to be carcinogenic. *Ratanjot* has colour pigments that are soluble in oil. It is mostly used for dyeing purposes.	It is not used as an ingredient directly in the food. It is kept in a strainer and hot oil is poured over it. The red coloured oil is collected in a bowl below and used for cooking.
Mowal	It is dried cock's comb flower. It is available whole or in a powder form and extensively used in Kashmiri dishes and curries.	It is boiled along with water and strained off. The water is then used for cooking curries and various gravies in Kashmir.

Sugar

As we know, sugar caramelizes upon the application of heat and changes its colour from white to golden to dark black colour. This property of sugar is also used in Indian cooking to add colour to few curries and dishes. This is quite popular in Bengal and Goa. Granulated sugar or jaggery is added to onions, while they become brown in the oil. This helps to impart a darker colour to the dish as in case of *kasha mangsho*—which is a dark coloured lamb dish from Bengal.

Flowers and Barks

Certain flowers are also used to provide colour to food. Marigold flowers are used in Indian cooking as imitation saffron. Certain other barks and flowers are used in Kashmiri cooking (refer to Table 26.10).

THICKENING AGENTS USED IN INDIAN COOKING

We read in the earlier chapters about the various thickening agents that are used in Western classical cooking such as roux, blood, etc. In Indian cooking curries and dishes are thickened on the basis of accompaniments that are served with them. If the dish is served with Indian breads, then the curries will be of thicker consistencies, but if they are to be eaten with rice, they will be more like stews. Many times a single spice has various uses—it can be used to add colour, flavour, thickening, aroma, piquancy, or even sweetness to the dish. Let us discuss various ingredients used as thickening agents in Indian cooking.

Onion Pastes

Both fried onion paste and boiled onion paste help provide thickness to the gravies. They indeed help to add base or body to the dish apart from acting as colouring agents too (refer to Table 25.1).

Nut Pastes

Various types of nuts ground into a fine paste are used for thickening Indian gravies. These are probably the influence of Mughal rulers and hence, mostly found in *Mughlai* kormas and curries often referred to as Indian royal cuisine. Many of the nut pastes, such as cashew nut, almond, and coconut pastes are discussed in Table 25.1. Apart from these, sometimes for special gravies, pistachio paste can also be used in thickening of gravy.

Seed Pastes

Many kinds of seeds are used in paste form to provide thickening to curries and dishes. Many of these seeds were discussed in Table 25.1, for example, poppy seed paste, *chironjee* paste, *char magaz* paste. Many other seeds such as sesame seeds are used in *salan* paste that also acts as a thickening agent. Mustard seeds are also used in the form of paste that acts as thickening and flavouring agents in many Bengali dishes.

Masala Pastes

Many types of dry masalas and wet masala pastes are used in the thickening of the dishes in Indian cuisine (refer to Chapter 25).

Lentils

Many Indian preparations use lentils for thickening purposes. It is used in various forms around India as shown in Table 26.11.

Dairy Products

Many types of dairy products, such as cream, etc., are used in Indian cooking to thicken some gravies, especially in *Mughlai* cuisine. Such use of dairy products is more prevalent in hotels and restaurants, and is rarely done in home style food. This is probably the influence of Western cooking. Table 26.12 shows some diary products used in thickening of gravies and curries.

Table 26.11   Lentils as thickening agents

Name	Description	Use
Lentil powder	It can be one lentil or a combination of lentils, such as *chana* and *urad*, which are roasted and ground into powder to thicken many south Indian dishes.	It is used for thickening some of the south Indian dishes.

Contd

Table 26.11 (Contd)

Name	Description	Use
Roasted *chana dal*	It is a commercially available product where the black gram is soaked in water and then roasted. It is available in both whole and powdered form.	It is popularly eaten as a snack in winters. It is also cooked with meat and provides binding for certain kebabs such as *shammi* kebab. *Chana dal* is also cooked with meat to provide thickening as in case of *haleem, dal gosht* from Hyderabad or *dhansak* from Parsee cuisine.

Table 26.12 Dairy products as thickening agent

Dairy Product	Description	Use
Cream	The cream that floats on top of milk is collected and used in various kebabs and curries in Lucknow. It is called as *balai* in Lucknow and in Hindi it is known as *malai*. The *malai kofta* is thickened with cream, and hence the name.	It is used for thickening in various *Mughlai* dishes such as kormas and *shahi* gravies.
Khoya	It is basically milk solid that is obtained by reducing milk. It is a commercially available product, that is used to thicken some curries and gravies.	It is available in many forms and is also used widely in the production of Indian desserts. In curries, it is grated and cooked with the gravies to thicken them. It is usually added in the last stage to the dish.

Vegetable Purees

Certain vegetable pastes, such as fresh turmeric, ginger, and garlic pastes, are also used for thickening of curries. Some green leafy vegetable pastes, such as spinach, fenugreek, etc., as discussed in Table 26.10 are used in thickening curries and dishes.

TENDERIZING AGENTS USED IN INDIAN COOKING

When talking about tenderizing agents, we refer to meat cookery and that too, specifically lamb. In the olden times, probably tenderizing of meat was done only for kebabs, etc. which were cooked with broiled heat. In curries and stews there was no need of tenderizing agents, as food was cooked on slow simmering charcoals, which cooked the meat to soft textures. With modernization and various types of fuels available, the cooking processes have become short and as a result the meat does not cook to its desired texture unless few tenderizers are added to it. Even today meats are tenderized for kebabs, biryanis, and sometimes for curries too. The essential principles of all tenderizers depend upon the acid content and enzymes present in them. These acids readily dissolve sinews and muscle fibres, making their use a necessity. Most of the tenderizing agents are used in the marination stage and the meat is allowed to rest in the marinade for some time. The time depends upon the type of tenderizer used. Some tenderizers, such as raw papaya paste, will tenderize the meat in couple of hours while some might require more time. Let us discuss about the common tenderizers used in Indian cuisine.

Curd

Curd is basically fresh milk inoculated with a culture of lactobacillus. It is the lactic acid in curd that helps to break down meat fibres and renders the product soft and succulent when cooked in *tandoor*. In curries it is usually added at the end. Apart from the curds mentioned in Table 26.3, the whey of the hung curd also known as *khatti lassi* is used for marinating the meat for tenderizing.

Lime

The citric acid content in this fruit is what causes the tenderizing action on the meat fibres. It is also used for marinating of meats for kebabs. The other citrus fruits, such as orange, *kokum*, and gamboge, can also be used as tenderizers.

Kachri

Kachri is a wild variety of melon-like fruit of cucumis family and is found in Rajasthan, Bengal, Punjab, parts of Maharashtra, the northern and western provinces, and the Sind area (now in Pakistan). It is used as a tenderizing agent for meats, especially for kebabs and biryanis (refer to Table 26.7).

Vinegar

Vinegar is produced when natural yeast in the air acts on the juice of the fruit used to make the vinegar. It converts the sugar to alcohol and then this alcohol is converted into acid. Acetic acid is the substance which gives it its tenderizing quality. The strength of acetic acid in vinegar varies depending on the fruit base. Every country produces vinegar made from locally abundant fruit. Grape and cider vinegar remain the most popular for cooking in the West. Refer to Table 26.1 on the various vinegars used in Indian cooking.

CHEF'S TIP
Avoid serving raw papaya to a pregnant woman, as it is believed that it aids in abortion in the initial stages of pregnancy.

Tamarind

Found mostly in south India, this is an extremely sour, beanlike fruit with larger seeds. The ripe fruit contains two acids—citric and tartaric acid—that give it a characteristic sour taste and its tenderizing properties (refer to Chapter 24).

Fruits

Many types of fruits in various forms are used as tenderizing agents especially for meats (refer to Table 26.13.)

Table 26.13 Fruits as tenderizing agents

Name	Description	Use
Raw papaya	Papaya is an oblong melon-like fruit that grows in most parts of India. The fruit is obtained from a fully grown tree and is used in the raw form. Papaya contains a protein digesting enzyme called 'papain' that gives it its tenderizing property. The enzyme is mostly present in the skin and not so much in the flesh. Thus while making paste, peel the papaya with only ½ cm of white flesh attached to it and grind into a paste. This paste can be made in advance and refrigerated for over a month.	Raw papaya paste is always used in marinations of meat and is never added as a tenderizing agent while making the curry. It is important to let the meat be marinated in raw papaya for only a couple of hours, otherwise it will destroy the protein structure of the meat and render it tough and chewy.

Contd

Table 26.13 (Contd)

Name	Description	Use
Raw pineapple	It is found in Assam and other north-eastern areas. This is an oval, spiky looking fruit with a pungent taste. The active enzyme found in pineapple is 'bromelain' which has a very similar tenderizing action to raw papaya. Here the skin is peeled and discarded and only the flesh is ground into paste.	Its usage is same as that of raw papaya.
Raw figs	Figs are found in many parts of India, but mainly in the north west and Punjab. The enzyme that acts as a tenderizer in figs is 'ficin'. It is interesting to see how certain enzymes and acidic content in raw fruits find their usage as tenderizing agents in cooking.	The whole figs are ground into a paste and used along with marinations for the tenderizing effect.

FLAVOURING AND AROMATIC AGENTS USED IN INDIAN COOKING

Indian food is full of flavours and life and this is achieved by using many agents such as spices and other condiments such as flowers, saps from certain plants, seeds, barks, and many others. We discussed in detail about many spices and condiments that give a flavour to food in one way or the other in Chapter 24. Apart from these spices mentioned in previous chapters, there are few other condiments that are used in Indian cuisine for flavour and aroma (refer to Table 26.14).

Table 26.14 Flowers as flavouring agents

Name	Description	Use
Rose petals	In Indian cooking rose has been closely associated with *Mughlai* cuisine.	Dried rose petal powder is used in Indian dishes such as kebabs and biryanis. It is also used as a condiment in the Awadhi garam masala. The fresh petals are used to garnish desserts. Rose petals are used in dried form, where they are powdered. The usage of rose petals is very common in *Mughlai* kebab. The famous *kakori* kebab from Lucknow is flavoured with rose petals.
Mogra	This is the Hindi name for jasmine flowers and again it is quite popular in some parts of Uttar Pradesh.	*Mogra* flowers are used to flavour certain pilafs from Uttar Pradesh.
Screw pine	These are plants that bear yellow flowers. An essence is extracted from the leaves of screw pine called *kewra* or *keora*. A concentrated form is also known as attar or *kewra* attar.	It is commonly used to flavour biryanis and kebabs. The usage is more prominent in Lucknow. A drop of attar is enough to flavour 1 kg of meat. It is often diluted and sold in bottles as *kewra* water.

SPICING AGENTS USED IN INDIAN COOKING

In Chapters 24 and 25 we dealt with many spices and condiments that are used as flavouring, aromatic, spicing, colouring, and souring agents. There is a thin line of difference between flavouring spices and aromatic spices. The aromatic spices are those which can be judged by the sense of smell through nose, while flavouring is a smell that is a combination of both taste and smell. Let us see the examples of flavouring, spicing, and aromatic agents in Table 26.15.

Table 26.15   Flavouring, spicing, and aromatic agents

Flavouring Agents		
• Ajowan	• Nigella seeds	• *Achari masala*
• Bay leaves	• Poppy seeds	• *Bhatti da masala*
• Cloves	• Turmeric	• *Goda masala*
• Coriander seeds	• Celery seeds	• *Sambhar masala*
• Cumin seeds	• *Marathi moggu*	• *Xacutti masala*
• Fenugreek seeds	• Stone flower	• *Tandoori masala*
• *Kokum*	• Black salt	• *Salan masala*
• Liquorice		
Spicing Agents		
• Chillies	• *Chana masala*	• *Rasam masala*
• Cubeb pepper	• *Pudina masala*	• *Chettinad masala*
• Mustard seeds	• Kebab masala	• *Goan masala*
• Black peppercorns	• Gunpowder masala	• *Rechado masala*
• White peppercorns	• *Bafat masala*	• *Balchao masala*
• Ginger powder	• *Kohlapuri masala*	• *Ver masala*
Aromatic Agents		
• Allspice	• Saffron	• *Paanch phoran masala*
• Aniseed	• Star anise	• *Dum ka masala*
• Asafoetida	• Royal cumin	• Malabari masala
• Green cardamom	• Curry leaf	• *Soola masala*
• Black cardamom	• Marjoram	• Rose petals
• Cinnamon	• Oregano	• Jasmine flowers
• Dill seeds	• Garam masala	• *Kewra/attar*
• Mace	• *Pudina masala*	
• Nutmeg	• *Potli masala*	

CONCLUSION

The previous two chapters listed various types of spices, their classification, and the ways of combining them to create various spice blends and pastes. This chapter brought us very close to the actual usage of the spices in an Indian dish and we have understood the role played by each spice in a dish or a curry. We have also seen that many spices play a dual role or sometimes multiple roles when it comes to cooking. The same spice can be a colouring, spicing, and a flavouring agent, for example, red chillies, while some act only as aromatic agents such as rose petals.

In this chapter we discussed about various other spices and condiments that are used as souring agents. Apart from the ones discussed earlier, we also discussed new condiments such as malted vinegar, palm vinegar, and synthetic vinegar, used as souring agents, tomatoes, curd and its by-products, lime and lemon, various forms of mango, tamarind, other vegetables such as *kachri* and gamboge and various chemical salts such as citric acid and tartaric acid, that lend souring and piquant taste to the Indian food. We discussed colouring agents and the colour that they give to the dish. We also discussed new condiments, such as *ratanjot* and *mowal*, that are used to add colour to the Kashmiri dishes. We dealt with various thickening agents used in Indian

curries and dishes. We also saw how lentils, onions, and some diary products, such as *malai* and *khoya*, are used for thickening of certain gravies. We discussed the use of spices and condiments as tenderizing agents and saw how acids present in them help to tenderize the meats for the preparation of kebabs and biryanis. We saw the unique usage of raw papaya, raw pineapple, and raw figs as major commodities for tenderizing purposes. We also listed the flavouring agents, spicing agents, and aromatic agents to understand how spices can be classified on the basis of their usage. This knowledge would now help us to understand the gravies and curries better which we shall discuss in the next chapter of this book.

KEY TERMS

Aam ka panna Savoury aperitif made with roasted *kairi*.

Aamchoor Hindi for dried raw mango.

Attar Concentrated form of *kewra*.

Balai Hindi name for *malai* or clotted cream.

Bathua Hindi for a green leaf called amaranth.

Bromelain Enzyme found in raw pineapple used as tenderizer.

Chironjee Type of nut used in Indian cooking.

Chonka dahi Dish made by tempering thick hung curd.

Desi tamatar Sour country tomatoes used in curries.

Dhansak Lamb and lentil dish from Parsi cuisine.

Ficin Enzyme found in raw figs used as tenderizer.

Gandharaj Bengali name for lemon.

Haleem Lamb dish from Hyderabad, where lamb is cooked to a paste consistency with lentils.

Kachri Berry found in Rajasthan used as tenderizing and souring agent.

Kairi Hindi for raw green mango.

Kakori kebab Famous lamb kebab from Lucknow, which has a soft texture.

Kasha magsho Lamb dish from Bengal.

Kewra Essence from the screw pine leaves.

Khoya Dried milk solids used for making desserts.

Kothmir Another name for fresh coriander leaves.

Lemon podlum Half a lime tied up in a cheese cloth and served as accompaniment.

Malt vinegar Fermented liquor of malted barley, used in marination of kebabs and pickles.

Methi tsaman Cottage cheese preparation with fenugreek leaves from Kashmir.

Milk cake Indian dessert made by reducing the milk and sugar and gradually curdling it along.

Mogra Hindi for jasmine flowers.

Mowal Hindi for cock's comb flower, used for colouring curries in Kashmir.

Nimona Vegetable preparation with spinach and dill leaves from Uttar Pradesh.

Papain Enzyme found in raw papaya used as tenderizer.

Plum tomato Type of tomato which is round in shape.

Raita Whisked curd flavoured with spices.

Ratanjot Bark from a tree in Kashmir, used for colouring the curries.

Roma tomatoes Type of tomatoes often used in Italian cooking. These are oval in shape and sweet.

Shammi Kebab Lamb kebabs from Hyderabad.

Shikanji Refreshing drink made with lime juice and water. Also known as *nimbu pani*, it can be sweet, salted or both.

Synthetic vinegar Acetic acid available in diluted form used as a condiment.

Thali A round metal plate used for eating food.

Toddy vinegar Fermented sap of the coconut palm, used in Goan food.

Unami Type of taste that is undefined. It is also referred as a kinaesthetic taste or the taste of sixth sense.

CONCEPT REVIEW QUESTIONS

1. How would the use of metal impact the colour of the dish?
2. Define souring agents and list at least five souring agents used in Indian cuisine and the method of their use.
3. When should the souring agents be added during the cooking process?
4. What is the sixth taste or kinaesthetic taste?
5. Name various types of vinegars used in Indian cooking.
6. How do you make curd?
7. What steps should you take to ensure that the curd should not curdle while making the gravy?
8. What is hung curd and what are its uses?
9. What are the various forms in which mango is used as a souring agent?
10. What is the difference between tartaric acid and citric acid?
11. List at least five colouring agents used in Indian cooking.
12. List the colour hues that different ingredients provide to the gravy.
13. What are the different kinds of herb pastes used for colouring and providing base to dishes? Give examples of the dishes.
14. How is sugar used as a colouring agent?
15. What is the difference between *ratanjot* and *mowal* and where are they used?
16. List at least five thickening agents used in Indian curries.
17. What are the various ways in which dairy products are used to thicken the gravies?
18. What roles do the tenderizers play in Indian cooking?
19. List at least five types of tenderizing agents and their use.
20. What is the difference between 'papain' and 'ficin'?
21. How does one make papaya paste?
22. Name at least three flower-based aromatic agents used in Indian cooking.
23. What is an Awadhi garam masala?
24. What is the difference between flavouring spices and aromatic spices?
25. Name at least 10 aromatic spices used in Indian cooking.

PROJECT WORK

1. In a group, visit the restaurants in your city and order for different dishes for food tasting. Analyse each dish and fill out the table below.

Dish	Colour	Flavour	Aroma	Taste	Texture

Judge the dish based on these elements and then critique in group and talk about the roles of spices used and how one can bring about changes in the dish by altering the ingredients.
2. Form groups of people from different regions of India and prepare few dishes cooked at your homes. Analyse the food from the culinary perspective and see the uniqueness of the dishes prepared. List the spices used for the same and try to categorize them into various groups, based upon their usage in the dish.
3. In groups of five, research about various souring, colouring, flavouring, thickening, and spicing agents used around the world in various countries. Now see if these ingredients can be used in Indian cooking and how they would be useful.

CHAPTER 27

BASIC INDIAN GRAVIES

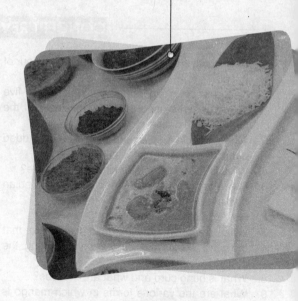

Learning Objectives

After reading this chapter, you should be able to:
- understand the various basic gravies used in Indian cooking
- know the difference between a curry and a gravy
- be aware of many regional gravies made in India
- know how to prepare various gravies, their usage, and storage
- make the *mise en place* flow for the preparation of various dishes in the hotel

INTRODUCTION

In the previous chapter we read about various kinds of condiments and spices that are used for specific purposes in the Indian kitchen. Some of these spices are used for colouring, flavouring, souring, tenderizing, etc. and many have dual or multiple roles to play to create a curry. We have read about the equipment used in Indian cooking and have discussed many of the commodities used to prepare Indian delicacies. It is quite sad that Indian food has still not found its place on the culinary map of the world. The reason is very simple—for a Westerner any Indian dish is a curry and any bread is a naan. It is only Indians who know the diversity and ethnicity of the Indian curries and gravies and it becomes very difficult to standardize the same. That is why, maybe, foreigners get confused when they eat as many different chicken curries as the places they visit.

This chapter will deal with basic gravies made in Indian cuisine. It is important to mention here that in Indian home style cooking, there is no concept of basic gravies. However, for bulk cooking of Indian food in hotels, it is important to make basic gravies from which various dishes can be prepared using various permutations and combinations. At homes, each dish is prepared separately, unlike Western food, where the stocks and sauces are ready and have to be tossed with the ingredients.

Let us now discuss various gravies used as a base for many Indian dishes.

GRAVIES AND CURRIES

In the last few chapters we talked about gravies and curries. There is very little difference between the gravy and the curry, and we can say that combinations of gravies are used to prepare curries. Curry is an English word and is adapted from South–East Asia where it is known by different names such as *kari*. In Indian cuisine gravies and curries are used interchangeably and this is so because it does not have any classification of gravies as Western cuisine has. Gravy, in Western cooking, is referred to thin juices that flow out of meat while roasting and those gravies are always meat based. In India gravy means various commodities, such as flavouring agents, thickening agents, and colouring agents, that have been simmered with liquid. We discussed various agents in Chapter 26 and now we shall see the usage of these agents in making gravy. Unlike in the West, India has vegetarian gravies as well. It can be said that gravies are central to Indian food. We must also realize that since there are certain agents that are limited to certain regions of India, there is another set of gravies called regional gravies. With so much of diversity and the regional influences it became important for chefs in hotels to start classifying gravies into certain categories, so that standardized products are available to the customers or guests.

There are basically four types of Indian gravies that are commonly used to prepare many dishes and curries around the country. However, there are some regional curries that are specific only to certain regions of India.

The four main gravies used in Indian cooking are white gravy, *hariyali* gravy, *makhni* gravy, and brown onion gravy. Some of these gravies have other names too. To make any gravy, there are certain procedures that need to be followed right from the selection of equipment to oil or fat, flavouring agent, souring agent, thickening agent, and colouring agent.

Let us discuss below the main gravies used in Indian cuisine under the parameters discussed above.

Onion Tomato Masala

This is also known as brown onion gravy or onion tomato gravy. This gravy is also known as *lababdar* gravy (refer to Table 27.1).

Table 27.1 Onion tomato masala

Equipment	Oil/Fat	Thickening Agent	Flavouring Agent	Spicing Agent	Colouring Agent	Souring Agent
Kadhai should be used only if darker coloured *lababdar* gravy is desired. Use tinned brass utensil, as an untinned utensil will react with tomatoes.	Any good quality vegetable oil is used.	Onions form the base and body of this gravy.	*Khada masala* is tempered in oil and garam masala is used in the finishing.	Red chilli powder is made into a paste and added after onions have turned brown.	Red chillies also act as colouring agent.	Chopped tomatoes also form the base of this gravy. Equal parts of onions and tomatoes are used for this gravy.

Preparation

Heat oil in the chosen utensil and temper with *khada masala* and slit green chillies. Add chopped onions and cook until slightly darker than golden brown. Take care not to burn the onions as they will impart a bitter taste. Add ginger and garlic paste and cook for a minute. Add red chilli powder made into a paste with water and cook for 30 seconds on a low flame. Add chopped tomatoes and cook. Add small amounts of hot water into the gravy and keep mashing the onions and tomatoes to form a base and thickening for the gravy. Cook this gravy covered on a hot plate. The covering will deepen the colour. Finish the gravy with chopped ginger and green chillies.

Usage and Storage

This gravy is used in north Indian cooking and forms the base for many curries and dishes. This can be used as tempering to boiled lentils or sautéed spinach can be tossed with it to make spinach dish. Small baby aubergine can be stuffed with *aamchoori masala* and simmered with this gravy. This gravy is also used of its own to make paneer *lababdar*, etc. This gravy is usually chunky in texture, but if desired it can be pureed and passed through a sieve. This gravy is usually made fresh for each dish in every Indian home and used in everyday cooking. Brown onion gravy can be made and stored up to one week in a refrigerator. It can also be vacuum packed and stored in the freezer for up to two months, but the philosophy of good Indian cooking is usage of fresh ingredients.

Makhni Gravy

This is a very popular gravy from north India and is used for many preparations such as *murgh makhni*, paneer *makhni*, etc. It is also added along with onion tomato masala to produce many other curries. As the name suggests, this dish is flavoured with butter (Table 27.2).

Table 27.2 Makhni gravy

Equipment	Oil/Fat	Thickening Agent	Flavouring Agent	Spicing Agent	Colouring Agent	Souring Agent
The gravy does not stick to the base of thick-bottomed utensils. Use tinned brass utensil, as an untinned utensil will react with tomatoes.	One could use the home churned white butter or salted butter.	Cashew nut paste, khoya, tomato, puree, and cream are used.	*Potli masala* consisting of green cardamom, mace, peppercorn, cinnamon, and bay leaf is used. *Kasoori methi* is used as an aromatic agent and sugar or honey is used for cutting down on the sourness of tomatoes.	Red chilli powder and slit green chillies are added as spicing agents.	Red chillies also act as colouring agent.	Pureed tomatoes also form the base of this gravy.

Preparation

Make a cross on the head of the tomatoes with a sharp knife, put them in a chosen pot, and add a little amount of water. Add crushed ginger and garlic and let the tomatoes stew until soft. Puree and stain the tomatoes and keep aside. Heat the butter, add red chilli powder and cook for 30 seconds. Add pureed tomatoes, salt, *potli masala*, green chillies and cook covered until the specks of melted butter are visible on the surface. Add cashew nut paste, toasted and powdered *kasoori methi*, and sugar. Add more butter if required and finish with cream.

Usage and Storage

This gravy is used in north Indian cooking and forms base for many curries and dishes. This gravy is paired with light flavoured foods such as chicken, fish, and cottage cheese. The *tandoor*-cooked chicken tikka is stewed with *makhni* gravy to make *murgh butter masala*. It is also used in conjunction with onion tomato masala to make curries. The dish prepared with this gravy will always have *makhan*, which means butter, mentioned with it such as *murgh makhni*, paneer *makhni*, and *subz makhan wala*.

If this gravy has to be made in bulk, then prepare only the base; which means till the time tomatoes are pureed and cooked with *potli masala*, when the oil specks appear on top. Finish the gravy with cashew nut paste and *kasoori methi* on a daily basis and use in dishes. The base gravy can be stored in a walk-in for one week and can be vacuum packed and stored for two months.

CHEF'S TIP
Add cream towards the end and do not boil after adding cream as it might result in curdling.

White Gravy

This gravy is white to blonde in colour. The base of this gravy is boiled onion paste and the gravy is thickened by nut pastes. It is usually used in *Mughlai* dishes (Table 27.3).

Table 27.3   White gravy

Equipment	Oil/Fat	Thickening Agent	Flavouring Agent	Spicing Agent	Colouring Agent	Souring Agent
The gravy does not stick to the base of thick-bottomed utensils. Use tinned brass utensil, as an untinned utensil will discolour the gravy and make it appear grey. Use a wooden *palta* to stir this gravy as metals could react to discolour the gravy.	Ghee is commonly used in *Mughlai* food. One can use oil too.	The boiled onion paste along with ginger, garlic, cashew nut, and *char magaz* provide the thickness to this gravy.	*Khada masala* is tempered in ghee.	Green chilli paste is used, one can also use slit green chillies that can be removed after cooking.	Green chillies also act as colouring agent.	Fresh curd needs to be used or else it will curdle. Add after the nut pastes have been added, as they will prevent the curdling.

Preparation

Heat ghee in the chosen utensil and temper with *khada masala* and slit green chillies. Add ginger–garlic paste and sauté for about a minute. Add boiled onion paste and cook until ghee comes out. Cook on a slow flame and ensure that the onions do not gain colour. Add cashew nut paste and *char magaz* paste and cook for another minute. If required, little hot water can be added at this stage. Now add whipped curd and cook until the gravy comes to a boil. Cover and cook until ghee comes on top. It usually takes 45 minutes for this gravy to cook.

Usage and Storage

This gravy is used in many dishes and curries. It is used as a base for kormas, where more curd and brown onion paste is added. This gravy is rarely used of its own, as it is very heavy. It is thus combined with *makhni, hariyali* gravy, etc. to create royal dishes such as *malai kofta, methi matar malai, navrattan* korma, etc.

This gravy should be made when required as the nut pastes and curd can make it sour and spoil rapidly. However, the *mise en place* such as the boiled onion paste can be kept ready in the refrigerator.

Hariyali Gravy

Hariyali in Hindi means green. This gravy is made by adding cooked spinach puree into brown gravy (Table 27.4).

Preparation

Make the onion tomato masala as shown in Table 27.1. When done, add spinach paste to the masala and cook without covering it until the oil floats on top.

Table 27.4 Hariyali gravy

Equipment	Oil/Fat	Thickening Agent	Flavouring Agent	Spicing Agent	Colouring Agent	Souring Agent
Iron equipment should never be used as it will discolour the gravy.	Any good quality vegetable oil or ghee can be used to make this gravy.	Spinach, tomatoes, and onions form the base and body of this gravy.	*Khada masala* is tempered in oil and ginger–garlic paste is cooked with onions and tomatoes.	Green chilli paste is added to maintain the green colour.	Green chilli paste also acts as a colouring agent.	Chopped tomatoes also form the base of this gravy.

Usage and Storage

This gravy is used in many dishes around India. The green paste used in the north can be of spinach, while in south India, this paste could be of curry leaves and coriander leaves. In Kashmir this green paste could be made of fresh fenugreek leaves to create *methi tsaman*. *Saag gosht* from north India is also made by combining *hariyali* gravy with lamb.

This gravy can be made instantly if the basic *mise en place*, such as brown onion gravy and spinach paste, is ready.

REGIONAL GRAVIES

The regional produce forms the base of the cooking of regional gravy. For example, the Malabar coast of Kerala uses liberal amounts of pepper and garam masala spices due to the abundance of these ingredients. The influence of religion and caste is also important. Religions such as Jainism advocate a strict code of vegetarianism. Geography also has a big part to play. In some coastal regions of India, climate, income levels, traditions, and beliefs influence cuisine. Even within the same state sometimes the preferences of the palate differ between communities. At different points in history, India was influenced by many foreign cuisines as a result of invasions and rule by invaders. Another major influence on Indian cuisine has been the ancient treatise on health, ayurveda, and the countless traditions that have developed over the centuries. It is interesting to note that there are thinner gravies made in rice-growing regions of India, whereas in the wheat belts, thicker gravies are made to be eaten with rotis and parathas. It would be difficult to mention in detail the cuisines of all the states, as the food of each state is an extensive study in itself. However, in the following tables we will talk about the most common curries and gravies from various regions in India.

We saw in the previous chapters that regional spice blends are influenced by the availability of the spices in that region. Similarly, regional gravies are also influenced by many factors such as availability of ingredients in that region, etc. The climatic conditions of the regions too play an important part in the regional gravies and curries. The hot and humid climates will have hot and spicy food that would aid in perspiration, thus cooling the body. In places that are cold, such as Kashmir and Himachal Pradesh, more ghee, oil, and spices will be used in the gravies so that they keep the body warm. Some of the regions were influenced by travellers and the ingredients available formed a main part of their cuisine. For example, in south India and Goa, coconut milk is used as a base and thickening agent for most of the gravies, whereas non-coconut producing states in the north do not use fresh coconut or its milk in curries. However, it is common to see the usage of dry coconut or *kopra*, which is used in some desserts as garnish. Let us discuss some of the famous regional curries.

Kadhai Gravy

As the name suggests, this gravy is made in a *kadhai*. This gravy is similar to onion tomato gravy; the only difference is that there is no onion in this gravy. It is used mostly for the vegetarian dishes such as *kadhai* paneer, *kadhai* vegetables, etc. Whole red chillies and pounded coriander seeds are the most commonly used flavours in the *kadhai* gravy (refer to Table 27.5).

Preparation

It is prepared by adding garlic paste to heated ghee, then adding coarsely pounded coriander and whole red chillies, followed by chopped green chillies and ginger. Finally, chopped tomatoes are added. The gravy is cooked till oil appears on the surface. This gravy forms the base for many vegetarian dishes such as *kadhai* paneer, *kadhai* vegetables, as well as chicken dishes. The gravy is usually finished with chunks of tomato, green capsicum, crushed *kasoori methi*, garam masala, and coriander leaves.

Usage and Storage

This gravy is used in north Indian cooking and forms the base for many stir-fried dishes such as *kadhai* paneer and vegetables such as *kadhai aloo, kadhai gobhi*, etc. These days it is not uncommon to see *kadhai* chicken as well, but essentially this gravy is used for vegetarian dishes.

Table 27.5 Kadhai gravy

Equipment	Oil/Fat	Thickening Agent	Flavouring Agent	Spicing Agent	Colouring Agent	Souring Agent
Kadhai should be used as the *kadhai* dishes are usually stirfried.	Any good quality vegetable oil is used.	Tomatoes used in gravy act as a thickening agent.	The masala of red chilli and coriander is also called *kadhai masala*. Diced peppers and tomatoes are added at last to garnish and add flavour to the dish.	Red chilli powder is made into a paste and added after the spices are tempered.	Red chilli powder also acts as a colouring agent.	Chopped tomatoes form the base of this gravy and act as a souring agent.

This gravy is usually made fresh for each dish but while making in bulk, it can be stored for a month if kept in a freezer. It can keep up to one week in a refrigerator, but since tomatoes are the only ingredient in this gravy it could spoil faster.

Achari Gravy

This is a regional gravy from Punjab and is used for many vegetarian and meat dishes. It gets its name from the *achari masala* used in it (Table 27.6).

Table 27.6 Achari gravy

Equipment	Oil/Fat	Thickening Agent	Flavouring Agent	Spicing Agent	Colouring Agent	Souring Agent
Use tinned brass utensil, as an untinned utensil will make the curd curdle.	Traditionally pickles are also made with mustard oil.	Onions and curd form the base and body of this gravy.	*Achari masala* contains seeds such as mustard, fennel, fenu-greek, *kalonji*, and cumin.	Red chilli powder is made into a paste and added after onions have been browned.	Red chilli powder also acts as a colouring agent.	Apart from chopped tomatoes and curd; sometimes chopped pickle is also added.

Preparation

Heat oil in the chosen utensil and temper with red chillies and broiled and roughly crushed *achari masala*. Add ginger–garlic paste and cook by *bhunao* method for 30 seconds. Add the paste of powdered spices such as turmeric, coriander, red chilli powder and stir for a minute. Add pureed tomatoes and cook covered until oil separates from the gravy. Stir in the beaten curd and keep stirring the gravy until it comes to a boil again. Adjust seasoning and cook it on *dum* for 20 minutes. One can also add chopped and mashed mango pickle to the above gravy to give a tang and a piquant *achari* flavour.

Usage and Storage

This gravy is used in north Indian cooking and forms the base for many curries and dishes. This gravy pairs up well with fish and meat items. Vegetables such as potatoes and cauliflower also pair up well. At homes the vegetable will be added to the gravy in the initial stage along with the tomatoes, but hotel cooking is different. We shall see the way an Indian kitchen would operate later in this chapter.

This gravy is curd and tomato based and hence, should be made fresh once in two days or preferably everyday. The other way of making this gravy as per order for an à la carte dish is discussed in Table 27.15.

Malai Kofta Gravy

This gravy is again very popular in north India and is used for making *malai kofta*. This is not regularly used in home cooking, but is made on festive occasions as a vegetarian delicacy (refer to Table 27.7).

Table 27.7   Malai kofta gravy

Equipment	Oil/Fat	Thickening Agent	Flavouring Agent	Gravy Spicing Agent	Colouring Agent	Souring Agent
Use tinned brass utensil, as an untinned utensil will react with tomatoes.	Any good quality vegetable oil is used.	Onions form the base and body of this gravy.	Cumin is used in half the quantity of coriander powder.	Red chilli powder is made into a paste and added after the onions are browned.	Red chilli powder also act as a colouring agent.	Pureed tomatoes form the base of this gravy.

Preparation

Prepare the boiled onion paste and keep aside. Make the tomato puree and cook it with small amount of oil for at least an hour.

Now heat oil in the chosen utensil and cook the boiled onion paste until slightly brown in colour. Add ginger–garlic paste and cook for another minute. Add the paste of spices and cook for another 30 seconds. Add cooked tomato puree and cook the gravy for few minutes. Add cashew nut paste and adjust seasoning. Lastly, add cream to this and use as required.

This is how one would make this gravy in a traditional manner. In hotels this gravy can be achieved by mixing white and *makhni* gravy. If we look closely at the ingredients, boiled onion paste and cashew nut paste are used for making white gravy and tomatoes are used for making *makhni*. Sometimes hotels adapt to necessary changes to get a standardized product that can be delivered in short period of time.

Usage and Storage

This gravy is used in north Indian cooking to make a dish called *malai kofta*. In this preparation deep-fried cottage cheese dumplings are often stuffed with dry fruits such as raisins and nuts.

This gravy can be used to make various other curries, but in the north it is commonly used for making *malai kofta* only.

In hotels as discussed above, this gravy is made as per order by combining white gravy and *makhni* gravy and hence the need of storage is same as that of white gravy and *makhni* gravy.

Yakhni Gravy

This is a regional gravy from Kashmir. It is a yoghurt-based gravy (Table 27.8). Many times a spice-flavoured lamb stock is also referred to as *yakhni*.

Preparation

Heat ghee in the chosen utensil and temper with *shahi jeera*. Add chilli paste, ginger–garlic paste and cook for 30 seconds. Add the spice powders except *(saunth* and cinnamon) made into a paste with little water and cook for a minute. Add cashew nut paste and cook for another 2 minutes on a slow flame. Add some water and bring to a boil. Now add the whisked curd and brown onion paste mixture and let the gravy cook on *dum* until the ghee floats on top. Season the gravy and add the *saunth* powder and cinnamon powder along with mint and saffron.

Table 27.8   Yakhni gravy

Equipment	Oil/Fat	Thickening Agent	Flavouring Agent	Spicing Agent	Colouring Agent	Souring Agent
Use tinned brass utensil as an untinned utensil will react with acid in the curd and the gravy will curdle.	Ghee is used traditionally in Kashmir.	*Besan* and brown onion paste is whisked together and added to the masala. Nut paste and ginger–garlic paste is also added for the thickening of this gravy.	Apart from *shahi jeera* and *saunf* which are used as tempering in ghee, all the other masalas are broiled and powdered finely.	Green chillies are made into a paste and added to the gravy. Peppercorn also helps to spice up the gravy.	Saffron is toasted and boiled in the gravy. Dried mint is added in the last and it also imparts some colour to the gravy.	Since *yakhni* is a curd-based gravy, it also acts as a souring agent.

Usage and Storage

This gravy is used to make various preparations in Kashmiri cooking. Dishes such as *goshtaba* are poached in this gravy. *Goshtaba* is used as a special dish during weddings. Other dishes, such as *dhaniwal* korma, *heddar*, also have the base of *yakhni* gravy. This gravy is also used to prepare many vegetarian delicacies such as *nadir yakhni* and *alyakhni*.

   This gravy is made fresh as it is curd based and the shelf life is limited.

Mughlai Yellow Gravy

This can be said to be a derivative of white gravy. The yellow gravy is coloured with turmeric. If the same is coloured with saffron, then it is known as *kesari* gravy or saffron gravy (Table 27.9).

Preparation

Cook curd along with turmeric and *besan* for around 40 minutes. In a separate utensil, heat oil or ghee and temper with *khada masala*. Add ginger–garlic paste and sauté. Add boiled onion paste and cook until oil comes on top. Add the powdered spices made into a paste with some water and cook for 30 seconds. Add puree of tomatoes and cook the gravy. Add some water if required. Now add the

Table 27.9 Mughlai yellow gravy

Equipment	Oil/Fat	Thickening Agent	Flavouring Agent	Spicing Agent	Colouring Agent	Souring Agent
Use tinned brass utensil, as an untinned utensil will react with curd and tomatoes and discolour the gravy.	Any good quality vegetable oil is used. Ghee is used in *Mughlai* cooking.	White gravy is made and combined with cooked curd and brown onion paste.	*Khada masala* is tempered in oil and mace and green cardamom powder is used in the finishing.	Red chilli powder is made into a paste and added into the oil with ginger–garlic paste.	Red chilli powder also acts as a colouring agent.	Tomato puree is used along with the whisked curd.

cashew nut paste, fried onion paste and the cooked curd. Add water, season the gravy, and put on dum for 15–20 minutes. Finish the gravy with cream and green cardamom and mace powder.

Usage and Storage

This gravy is used in the preparation of many *Mughlai* dishes which need to be given yellow colour. *Guchhi matar, Mughlai* paneer, *dum ki subziyan,* etc. use this gravy. This gravy can be combined with other gravies to create other curries.

This gravy is made fresh and can be stored up to one day in the refrigerator.

Rajasthani Yellow Gravy

This is a regional gravy from Rajasthan and is very different from *Mughlai yellow* gravy. Unlike *Mughlai* gravy, this gravy does not use the base of white gravy (refer to Table 27.10).

Table 27.10 Rajasthani yellow gravy

Equipment	Oil/Fat	Thickening Agent	Flavouring Agent	Spicing Agent	Colouring Agent	Souring Agent
Use tinned brass utensil, as an untinned utensil will react with curd.	Any good quality vegetable oil is used.	Onions form the base and body of this gravy.	*Khada masala* is tempered in oil and powdered masalas are added to sliced onions. *Hing* is dissolved in water and tempered in oil.	Whole red chillies are used in tempering. Red chilli powder is made into a paste and cooked with onions. Few slit green chillies are also added to make the gravy hot.	Turmeric adds a bright yellow colour to the gravy.	Curd is also used as a base and as a souring agent for this gravy.

Preparation

Heat oil in the chosen utensil and temper with *khada masala* and *hing*. Add sliced onions and cook until translucent. Add ginger–garlic paste and cook for a minute. Add powdered spices made into a paste with water and cook for a minute. Add curd boiled with *besan* and cook the gravy in *dum* for 20 minutes or until the oil floats on top. Blend the gravy until smooth and strain through a sieve.

Usage and Storage

This gravy is used in many Rajasthani preparations such as *maas ki kadhi*, *makki ka soweta*, Rajasthani *gatta* curry, etc. Just like onion tomato masala is used in the north, yellow gravy is used in Rajasthan. Since curd is abundantly available in Rajasthan, it is used in many dishes. This gravy has a limited shelf life and should be made fresh everyday.

Rizala Gravy

This gravy is a regional gravy from Awadh (refer to Table 27.11). However, when Nawab Wajid Ali Shah went to Kolkata, it got popularized there and today Rizala is associated with the *Mughlai* cuisine of Kolkata. It can be seen that most of the regional *Mughlai* gravies are in a way derived from white gravy.

Preparation

Cook curd along with *besan* for around 40 minutes. In a separate utensil, heat oil or ghee and temper with *khada masala*. Add ginger–garlic paste and sauté. Add boiled onion paste and cook until oil comes on top. Add the powdered spices made into a paste with some water and cook for 30 seconds. Add some water if required. Now add the cashew nut paste and almond paste and cook the gravy for 5 minutes. Add the cooked curd and saffron. Add water, season the gravy and put on *dum* for 15–20 minutes. Finish the gravy with green cardamom, nutmeg, mace powder, and *kasoori methi* powder.

Table 27.11 Rizala gravy

Equipment	Oil/Fat	Thickening Agent	Flavouring Agent	Spicing Agent	Colouring Agent	Souring Agent
Use tinned brass utensil, as an untinned utensil will react with curd and discolour the gravy.	Any good quality vegetable oil is used. Ghee is traditionally used in *Mughlai* cooking.	White gravy is made and combined with cooked curd flavourings. Cashew nuts and almond paste is also used for thickening of this gravy.	*Khada masala* is tempered in oil and nutmeg powder. Green cardamom powder and *kasoori methi* powder is used in the finishing.	Green chillies are made into a paste and few slit ones are also added for making the gravy hot.	Saffron is broiled and powdered. This way it gives a deeper colour.	Whisked curd acts as a souring agent.

Usage and Storage

This gravy is used in making many dishes from Awadh such as *murg rizala*, *gosht rizala*, etc. This gravy is mostly used for making meat curries. This gravy is not very thick and is mostly served as a stew with Indian tandoori breads.

Rizala gravy is usually made *à la minute* by combining white gravy with spices and cooked curd. It is made fresh as it has a limited shelf life because of curd in it.

Korma Gravy

This is a regional gravy from the Mughal *gharanas* of Uttar Pradesh. Most of the Mughal rulers settled around Delhi, Lucknow, and Agra and hence, it is very common to see the influence of white gravies in Uttar Pradesh (refer to Table 27.12).

Preparation

Heat oil and butter in the chosen utensil and temper with *khada masala*. When it crackles, add ginger–garlic paste and cook for 30 seconds. Add coriander and cumin powder and turmeric powder made into a paste with some water. Then cook the gravy for 2 minutes. Add tomato puree and green chilli paste and cook until the oil floats on top. Add the cooked curd, brown onion paste and add some water and cook the gravy on *dum* for 20 minutes. Finish the gravy with mace, cardamom powder, and *kewra* water.

Usage and Storage

This gravy is used in *Mughlai* preparations to make many curries such as kormas, *murgh handee lazeez,* etc. This gravy has a smooth texture and can be strained to give the extra velvety touch to it. It is again a very traditional curry made for festivities.

This gravy is made *à la minute* if the white gravy is ready as a *mise en place* along with other ingredients. This gravy cannot be stored for long due to the presence of curd in it.

Table 27.12   Korma gravy

Equipment	Oil/Fat	Thickening Agent	Flavouring Agent	Spicing Agent	Colouring Agent	Souring Agent
Use tinned brass utensil, as an untinned utensil will react with tomatoes.	Any good quality vegetable oil along with butter is used. One can also use ghee.	Brown onion and curd paste form the base and thickening of this gravy.	*Khada masala* is tempered in oil and garam masala, mace, cardamom powder, and *kewra* is used in the finishing.	Green chilli paste is added so that the beige or the blonde colour of the gravy is maintained.	Green chilli paste also acts as a colouring agent.	Cooked tomato puree is used along with curd as a souring agent.

Mustard Gravy

This is also known as *sorshe* gravy in Bengal. It is a regional gravy typically used in Bengali cooking. Let us discuss this gravy in Table 27.13.

Preparation

Soak the mustard seeds, ginger, poppy seeds, and ginger in water overnight and grind to a coarse, smooth paste on a *sil batta*. Heat oil in the chosen utensil until it smokes. Remove the pan from the fire and let the temperature of the oil come down. Now add the paste and heat it slowly. When the mixture starts to splutter, add turmeric and cook for 1 minute. Add water and cook the gravy until oil comes on top. Add coconut milk, slit green chillies, and adjust seasoning.

Table 27.13 Mustard gravy

Equipment	Oil/Fat	Thickening Agent	Flavouring Agent	Spicing Agent	Colouring Agent	Souring Agent
Use tinned brass utensil, as an untinned utensil will make the gravy lose its colour.	Mustard oil is smoked and cooled to get rid of pungency.	Both yellow and black mustard seeds, poppy seeds, green chillies, and ginger are soaked and ground to a paste on *sil batta*.	Mustard oil and mustard paste form the main flavours of this gravy.	Green chillies and also mustard seeds lend some spiciness to the gravy.	Green chillies and mustard also act as colouring agents.	There is no particular souring agent used in this gravy, but some pungency from the mustard paste does the job.

Usage and Storage

This gravy is usually used with a fish called *hilsa* and it is very popular in Bengal. This gravy is usually made at homes or even used in special occasions. This mustard gravy can also be used as gravy for many fish curries, where it will be diluted further by water. This gravy should be made fresh. Only the paste can be ground and refrigerated for a day as the poppy seeds are perishable and can ferment if kept for more than a day or two. The flavours are better if made fresh.

Salan Gravy

This is a regional gravy from Hyderabad. It is commonly served as an accompaniment with biryani and usually it is made of long banana peppers also known as *salan mirchi* (Table 27.14).

Preparation

Heat oil in the chosen utensil and add ginger–garlic paste and cook for 30 seconds. Now add powdered spices made into a paste with little amount of water and add *salan* paste. Cook on a slow flame until the oil comes to the surface, which will take time. This gravy will splutter a lot when being cooked.

Table 27.14 Salan gravy

Equipment	Oil/Fat	Thickening Agent	Flavouring Agent	Spicing Agent	Colouring Agent	Souring Agent
Use tinned brass utensil, as an untinned utensil will discolour the gravy.	Any good quality vegetable oil is used.	The *salan* paste adds the thickness to the gravy and also forms the base.	*Salan* gravy is tempered with the above spices which impart the flavour to the gravy. Spices, such as whole red chillies, mustard seeds, and fenugreek, are used in tempering of this gravy.	Red chilli powder is made into a paste and added after the paste has been cooked in oil.	Red chilli powder also acts as a colouring agent.	Tamarind pulp is added to the gravy. This also helps to add sourness. It also thickens the gravy and adds colour.

When done, add the tamarind pulp and season the gravy. Tamarind is added last as the acid will increase the time of cooking if added in the initial stages. Now heat some more oil and add mustard seeds, whole red chillies, fenugreek seeds, and add to the gravy. This is called tempering.

Usage and Storage

This gravy is used in Hyderabadi *salans*, vegetarian curries served with biryanis. *Salan* is a word that is often related to meats, for example, *salan* in Kashmir would refer to a meat preparation, but in Hyderabad it refers to this gravy. This gravy can be used for making various *salans* such as *baingan ka salan, mirchi ka salan, arbi ka salan*, etc. To make any vegetable *salan*, fry the vegetables and add to the gravy.

The base of this gravy can be made in bulk and stored in the refrigerator for a week. The gravy should be finished with tamarind and tempering only while being used. It can also be vacuum packed and stored in the freezer for a month.

Meen Moilee Gravy

This is the regional gravy from Kerala and probably has the influence of Oriental curries such as Thai curry. *Meen* means fish in Malayalam and as this gravy is paired with fish (Table 27.15) it is called *meen moilee* gravy.

Preparation

Heat oil in the chosen utensil and add *methi* seeds and cook for some time until it colours. Then add mustard seed; when they crackle add curry leaves and stir-fry for a few seconds. Now add sliced onions, julienne of ginger, chopped garlic, and slit green chillies. Sauté for a minute and add turmeric, coconut milk, and bring to a boil. Add diced tomatoes and cook the gravy for 10 minutes. One can now add pan-fried fish cubes and season it.

Usage and Storage

This gravy is used in Kerala for preparing fish *moilee*. It is a coconut milk based gravy and hence, pairs up well with steamed rice or *appams. Appams* are fermented rice pancakes that are made in a special *appam kadhai*. The usage of this gravy is limited to preparing fish in Kerala.

Moilee gravy is always made *à la minute* when the order comes. Because of the use of coconut milk, the shelf life of this gravy is limited.

Table 27.15   Meen moilee gravy

Equipment	Oil/Fat	Thickening Agent	Flavouring Agent	Spicing Agent	Colouring Agent	Souring Agent
Use tinned brass utensil, as an untinned utensil will discolour the gravy.	It is traditional to use coconut oil.	Traditionally thin and thick extracts of coconut milk are used, but in hotels canned coconut milk is used.	Fenugreek seed, mustard seeds, and curry leaves are tempered in oil, which give the flavours. Onion, ginger, and garlic are not pureed here so they do not provide thickening in this case.	Slit green chillies and crushed peppercorn are used for spicing.	Turmeric gives the bright yellow colour to this gravy. Tomatoes are also diced and added for colour.	Chopped tomatoes give the sourness to the gravy.

There are many other gravies made in Indian cuisine apart from the ones mentioned above. But most of them are specific to one particular dish and are derivatives of the ones mentioned above. For example, a Goan fish curry would be the combination of pureed onion tomato masala with Goan masala paste and coconut milk. Similarly, many other types of gravy rely upon the pastes that are cooked in coconut milk base if they are from south India and with onion tomato masala if they are from north India.

Indian cooking at homes is very different from the cooking in hotels. At home each ingredient is freshly processed and the food is not carried over to the next day. The food in the hotels has to be modified for various reasons such as:

- Certain amount of dishes has to be achieved in a fixed time.
- Guests want variety and specialities and do not want to eat what they eat at home everyday.
- Technically some dishes would take enormous amount of time if started from scratch. The basic *mise en place* is kept so that it is easier to turn out dishes in more authentic way.

For example, if there are four dishes on the menu which use the base of onions and tomatoes with spices, it will be a good idea to make the base common and then proceed with each dish individually. This would save time, efforts, and also yield a standardized taste.

Fig. 27.1 shows how an Indian curry section is set up in a hotel and how varieties of curries and dishes can be made in a stipulated time.

The following set-up is shown in Fig. 27.1.

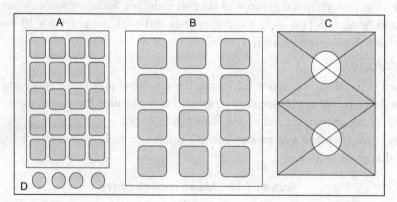

Fig. 27.1 Workstation of Indian curry section
A: Spice box, B: Gravy tray, C: Cooking ranges, D: Fats and oils

A: Spice box containing various whole and powdered spices, such as red chilli powder, turmeric, coriander powder, cumin powder, whole cinnamon, red chillies, peppercorns, cloves, mace, bay leaf, nutmeg, cumin seeds, coriander seeds, *saunf* seeds, mustard seeds, fenugreek seeds, *kalonji*, curry leaves, broiled saffron, salt, *kasoori methi* powder, garam masala, green cardamom powder, mace powder, *kewra, hing*, dissolved in water.

B: Gravy tray which may have things depending upon the menu, but the most common ones will be onion tomato masala, *makhni*, white gravy, *kadhai* gravy, cooked spinach paste, cooked tomato puree, cooked curd with *besan*, cream, cashew nut paste, brown onion paste, ginger paste, garlic paste, chopped tomatoes, slit green chillies, chopped coriander, diced capsicum, diced tomatoes, and diced onions.

C: Cooking ranges with assortment of utensils such as *kadhai*, brass *handi, degchi*, and *lagan*.

D: Fats and oils such as vegetable oil, ghee, coconut oil, and mustard oil.

With the set-up mentioned above one will be able to make all kinds of Indian curries and even home style dishes. Apart from these, some vegetables are also kept in a precooked state so that it will be easier to finish cooking in less time, without compromising upon the taste and flavours. Many people keep boiled vegetables, but there is a small trick that shall improve the quality of Indian food—*dum* cook individual vegetable with small amount of cumin tempering, red chillies, and turmeric and keep it as a *mise en place*. Cauliflower, potato cubes, diced carrots, beans, etc. can be prepared in this way and set up as *mise en place*.

Fig. 27.2 shows the curry section of the menu of a speciality Indian restaurant.

INDIAN MENU OF THE DAY

Adraki gobhi	**Achari machli**
Cauliflower and potatoes flavoured with ginger	*Fish simmered in achari gravy*
Kadhai paneer	**Murgh handee lazeez**
Cottage cheese stir-fried with tomatoes and capsicum	*Chicken cooked with korma style gravy*
Dakhini miloni tarkari	**Gosht rizala**
Vegetables cooked in a south Indian style	*Famous dish from Awadh, made with yoghurt-based gravy*
Guchi malai matar	**Chicken curry**
Morels and green peas cooked in a nut-based gravy	*Home style chicken curry from Punjab*
Lasooni baingan	**Bhunna gosht**
Baby brinjal cooked with garlic flavoured masala	*A dry preparation of lamb stir-fired with spices*

Fig. 27.2   Curry menu of an Indian restaurant

Table 27.16 shows how the dishes in Fig. 27.2 can be made if the respective *mise en place* is there.

Table 27.16   Style of making curries in hotels and restaurants

Dish	Ingredients from the *mise en place*
Adraki gobhi	Heat oil, add cumin and slit green chillies, add onion tomato masala and *dum* cooked cauliflower. Add some water and cook covered. Now add garam masala, touch of *makhni* gravy and stir-fry. Finish with shredded ginger and season. Serve garnished with chopped coriander and julienne ginger.
Kadhai paneer	Heat oil, add cumin seeds, chopped garlic, diced onions, capsicum, and tomatoes. Stir-fry and add *kadhai* gravy. Add cottage cheese, little *makhni* gravy, and *kasoori methi*. Stir-fry and finish with garam masala.

Contd

Table 27.16 (Contd)

Dish	Ingredients from the *mise en place*
Dakhini miloni tarkari	Heat oil, temper with whole red chillies, mustard seeds, and curry leaves. Add *dum*-cooked diced vegetables such as beans, carrots, cauliflower. Add turmeric, cumin powder, and onion tomato masala. Add a small amount of spinach paste and cook. Add the cooked curd gravy and stir-fry. Finish with ginger julienne and chopped coriander.
Guchi malai matar	Heat ghee, add *shahi jeera* and chopped garlic, and sauté. Add morels and green peas and add turmeric and red chilly powder. Add cooked tomato puree and white gravy. Cook covered for 5 minutes. Add cooked curd and finish the gravy with cardamom and mace powder. Add cream to the dish and stir along with a dash of saffron.
Lasooni baingan	Heat oil, add cumin seeds and chopped garlic. Add red chilli powder and *kadhai* gravy. Cook covered and add *aamchoori masala* for giving piquancy. Serve garnished with shredded ginger and slit green chillies.
Achari machli	Heat oil, add *achari masala* (refer to Chapter 25), red chillies (whole), and ginger-garlic paste. Sautè for a minute and add turmeric, coriander, and red chilli powder. Cook for 30 seconds and add cooked tomato puree. Add cooked curd and chopped pickles. Now add some water and pieces of pan-fried fish. Cook on *dum* and serve.
Murgh handee lazeez	Heat oil, sauté the chicken with turmeric, ginger–garlic paste, coriander powder, and cumin powder. Add white gravy and cook. Add brown onion paste and cooked tomato puree and bring to a boil. Finish with cream, garam masala, green cardamom powder, mace powder, and *kewra*.
Gosht rizala	Heat ghee, sauté the chicken with coriander powder, and add almond paste. Add white gravy and some water and cook covered. Add cooked curd and more water or stock and adjust the consistency of the gravy. It is a thin gravy. Finish with green cardamom, mace, and *kasoori methi*.
Chicken curry	Heat oil, add cumin seeds, ginger–garlic paste and then add chicken. Sauté and add red chilli powder, turmeric, cumin powder, and coriander powder. Add onion tomato masala and water. Cook until the chicken is tender. Add garam masala and chopped coriander.
Bhunna gosht	Heat oil in a *kadhai*. Add lamb and sauté. Add red chilli powder and onion tomato masala. Add water and cook until the lamb is tender. Now keep stir-frying the lamb so that the water evaporates. Add garam masala, chopped coriander, dried mint powder, and *aamchoor*.

We saw in Table 27.16 how the same curries that are made individually at homes, can be combined with other base gravies to achieve the desired results. They are combined in a particular way to maintain the ingenuity of the dish. It is difficult to keep the *mise en place* of so many individual dishes so one has to prepare a common *mise en place* list which can be combined to produce various kinds of dishes and curries.

PREPARATION OF GRAVY

There are many basic and regional types of gravies that we discussed in this chapter and all the gravies have the same principles of cooking, i.e. sautéing, stewing, or cooking on *dum* in some cases. Let us discuss some important points for making gravies.

- The cuts of vegetables, such as onions, tomatoes, etc., should be of the same size, as they will cook alike. If some of the onions are thinly sliced than others, they will burn while the rest cook, thereby imparting a burnt flavour to the dish.
- Always smoke the mustard oil and take it off the fire and bring to a stable temperature before adding any spices for tempering.
- Let the onions turn brown first and then add the ginger–garlic paste. In case of boiled onion paste, add ginger and garlic first as the boiled onions will not take long to cook.
- Make a paste of spice powder with water and then add to the browned onions. The powder, if directly added, might burn, thereby losing its flavour and giving an unpleasant flavour to the dish.
- When using chopped tomatoes in the gravy, add browned onions and keep mashing them in with small amounts of water. This will help the gravy to thicken and deepen the colour of the gravy.
- Cook the gravy on *dum* for some time. This is done by covering it and keeping it on a hot plate to let the oil come to the surface.
- Add small amounts of salt in the beginning and also season the food towards the end.

CONCLUSION

In this chapter we read about making gravies and curries. We saw the usage of various kinds of utensils and understood why a particular utensil is used for a type of gravy. We saw various kinds of spices, blends, and pastes and their specific roles in making gravies in Indian cuisine. The thickening agents in the gravy have a clear role to play and they also have other uses such as flavouring and colouring. Some spices are used for their colouring properties and some are used for flavouring and souring. We saw how each gravy comprised all these elements.

We also touched on the *ghar ka khana* concept, where all the dishes are made from scratch, but in hotels, chefs have devised and categorized gravies into a classification that helps them produce a variety of dishes in the stipulated time. The combination and peculiar eating habits of the people of a diverse country like India has also given birth to regional gravies. Some of these regional gravies are used as a base gravy for many curries, whereas in some cases they are used for a particular dish only, for example, *meen moilee*. We discussed various kinds of base gravies and regional gravies, their ingredients, preparation, and also their usage and storage. We also discussed the ways in which traditional gravies are made in hotels by combining various bases and gravies to produce dishes without compromising the flavours and the taste of the dish. We also discussed the salient features of making gravies. This knowledge will enable us to produce good quality Indian curries and gravies suitable for home and hotels.

KEY TERMS

Achari gravy Curd and tomato based gravy tempered with *achari* spices.

Al yakhni White pumpkin cooked with *yakhni* from Kashmir.

Appam Fermented rice pancakes from south India.

Curry Indian stew made from single or combination of base gravies.

Dhaniwal korma Dish from Kashmir, where *yakhni* gravy is flavoured with fresh coriander paste.

Gatta Poached dumplings of gram flour stewed in Rajasthani yellow gravy.

Goshtaba Pounded meat with spices, made into dumplings and poached in *yakhni* gravy from Kashmir.

Gravy Cooked mixture of thickening, colouring, flavouring, and souring agents used as a base of Indian stews.

Guchi matar Indian dish of morsel and green peas cooked in *Mughlai* yellow gravy.

Hariyali gravy Spinach-based gravy cooked with onions, tomatoes, and Indian spices.

Heddar Dish from Kashmir, where mushrooms are cooked with *yakhni* gravy.

Kadhai gravy Tomato-based gravy flavoured with red chillies, coriander, and *kasoori methi*.

Kesari gravy Type of *Mughlai* gravy which is coloured and flavoured with saffron.

Kopra Hindi for dried coconut.

Korma White gravy flavoured with brown onion paste and tomatoes.

Lababdar gravy Another name for onion tomato masala.

Maas ki kadhi Lamb preparation from Rajasthan, where lamb is cooked in Rajasthani yellow gravy.

Makhni gravy Tomato-based gravy flavoured with butter and Indian spices.

Makki ka soweta Speciality of Rajasthan, where lamb is cooked with fresh corn.

Malai kofta Cottage cheese dumplings stewed in tomato and nut-based gravy flavoured with cream.

Meen Malayalam for fish.

Methi tsaman Cottage cheese and fresh fenugreek preparation from Kashmir.

Murgh butter masala Tandoori chicken stewed in *makhni* gravy.

Nadir yakhni Lotus roots cooked with *yakhni* gravy from Kashmir.

Navrattan korma Blend of nine *rattans* or vegetables cooked in a tomato-based gravy flavoured with cream and thickened with nut paste

Onion tomato masala Also known as brown onion gravy, OTM, or lababdar gravy.

Rizala Curd-based gravy from Kolkata, flavoured with *kasoori methi* and saffron.

Sorshe gravy Mustard paste based gravy from Bengal.

White gravy Indian base gravy made with boiled onion paste, nut paste, and Indian spices.

Yakhni Curd-based gravy from Kashmir, also referred to lamb stock with flavouring spices.

CONCEPT REVIEW QUESTIONS

1. Differentiate between a gravy and a curry.
2. How is Indian gravy different from the Western gravy?
3. What is the difference between a base gravy and a regional gravy?
4. What are the basic gravies used in Indian cooking?
5. How would the choice of cooking utensil affect the colour and usage of onion tomato masala?
6. List the preparation of *lababdar* gravy.
7. Describe a *makhni* gravy and list three preparations that you can make from it.
8. What are the thickening agents used in white gravy?
9. List the various gravies that can be derived from white gravy.
10. What is *hariyali* gravy and where is it used in Indian cooking?
11. List at least three regional gravies from north India.
12. What are the different types of *yakhni*?
13. List at least three Kashmiri dishes that can be made from *yakhni* gravy.
14. What is *kadhai* gravy and how is it different from *lababdar* gravy?
15. What are the thickening agents used in kadhai gravy?
16. List the preparation of *achari* gravy. How would you make it differently in an à la carte kitchen?
17. Define *malai kofta* gravy and list the traditional procedure of making the same.
18. Which gravies would you combine to create *malai kofta* for an à la carte order?

19. What is the difference between *Mughlai* gravy and Rajasthani yellow gravy?
20. List at least three preparations made by using *Mughlai* yellow gravy.
21. What is the souring and colouring agent used in Rajasthani yellow gravy?
22. Describe *rizala* gravy.
23. What is korma gravy and how is it different from white gravy?
24. What is *sorshe* gravy and how is it used?
25. What is the thickening agent used in *sorshe* gravy?
26. List the salient features of making *salan* gravy.
27. List the souring agents used in *salan* gravy.
28. What is *meen moilee* and how is it made?
29. List the salient features of making basic gravies.
30. Draw the basic *mise en place* set-up of the curry section of an Indian kitchen.

PROJECT WORK

1. The recipes of the spice gravies are provided in the ORC. Divide yourselves into two groups. One group should make the standard recipe and the other group should make the deviations in the thickening agents, souring agents, and colouring agents. Compare the results with the standard recipes and record your observations.
2. Visit various hotels and restaurants and list the curries and gravy dishes served on the menu. Now design a *mise en place* and workstation set-up, so that there is a good flow of work. The amount of *mise en place* should be limited.
3. In groups, visit the various hotels and restaurants and sample the food. Analyse the kinds of gravies used and critique the usage of spices as mentioned in the chart below.

Gravies	Thickening Agent	Souring Agent	Flavouring Agent	Colouring Agent	Souring Agent

CHAPTER 28

CHEESE

Learning Objectives

After reading this chapter, you should be able to:
- understand the process of making cheese with regards to various criteria
- list the famous cheese of different parts of the world
- comprehend the use of cheese in cooking and cooking related processes
- know the parameters of selecting and storing cheese
- understand the nuances of making a cheese board

INTRODUCTION

Cheese is a by-product of milk which is obtained from the separation of fats and proteins from the milk, through the process of coagulation that leaves behind a liquid known as whey. This process of coagulation can be natural or artificially induced.

Cheese has always been the basic dish on every table of the European continent. The French, Italians, Swiss, and the Dutch boast about a large selection of cheese and this is undoubtedly true. The manufacturers prepare cheese that reflects the region and the place of its origin. Many of the cheeses are DOP (*Denominazione di Origine Protetta*), a classification given to a food item that denotes the origin of the product from that particular region. A large variety of cheeses are available in the world.

It would actually be impossible to say when cheese was first invented, but a few stories say that an Arab traveller carrying milk in a saddle bag made from camel skin made cheese. Due to the hot weather and the movement of the milk inside the saddle bag the milk separated into cheese and whey. The modern science confirms that a special enzyme, known as 'rennet', which is obtained from the stomach of animals such as calves helps in the coagulation of cheese. Even today, cheese is prepared by separating the coagulated milk proteins known as curd through enzymatic activity.

Thousands of years ago, man learnt the art of domesticating animals for their milk and meat. A range of animals such as wild cattle, sheep, goat, and camel were reared by farmers, during which they also learnt the art of converting the surplus milk from these animals into various by-products that could add variety to their meal and help in the preservation of the milk. The cheese industry, in real terms was popularized by the monks of Italy who had to look for various vegetarian options to include in their diet.

The art of making cheese was handed over from one generation to another and a few families guarded the secret recipes of cheese production—a practice that is still prevalent in many countries. In the modern times, cheese can be made from the milk of various animals such as cows, sheep, buffalo, yak, and reindeer, but the most popular are cow's and sheep's milk for making cheese.

Innumerable kinds of cheeses are made around the world. In India we are only limited to cottage cheese also known as paneer, but in an international market cheese holds a very special place in the meals. Each country boasts about its speciality cheese; these cheeses are commonly made from the milk of animals such as cow, buffalo, sheep, goat and even yak, which lends a typical taste to the particular cheese. Cheeses can be classified into various categories as follows:

- Country of origin
- Method of production
- Types of cheese
  o Fresh cheese
  o Semi soft cheese
  o Hard cheese
  o Blue cheese
  o Speciality cheese

PRODUCTION OF CHEESE

Generally, the following two methods are used for producing cheese.

- Sour milk cheese method
- Sweet milk cheese method

In the production of sour milk cheese, the *lactobacillus bacteria* present in the environment react with the milk to make it sour. When this milk is heated, the milk solids coagulate, leaving the whey behind. The types of cheese produced through this method acquire a sour taste and hence, this is not the most popular method for producing cheese among the commercial cheese makers.

Commercially, the sweet milk cheese method is preferred by the cheese makers across the world. In this method, the rennet enzyme is added to the milk which helps in separating the solids in the milk from the whey. The rennet helps in coagulating the milk proteins without letting the milk go sour. The resulting coagulated proteins are then drained, leading to the formation of solid curds.

Many kinds of cheese are made by this method. Such cheeses can vary in taste, texture, flavour, and smell. The taste can vary from mild to sharp and the texture can vary from soft to hard. The in-between range of the cheese can be semi soft, firm, or hard. The cheese can be ripened and aged for a suitable period of time or it can be an unripened variety, which is known as fresh cheese. Cottage cheese or paneer is an example of fresh cheese. Let us now read about the process of production of cheese.

Process of Production

The process of making cheese is ideally accomplished in the following seven major steps.

1. **Preparation of Milk:** The milk needs to be pasteurized and even homogenized in a few cases. The preparation of milk really depends on a particular cheese. For a few cheeses, the temperature of the milk also plays an important part in their texture. In this stage certain useful bacteria that produce lactic acid are infused into the milk so that the right acidity in the milk is achieved, which will help to precipitate the milk proteins known as *casein,* when the rennet is added.

   Milk can be used when it is freshly extracted or it may even be skimmed to produce certain types of cheese. Many kinds of milk are enriched with cream to form cheeses. Each of these concoctions of milk will produce a cheese with a distinctive taste, texture, and colour. Even the milk from a particular animal will give a distinctive aroma and flavour to the cheese. The milk of sheep and goats has more fatty acids than the cow's milk and this property of milk gives distinctive characteristics to the cheese.

2. **Addition of Starters and Rennet:** This is the most important aspect of cheese making and the final flavour and texture of the cheese largely depends upon this step. Traditionally an enzyme called rennet, obtained from the belly of calves is used as a starter to start the process of curdling the milk. Rennet helps to bring together the casein molecules trapping the milk proteins and the fats in the suspended whey.

   The milk is heated to around 41°C and held for around 30 minutes to 1 hour. This helps in the formation of a gel like structure, where the casein and fats come together to form unique bonds. The gel once set is then cut into small pieces by a special tool made of thin wires. The cutting of the gel exposes the interiors thereby allowing more whey to spate out from the proteins. These days to keep up with the growing demands of vegetarians, substitutes like vegetarian rennet are commonly used. The benefit of using rennet over lactobacillus starters is that the cheese will not have a sour taste. After this step, the milk is allowed to rest for coagulation to start.

3. **Coagulation or Curd Formation:** The third step is to cook the gel between 41–45°C, whilst agitating the gel mixtures to form the chunky substance called curds. The rennet continues to do its job and more and more casein is precipitated resulting in denser and more congealed curds. The cooking temperature between 41–45°C depends on whether one wants a soft cheese or a hard cheese. Heating gel at 41°C results in soft cheese whereas cooking the gel to 45°C results in hard cheese. Once the pieces of curd have become firm and somewhat rubbery to touch, the stirring of the same is stopped and then the thick curds start to settle at the bottom leaving the whey.

   In this process this thick mass, also known as young curd, and the proteins, fats, and other substances present in the milk come together to form a congealed mass leaving behind a colourless liquid—the whey. This whey is often discarded, but in few processes like in the parmesan cheese making industry, it is fed to the pigs, which are then reared for making Parma ham.

4. **Salting:** The curds so obtained are cut into blocks and are piled on top of each other. This process often known as *cheddaring* is crucial for the taste and texture of the cheese. The piles are kept for around 2–3 hours depending upon the type of cheese. These curds go through many chemical and physical changes at this stage due to the continued action of the rennet and the force that the curds experience when piled upon each other.

Salt is added into the cheese by wet or dry method as per the recipe. The curds can be dipped in brine solution or dry salt can be rubbed on top of them. This process of salting helps the cheese in two ways—it stops the action of bacteria on the cheese to maintain its texture and taste and the cheese is dehydrated releasing more whey from the curds. Brining some cheeses also leads to longevity in their shelf life.

5. **Moulding or Vatting and Pressing:** After salting, cheese is put in moulds for it to acquire a particular shape. This can be done in plastic or wooden moulds. The cheese is pressed to give it a definite shape. In few cases the cheese is not pressed. Pressing is done to give a unique and uniform shape and size to a particular cheese. Also, pressing will determine the final texture of the cheese. For harder cheese more pressure is applied to expel more moisture from the cheese and vice versa.

This pressing of cheese is done at controlled temperatures of around 10°C to avoid unnecessary microbial growth. The cheese at this stage is also known as 'green cheese' and is matured in controlled humid conditions. The humidity for hard cheeses is less as we would want a harder and drier cheese.

6. **Ripening/Maturing:** Cheese is then matured in temperature controlled chillers or traditionally in dark places called cheese caves. The maturity of the cheese determines its final texture. The ripening or maturing of cheese can take anywhere between a few weeks to up to a year. It is a very complex process, whereby the good bacteria developed in curds allow them to go through complex chemical changes that result in a unique taste and final texture of the cheese.

7. **Finishing:** The moulded cheese is left for a few weeks in controlled temperatures until it acquires a stable shape. It is then de-moulded and coated with rind or any other coating such as herb crust, grape must, and plastic wrapping. In few cases the softer cheese gets a natural rind on itself. The coating is done for various reasons. It is used for branding purpose and also helps in preventing moisture loss, spoilage, or any physical damage to the young cheese.

SELECTING A GOOD CHEESE

Each of the cheese manufactured has its own unique taste and identity and it is a matter of choice and personal preference. The cheese made from unpasteurized fresh milk is always better and more superior to the one that is made with pasteurized milk; however the modern cheese making industry cannot take the risk of preparing cheese with unpasteurized milk due to health risks. In Europe, the farmers at home can however easily make the cheese from fresh milk and such cheeses are known as artisanal cheeses and they fetch a higher price in the market than other cheeses. However the general rule is to select a good cheese by doing the organoleptic checks such as that of sight, smell, and taste. Odour from the cheese is one of the most important parameters for selecting a good cheese. Fresh cheeses should have a pleasant sweet smell and the hard cheeses should smell of earthy flavours. One should avoid buying cheese that smells acidic or has a smell of ammonia or has no smell at all. The cheese must have a distinctive aroma.

Just by looking at a cheese one can tell if the cheese is of good quality or not. The rind of the cheese is the biggest indicator of its quality. An old cheese will have a dried rind, but rinds that are cracked, chipped, or have patches are not good indicators. Also as the cheese ages it needs to breathe through its rind. Many cheeses are coated with synthetic materials such as plastics that do not allow the cheese to mature in the right manner. So avoid cheeses that are covered with synthetic materials and are aged for a long time.

Lastly, select the whole cheese block rather than a pre-cut one. The cheese when cut into smaller pieces and stored develops dried edges and its flavour is spoiled.

SERVICE OF CHEESE

In Classical French 17 course menu, cheese features at the 15th place under the heading *fromage*. Cheese is used in food in many ways, but when it is served as a course by itself, it is traditionally served with dried fruits such as walnuts, apricots, and almonds, fresh fruits such as grapes and pickled olives, and cheese crackers. Some salient features that should be observed whilst serving cheese are:

- In most hotels cheese is displayed during all three meals on buffet.
- In specialty restaurants guest can choose from a cheese trolley.
- Cheeses should be served at room temperature.
- Slices and bite-sized pieces should be cut just before service. Once dried they are useless for service.
- The cover for cheese should be a dessert plate with appetizer fork and appetizer knife.

Creating a Cheese Board

When cheese is served on buffet, it should be arranged elegantly where an assortment of cheeses varying in taste and texture are arranged on a platter or traditionally on a wooden plank, also known as cheese board. The cheese board is served with a special cheese knife which has a forked tip in the front to pick up the cut cheese. Following criteria should be kept in mind whilst presenting cheese on cheese board.

- Cheese boards are usually made of dark wood with a rustic look.
- One must consider a good mix of 5–6 types of cheeses with different textures, colours, rinds, and countries of origin.
- The board should also have accompaniments of cheese such as crackers, grapes, walnuts, watercress, and celery stick.

HOLDING AND STORAGE

One has to take special care in storing cheese. A few hotels have a separate refrigerator designated for storage of cheese as they can acquire the smell of other ingredients. Also cheese is the richest source of proteins and hence, a favourite food for bacteria and moulds. One must take extreme care to hold it with gloves and individually wrap the cut pieces. The shelf life of cheese will depend on its type and can range from a few days to a few weeks.

COOKING WITH CHEESE

Cheese cookery poses no problem as long as you keep the temperature low or the cooking time short or both. The high protein content of cheese means that it becomes tough and stringy with high temperatures and prolonged cooking. Fat may also get separated if the cheese is subjected to high temperatures.

Cheese used in starch infused sauces must not be added until the thickening process is completed. Cheese for gratinating should be added near the end of the cooking process.

Generally, cheese in cooking is used in the following ways.

- **Stuffing:** Many kinds of preparations are used by stuffing cheese either by itself or mixed with another ingredient. For example, a dish called spanakopita in Greece is made by stuffing a phyllo pastry with a mixture of feta cheese and spinach. Cheese may also be stuffed in creamy buns that are eaten as snacks. For example, a very popular snack from Italy known as arancini is made from risotto that is stuffed with mozzarella cheese and is crumb fried.

- **Topping:** All of us can relate to the rich toppings of grated mozzarella on pizzas. Apart from pizzas, cheese can also be used as a topping on many Mexican dishes such as nachos and tacos.

- **Flavouring:** Cheese is often added as a flavouring agent, where it is cooked along with sauces such as cream cheese sauce flavoured with blue cheese. Grated sharp cheese such as *pecorino* is often grated on top of risottos to add the flavour.

- **Base:** Cheese can be used in many other ways all by itself for several dishes. Cheeses such as *mascarpone* are the base for making a popular dessert called *tiramisu*. Many cheeses such as *haloumi*, *kasseri*, and *saganaki* are often served grilled in Turkish and Greek cuisines. Crumbed and fried camembert cheese fingers are a very popular snack in France.

POPULAR CHEESES IN THE WORLD

All around Europe cheese is the focal point on every table. Let us discuss some of the famous cheeses available around the world and commonly used in hotels worldwide.

Cheeses from Italy

The Italians have a larger selection of cheese than French and they prepare cheeses that reflect the region and the place of their origin. Hence, many of the Italian cheeses are DOP classified. An innumerable variety of cheeses are available in Italy, and in Table 28.1 we discuss the most commonly used Italian cheeses in hotels around the world.

Table 28.1   Popular cheeses from Italy

Cheese	Description
Parmesan	This is one of the oldest cheeses of Italy and is particularly used for cooking. It is also the biggest of cheeses and is aged for a duration of 12 months to 2 years. Parmesan cheese is commonly known as *Parmigiano Reggiano* and is made in Emilia Romagna. It is a DOP cheese and is made from cow's milk. The cheese is called *Grana Padano* if it is made in the Lombardy region. This cheese is made in large wheel shapes and it weighs around 30 kilos.
Pecorino	This is a pungent cheese made from sheep's milk. This cheese is produced in almost every province of central and southern Italy. It matures faster than parmesan cheese and becomes hard in 7–8 months only. This is a cooking cheese and is used in making sauces and pastas. The pecorino that is made in Emilia Romagna is called *Pecorino Romano* and has a peppered flavour while the one from Sardinia is called *Pecorino Sardo* and is slightly more acidic and sharp.

Contd

Table 28.1 (Contd)

Cheese	Description
Ricotta	It is a soft creamy cheese made from ewe's milk. The unique thing about this cheese is that it is made from the whey which is low in fat content. Ricotta cheese is drained in special baskets and the marks of the baskets can be easily seen on the surface of the cheese. Ricotta cheese is often used as stuffing for pastas and is also commonly used for making desserts.
Mascarpone	This creamy cheese comes from the Lombardy region of Italy and is made from cow's milk. It is prepared by adding tartaric acid to warm milk, which is then allowed to curdle. It is then drained in cheesecloth and allowed to ripen for a few days. Mascarpone is commonly used for preparing a famous Italian dessert called *Tiramisu* but it can also be used as stuffing for pasta.
Gorgonzola	Gorgonzola is a famous blue veined cheese from regions of Piedmont and Lombard in Italy. Since these regions border France, it is quite possible that the French blue cheese called *Roquefort* was also inspired by this cheese. Made from cow's milk, this cheese is curdled with starter bacteria and penicillium glaucoma, which are responsible for creating the traditional greenish blue streaks in the cheese. Gorgonzola melts easily and it is thus commonly used in cooking.
Mozzarella	Mozzarella is a native of Campania in Italy and is made from the milk of water buffalo. It is a fresh cheese and is low in fat content. Mozzarella is always associated with pizza and the unique property of this stringy melting cheese makes it ideal to be used as topping on pizzas and other dishes.
Caciocavallo	This cheese can be prepared from cow's or sheep's milk and is predominantly produced in Sicily. The word comes from the word *cavallo* which means horseback and this is so because the cheese is linked together like saddlebags. The cheese is kneaded in lukewarm water and shaped by hands into traditional pouch shapes.
Scarmorza	This cheese is quite similar to the mozzarella cheese and its unique pear shape comes from the fact that it is tied in plastic and then tied with string and hung to dry. This cheese is also commonly served smoked and can be used instead of mozzarella.
Provolone	Provolone is made in Veneto and Lombardy regions of Southern Italy. It is a semi hard cheese that is aged for at least 3–4 months. Provolone comes in fancy shapes as it is sometimes tied up with strings and that is how it gets its traditional oval shape. Three different types of Provolone are quite popular—mild called *dolce* which means sweet, *picante* which means sharp, and *affumicato* which means smoked.
Bocconcini	This is a small bundled cheese prepared from the milk of water buffalos. It is a *pasta filata* cheese, which means that it is prepared by dipping the curd cheese in warm salted water and then stretched and folded to give smooth shapes. They are small dumplings of fresh cheese packed in their own whey. This cheese is used in salads to prepare tomato and mozzarella salad or is used in bruschetta and sandwiches. It can also be coated in Anglaise style and deep fried and served as a popular snack.

Greek Cheese

Greek food has influenced most of the cuisines of the Western world, as it has a culinary tradition of more than 4000 years. It spread its culture and cuisine through Rome to Europe and other parts of the world. The Greeks traded their olive oils, wheat, and wine–all the three things that are most common in the Mediterranean cuisine. Fish and seafood have always played a major role in the Greek cuisine. Seafood such as tuna, sea bream, octopus, and squids are very popular till date in Greek cuisine and the art of pickling the fish is still prominent. Since cattle was hardly present in this region due to scarcity of green pastures, goat and sheep were reared not so much for consumption, but for milk and cheese. Let us discuss the most common Greek cheeses in the Table 28.2.

Table 28.2 Common Greek cheeses

Cheese	Description
Feta	Feta is an aged cheese that is prepared from goat's or sheep's milk. It has a granular texture and is used as a table cheese or in salads such as Greek salad. This cheese is also crumbled and used as stuffing for many baked pastries like spanakopita. Feta cheese is salty in taste and should be used accordingly.
Kasseri	This cheese is prepared in Greece as well as Turkey. It is a hard pale yellow coloured cheese prepared from sheep's milk. The texture is soft but stringy like mozzarella. It is aged for at least 3–4 months to get the full flavours and texture. This cheese is served in sandwiches and can also be pan fried in olive oil and served drizzled with lemon juice.
Graviera	This cheese is considered to be the *Gruyere* of Greece and is the second most important and famous cheese in Greece. The unique thing about this cheese is that it is made from a blend of the milk of cow, sheep, and goat. It is used in cooking or simply served grated on top of dishes.
Kefalotiri	It is a hard cheese and more salty than the usual cheese. It is made from goat's or sheep's milk and is traditionally used for preparing a fried snack called *saganaki* or is simply served with grapes and red wines.

Cheeses from France

France is one of the largest producers of the most popular cheeses in the world after Italy. The various regions of France specialize in a particular type of cheese that is also controlled by AOC or *Appellation d'origine contrôlée*, which literally translates to controlled designation of origin. It means that an AOC product has to originate from one type of region only and if produced elsewhere it would be known by a different name. For example, the wine *Champagne* can only come from Champagne region and when produced elsewhere in France it will be known as sparkling wine, but not Champagne, even though the method and procedure for making the product is similar.

The availability of the ingredient at a particular place has a major impact upon the cuisine of that region and this can be validated with the examples of apples from Normandy and butter and cream from the Loire Valley. Table 28.3 discusses some of the popular cheeses used in French cuisine.

Table 28.3   Popular cheeses from France

Cheese	Description
Brie	Brie is like camembert, the only difference being that brie comes from Champagne region, whereas camembert comes from Normandy. Brie is a soft pale white cheese made from cow's milk. The whitish mouldy rind with slight ammonia flavour is edible. Brie is also available in large wheels.
Roquefort	Roquefort is a blue veined cheese made from sheep's milk. It is one of the best known cheeses in the world. It is also one of the cheeses with AOC. The cheese is white in colour with green moulds. It is a slightly tangy cheese and commonly served on cheese platters.
Saint nectaire	This cheese is made from cow's milk and is from the Auvergne region of France. The texture of this cheese is semi soft and it is aged for around 8 weeks. This cheese is also an AOC cheese.
Petit basque	This cheese is made in the Pyrenees mountain and is made from sheep's milk. Petit basque is buttery yellow in colour and is aged for at least 70 days.
Rocamadour	This cheese comes from the Perigord region of France and is one of the AOC cheeses. Rocamadour is made from goat's milk and is aged only for 12–15 days. The soft texture of this cheese makes it a popular cheese for salads and sandwiches.
Brillat savarin	This is a soft white cheese made from cow's milk. It is mainly produced in the Normandy region and comes in 12–13 cm wheels. It is aged only for a few weeks and is also available as fresh cheese; as fresh cheese it resembles cream cheese.
Neufchatel	This cheese is also made in the Normandy region of France and is soft to crumbly cheese. The outer look of the cheese resembles camembert but the taste is much saltier and sharp as compared to camembert. Neufchatel has a typical mushroom flavour and it is usually moulded in shape of hearts.
Munster cheese/ *Munster gerome*	This is a strong flavoured soft cheese made from cow's milk. Its name comes from the little town of Munster in Alsace, it was aged in the monk's cellars. It is one of the AOC cheeses and should not be confused with the American cheese called *Muenster*.
Camembert	A soft white mould cheese from Camembert village of Normandy, it is made from unpasteurized cow's milk. This is one of the most famous cheeses of France and is an AOC cheese. Camembert can be served on the cheese display or can be commonly eaten pan fried as a hot hors d'oeuvre.
Boursin	This is a creamy cheese with almost 75% fat content and can be flavoured with many ingredients such as herbs, pepper, or even garlic.
Chavignol	A popular cheese of France, it is made from goat or sheep's milk and has a fat content of around 45%. It is sold in small cylindrical shapes. In many places this cheese is stored in glass jars immersed in olive oil and herbs for pronounced flavour.

Cheeses from Great Britain

The British cuisine has been influenced by its neighbouring countries. Being an island it had also been a favourite spot for traders and travellers, who brought in various ingredients into its cuisine. British prefer beef over any meat, though other meats such as lamb, pig, and game birds such as geese, duck, and pigeons, called squabs, are popular as well. Meats are roasted, boiled, and stewed. They can also be made into sausages. The famous Cumberland sausage from England is quite popular around the world. The sausages are popularly known as *bangers* in England. England is also known for its cheese that gets its name from the region it comes from. A few examples of such famous cheeses are stilton, cheddar, and derby. Many of the ingredients such as cheese are controlled by PDO, which refers to the *protected designation of origin*. Table 28.4 discusses few of the commonly used cheeses in British cuisine.

Table 28.4   Common cheeses from Britain

Cheese	Description
Stilton	Stilton is one of the famous blue cheeses from England. It is known for its characteristic strong smell and taste. This cheese is guarded by the PDO status and it can only be called stilton if it is made in Derby, Leicester, or Nottingham following a particular recipe. This cow milk's cheese is soft and crumbly and is usually aged for at least 2–3 months.
Red Leicester	This is so called because of its reddish to orange tinge. It is a hard cheese made from cow's milk and is aged for a duration of 5–9 months. This cheese with a nutty taste hails from Leicestershire in England. The cheese at the ageing of 5 months is mild in taste and as the ageing is prolonged it develops a tangy taste.
Cheddar	This is the most popular cheese of England. It is a hard cheese with creamy to off white colour. It is made in the cheddar town of Somerset and hence the name. It is one of the largest exported cheeses of UK. The ageing time of this cheese could vary between 3 months to 5 years depending upon the texture and flavour required. It is commonly used in cooking as it melts easily.
Gloucester	This cheese is prepared from the milk of cows of Gloucestershire region and is a semi hard cheese that is aged for at least 9 months. This cheese is of two types– single Gloucester and double Gloucester. Double Gloucester is allowed to ripen for a longer time yielding a sharper and more savoury flavour.
Derby	This is a medium hard cheese made from cow's milk and is known for its buttery flavour. The distinctive pale golden orange colour of this cheese differentiates it from cheddar as it is very similar in texture and taste to cheddar. It is aged between 1–6 months.
Lancashire	This cheese is made from cow's milk and its texture is creamy to hard to crumbly depending upon its ageing time. It is aged between 1 month–2 years. It is a popular cheese used in cooking and is given a PDO accreditation.

Cheeses from Denmark

The cuisine of Denmark is known as Danish cuisine. Danes' way of living has been dependent on agriculture, which still forms a large part of the Danish economy. The cuisine is influenced by its neighbouring countries such as Holland and Germany and so the food is heavy and rich in fat. The reason for this type of food is the cold climate of Denmark. Its long winters like other Scandinavian countries have created the need for pickling and preserving food, which can be seen quite frequently in its cuisine. Though during season, emphasis is laid on fresh agricultural produce, in off seasons people have to rely largely on smoked and cured meats, pickled fish, cheese, and vegetables.

Denmark is famous for its cheese. The most famous part of Denmark is its Jutland peninsula that touches the north of Germany. The south of Jutland is famous for its smoked meats and charcuterie products as well as many kinds of popular cheeses that we will discuss in Table 28.5.

Table 28.5 Popular cheeses from Denmark

Cheese	Description
Danish blue	This blue cheese from Denmark is made from cow's milk and is often known as *Danablu*. This cheese like other blue cheeses has a sharp and sour taste with a salty flavour. The texture is creamy to taste. This cheese is ripened for 2–3 months, where it develops the typical blue to black mould. It has also been awarded the PDO accreditation. It is one of the two cheeses to get a PDO in Denmark. The other is Esrom.
Esrom	The word Esrom comes from the famous monastery that was founded in Denmark and hence the name. It is one of the two Danish cheeses to have been accredited with PDO. This is a semi soft cheese made from cow's milk and has a traditional sticky texture with yellow colour and holes scattered all around it. It is ripened for 10–12 weeks, but the longer it matures the better it becomes.
Havarti	This semi soft creamy cheese from Denmark is probably the most popular cheese in Denmark. It has distinctive small holes scattered throughout. It is served as a table cheese and is often served in desserts with wine.

Cheeses from Holland

The Dutch are very famous for their cheeses and are also one of the largest consumers of cheese in the world. Holland exports around 2/3rd of its cheese production thereby becoming one of the largest exporters of cheese in the world. Apart from the most popular Dutch cheeses such as gouda and Edam ball, Holland produces some unique cheeses that we will discuss in Table 28.6.

Table 28.6 Popular cheeses from Holland

Cheese	Description
Gouda	It is one of the most popular and the largest produced cheeses in Holland. The unique round shape of this semi soft cheese coated with a unique wax coating is prepared from cow's milk.

Contd

Table 28.6   (Contd)

Cheese	Description
Edam	Also known as Edam ball, it is a very distinctive Dutch cheese with its unique ball like shape and slightly flattened edges. It is covered with a bright red cultured wax coating. It is made in the Edam town of Holland and hence the name. Both Edam and gouda have a similar process of being made and only the quantity of fat in these cheeses differentiates one from the other. Edam is made from 40% partially skimmed milk and gouda is made from 48% butterfat milk.
Leyden	Like most of the Dutch cheeses, which get their name from the town they are made in, Leyden is made in the Leiden town of Holland. The unique cumin flavoured cheese is sharp and tangy in taste and is made from 20% butterfat cow's milk.
Maasdam	Maasdam is a typical Dutch cheese that is inspired from the Emmental in Switzerland. It is made in the size of around 13–14 kg wheels that develop peculiar holes during the maturing process. It is a cheese with a sweet and nutty flavour.

Cheeses from Switzerland

Switzerland is popular for its cows that graze in its lush green meadows thereby producing high quality milk, which has allowed Switzerland to make one of the finest variety of cheese and milk based products including milk chocolates. Due to its unique geographical location, nearly 80% of its land is not used for farming and hence is used for livestock agriculture. Cheese making in Switzerland is one of the oldest methods of livelihood for the people living there. Around 750 varieties of cheese are produced in Switzerland and many of these cheeses are accorded the status of AOC and are used all across the world. Switzerland makes one of the finest cheeses that can be used in cooking and one very popular dish made from its cheeses is *Fondue*. In this dish, a variety of cheeses are melted and it's customary to dip crusty bread in the melted pot of cheese and enjoy it with roasted potatoes and olives. Swiss cheeses as they are popularly known as are mostly made from cow's milk, however there are some Swiss cheeses that are also made from sheep or goat's milk. Table 28.7 discusses a few famous Swiss cheeses.

Table 28.7   Popular cheeses from Switzerland

Cheese	Description
Appenzeller	Appenzeller is produced in the Appenzell region of Switzerland that lies in the north eastern part of the country, which is close to the Alps Mountains. This hard Swiss cheese is very popular for its nutty and earthy flavour as it is brined in a typical secret recipe that is guarded by the producers. Few producers use a combination of wine and cider vinegar to brine the cheese, whereas few brine this cheese in an herbal solution. The cheese is packed in three kinds of wrappers– the silver label indicates that the cheese has been aged for 3–4 months, whereas the gold label denotes that it has been aged for 6 months, the black label is the strongest of all and more expensive than others, it indicates that the cheese is aged for more than 6 months to develop robust and nutty flavours.

Contd

Table 28.7 (Contd)

Cheese	Description
Emmental	Emmental cheese is one of the most popular Swiss cheeses across the world. It is often used in making sandwiches due to its unique melting properties and distinctive holes that are created by certain types of bacteria that cause carbon dioxide molecules to get trapped inside the cheese and create a network of holes and tunnels in the cheese. Like Appenzeller, it gets its name from the Emmental town in Switzerland where it is produced.
Gruyère	Gruyère is similar to Emmental in taste and melting properties, but it is without any holes. It is one of the cheeses used in fondue and also for baking savoury dishes. This Swiss cheese gets its name from the town of Gruyere where it was traditionally produced. Due to its melting and gratinating properties, Gruyere is even used in certain dishes in France. This cheese can be aged for 2 months to up to 10 months and the flavour becomes more pronounced with aging.
Raclette	Raclette is a unique cheese with a distinctive salty crust and flavour that reminds one of smoky bacon, cream, and nuts. This cheese has a unique ability to melt under a heat source such as griller and then it's scraped over an ingredient that has neutral flavour, such as boiled potatoes or plain bread. There is a special raclette melting equipment and in many food outlets, chefs melt the raclette and serve it in front of the guest as a live counter.
Tête de Moine	Tête de Moine, literally translated as monk's head is a very popular Swiss cheese made in the French speaking region of Jura mountains. The cheese was traditionally scraped to make shavings as the people believed that shaving this cheese exposes a larger surface area of the cheese to come in contact with oxygen, which in turn enhances its flavour. Recently, equipment known as *girolle* has been created to be used for shaving the cheese in such a way that the shaved cheese resembles a flower.

CONCLUSION

Cheese is one of the most consumed dairy products in the world. In this chapter we discussed about the origins of cheese and how it became so popular all across the world. Cheeses are very unique to each country and some of these cheeses are also given special status that is the mark of their quality by the government of that country. For example, in Britain the PDO status is accorded to the cheese that comes from one particular region and is made with the traditional method that was used in olden times.

AOC in France, DOP in Italy, and AOP in Switzerland all mean the same thing, that is, special status given by the government to indicate a product's place of origin, but in their own languages.

In this chapter we discussed about how cheese is processed from the milk of cow, goat, and sheep. We discussed all the major seven steps that are used for producing particular types of cheese. From preparation of milk to the addition of rennet and starters to coagulation of curds and then moulding the cheese and pressing it to determine if the cheese will be soft or hard, were discussed under the section process of production.

We also discussed some salient features of selection of good cheese and how a cheese is served with different accompaniments on a cheese board. In this chapter we also discussed how to hold and store the cheese so that there is no impact on its shelf life and flavour. Cheese has been used in cooking since time immemorial and we also discussed the application of cheese in cooking.

Lastly we discussed various famous cheeses from Italy, Greece, France, Great Britain, Denmark, Holland, and Switzerland. In these tables, we discussed a range of soft, creamy, and blue cheeses typical to these countries.

KEY TERMS

Affumicato   Italian for smoked, often referred to cheese and meats.

AOC   *Appellation d'origine contrôlée,* which literally translates to controlled designation of origin, is the French classification for high quality ingredients that come from a designated origin.

Arancini   A traditional Italian snack made from cooked risotto.

Artisanal cheese   Cheese made on farm by using traditional methods, also can be referred to as hand crafted.

Bocconcini   Fresh mozzarella cheese from Campania region of Italy.

Brining   A process where the substance is immersed into a mixture of salt or salt and sugar.

Caciocavallo   Cow or sheep milk cheese from Italy.

Casein   Milk proteins present in milk.

Cheddaring   The cutting of curd cheese into slabs and placing on top of each other.

Crackers   Savoury crisp biscuits served as an accompaniment with cheese.

DOP   Denominazione di Origine Protetta is a classification given to a food item which denotes the origin of the product from that particular region.

Feta   Sheep or goat's milk cheese used as table cheese or for cooking.

Fromage   French for cheese and also the 15th course of traditional French menu.

Grana Padano   Italian hard cheese from Lombardy region in Italy made from cow's milk.

Graviera   Semi hard cheese from Greece often used in cooking.

Green cheese   Unripened cheese which has just been moulded into shapes.

Kasseri   Sheep's milk cheese from Greece that resembles Italian mozzarella.

Kefalotiri   A hard and saltier variety of Greek cheese made from goat or sheep's milk.

Mascarpone   Cream cheese from Lombardy region of Italy, used for making tiramisu.

Mozzarella   Italian stringy cheese made from the milk of water buffalo.

Parmigiano Reggiano   Italian hard cheese from Emilia Romagna in Italy made from cow's milk.

PDO   Protected designation of origin is the English classification for high quality ingredients that come from a designated origin.

Pecorino Romano   Italian DOP cheese made from sheep's milk in Emila Romagna district.

Pecorino Sardo   Italian DOP cheese made from sheep's milk in Sardinia district.

Phyllo pastry   Thin sheets of flour based pastry from Greece.

Provolone   Semi hard cheese from Italy.

Rennet   An enzyme present in the stomach of young calves that is used as a coagulant in cheese making.

Ricotta   Soft cheese made from the whey of sheep's milk.

Saganaki   A Greek snack made by crumbed and deep fried Kefalotiri cheese.

Scarmorza   Smoked cheese from Italy.

Spanakopita   Greek dish made with phyllo pastry that is stuffed with feta cheese and spinach.

Tiramisu   Italian dessert made from coffee and mascarpone cheese.

Whey   Green to yellow colour liquid remaining after the milk proteins get coagulated from milk.

1. Define cheese.
2. Name the countries that are famous for cheese production.
3. Name the enzyme and its source that is responsible for production of cheese.
4. Which community was responsible for producing and popularizing cheese?
5. Name 4 animals that are reared for cheese manufacturing.
6. List down the factors that you will keep in mind whilst preparing milk for cheese.
7. What is the role of a starter and rennet in cheese manufacturing?
8. What is the temperature at which milk forms curd?
9. What are the salient features of salting cheese in the production cycle?
10. Describe green cheese and how is it different from other cheeses.
11. List down the factors to be kept in mind whilst maturing cheese.
12. How do you serve cheese on the board?
13. What are the ways in which cheese is used in cooking? Name at least 4 cheeses that can be used for cooking.
14. Differentiate between parmesan cheese and pecorino cheese.
15. Which Italian cheese is used for making a dessert?
16. What do you understand by the term pasta filata?
17. Name at least 3 Greek cheeses and their uses.
18. What do you understand by the term AOC in France?
19. What is the difference between camembert and brie?
20. Name at least 4 blue veined cheeses that are famous around the world.
21. Differentiate between Edam and Gouda cheese.
22. What is fondue and what is its country of origin?
23. Name at least 4 famous cheeses from Switzerland.
24. Explain the unique features of raclette cheese and how it is served.
25. What do you understand by the word cheddaring?

1. In groups of 3–4, research about few dishes made for at least 8 countries famous for their cheese and then along with your instructor, create the dishes and present for tasting and evaluation
2. In groups of 3–4, visit various pastry shops and food markets that sell cheese. Write down your observations with regards to its colour, packaging, and price.
3. In groups of 3–4, create a cheese board with at least 5 cheeses of various textures like soft, hard, and creamy. Prepare a card for each cheese, which lists down the peculiar features of the cheese such as origin, milk type, process, and taste. Present the cheese board with accompaniments. Now do the food tasting and as people are tasting the cheese, read out the card and engage in a question answer session for better learning.

Index

About the Author

Parvinder S. Bali is Corporate Chef–Learning and Development, at the Oberoi Centre of Learning and Development (OCLD), New Delhi. He is a certified hospitality educator from the American Hotel and Lodging Association (AHLA), a certified professional chef from the Culinary Institute of America, and also a certified chef de cuisine from the American Culinary Federation. Chef Bali is the recipient of the prestigious Gourmand World Cookbook Awards held in Spain for *Quantity Food Production Operations and Indian Cuisine.* His other book *International Cuisine and Food Production Management* is the winner of the First Prize for Excellence in Book Production 2012 awarded by The Federation of Indian Publishers.

Related Titles

Food and Beverage Service
[9780199464685]
R. Singaravelavan, *State Institute of Hospitality Management, Kozhikode*
The second edition of *Food and Beverage Service* is specifically tailored to meet the requirements of the students of hotel management courses. Each of the six sections—introduction to food and beverage service, menu knowledge and planning, food service, beverages and tobacco, bar operations and control, and ancillary functions—have been thoroughly updated to cover all the aspects of the food service industry.

Key Features
- Illustrates the key concepts with the help of photographs of various table layouts and other services, sample menus, and side bars
- Provides a detailed description of the various types of wines, non-alcoholic beverages, guéridon service, and specialized service skills for breakfast, afternoon tea, brunch, and so on
- Includes the French terms used for the various staff members, menu, and dishes

Hotel Housekeeping: Operations & Management
[9780199451746]
G. Raghubalan & Smritee Raghubalan, *GR: Hospitality Consultant and Trainer SR: Garden City College, Bengaluru*
The third edition of the book continues to discuss all important aspects of housekeeping such as role of housekeeping in hospitality operations; composition, care and cleaning of different surfaces; room layout and guest supplies; area cleaning; routine systems and records; as well as pest and odour control, uniforms, laundry, flower arrangement, interior decoration, indoor plants, lighting, and contract services.

Key Features
- Includes case studies that discuss the challenges faced by housekeeping personnel
- New sections on floor pantry, contract specification, access equipment, basics of room layout, Wi-Fi and Internet devices, pillow menu, gate pass procedure, and more

Hotel Front Office: Operations and Management
2E [9780199464692]
Jatashankar R. Tewari, *Uttarakhand Open University*
The second edition of Hotel Front Office is specifically tailored to meet the requirements of the students pursuing hotel management courses. The book aims to explore all the relevant aspects and issues related to front office operations and management with the help of numerous industry-related examples, cases, and project assignments.

Key Features
- Discusses the functions of front office operations, and suggests ways and means to make them more effective
- Includes well-illustrated chapters with numerous photographs, flowcharts, illustrations, tables, and examples
- Contains cases to enhance critical thinking and relate concepts to real-life situations

Food Science and Nutrition 3E
[9780199489084]
Sunetra Roday, *Maharashtra State Institute of Hotel Management and Catering Technology (MSIHMCT), Pune*
The third edition of Food Science and Nutrition provides complete and exhaustive coverage of topics related to food science, food safety, and nutrition. It is aimed at students of undergraduate, diploma, or certificate courses in hotel management, hospitality studies, and catering technology.

Key Features
- Covers subjects taught in hospitality and hotel administration, food technology, applied sciences, home science, and nursing courses
- Provides ample examples, review questions, analytical thinking exercises, and updated reference charts and tables

Books by the Same Author

9780198073895 Bali: *International Cuisine and Food Production Management*
9780198068495 Bali: *Quantity Food Production Operations and Indian Cuisine*
9780199474448 Bali: *Theory of Cookery*
9780199488797 Bali: *Theory of Bakery*

Other Related Titles

9780198062912 Ghoshal: *Hotel Engineering*
9780199458844 Devendra: *Soft Skills for Hospitality*
9780199469833 Seal: *Food & Beverage Management*
9780198084013 Devendra: *Hotel Law*
9780195694468 Iyengar: *Hotel Finance*
9780198084013 Devendra: *Hotel Law*
9780198064633 Bansal: *Hotel Facility Planning*
9780198084006 Seal: *Computers in Hotels: Concepts and Applications*